Introduction to the Study of Dinosaurs

SECOND EDITION

Introduction to the Study of
Dinosaurs

Anthony J. Martin

Department of Environmental Sciences
Emory University
Atlanta, Georgia

Blackwell
Publishing

BLACKWELL PUBLISHING
350 Main Street, Malden, MA 02148-5020, USA
9600 Garsington Road, Oxford OX4 2DQ, UK
550 Swanston Street, Carlton, Victoria 3053, Australia

First edition published 2001 by Blackwell Publishing Ltd
Second edition published 2006

1   2006

Library of Congress Cataloging-in-Publication Data

Martin, Anthony J.
  Introduction to the study of dinosaurs / Anthony J. Martin. – 2nd ed.
    p. cm.
  Includes bibliographical references and index.
  ISBN-13: 978-1-4051-3413-2 (pbk. : acid-free paper)
  ISBN-10: 1-4051-3413-5 (pbk. : acid-free paper)   1. Dinosaurs.   I. Title.
  QE861.4.M37 2006
  567.9–dc22

                                                    2005020441

A catalogue record for this title is available from the British Library.

Set in 10.5/12.5pt Stone Serif
by Graphicraft Limited, Hong Kong
Printed and bound in Italy
by Rotolito Lombarda SPI

For further information on
Blackwell Publishing, visit our website:
www.blackwellpublishing.com

Contents

Preface

While this is indeed another dinosaur book, it is also a book about basic science that just happens to be about dinosaurs. In other words, the primary goal of this book is to teach basic scientific methods through the theme of dinosaur paleontology. My expectation in this respect is that dinosaurs provide a tempting hook for undergraduate non-science majors, who may already be enthused about dinosaurs but perhaps need some encouragement to learn basic science.

Learning about science has two approaches, both of which are followed throughout the book: (i) science literacy, which is fundamental knowledge about facts in science; and (ii) scientific literacy, which is the ability to apply scientific methods in everyday life. The study of dinosaurs requires both types of literacy, as well as the use of geology, biology, ecology, chemistry, physics, and mathematics. Accordingly, facets of these fields of study are woven throughout this book. The process of science is thus united by a journey into the geologic past with dinosaurs, which hopefully will inspire many exciting learning opportunities and lead to reevaluations of the assumption that science is just a dull recitation of facts. Realistically, very few of the undergraduate students taking a dinosaur course for non-science majors will become professional scientists (let alone professional paleontologists), but all of them will have opportunities to appreciate science in their lives long after college.

The first edition of this book certainly aspired to these lofty goals and sentiments. Nevertheless, in the spirit of how science progresses because of peer review (Chapter 2), it has been revised in ways both great and small on the basis of valuable feedback while retaining its original theme. Helpful advice came from instructors who used the first edition in their classes, independent reviewers, paleontologists, and others who had useful insights.

Organization

The book as a whole is broadly divisible into two parts composed of an equal number of chapters. Chapters 1–8 introduce the major concepts associated with the study of dinosaurs, and provide an understanding of factual information in the basic sciences surrounding dinosaur studies (science literacy), as well as scientific methods used to investigate dinosaurs (scientific literacy). By the time a student finishes these chapters, he or she should be able to speak the language of science by asking the right questions. A student will also be familiar with the terminology used in dinosaur paleontology.

Chapters 9–13 delve into the major clades of dinosaurs, while also expanding on additional important topics associated with dinosaurs. Each of the clade chapters then uses the same template of subheadings:

Why Study [This Clade]?;
Definitions and Unique Characteristics;
Clades and Species;
Paleobiogeography and Evolutionary History;
[This Clade] as Living Animals.

This parallel structure should make learning about each clade routine for students, while retaining interest through the comparisons and contrasts it creates. The remainder of the book elaborates on the ichnology of dinosaurs (Chapter 14), the evolutionary history of birds and their role as modern dinosaurs (Chapter 15), and hypotheses for dinosaur extinctions (Chapter 16). Trace fossil evidence continues to be threaded throughout nearly every chapter as a normal form of paleontological data parallel to or exceeding body fossil evidence. The topic of extinctions in general closes the book, with some thought-provoking questions regarding modern extinctions and associated environmental issues. I anticipate that this material will impart lessons about how the lives and deaths of dinosaurs relate to our world today and to the future.

Level and Use

The text is suitable for introductory-level undergraduate geology or biology classes. It also provides enough information on advanced topics that it could be supplemented by primary sources (such as journal articles) for an upper-level undergraduate course. The book is designed for a one-semester course, but it contains sufficient depth that it could be expanded to two semesters if supplemented by other materials. If you are a dinosaur enthusiast not taking a college class but just wanting to learn more about dinosaurs, I am very happy that you chose this book, because in many ways it is written especially for you. Although the chapters are connected to one another in sequence, an instructor or student could certainly skip to specific chapters in the book out of sequence and not be completely lost. Cross-references between chapters serve as small signposts along the way guiding readers to important concepts in preceding chapters.

In every equation presented in this textbook, I define the terms presented in it and try to go through their solutions step-by-step and with examples, so that students can see metaphorically that what they initially thought was a *Velociraptor* (Chapter 9) is actually a parakeet (Chapter 15). This edition does have slightly fewer of these applications than the first edition, but instructors can use their discretion in how much emphasis they might place on them.

The long-term goal of the approach taken in this book is to provoke inquiry about the natural world long after the semester is over. Accordingly, macroscopic phenomena that can be observed in natural settings, museums, or zoos are emphasized throughout the book, simply because a reader of this textbook is much more likely to travel through a national park containing spectacular geology and fossils than to look through a petrographic microscope or use a mass spectrometer (Chapter 4). This book is about learning how to learn, appreciating the integrative nature of science, noting the humanity that shines throughout its endeavors (Chapter 3), and marveling at the beauty of the interwoven web of life and how it changes through time.

Last but not least, the scientific methods repeated throughout this book, albeit using dinosaur paleontology as a uniting theme, should cultivate a healthy skepticism of any ideas, scientific and otherwise. In our culture, where phrases such as "There are two sides to every argument" and "Everyone is entitled to their opinion" are accepted at face value, I encourage all students to critically examine these and other kinds of statements for factual content, reasoning, and accuracy. Once an idea has been subjected to critical reasoning, it can be then better judged for its veracity.

Special Features

To provoke inquiry about the main topics, each chapter begins with an imagined situation in which some facet of dinosaurs is placed in the context of an everyday experience for the student. From this premise, questions are formulated, such as "What was or was not a dinosaur?" (Chapter 1), "Who made some of the original discoveries of dinosaurs?" (Chapter 3), "How do people know the ages of rocks?" (Chapter 4), "What did different dinosaurs eat?" (Chapters 9–13), and "How could crocodiles, birds, and dinosaurs be related to one another?" (Chapters 6, 9, 15). To answer the questions, the student then must read and study the chapter that follows. The answers are not given in an answer key, although the Summary statement at the end of each chapter may provide some clues.

In another attempt to prompt inquiry-based learning, major concepts of the chapters are then explored further through Discussion Questions at the end of each chapter. The title is self-explanatory in that instructors can use them for either written assignments or in-class discussions. Students may find that some of these questions remain unanswered; indeed, the lack of an answer key again may lead to their asking more questions. In this respect, they learn a realistic aspect of science: it does not always provide answers. Nevertheless, the process of science always involves asking questions and operates on the principle that answers can be found to questions if we ask the right questions.

The special features include:

- Chapter-Opening Scenarios;
- Website icons within text indicate relevant material on website;
- Chapter Summaries;
- Discussion Questions;
- Accompanying website with sample syllabi, instructor notes, interactive cladograms, and links to additional resources at www.blackwellpublishing.com/dinosaurs;
- All art from the book available in JPEG format on the website and on CD-ROM for use by instructors.

Summary of New Features

- Many more new illustrations, most of which are photographs. Moreover, the quality of these illustrations has been improved to better augment text descriptions.
- Updated scientific information, including peer-reviewed dinosaur discoveries through 2005, folded into the text wherever appropriate.
- Inclusion of these new discoveries into appropriate chapters of the major dinosaur clades: Theropoda (Chapter 9), Sauropodopmorpha (Chapter 10), Ornithopoda (Chapter 11), Thyreophora (Chapter 12), and Marginocephalia (Chapter 13).
- Cladograms revised to incorporate the latest information, analyses, and hypotheses of dinosaur phylogeny.
- Chapters reorganized so that dinosaur evolution (Chapter 6) and dinosaur clades (Chapters 9–13) are covered earlier than in the first edition.
- Some chapters split and expanded for better coverage: new chapters on dinosaur physiology (Chapter 8), dinosaur ichnology (Chapter 14), birds as dinosaurs (Chapter 15), and dinosaur extinctions (Chapter 16).

■ Chapters are arranged in an order that follows a building of basic scientific knowledge (e.g., scientific methods, plate tectonics, evolutionary theory), but subjects taught after this foundation can be taught or rearranged at the discretion of the instructor.

Acknowledgments

Nancy Whilton of Blackwell Publishing deserves credit for urging me to write a prospectus for the first edition of this textbook, which I was pleased to learn was accepted and lauded by the editorial staff at Blackwell. My energetic, cheerful, and enthusiastic assistants in editorial development and production at Blackwell included Elizabeth Frank, Rosie Hayden, and Sarah Edwards. They deserve not only raves, but raises. This book would be much less educationally valuable without the illustrative talents of Caryln Iverson, who provided the majority of illustrations in the book. I am extremely happy that she chose my book to showcase her abilities. I was continually amazed when I saw proofs of her illustrations that perfectly matched my sketches and descriptions of the educational concepts I envisaged for the figures.

Additional and invaluable editorial assistance from outside of Blackwell came from Janey Fisher and Philip Aslett (Cornwall and Sussex, UK, respectively), who were extraordinarily helpful with noticing and fixing errors that crept into the text throughout the process. Debbie Maizels of Zoobotanica provided additional artwork for the new edition.

Reviewers added considerable insights and corrections through their feedback on the drafts of chapters in both the first and second editions. Their guidance helped to tighten the text, better define the educational objectives, and teach me about what methods and subjects would work best for their students. In a few cases I respectfully disagreed with their suggestions, but their intentions were good, so I listened carefully to what they had to say and tried to respond to their concerns as well as possible or practicable. I take full responsibility for any errors, but will be delighted if you find them. Some stayed anonymous, while others did not. The latter people are:

First edition: James Albanese, SUNY-Oneonta; Stephen W. Henderson, Oxford College of Emory University; Stephen Leslie, University of Arkansas at Little Rock; Scott Lilienfeld, Emory University; Franco Medioli, Dalhousie University; Charles Messing, Nova Southeastern University Oceanographic Center; Mark Messonnier, Centers for Disease Control and Prevention; David Meyer, University of Cincinnati; Andrew K. Rindsberg, Alabama Geological Survey; David Schwimmer, Columbus State University; Roy Scudder-Davis, Berea College; and William Zinsmeister, Purdue University.

Second edition: Sandra Carlson, University of California at Davis; Raymond Freeman-Lynde, University of Georgia; Kathryn Hoppe, University of Washington, Seattle; Roger L. Kaesler, Kansas University; Thomas B. Kellogg, University of Maine; Ronald Parsley, Tulane University; and Gustav Winterfeld, Idaho State University.

A few other paleontologists, through research done in the field or in conversations with me, helped by inspiring many thoughts and insights about dinosaurs and the Mesozoic world. David Varricchio (Montana State University) gave me an all-too-brief tour of the Late Cretaceous Two Medicine Formation of western Montana and its dinosaur fossils (both bodies and traces) soon after the first draft of the book was completed. This visit helped to give me a perspective on how much I had learned about dinosaurs and how much I still needed to learn; our subsequent correspondence and research has been very helpful in this respect. Emma

Rainforth provided many insights on Triassic and Jurassic dinosaur track ichnotaxonomy and preservation that assisted with testing hypotheses I had assumed about these important fossils. Jorge Genise (Museo Paleontológica Egidio Feruglio, Trelew, Argentina) was a source of important information about the Mesozoic environments and faunas (especially insects) of South America: *gracias mucho por ayudarme*, Jorge! Christine Bean and Nancy Huebner (Fernbank Museum of Natural History, Atlanta, Georgia), who both teach K-12 students and deal with the everyday public thirsting for more knowledge about dinosaurs, were extremely helpful in providing hints about what interested people about dinosaurs.

Three dinosaur paleontologists are to be thanked for not only being world-class scientists but also for taking time out to teach students in a dinosaur field course co-taught by Stephen Henderson and me. These paleontologists are James Kirkland (Utah Geological Survey), Martin Lockley (University of Colorado, Denver, Colorado), and Don Burge (College of Eastern Utah Prehistoric Museum, Price, Utah). The students were thrilled to interact with them and we were very happy to have our students gain personal perspectives on paleontology from them that they might not have gained from, say, only reading a textbook.

Additionally, I must thank Michael Parrish (Northern Illinois University, DeKalb, Illinois), who provided a "foundation for the foundation" of this book through his teaching a National Science Foundation-sponsored Chautauqua course for college teachers on the paleobiology of dinosaurs in 1996. Soon after I took this course, I composed a website about dinosaur trace fossils that was my first attempt at public outreach in scientific literacy through the subject of dinosaurs. Thanks to him and everyone else for their encouragement, and I look forward to our continued learning about science and dinosaurs.

Last but not least, my heartfelt appreciation goes to my wife, Ruth Schowalter, for her quiet encouragement, as well as her patient endurance of the many hours and considerable work required for this second edition. May you be proud when you finally hold this book in your hands and say, "I helped to make this happen."

Timeline

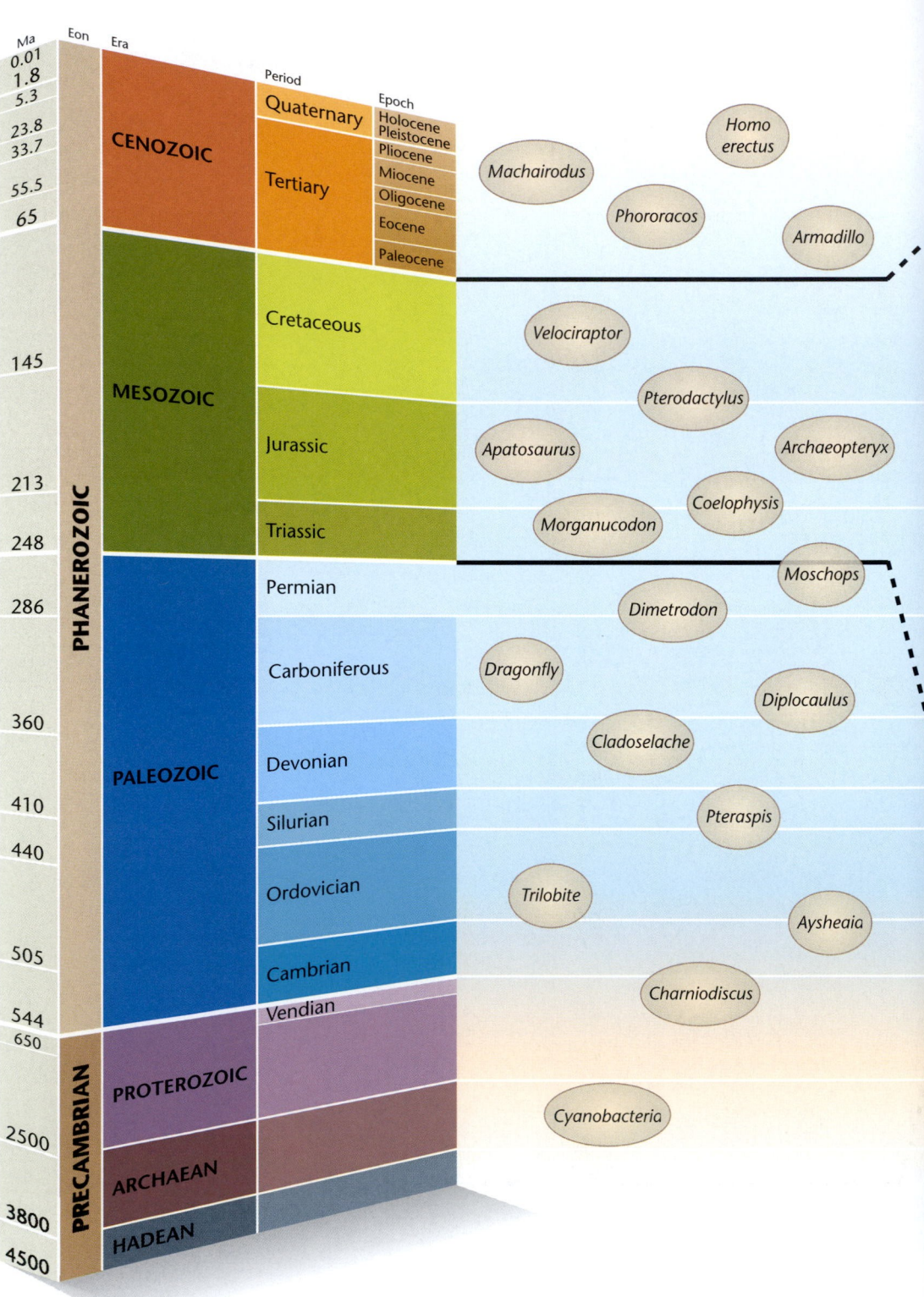

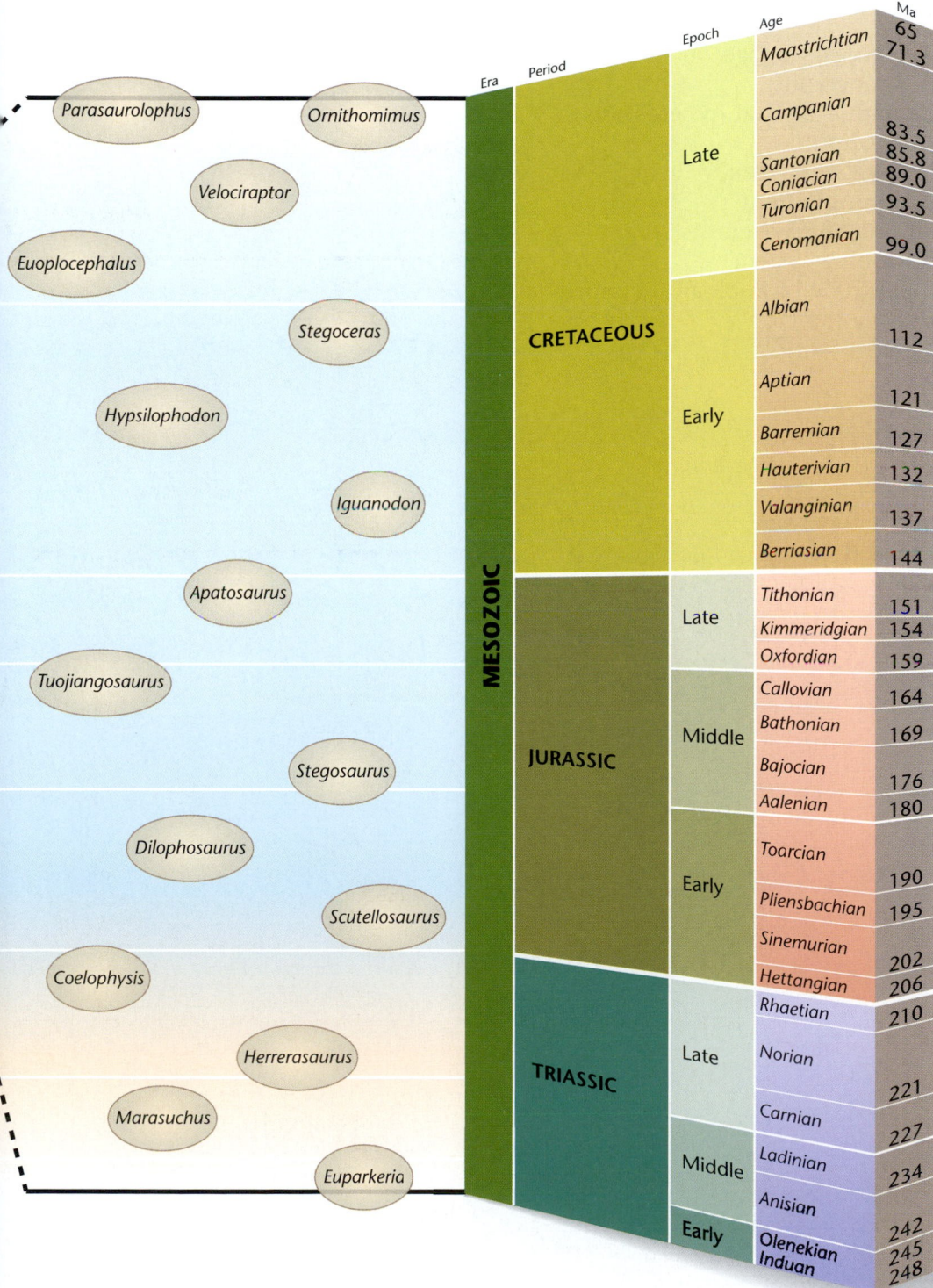

Abbreviations

atm	atmosphere
AV	Avogadro
CGI	computer-generated image
CT	computer tomography
DISH	diffuse idiopathic skeletal hyperostosis
d/p	daughter to parent
EPA	European Protection Agency
EQ	encephalization quotient
GIS	Geological Imaging System
GPS	global positioning system
ICZN	International Code of Zoological Nomenclature
IQ	intelligence quotient
J	joule
kcal	kilocalorie
kJ	kilojoule
LAG	lines of arrested growth
Ma	million years (Latin *mega annus*)
N	newton
PDA	personal digital assistant
ppb	parts per billion
USGS	United States Geological Survey

Chapter 1

Your nine-year-old nephew draws a picture of a plesiosaur, which is a large, extinct marine reptile, some of which had long necks and well-developed fins. This plesiosaur is accurately depicted as swimming in an ocean, and in the sky above are a few pterosaurs, which were flying reptiles. One of the pterosaurs, however, is carrying a cow in its claws. Your nephew patiently explains to you that the "dinosaur" in the water is like the Loch Ness monster, and the "dinosaurs" flying overhead saw some cows in a field. One of them was hungry and wanted to feed its babies, so it captured the cow and was carrying it off to its nest.

How do you explain to him, without crushing his imagination or ego, some of the scientific inaccuracies of what he has illustrated and told you?

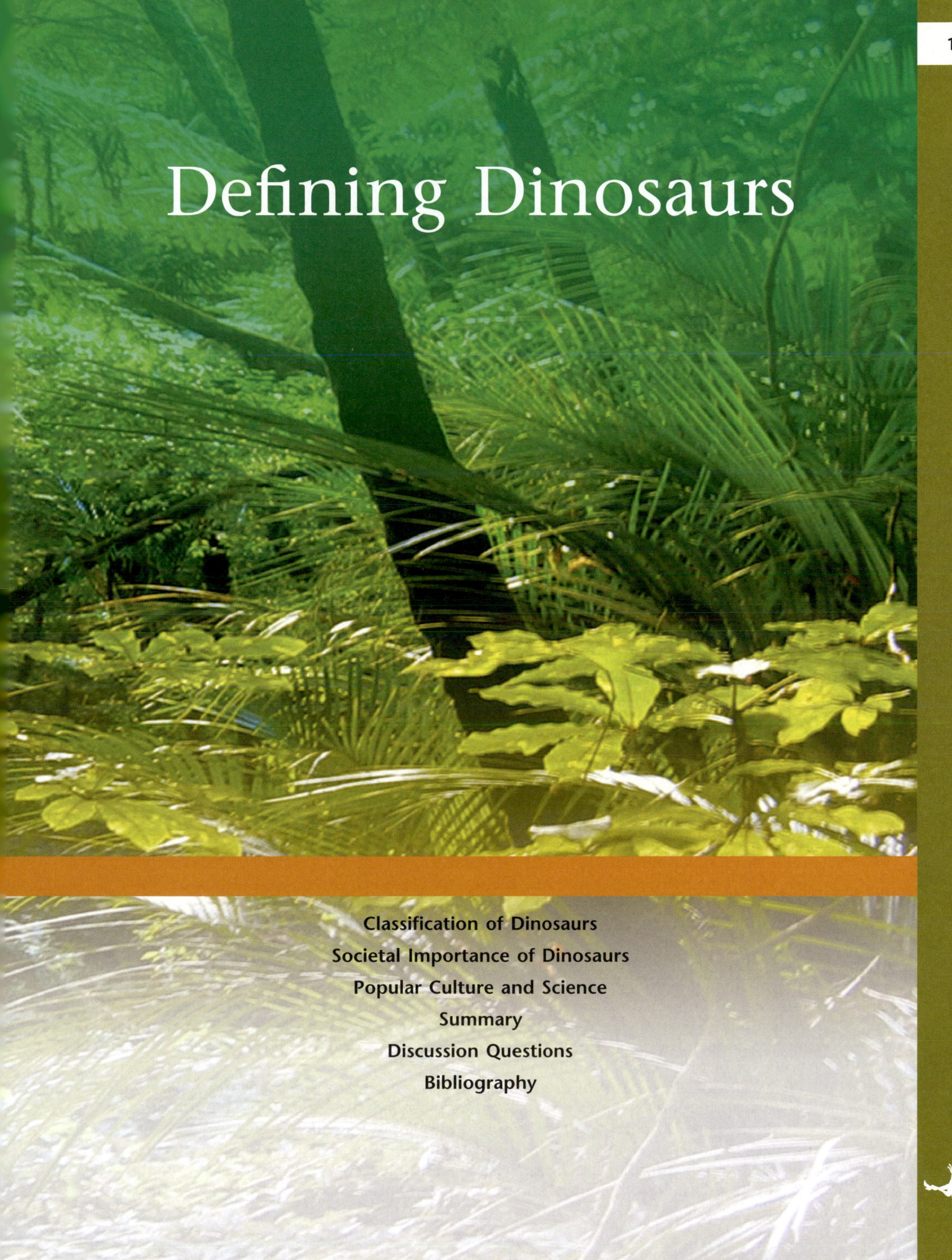

Defining Dinosaurs

1

Definition of "Dinosaur"

Because this book is about **dinosaurs**, probably the most appropriate way to start is by defining them. This is not an easy task, even for dinosaur experts, so here is a preliminary attempt:

> *A dinosaur was a reptile- or bird-like animal with an upright posture that spent most (perhaps all) of its life on land.*

The term "reptile-like" is applied because dinosaurs evolved from reptilian ancestors, yet they were clearly different from present-day reptiles such as crocodiles, alligators, and lizards. Hence these modern animals are not "living dinosaurs," nor were their ancient counterparts. Therefore, anatomical distinctions and differing lineages separate modern reptiles and dinosaurs, even though both groups had common ancestors. However, dinosaurs had many features similar to those of modern reptiles, which warranted their original classification as such (Chapters 3 and 5). Yet some dinosaurs also had anatomical and attributed behavioral characteristics similar to modern birds (Chapter 15). So dinosaurs would appear as a diverse group of organisms that were transitional between certain ancestral reptiles and modern birds, although these relations will be expanded upon, clarified, and corrected later.

Upright posture, also known as an **erect** posture, is important when defining dinosaurs. "Upright" means that an animal stands and walks with its legs directly underneath its torso. This posture is distinguished from **sprawling** or **semi-erect** postures, where the legs project outside the plane of the torso. Sprawling postures are seen in most modern amphibians and reptiles (Fig. 1.1). With only a few exceptions, dinosaurs were among the first animals to be **bipedal**, or habitually walk on two legs. This is indicated by both the anatomy and tracks of early dinosaurs or dinosaur-like animals (Chapter 6). A bipedal stance that is not upright does not result in effective movement. Four-legged (**quadrupedal**) dinosaurs also had an upright posture, as can be seen from their anatomy and tracks (Chapters 5 and 14). In the nineteenth century, dinosaurs were interpreted as large lizards, so older illustrations depict sprawling, reptile-like stances (Chapter 3). Nowadays, modern museum mounts of dinosaurs and better-informed illustrators reconstruct nearly all dinosaurs with their legs underneath their torsos. Why dinosaurs developed an upright posture is not yet fully understood, but current evidence points toward the evolution of more efficient movement on land (Chapter 6).

The land-dwelling habit of dinosaurs is also important in their definition. Based on all information to date, dinosaurs that preceded the evolution of birds did not fly as part of their normal lifestyle, although some may have been gliders (Chapters 9 and 15). Likewise, no conclusive evidence indicates that dinosaurs swam, although a few of their tracks suggest swimming abilities (Chapter 14). Their remains in deposits from ancient aquatic environments suggest that they sometimes

FIGURE 1.1 Differences in postures of a dinosaur and a large modern reptile. (A) Skeleton of the Late Cretaceous ornithopod *Edmontosaurus annectus* from Alberta, Canada. Posterior view of the rear limbs leaving a trackway, showing the typical dinosaurian trait of legs held underneath its body (erect posture). Specimen in the Royal Ontario Museum of Toronto, Ontario. (B) American crocodile, *Crocodylus acutus*, in Costa Rica, showing a sprawling posture and also leaving a trackway. This same typical reptilian posture can change to a semi-erect posture by the crocodile standing up or walking. Photo by Nada Pecnik, from Visuals Unlimited.

(A)

(B)

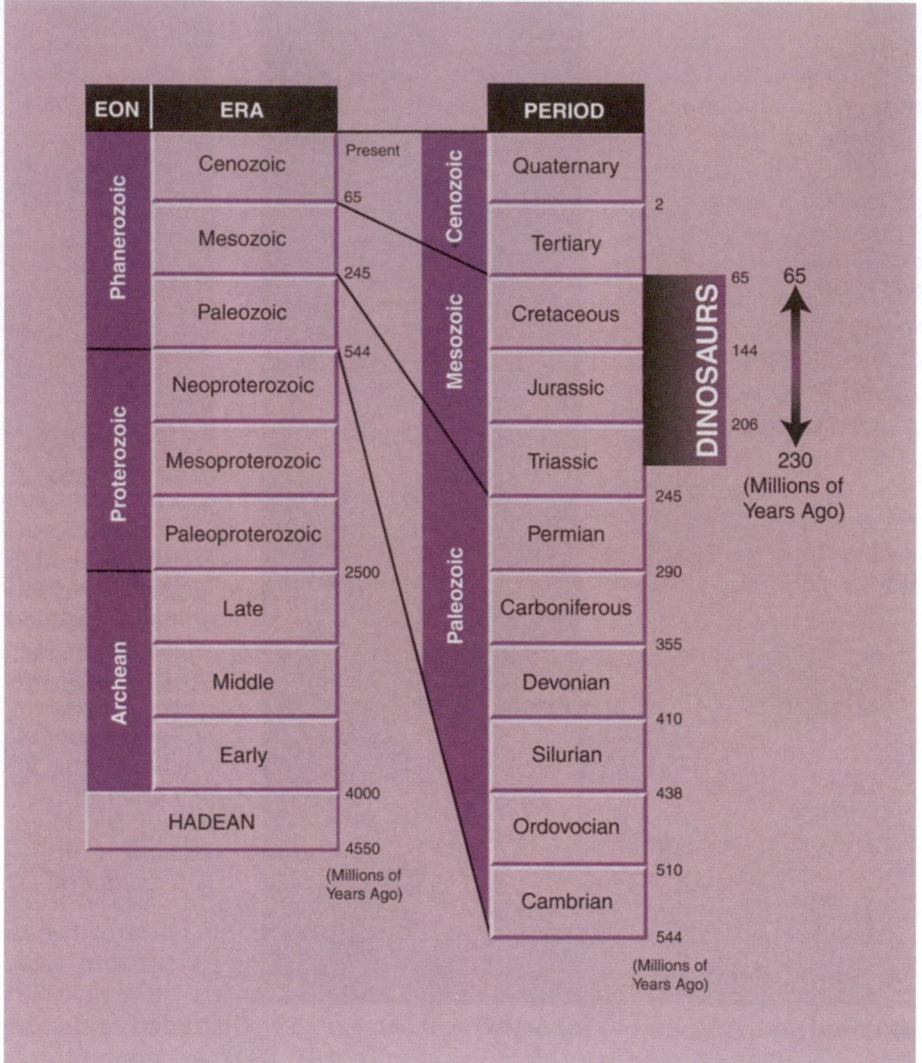

FIGURE 1.2 Geologic time scale used as a standard by geologists and paleontologists worldwide. Largest units of geologic time are eons, followed (in order of most inclusive to least inclusive) by subdivisions eras, periods, and epochs. Figure is not scaled according to amounts of time.

drowned while attempting to swim across lakes or streams (Chapter 7). However, dinosaurs did have some reptilian contemporaries, pterosaurs and plesiosaurs, which flew and swam, respectively (Chapter 6). These were not dinosaurs, although all three groups had common ancestors. Furthermore, no convincing evidence has revealed that dinosaurs lived underground because no dinosaur has ever been found in a burrow, nor have any burrows been attributed to them. Some anatomical evidence indicates that a few small dinosaurs were capable of climbing trees (Chapter 9), but no skeletal remains have been found in direct association with a fossil tree. Dinosaurs appear to have been well adapted to living in the many environments associated with land surfaces, which obviously worked very well for them during their 165-million-year existence.

Dinosaurs are well known for the enormous size of some individual species, in comparison to modern land-dwelling animals. Indeed, some dinosaurs were the largest land animals that ever left footprints on the face of the Earth (Chapter 10).

To date, no indisputable scientific evidence has established the existence of dinosaurs from earlier than 230 million years ago. Furthermore, no living dinosaurs have been discovered in recent times, contrary to claims of some tabloid headlines and Web pages.

However, many adult dinosaurs were smaller than the average human and some were even smaller than present-day chickens (Chapters 9 and 15). To say that large dinosaurs were the most abundant animals in their environments is a misconception. It is analogous to saying that elephant herds in modern Africa dominate those environments, when actually more than half of the mammal species in that environment are smaller than domestic dogs.

Dinosaurs lived only during a time in the geologic past called the **Mesozoic Era**, which is divided into three smaller times periods: the **Triassic Period**, **Jurassic Period**, and **Cretaceous Period**, in order from oldest to youngest, respectively (Fig. 1.2). Because the geologic record for human-like animals only extends to about 4 million years ago, we can be certain that no human has ever seen a living dinosaur. The only record of a human being "killed by a dinosaur," happened in 1969 when a coal miner fatally hit his head against a dinosaur track on the roof of a coal mine. Nevertheless, the formation of the track and the unfortunate miner's death were separated by about 75 million years.

A more precise definition of what constitutes a dinosaur, based on detailed aspects of its skeleton, is covered later (Chapter 5). Some people insist that modern birds are dinosaurs and so do not fit this initial definition (Chapter 15), an objection that is reasonable. Thus, this book is mostly about non-avian ("non-bird") dinosaurs and, from now on, the term "dinosaur" will refer to those same animals limited to the Mesozoic Era.

Once this definition and all of its amendments are formed into a conceptual framework, think about what extinct or living animals are *not* dinosaurs and test the definition whenever possible. For the purposes of this book, a familiarity with the names given for different dinosaur groups and their general characteristics will also help to reinforce the identification of certain names with dinosaurs.

Classification of Dinosaurs

The method by which organisms or traces of their activities are named, which provides a framework for communicating through a classification system, is **taxonomy**. Thus, a name given to a group of organisms in a classification system is called a **taxon** (plural **taxa**). Dinosaurs can be classified in two ways. The more up-to-date of those two methods, **cladistics** (explained below), is the preferred one used worldwide by **paleontologists** (people who study the fossil record). The older, traditional method is the **Linnaean classification**, named after the Swedish botanist, **Carl von Linné** (1707–78), better known by his pen name **Carolus Linnaeus.** In his botanical studies, Linné realized that a standard method was needed to name organisms, which he presented in 1758. The Linnaean method is based on hierarchical **grades** of classification, meaning that organisms are fitted into increasingly more exclusive categories, based on a standard set of anatomical attributes of members in that category. The higher grades become more stringent about which organisms belong to them on the basis of an arbitrary number of characteristics that an organism might have or not have. Such a classification system is typically stratified, starting with groups that contain many members, then progressing to groups with fewer members, such as, in order of largest to smallest group, **kingdom**, **phylum**, **class**, **order**, **family**, **genus**, and **species**. In botany, the equivalent grade to a phylum is a **division**, otherwise the categories are the same. Under this classification scheme, dinosaurs are categorized as below, with the more exclusive grades descending to the right:

Phylum Chordata
    Subphylum Vertebrata
        Class Reptilia
            Subclass Diapsida
                Infraclass Archosauria
                    Superorder Dinosauria
                        Order Saurischia
                        Order Ornithischia

For humans, the categories would be: Phylum Chordata, Subphylum Vertebrata, Class Mammalia, Order Primates, Family Hominidae, Genus *Homo*, and with *Homo sapiens* as the species.

The modern and more commonly used classification method applied to dinosaurs, began in 1984, is the **phylogenetic** classification. This classification is also known as cladistics because it is based on placing organisms into units called **clades**, which are supposed to represent their evolutionary history, or **phylogeny**. Thus, clades are groups of organisms composed of an ancestor and all of its descendants. They are defined on the basis of **synapomorphies**, which are shared, evolutionarily derived anatomical characteristics, also known as **characters**. For example, all mammals have synapomorphies of hair and mammary glands, which they share with ancestral mammals. Cladistic classifications are basically explanations of evolutionary relationships between organisms and are best summarized in a diagram called a **cladogram** (Fig. 1.3).

Cladistics produces a bush with many branches, rather than a ladder with many rungs.

A cladistic classification for dinosaurs based on characters, where one clade branches to another to show descent to the lower right, is:

Chordata
    Tetrapoda
        Amniota
            Reptilia
                Diapsida
                    Archosauriformes
                      Archosauria
                        Ornithodira
                          Dinosauria
                            Saurischia
                            Ornithischia

This may look like a "line of descent," but is not because:

1 it does not include the many branches that emanate from each clade; and
2 it does not show the timing for the evolution of a new clade (Chapters 5 and 6).

In other words, a clade did not have to become extinct in order for the next clade to evolve. Because verbal descriptions of phylogenetically-based classifications can become confusing, cladograms are more commonly used to explain them instead.

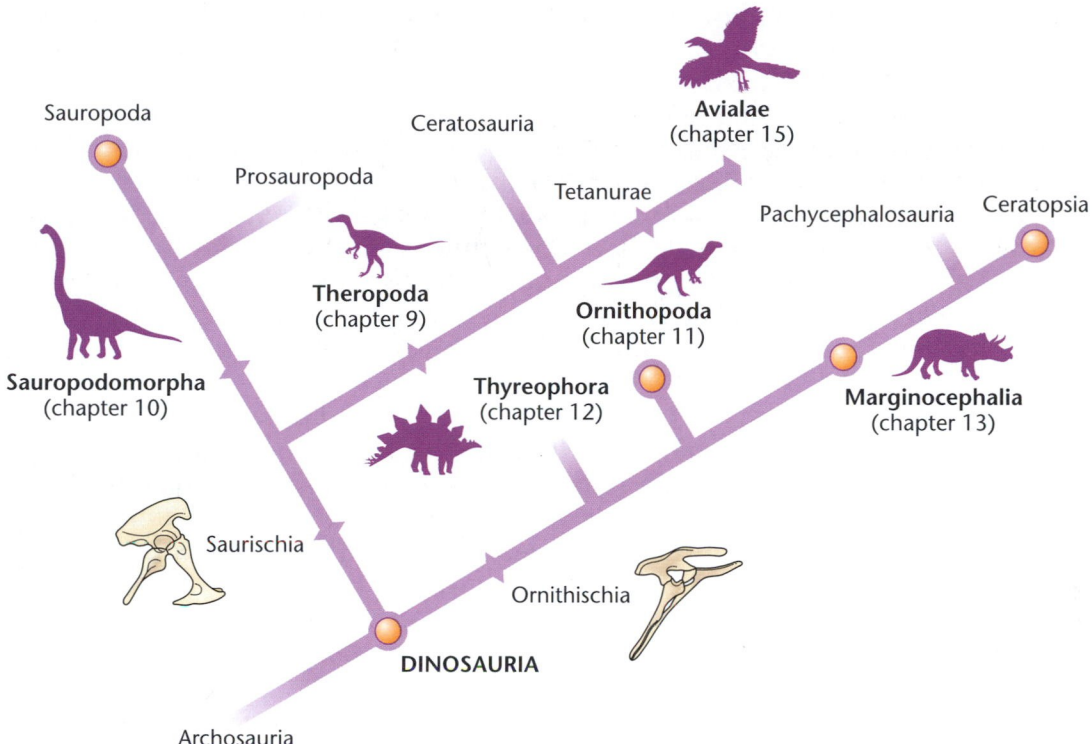

FIGURE 1.3 Cladogram of the major dinosaur clades covered in this text, using Saurischia and Ornithischia hip structures as a basis for dinosaur classification.

Although Linnaean and phylogenetic classification methods differ from one another, a comparison of categories used in each dinosaur classification shows that they use many of the same names. Unless antecedents such as "clade" or "Order" are used, confusion may result from not knowing which scheme a paleontologist is using. Consequently, many dinosaur paleontologists will merely abbreviate references to certain groups of dinosaurs through general categorical names, such as "theropods" (Chapter 9), "sauropods" and "prosauropods" (Chapter 10), or "ornithopods" (Chapter 11), although nowadays these designations implicitly refer to clades. Cladistics is used in this book because dinosaur paleontologists mostly use this method, and it is based on evolutionary relatedness. However, an awareness of the Linnaean system is helpful for understanding the extensive literature on dinosaurs published prior to the 1980s, and some even later.

One aspect of classifying dinosaurs, unchanged since Linnaean times is the tradition of naming species. The species name of a dinosaur or any other organism is based on the **biological species** concept, where a species is a population of organisms that can interbreed and produce offspring that can also reproduce with one another (Chapter 6). The species name was an elegant solution devised by Linné for problems associated with the common practice of applying numerous names to the same organism. The species name uses a **binomial nomenclature**, meaning that two italicized names are used together, a capitalized **genus** name followed by a lowercase **trivial** name, to name a species (i.e., *Tyrannosaurus rex* for a specific dinosaur, *Homo sapiens* for modern humans). The trivial name is "trivial" in the sense that it cannot be used by itself to identify an organism and must always be used in combination with and preceded by a genus name. However, the genus name can be used alone and represents a broader category that may include several species.

This principle is similar to that used by some Asian societies, who place the family name first and the surname second. For example, in Korea, the names Moon Jai-Woon and Moon Hyun-Soo both have the Moon family name (a general category) followed by their surnames, which identify specific individuals when used in combination. Species and other categories in the Linnaean classification originated with Latin and Greek roots for the sake of universal standards, which prompted such well-known dinosaur genus names such as *Stegosaurus*, *Triceratops*, *Allosaurus*, and *Tyrannosaurus*. Since Linné's time, many languages have contributed roots for taxonomic categories, a practice that is especially evident in species names seen throughout this book. For example, French, Spanish, German, Swahili, Mandarin Chinese, and Japanese, among others, have contributed to dinosaur species names.

Using cladistics as a framework, the names of major dinosaur groups, such as ceratopsians (Chapter 13), ceratosaurs (Chapter 9), hadrosaurs (Chapter 11), and prosauropods (Chapter 10), will be repeated throughout this book. Likewise, association of these groups with certain well-studied or otherwise famous dinosaur genera or species will provide an outline of general anatomical characteristics shared within such groups (Table 1.1), which will suffice for discussion of what information can be discerned from dinosaurs. Information about synapomorphies that define each clade will be given in greater detail in later chapters (Chapters 9 to 13).

TABLE 1.1 Summary of different major clade groups used to classify dinosaurs, general descriptions of anatomical characteristics for each group, and genus examples. Detailed classifications, less represented groups, and interrelationships are presented in Chapters 5 and 11 to 15.

Saurischia ("lizard-hipped" dinosaurs)

Theropoda: Late Triassic to Late Cretaceous; feet and legs reflect bipedal habit; hands able to grasp; hollow limb bones; teeth indicate meat eating; 1–16 m long.

Ceratosauria	*Abelisaurus*
	Ceratosaurus
	Coelophysis
	Dilophosaurus
Tetanurae	*Allosaurus*
	Compsognathus
	Oviraptor
	Tyrannosaurus

Sauropodomorpha: Late Triassic to Late Cretaceous; feet and legs reflect bipedal habit in some forms, quadrupedal in most others; often characterized by small head in proportion to rest of body and long necks; teeth indicate plant eating; 2–38 m long.

Prosauropoda	*Plateosaurus*
	Lufengosaurus
	Coloradisaurus
	Riojasaurus
Sauropoda	*Apatosaurus*
	Argentinosaurus
	Brachiosaurus
	Camarasaurus

TABLE 1.1 Continued

Ornithischia ("bird-hipped" dinosaurs)

Ornithopoda: Early Jurassic to Late Cretaceous; feet and legs reflect mostly bipedal habit; teeth indicate plant eating with multiple rows of teeth; 1–15 m long.

Hypsilophodontidae	*Hypsilophodon*
	Orodromeus
	Othniella
Iguanodontidae	*Camptosaurus*
	Dryosaurus
	Iguanodon
	Ouranosaurus
Hadrosauridae	*Corythosaurus*
	Edmontosaurus
	Hadrosaurus
	Saurolophus

Thyreophora: Early Jurassic to Late Cretaceous; feet and legs reflect quadrupedal habit; armored with plates or spines; teeth indicate plant eating; 3–12 m long.

Ankylosauria	*Ankylosaurus*
	Hylaeosaurus
	Nodosaurus
	Pinacosaurus
Stegosauria	*Huayangosaurus*
	Kentrosaurus
	Stegosaurus
	Tuojiangosaurus

Marginocephalia: Cretaceous only; feet and legs reflect bipedal habit in one group (Pachycephalosauria), quadrupedal habit in other group (Ceratopsia); enlarged or thick skull in proportion to rest of body, in some cases with prominent horns; teeth indicate plant eating; 2–12 m long.

Pachycephalosauria	*Homocephale*
	Pachycephalosaurus
	Prenocephale
	Stegoceras
Ceratopsia	*Chasmosaurus*
	Protoceratops
	Torosaurus
	Triceratops

Societal Importance of Dinosaurs

Dinosaurs as an Example of Scientific Inquiry

The main purpose of this book is to introduce the study of dinosaurs as a scientific endeavor. What is and is not science is a major theme of this book, and the study of dinosaurs is an appropriate way to show how scientific methods are applied to real-world situations (Chapter 2). Because dinosaurs have been studied through

scientific methods since at least the early part of the nineteenth century (Chapter 3), many examples are given of how these methods increased knowledge of dinosaurs. Furthermore, subjects in the various chapters are covered to provide a sense of the historical continuity of the science. Science, by design, is always changing and updating itself, and the nearly unprecedented new discoveries and subsequent insights about dinosaurs, in just the past 30 years, have provided an exhilarating example of this dynamism. In fact, research on dinosaurs published in only the four years since the first edition of this book necessitated some major revisions for this second edition (e.g., Chapters 8, 9, and 15).

Although the study of dinosaurs is interesting and fun, it is not easy. Those who think that reading this book and maybe a few other references will be adequate preparation for "going on a dig" and discovering new dinosaur species are probably being overly romantic and naïve. For example, people who are interested in serious study of dinosaurs may need to, at various times, apply geology, biology, chemistry, physics, math, or computer science. All of these fields (and more) are not only used but are necessary in order to make any meaningful sense out of the fossil record. An integrative use of these sciences can help in gaining an appreciation for application through a common theme of dinosaurs, as well as reaching a better understanding of the eclectic and integrative nature of science in general.

The best-known sciences connected to dinosaur studies are geology and biology, which are sometimes united through **paleontology**, the study of ancient life. In fact, many paleontologists who study dinosaurs also call themselves geologists, whereas others were trained as biologists. As a result, distinctions between these two seemingly separate fields are sometimes blurred. Paleontology is studied mostly through the examination of **fossils**, any evidence of ancient life, which can consist of **body fossils** or **trace fossils**. A body fossil is any evidence of ancient life as represented by preserved body parts, such as shells, bones, eggs, or skin impressions. In contrast, a trace fossil is any evidence of ancient life other than body parts that reflects behavior by the animal while it was still alive, such as tracks, nests, or toothmarks. How fossils are preserved in the geologic record is the science of **taphonomy**, important when appraising any dinosaur body fossil or trace fossil (Chapter 7).

Many paleontologists have considerable knowledge of biological principles or perform experiments and field study of modern organisms to gain better insights into their long-dead subjects. Paleontologists tend to study a specific group of organisms and some of the most common subdivisions are:

1 **invertebrate paleontology**, the study of fossil animals without backbones, such as insects;
2 **vertebrate paleontology**, the study of fossil animals with backbones;
3 **micropaleontology**, the study of fossil one-celled organisms and other microscopic fossils; and
4 **paleobotany**, the study of fossil plants.

With these categories in mind, dinosaur paleontologists will often call themselves vertebrate paleontologists. Nevertheless, not all vertebrate paleontologists are dinosaur specialists – some study fish, amphibians, reptiles, and mammals.

For a paleontologist, a more complete understanding of organisms, fossil or living, can be gained by studying them in the context of their environments, which includes all biological, chemical, and physical factors, such as other organisms, nutrients, and sunlight. The study of organisms and their interactions with environments is **ecology**. Ecologists specifically examine a group of organisms as an **ecological community** that interacts with a habitat, called an **ecosystem**. The equivalent practiced by

TABLE 1.2 Commonly encountered elements and compounds in geology, with their chemical symbols and formulas.

ELEMENTS

Ag	Silver	Mn	Manganese
Al	Aluminum	N	Nitrogen
Ar	Argon	Na	Sodium
Au	Gold	Ni	Nickel
C	Carbon	O	Oxygen
Ca	Calcium	P	Phosphorus
Cl	Chlorine	Pb	Lead
Co	Cobalt	Pt	Platinum
Cr	Chromium	S	Sulfur
Cu	Copper	Si	Silicon
F	Fluorine	Sn	Tin
Fe	Iron	Ti	Titanium
H	Hydrogen	U	Uranium
Hg	Mercury	W	Tungsten
K	Potassium	Zn	Zinc
Mg	Magnesium		

COMPOUNDS

CO_2	Carbon dioxide	H_2SO_4	Sulfuric acid
$Ca_5(PO_4)_3(OH, F, CO_3)$	Apatite	Fe_2O_3	Hematite
$NaCl$	Sodium chloride (salt)	Fe_3O_4	Magnetite
$CaSO_4 * H_2O$	Gypsum	$CaCO_3$	Calcium carbonate (calcite or aragonite)
FeS_2	Pyrite	H_2CO_3	Carbonic acid
H_2O	Water	SiO_2	Silicon dioxide (quartz)
$KAlSi_3O_8$	Potassium feldspar		

paleontologists is **paleoecology**, where they attempt to reconstruct the biological and physical factors that affected ancient ecosystems, based on clues left in rocks.

Although the connections of dinosaur studies to geology and biology are well known, the relationship of chemistry, physics, math, and computer science to dinosaur studies may be less clear. These sciences are essential to dinosaur studies and definitions of these sciences and their applications may clarify why these subjects relate to dinosaurs.

In chemistry, properties and changes in materials involve the interactions of atoms of elements, as listed in the periodic table. Chemistry is important to dinosaur studies because dinosaur bones and eggs, as well as their associated sediments, are made of chemicals, which potentially contain information pertinent to the life, death, and after-death history of dinosaurs, as well as their extinction. Consequently, chemical formulas and reactions are used throughout this book. A summary of the most commonly encountered elements in geology and the compounds they compose is listed in Table 1.2. Some chemical formulas and reactions used in dinosaur studies are under the realm of **geochemistry**, the study of chemistry pertaining to the Earth, and **biochemistry**, the chemistry of life. **Microbiology**, which is related to biochemistry, is the study of one-celled organisms (often called **microbes**) and their interactions with their environments.

TABLE 1.3 Units of measurement used in dinosaur studies.

A. DISTANCE UNITS IN COMPARISON TO 1 METER

Unit	Decimal	Fraction	Scientific Notation
micron	0.00001	1/1,000,000	1.0×10^{-6}
millimeter	0.001	1/1000	1.0×10^{-3}
centimeter	0.01	1/100	1.0×10^{-2}
decimeter	0.1	1/10	1.0×10^{-1}
meter	1.0	1/1	1.0×10^{0}
kilometer	1000	1000/1	1.0×10^{3}

B. TIME UNITS IN COMPARISON TO 1 SECOND

Unit	Decimal	Fraction	Scientific Notation
second	1.0	1/1	1.0×10^{0}
minute	0.017	1/60	1.7×10^{-2}
hour	0.00028	1/3600	2.8×10^{-4}

C. MASS MEASUREMENTS IN COMPARISON TO 1 KILOGRAM

Unit	Decimal	Fraction	Scientific Notation
gram	0.001	1/1000	1.0×10^{-3}
kilogram	1.0	1/1	1.0×10^{0}
ton	1000	1000/1	1.0×10^{3}

D. TEMPERATURE UNITS (CELSIUS SCALE)

Temperature	Freezing Point of Water	Boiling Point of Water
Centigrade	0°C	100°C
Kelvin	273K	373K

E. COMBINATIONS OF DISTANCE, TIME, AND MASS FOR OTHER COMMON UNITS

Measurement	Unit	Formula
Area (square or rectangle)	centimeters^2	length^1 × length^2
Volume	centimeters^3 (cm^3)	length^1 × length^2 × length^3
Density	g/cm^3	mass/volume
Velocity	m/s	distance/time
Acceleration	m/s^2	distance/time^2
Force	kg/m/s^2 (newton)	mass × acceleration
Pressure	n/m^2	force (newton)/area

F. CONVERSIONS FROM ENGLISH TO METRIC SYSTEMS AND VICE VERSA

English Unit	Metric Conversion	Metric Unit	English Conversion
Inch	2.54 cm	Centimeter	0.39 inches
Foot	30.48 cm	Meter	39.37 inches
Yard	91.44 cm	Meter	3.28 feet
Mile	1.61 km	Kilometer	0.62 miles
Ounce	28.3 g	Gram	0.035 ounces
Pound	452.8 g	Kilogram	2.21 pounds
1°F	−31.4°C	1°C	33.8°F

Biogeochemistry, the study of chemical processes caused by organisms in geologic media and how elements are cycled in the biosphere, is typified by microbe-mediated reactions in soils (Chapter 7).

The interaction of matter and energy explored in physics is exemplified in dinosaurs through the applications of **biomechanics** and **thermodynamics**. Biomechanics is the study of how living systems, such as animal bodies, perform work. Thermodynamics is the study of heat and its relationship with work, an important aspect of dinosaur physiology (Chapter 8). Physics can also be applied to understand how dinosaurs related to their world through physical properties, such as mass, density, and motion. Furthermore, dinosaurs sensed certain aspects of their environments through their vision or produced sounds with certain frequencies and pitches (Chapter 11). **Geophysics** combines geology and physics, where basic principles of physics are used to understand the Earth, particularly its interior. Some geophysical methods are used to interpret the subsurface distribution of rocks, providing information on the geologic history of an area where dinosaurs lived (Chapters 4 and 6).

In terms of mathematics, this book primarily will use numbers as they are applied, to better understand dinosaurs through measurements and models. Examples of this include **biometry** and **allometry**. Biometry is the study of life through measurements and statistical methods, whereas allometry is the study of size and how it changes with growth of an organism in various dimensions (Chapters 8 to 13). All dinosaur fossils have involved or could involve measurements of some sort. Thus, statistical methods in particular are important in describing dinosaurs and testing data sets for similarities or differences (Chapter 2).

Computers are now essential tools for most paleontologists and are used for cladistics and analyzing results of experimental work. They are also important for communication among scientists, and between scientists and the general public, whether through e-mail or the Web. Computer-generated simulations, in conjunction with hypothesized environmental parameters, are now quite common. They are also used for documentation and interpretation of field sites containing dinosaur fossils, especially through **geographic information systems (GIS)** (Chapter 4). These are programs that integrate spatial data with other forms of information. Consequently, map-reading skills are also needed in dinosaur studies. Geographic methods can be extended to the geologic past through maps that show the distribution of ancient landmasses in association with fossils, a practice called **paleobiogeography**.

*Three-dimensional imaging, using **computer tomography** (CT), and animation of dinosaur fossils is yet another use of computers in dinosaur studies.*

An integration of the preceding subjects is therefore necessary for a fuller understanding of dinosaurs and to appreciate how each subject is an important tool for better understanding the ancient and modern worlds. Only a small amount of previous knowledge of these subjects is needed to understand this book, and the math uses the standard system of measurement in the scientific world and its units: the metric system (Table 1.3).

Keep in mind that this book was written by using words in connected phrases, punctuated by line drawings and photographs, all of which hopefully communicate basic concepts about dinosaurs. As a result, good communication skills expressed through writing, illustrating, or speaking are extremely important to the study of dinosaurs. In other words, the most brilliant paleontological discovery of the century can remain unnoticed if the results are not communicated in a clear and understandable manner. Formal education is not necessary for an extraordinary discovery in paleontology. Some people who study dinosaurs are not associated with prestigious universities and museums. Rather, they may simply have

FIGURE 1.4 Triassic and Jurassic formations of Canyonlands National Park, eastern Utah, an area well known for both dinosaur body fossils and trace fossils. Notice the lack of convenience stores and coffee shops in the field area.

much field, museum, and laboratory experience that they can also relate through excellent communication skills, such as artwork, photography, computer applications, and public speaking. In short, paleontologists should be good teachers in order to be effective.

To these intellectual requirements of dinosaur studies, add the physical demands. Such studies often require fieldwork in remote areas that do not have running water and room service (Fig. 1.4). Similarly, dinosaur studies might involve rummaging through museum drawers for years, with little or no pay. Fieldwork also may require securing funds and logistical planning through hostile (or worse, bureaucratic) institutions, long days filled with physical exertion in the aforementioned remote areas, and saintly patience. Fulfilment of all these may or may not result in any significant dinosaur discoveries. The risk of disappointment caused by looking for something that apparently is not there can be personally discouraging. However, a love for the work and the joys of discoveries, or just the promise of discoveries, are often enough reward for people who study dinosaurs.

Dinosaurs as a Part of Popular Culture in Fiction

For reasons that perhaps can only be explained by psychologists, dinosaurs have always had a large popular appeal. This is evidenced by them being the subject of numerous books, comics, movies, television shows, Web pages, toys, models, and works of art in nearly every industrialized nation of the world. Recognition of this pervasive celebration of everything dinosaurian leads to a sociological observation: dinosaur images in popularized media serve as the most direct source of many public ideas about dinosaurs. Consequently, acknowledgement of mainstream influences, especially in works of fiction, is warranted in order to correct or confirm commonly held notions about dinosaurs.

Dinosaurs were portrayed in fiction relatively soon after their scientific descriptions in the early to mid-nineteenth century. **Charles Dickens** (1812–70) mentions the dinosaur *Megalosaurus* (Chapter 9) in the beginning of *Bleak House* in 1853, only 29 years after the name for that dinosaur was formally proposed (Chapter 3). Other uses of dinosaurs in fiction were apparently uncommon until 1912, when Sir **Arthur Conan Doyle** (the creator of Sherlock Holmes) published his seminal novel *The Lost World*. This book dealt with the experiences of five explorers who discover the existence of live dinosaurs, such as *Megalosaurus* and *Iguanodon* (Chapter 11), in a

FIGURE 1.5 Photograph from the film *The Valley of the Gwangi* (1969), set in the early twentieth century western USA, with cowboys attempting to capture a large theropod. From Horner and Lessem (1993), *The Complete T. Rex*, Simon & Schuster, NY, p. 87. (Dave Allen/PhotoFest).

remote area of South America. Similar portrayals of modern dinosaurs in remote places were written from 1915 to 1944 by Tarzan creator **Edgar Rice Burroughs** (1875–1950). Among the dinosaurs were well-known favorites, *Stegosaurus* (Chapter 12) and *Triceratops* (Chapter 13). From the 1940s through to the present day, science-fiction magazines and comic books also continued this imaginative theme of humans in conflict with dinosaurs. Some contemporary writers have attempted to incorporate scientific knowledge about dinosaurs in their fictionalized accounts, such as Michael Crichton's *Jurassic Park* (1990) and *The Lost World* (1995), and Robert Bakker's *Raptor Red* (1996).

The long and successful use of dinosaurs as subjects in film began less than 20 years after the invention of this entertainment medium in 1890. Of these films, the most important for its adherence to what was known about dinosaurs then and its influence on future dinosaur-themed films was *The Lost World* (1925). This movie, based on the previously mentioned work by Doyle, presented *Allosaurus*, *Tyrannosaurus*, *Triceratops*, and other Mesozoic animals as either individuals or in groups. The portrayal of this assemblage departed from a standard cinematic formula of having a single dinosaur responsible for virtually all on-screen action and carnage. Other movies that showed dinosaurs based on actual species were *King Kong* (1933), *One Million BC* (1940), *Journey to the Beginning of Time* (1954), *The Valley of Gwangi* (1969; Fig. 1.5), *Jurassic Park* (1993), *The Lost World: Jurassic Park* (1997), *Dinosaur* (2000), *Jurassic Park III* (2001), and the large-format IMAX film *T. Rex: Back to the Cretaceous* (2002). Many other dinosaur movies have animals that superficially resemble some known dinosaur species or are exaggerated and embellished conglomerations based on various traits from several known dinosaurs (i.e., all of the *Godzilla* films).

Cinematic treatments of dinosaurs thus provide a good opportunity for critical reviews. For example, the intriguing titles of some films (e.g., the 1991 film *A Nymphoid Barbarian in Dinosaur Hell*) tell how entertainment was their intent, not information. The recurring words in the movie list to note are "lost," "unknown," "prehistoric," or some variation on the theme of "beast" or "monster." The frequency of these words in movie titles is probably the result of perceived favorable reactions of the audiences. After all, these films were made with financial profit in mind. Nevertheless, the viewing of any dinosaur-themed films, especially the older ones, allows for a critical examination of their scientific content. Important questions to ask include:

1 Did the film use scientific information that was known at the time; or
2 Was the scientific information known, but ignored for the sheer entertainment value of seeing live dinosaurs on the screen?

Compared to the motion-picture industry, television had limited production budgets for special effects, which meant that dinosaurs were less common and usually took the form of cartoons or actors in clumsy costumes. However, dinosaurs began appearing more frequently on television within several years of the necessary computer technology becoming commercially viable. With the improvement and economic feasibility of such computer-generated images (CGI) in recent years, the increased integration of dinosaurs into the plots of television episodes has begun. For example, the syndicated TV series *The Lost World*, again reprising the characters and general plot of Conan Doyle's seminal work, premiered in the late 1990s and featured dinosaurs as recurring plot devices. *Dinotopia*, an imaginatively illustrated book that depicts a place where humans and dinosaurs co-exist in near-peaceful harmony, was also produced as a TV mini-series in 2002. Aside from such overt attempts at entertainment, the 2001 BBC-produced documentary series *Walking With Dinosaurs* set a new standard by combining scientific information with startlingly realistic CGI dinosaurs dropped into real, natural environments. The overall effect was to emulate wildlife documentaries. An added twist, however, was to use intermittent brief interviews with dinosaur paleontologists to discuss scientific evidence that supported or refuted some of the dinosaur behaviors depicted in preceding scenes.

On the Internet, Web pages with dinosaur themes are exceedingly abundant and now rival print literature in some respects.

Many web pages with dinosaur themes are non-fiction and attempt to be educational, and some succeed in that goal. However, an increasing number of these pages not only have written material but also showcase works of art as scanned images of drawings, paintings, or sculptures. CGI artwork or computer animations are also more common as people creatively employ sophisticated hardware and software at home. In many cases, the interpretations of some dinosaur behavior in Web pages blend both fiction and self-expression. The Web authors may not be so concerned with scientific accuracy but with entertainment and voicing their speculations on dinosaur behavior. In this sense, fiction is being created without the authors necessarily realizing it, although the same might be said for every scientist who has ever been wrong about an expressed hypothesis.

Dinosaurs as Objects of Art and Artistic Inspiration

The first drawing of a dinosaur bone was in the seventeenth century, but it was interpreted as something entirely different at the time (Chapter 3). Much later, after their public recognition as formerly reptile-like animals, dinosaurs were depicted as dynamic creatures by many nineteenth-century artists. Dinosaurs have been a popular theme in art ever since, portrayed worldwide in drawings, paintings, and sculptures. More recently, multimedia approaches use photography (particularly digital) and computer applications as the means for expressing the artistic qualities

FIGURE 1.6 Comparison of photograph and line drawing of a skull of the Late Jurassic theropod *Allosaurus fragilis* from the Morrison Formation (Late Jurassic) of Utah, USA, showing more easily discernable anatomical details in line drawing. Skull is a replica, formerly on display in the Western Colorado Museum of Paleontology, Grand Junction, Colorado, USA.

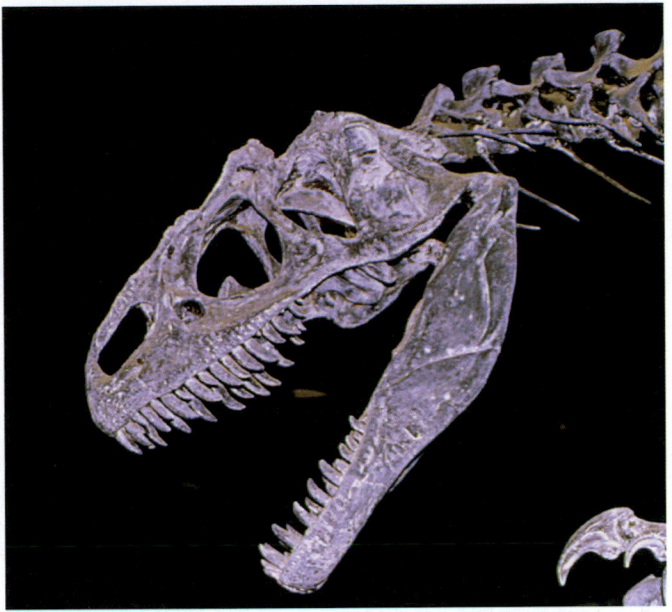

of dinosaur fossils. Artistic renditions of dinosaur appearances and behavior are note-worthy because, like films and television, they reflect basic popular conceptions of dinosaurs. These views of how ancient life and environments have changed through time often accord with scientific progress. Depictions of dinosaurs have been affected by two broad, but often overlapping, influences:

1 science, in the form of **scientific illustration**, which is typically in associ-ation with a scientific text; and
2 **aestheticism**, which is simply the expression of their wonder, beauty, or awe-inspiring power.

Drawings accompanied the first scientific descriptions of their bones in nineteenth-century Europe (Chapter 3). Despite the advent of digital photography and computer graphics, drawings are still a necessary part of dinosaur studies (Fig. 1.6). Some artists depicting dinosaurs are professional scientific illustrators, whose artistic talents lie in combining fossils with living animals while working within the prescribed bound-aries of fact. Serious scientific illustration of dinosaur fossils requires much study of the anatomy, inferred physiology, and behavior of dinosaurs. Not coincidentally, some illustrators are professional paleontologists who honed their observational skills

FIGURE 1.7 A classic painting by Charles R. Knight of the Late Jurassic sauropod *Apatosaurus* (more popularly known as *Brontosaurus*) in an aquatic habitat. First published in *The Century Magazine* (1904) in the article "Fossil Wonders of the West: The Dinosaurs of the Bone-Cabin Quarry, Being the First Description of the Greatest Find of Extinct Animals Ever Made," written by Henry Fairfield Osborn. Transparency No. 2417(5), courtesy of the Library, American Museum of Natural History.

through meticulous drawings of their subjects (Chapter 2). An artist's knowledge base is expanded considerably if the dinosaurs are to be re-created in their original natural environments. Such illustrations necessitate study of non-dinosaurian animals, plants, ecosystems, and landscapes that probably accompanied them. For example, the illustrations of paleontologists Gregory Paul and Robert Bakker often show dinosaurs in their interpreted environmental context. Such works demonstrate that these illustrators are well acquainted with the anatomical traits of their subjects, and are also familiar with evidence for ancient environments.

*The most revered of artists who depicted dinosaurs was **Charles R. Knight** (1874–1953), an American who worked with drawing, painting, and sculpting to fashion portraits of dinosaurs.*

The works of artist Charles R. Knight were so evocative and influential that they arguably constituted the foundation of the popularity that surrounds some of the most famous dinosaurs today, such as *Allosaurus*, *Tyrannosaurus*, *Triceratops*, *Stegosaurus*, and the sauropod *Apatosaurus* (previously known as *Brontosaurus*; Fig. 1.7). Knight's attempts at realistically illustrating dinosaurs as living, active animals were facilitated by his consultations with professional paleontologists and intensive study of his subjects. Knight's enduring images of *Apatosaurus* immersed in bodies of water, and *Tyrannosaurus* confronting *Triceratops*, have served as icons for the popular conception of these dinosaurs, although some of these interpretations of dinosaurs' behaviors changed over the ensuing years. Some of Knight's illustrations reflect hypotheses about dinosaurs that were surprisingly ahead of his time, such as active and agile carnivorous dinosaurs (Chapter 9) and extremely large dinosaurs raising their front feet off the ground (Chapter 10).

The evolution of dinosaur illustrations also reflects the evolution of dinosaur studies, particularly in the past 35 years. Dinosaurs were initially shown as slow, dull-witted, "cold-blooded" reptiles (Chapter 3). However, they are now frequently illustrated as dynamic, reasonably intelligent, and "warm-blooded" bird-like creatures that were unique animals in the history of vertebrate life (Chapters 6 and 8). Not all of the latter presumptions are firmly grounded in science. For example, one prominent and potentially sensory-assaulting genre in dinosaur art, which resulted from the re-interpretation of dinosaurs as bird-like, is the use of garish, near-fluorescent color schemes, as well as inclusion of feathers in dinosaurs not known to have them. In these cases, one must realize that art sometimes fails to imitate life (or death, in the case of fossils). After all, interpretations of dinosaur coloration are based on scanty evidence, and feathers that show some of that evidence for coloration have only recently been reported for relatively few dinosaurs (Chapters 5 and 9). However, recent discoveries of numerous species of feathered dinosaurs in Cretaceous deposits of China now lend some credibility to such fanciful portraits (Chapter 9).

The scientific bases of some dinosaur depictions in art can be questionable in other respects, just as in other aspects of popular culture that attempt to mirror reality. Among the most common mistakes made by illustrators is the inclusion of **anachronisms**, which have dinosaurs or other organisms from different times together. An example of an anachronism is *Stegosaurus* of the Jurassic Period and *Tyrannosaurus* of the Cretaceous Period fighting one another, as shown in the animated film *Fantasia*. *Stegosaurus* died out millions of years before *Tyrannosaurus*. Another error is the juxtaposition of inappropriate environmental vistas surrounding dinosaurs, such as volcanoes in areas where there is no scientific evidence that volcanism occurred. Nevertheless, such unscientific portrayals are still potentially valuable for application of the scientific method and critical reasoning skills (Chapter 2). Simply because a dinosaur is shown behaving a certain way in an illustration can promote inquiry into what evidence may support such a depiction.

Throughout this book, there are many opportunities to critically examine the fossil evidence for dinosaurs with regard to their behavior and evolution. Such analyses then can be compared with previous conceptions of dinosaurs and how dinosaurs are depicted in popular culture. Some depictions may actually reflect current scientific knowledge about dinosaurs, but such accuracy may have been unintentional. Just because dinosaurs in a movie, television show, fictional book, or artwork are shown behaving in agreement with modern scientific knowledge does not mean that the producers of these works did their homework. Nevertheless, with all of these scientific caveats in mind, one can still appreciate the beauty of a well-done dinosaur illustration. This is regardless of the fact that the dinosaurs are reconstructed as living, breathing animals or portrayed through the earthy, static realism of their fossils.

Popular Culture and Science

Dinosaur Models and the Estimation of Dinosaur Weights

An example of how science, art, and popular culture can be combined is through information derived from models of dinosaurs. Dinosaurs are often associated with huge sizes, but how can the question "How big were dinosaurs?" be answered? This book refers to the kilograms or metric tonnage (1000 kg, which equals 2200 pounds) of a particular dinosaur, even though no one has actually weighed a living (or even recently dead) one. Arriving at such figures requires a few simple

principles of physics, a little bit of math in the form of biometry, and some help from the dinosaur models.

Dinosaur models, usually encountered in toy stores or gift shops of natural history museums, are a form of mass-produced "artwork" for which the artists are usually not credited. Nonetheless, many of the models are based on at least some scientifically-derived estimates for dinosaur morphology. Moreover, they are sometimes scaled to a standard size in relation to a full-sized species of dinosaur. Armed with these models, a vessel containing water, some measuring tools, and a little bit of knowledge, the approximate weight of a dinosaur can be calculated.

Weight is a measurement of the amount of force exerted by **gravity**, which is caused by the attraction of the matter for matter. In the case of the Earth, the force of gravity is expressed by the following equation:

$$F = Gm_1m_2/d^2 \tag{1.1}$$

where G is the **gravitational constant** (9.8 meters/second^2); m_1 and m_2 are the masses of the objects attracted to one another (one of them being the Earth, the other being any other object); and d is the distance separating the two objects. The force is measured in **newtons** (**N**), expressed as kg/m/s^2. This shows that weight, in this case, is a force expressed by the mass of an object multiplied by the acceleration that is imparted to it from its attraction to the Earth. As a force, a person's weight will vary very slightly on the Earth's surface. This variation depends on whether a person is directly over an area of the Earth with slightly more or less mass interacting with their mass, as well as the distance between those two masses. For dinosaurs that had much mass, which we have interpreted on the basis of the large size of their skeletal parts and inferred musculature, a logical conclusion is that they correspondingly had much weight.

If a scale was not to hand to measure someone's weight, it could still be estimated on the basis of two parameters:

1 **volume**, which is the three-dimensional space occupied by a certain amount of matter and normally expressed in cubic centimeters (cm^3); and
2 **density**, which is the mass of that matter divided by volume and expressed in grams per cubic centimeter (g/cm^3).

Dipping someone into a bathtub and measuring the volume of water displaced could measure the volume. For example, once immersed, the person might displace 72.0 liters of water, which converts to 72,000 cm^3 (because 1.0 ml = 1.0 cm^3 = 1.0 g, with pure water as a standard). Because the human body is mostly composed of water, its density is also close to that of water, about 0.9 g/cm^3. To find out the weight, simply multiply mass by volume, where **W** is weight, **d** is density, and **v** is volume:

$$W = dv \tag{1.2}$$

Step 1. $W = 0.9$ g/cc $\times$ 72,000 cc = 64,800 g
(converting to kilograms)
Step 2. = 64,800 g $\div$ 1000 g/kg = 64.8 kg

The present mass of the Earth is assumed to be identical to that in the Mesozoic Era. A model of a tyrannosaur, scaled at 0.033 (3.3%) of the original size of the dinosaur, would displace 235 ml (235 cm^3) of water, if fully immersed. However, the assumed density for the tyrannosaur is 0.8 g/cc, which is less dense than a person because of the degree of "hollowness" in some dinosaur bones (Chapter 8). Is 0.8 g/cm^3 then multiplied by 235 cm^3? No, because the tyrannosaur must be made "larger" by scaling it to life-size. This means recognizing that 3.3% is about equal to 1/30 and that it had three dimensions (length, width, height), which corresponds approximately to its original volume. Thus, scaling involves making the tyrannosaur 30 times longer, wider, and higher than the model, which results in the following volume change, where V is volume, l is length, w is width, and h is height:

$$V = lwh \tag{1.3}$$

Step 1. $V = 30 \times 30 \times 30 = 27{,}000$ times the volume of the model

Using this volume increase and multiplying it by the density and the measured volume yields the following results for the tyrannosaur:

$$W = dv_1v_2 \tag{1.4}$$

Step 1. $W = 0.8$ g/cm$^3 \times 235$ cm$^3 \times 27{,}000 = 5{,}076{,}000$
           (Converting to kilograms)
Step 2. $= 5{,}076{,}000$ g $\div$ 1000 g/kg $= 5076$ kg
           (Converting to metric tons)
Step 3. $= 5076$ kg $\div$ 1000 kg/ton $= 5.076$ metric tons

where W is weight, d is density, v_1 is measured volume, v_2 is the volume increase.

Hence an initial estimate of how much a particular dinosaur weighed can be calculated. This is probably not accurate, because the first assumption is that the model is an accurate representation of the dinosaur. This assumption is made despite the fact that many species of dinosaurs are known from less than 90% complete skeletons. As a result, their reconstruction is sometimes sketchy (Chapters 6 and 7). Furthermore, not all model-makers are concerned with constructing scientifically accurate figures. Another assumption is that the density was 0.8 g/cm^3, whereas other researchers have made estimates of 0.9–1.1 g/cm^3.

Alternative methods have been used for estimating dinosaur weight. One method uses measurements of leg-bone circumferences of extant mammal species and correlates these data with animal weights. This results in different values for dinosaurs, suggesting that either method might work, or not.

The important point here is that some artistic interpretations of dinosaurs, which are based on at least some available scientific information, can be tested in a scientific manner for their feasibility. Such tests can demonstrate that any supposed gap between science and popular art is not as wide as we sometimes think. These weight estimates derived from models also help us to better appreciate the possible weights of some dinosaurs relative to living animals. For perspective, an adult African elephant can weigh 5 metric tons, which is about the same weight as our hypothetical tyrannosaur. Realizing that a carnivore, such as *T. rex*, may have weighed as much as an African elephant adds a sense of realism to it that transcends models, paintings, or photographs of its remains, and brings it more to life.

SUMMARY

Because dinosaurs are an important part of popular culture and hence are easily recognizable, the study of them serves as an apt vehicle for understanding how science is applied to their study. A starting point for applying the science of dinosaur studies is to understand what is or is not a dinosaur, using a definition as a prompt for asking questions, as a large number of animals regarded as dinosaurs actually are not. A dinosaur is defined initially as a reptile-like or bird-like animal, with an upright posture, that spent most (perhaps all) of its life on land and lived from about 230–65 million years ago. Dinosaurs then can be classified by either a Linnaean or phylogenetic (cladistic) classification system. The cladistic method is preferred because it better expresses hypotheses about evolutionary relatedness within dinosaurs as a group. These hypotheses are best described through a cladogram, a diagram that shows ancestor-descendant relationships.

The two sciences most commonly associated with dinosaur studies are geology and biology, which are also augmented by other sciences, such as chemistry, physics, math, and computer science. Their use illustrates how the interrelation of all sciences can contribute to a field of study. Despite the apprehension of many people about the sciences, especially those that frequently use symbols and numbers, it is necessary to know a minimal amount about them to better understand dinosaurs. Professional paleontologists typically have to know some facets of all scientific disciplines. In many cases they also must be illustrators, writers, public speakers, and deal with the physical and logistical difficulties of performing fieldwork in remote locations.

Popular culture, such as books, TV shows, movies, artwork, and Web pages, reflect public ideas about dinosaurs that may or may not be based on scientific reality but they can follow general scientific trends. Whenever encountering these images of dinosaurs, the question of "What evidence justifies these depictions?" should be asked. However, of all dinosaur artwork, scientific illustration is the most important with regard to dinosaur studies and combines scientific knowledge with artistic abilities to convey accurate information.

Math is an essential tool for dinosaur studies and is expressed mostly through measurements, which are made through the international standard of the metric system. Math can be used in nearly every aspect of dinosaur studies, as demonstrated by the use of some simple calculations of estimated dinosaur weights based on their models. Such step-by-step methods help to show that math has practical uses in dinosaur studies and can be made more understandable in an applicative context.

DISCUSSION QUESTIONS

1. Name three examples of animals that you thought were "dinosaurs" before reading the definition given at the beginning of the chapter. How did application of this definition help you to change your mind?

2. Think of what animals today are bipedal and have an erect posture. What might be an advantage or disadvantage of bipedalism and erect posture, as a direct response to some change in the animals' environments?

3. Think of what large animals today live most of their lives on land but are also capable of swimming. What evidence would you need to show that some large dinosaurs sometimes swam?

4. Current archaeological evidence indicates that the oldest Egyptian pyramid was built about 2800 BCE. Calculate the percentage its age comprises out of 65 million years, which is about when the last dinosaur died.

5. Is a fossil egg a body fossil or a trace fossil? Explain why.

6. Which of the sciences outlined in the chapter would you use to answer these questions about dinosaurs and why? (You may choose more than one.)
   a. How high could they have jumped?
   b. How did they mate?
   c. Did dinosaur bones get buried quickly or did they lie out in the open for a long time?
   d. Does a deep track left by a dinosaur mean that it weighed a lot?
   e. How long was a particular dinosaur if you only find its leg bones?
   f. What food did dinosaurs eat and how did they digest it?
   g. Did dinosaurs make sounds and, if so, what did they sound like?
   h. What did dinosaurs smell like?
   i. Where did dinosaurs live?

7. Give an example of what you think is an ecosystem and outline some of its physical characteristics, such as vegetation, rainfall, and temperature. How might dinosaurs have interacted with or been affected by some of these environmental factors?

8. Think of two examples of how you learned about some concept of dinosaur behavior through reading a book, seeing a TV show, or watching a movie. Did you assume that these portrayed behaviors were based on actual research? If so, what aspects of the portrayal convinced you of that?

9. Flip through this entire book and pick out your top five favorite illustrations (either photographs or drawings). Why did you make this choice?

10. Estimate the weights of the following dinosaurs using the described models.
    a. *Camarasaurus*: 172 cm^3, 2.5% scale, density of 0.9 g/cm^3
    b. *Allosaurus*: 220 cm^3, 3.3% scale, density of 0.8 g/cm^3
    c. *Pachycephalosaurus*: 215 cm^3, 5.0% scale, density of 0.95 g/cm^3
    d. *Brachiosaurus*: 546 cm^3, 2.5% scale, density of 1.05 g/cm^3

Bibliography

Alexander, R. M. 1989. *Dynamics of Dinosaurs and Other Extinct Giants*. New York: Columbia University Press.

Bakker, R. T. 1986. *The Dinosaur Heresies*. New York: Kensington Publishing.

Crichton, M. 1990. *Jurassic Park*. New York: Alfred A. Knopf.

Currie, P. J. and Padian, K. (Eds) 1997. *Encyclopedia of Dinosaurs*. San Diego, California: Academic Press.

Currie, P. J. and Tropea, M. 2000. *Dinosaur Imagery: The Science of Lost Worlds and Jurassic Art (The Lanzendorf Collection)*. New York: Academic Press.

Czerkas, S. and Glut, D. F. 1982. *Dinosaurs, Mammoths, and Cavemen: The Art of Charles R. Knight*. New York: E. P. Dutton, Inc.

Farlow, J. O. and Brett-Surman, M. K. (Eds) *The Complete Dinosaur*. Bloomington, Indiana: Indiana University Press.

Feldman, H. R. and Wilson, J. 1998. The godzilla syndrome – scientific inaccuracies of prehistoric animals in the movies. *Journal of Geoscience Education* **46**: 456–459.

Glut, D. F. and Brett-Surman, M. K. 1997. "Dinosaurs and the media". *In* Farlow, J. O. and Brett-Surman, M. K. (Eds), *The Complete Dinosaur*. Bloomington, Indiana: Indiana University Press. pp. 675–706.

Gould, S. J. 1991. "The dinosaur rip-off". *In* Gould, S. J., *Bully for Brontosaurus*: New York: W. W. Norton and Company. pp. 94–106.

Haste, H. 1993. Dinosaur as metaphor. *Modern Geology* **18**: 349–370.

Henderson, D. 1997. "Restoring dinosaurs as living animals". *In* Farlow, J. O. and Brett-Surman, M. K. (Eds), *The Complete Dinosaur*. Bloomington, Indiana: Indiana University Press. pp. 165–172.

Jones, S. 1993. *The Illustrated Dinosaur Movie Guide*. London: Titan Books.

Lockley, M. G. and Wright, J. L. 2000. Reading about dinosaurs – an annotated bibliography of books. *Journal of Geoscience Education* **48**: 167–178.

Norell, M. A., Gaffney, E. S. and Dingus, L. 1995. *Discovering Dinosaurs in the American Museum of Natural History*. New York: Alfred A. Knopf, Inc.

Norman, D. B. 1985. *The Illustrated Encyclopedia of Dinosaurs*. London: Salamander Books, Ltd.

Psihoyos, L. and Knoebber, J. 1994. *Hunting Dinosaurs*. New York: Random House.

Rossbach, T. J. 1996. *Fantasia* and our changing views of dinosaurs. *Journal of Geoscience Education* **44**: 13–17.

Torrens, H. S. 1993. The dinosaurs and dinomania over 150 years. *Modern Geology* **18**: 257–286.

Weishampel, D. B., Dodson, P. and Osmólska, H. (Eds). 2004. *The Dinosauria* (2nd Edition). Berkeley, California: University of California Press.

Chapter 2

While you are on a plane, the man in the seat next to you notices that you are reading this book and starts a conversation with you about dinosaurs. Soon he begins to tell you that he heard a "theory" that dinosaurs and people actually lived at the same time, and that the "proof" is represented by some tracks in east Texas that show that dinosaurs and people were walking in the same area at the same time. When you ask where he heard this, he replies that he "read it on the Internet," but he knows lots of other people who also believe it. When you express your skepticism about his claim, he says, "Well, that's what I believe. Besides, you don't have any proof that dinosaurs and people didn't live at the same time. That's just your opinion."

In what ways can you use scientific methods to comment on your traveling companion's methodology?

Overview of Scientific Methods

2

Importance of Scientific Methods

Science in Both Paleontology and Everyday Life

Paleontology is a science and, as with any science, paleontologists test hypotheses to see if they are wrong. If an alternative hypothesis is better supported by the evidence and analysis than a previously accepted hypothesis, or if the previous hypothesis is shown to be false, then it is rejected. Examples of rejected hypotheses in dinosaur paleontology include:

1 some very large dinosaurs lived most of their lives immersed in water (e.g., Fig. 1.7; Chapter 10);
2 bipedal dinosaurs stood upright and walked with their tails dragging on the ground (Chapters 9, 11, and 14); and
3 dinosaurs were large reptiles and behaved like modern reptiles (Chapter 8).

Paleontologists who have concluded that a currently accepted hypothesis is wrong must write their results coherently, in many cases accompanying their written evaluation with photographs or other illustrations. They then send these reports to colleagues, who they know will be honest in their evaluation of them. After the reviews are completed, they will discuss any criticisms with those colleagues. They may send the revised draft of their report, perhaps with new or changed illustrations, to other reviewers. Finally, they may need to present their hypothesis, either written or orally, in a public forum to other paleontologists who are well acquainted with their subject. If, at the end of this process, the evidence still supports their new hypothesis, it will be conditionally accepted – that is, until someone else provides sufficient evidence to persuade the paleontological community otherwise.

The study of dinosaurs is largely a science, so knowing how it works gives us an appreciation of science in general. This knowledge is useful, even if one does not intend to become one of the few hundred professional dinosaur paleontologists distributed worldwide. For example, deciding whether to take an umbrella before leaving home in the morning may involve the use of scientific methods. Evaluating a potential home before deciding whether to move into it can also use scientific methods. Deciding who receives a vote in an election may use scientific methods. Properly assessing the factual content of a news story necessitates scientific methods. If applied properly, they constitute an excellent way to make informed decisions. This is why many students who perceive themselves as non-scientists are actually scientists (albeit non-professional) in the

Scientific methods constitute a form of evidence-based reasoning that everyone uses in their everyday life.

sense that they have been actively applying scientific methods in some facet of their lives. If they look closely enough, they will find sufficient evidence to disprove their initial perception that they are non-scientific.

What Is a Fact?

A **fact** is a phenomenon that has an actual, objective existence. For humans to understand facts, observations must be made of them. Contrary to the old adage "seeing is believing," these observations are not necessarily visual, but might be gathered through other senses. Observations regarding everyday facts could include seeing a sunset, smelling a flower, or hearing thunder. However, what is considered as a fact can change in terms of how it is interpreted. For example, the sun can be observed to move through the sky, which was at one time interpreted as evidence that the sun moves around the Earth. Of course, we now know that the Earth revolves around the sun, and it is the Earth that is moving. The Earth's rotation and its revolution around the sun, however, did not suddenly become real as soon as humans realized that these were the actual processes. So, to qualify as facts, actual and objective phenomena should exist independent of human perceptions.

Observations do not have to be *direct* to provide facts. After all, no one has actually seen an atom or a bacterium without the aid of instruments, yet no rational person doubts their factual existence. Phenomena detected by animals (see box) have no less existence just because humans cannot perceive them. Those that lie beyond the unaided sensory realm of humans can be detected through tools that amplify their effects. Examples of these are mass spectrometers that count atoms (Chapter 4) and microscopes that provide magnified images of bacteria. Direct observations that deal with dinosaurs might include seeing a footprint made by one or feeling a dinosaur skeleton, but *indirect* observations might include detecting what chemicals compose their eggs or looking at dinosaur bone structures through a microscope. Regardless of whether these observations are direct or indirect, they qualify as facts because they are based on objects that actually exist. Facts are ideally undeniable, although some observations can lead to different interpretations. Consequently, explanations for those facts are subject to debate and are malleable, but facts constitute evidence, which is the foundation of scientific methods.

Unlike humans, canines can hear in frequencies beyond normal human hearing, and some birds can see in the ultraviolet part of the electromagnetic spectrum.

Interestingly, different forms of evidence in paleontology are treated as being less or more direct evidence of ancient life. Body fossils, such as shells, bones, eggs, feathers, and skin impressions, are often considered as more directly relating to ancient life than trace fossils, such as burrows, tracks, trails, nests, toothmarks, and feces (Chapter 14). An analysis of which type of fossil evidence is held in the higher regard by paleontologists can be conducted by simply examining cover photographs or illustrations of science journals. The clear and overwhelming favorite is body fossils, and the majority of these are dinosaurs or fossil humans. An independent test of this favoritism can then be applied to the articles in the journals. Again, those that deal with body fossils are much more common than those about trace fossils, despite the fact that trace fossils made by these same organisms may be much more common in the geologic record. Nevertheless, trace fossils are now more highly regarded than in the past because paleontologists who study them are promoting their intrinsic value in interpreting, for example, ancient behavior (Chapter 14). In paleoanthropology, controversy raged for a long time over whether ancient hominids from 3.5 million years ago walked upright or not, and the conflicts were all based on interpretations of a few fragmentary skeletal

FIGURE 2.1 Hominid footprints overprinting other hominid footprints, suggesting that one hominid purposefully stepped into the footprints of another preceding it. Cast of original trackways, which were preserved in 3.5 million-year-old volcanic ash in Laetoli, Tanzania.

remains. But the controversy was mostly put to rest when paleoanthropologists found footprints attributed to these same hominids in a 3.5 million-year-old deposit. The tracks not only showed upright walking, but an ease with it, as one individual had purposefully stepped in the tracks of a preceding one (Fig. 2.1). Moreover, a nearby smaller individual, possibly a juvenile, also showed the same evidence of bipedalism, meaning that three individuals were all walking in the same way. In this case, and in many others, body fossils and trace fossils certainly constitute different forms of fossil evidence, with one not being necessarily better than the other. The scientific significance of the evidence depends on the factually-based quality of the fossils themselves and how carefully the associated observations are recorded and interpreted.

Facts are described through collecting **data** ("data" being the plural form of "**datum**"), which comprise the recording of observations. Not all data are created equally, however, and the quality of the descriptive methods and the classification of data are important for distinguishing what is useful for science and what is not. For example, a dinosaur bone might be seen protruding from the ground by two different observers, who respectively record their data as follows:

OBSERVER A: There was this big dinosaur bone, but not too big, which looked gray, like my grandmother's hair, and it was sticking out of the dirt.

OBSERVER B: The object had a linear trend and was 12.5 cm wide with an exposed length of 24.7 cm. It also had millimeter-wide parallel striations running the length of it, a light to medium gray overall color, and a noticeable but slight widening to its distal, rounded end. The host sediment was fine-grained sand mixed with hematitic clay, and the object was protruding at about a 20-degree angle with respect to the horizontal plane of the ground surface.

Observer A showed some promise and laudable enthusiasm, but did a poor job overall of collecting any meaningful data that could be classified or communicated readily to others who did not observe the bone. Observer B used a combination of verbal description and numbers in the data collection, and used a minimum of interpretation (the object was not even identified as a "dinosaur bone" or any other type of bone). Note that the fact of the dinosaur bone's existence does not change with either description. As the preceding example shows, however, the way the bone is described can differ considerably, and if done inadequately can inspire doubt in other potential observers about the factual existence of the bone.

The example also shows some methods of data collection and how data are classified. Data can be collected through either qualitative or quantitative methods. **Qualitative methods** typically include using oral or written descriptions of the observed phenomena, as well as illustrations. The latter can be diagrams, sketches, or photographs, which are particularly useful for summarizing a large amount of information without added verbosity. **Quantitative methods** involve the use of measurements and the recording of the numbers associated with them; such measurements may be then described further through statistics and equations. Qualitative and quantitative methods can reinforce one another, such as when a diagram depicts visually what otherwise may be complex mathematical relationships (Fig. 2.2). A cladogram (see Fig. 1.3) is an example of a diagram that combines the results of qualitative and quantitative methods. It is based on observations of anatomical traits, then statistical analyses of the data are used to hypothesize which organisms are the most closely related to one another (Chapter 5).

Once qualitative and quantitative data are carefully collected and communicated to other people, facts become clearer to observers. For example, people have repeatedly observed falling objects and have collected data from these observations, leading them to conclude that gravity is a fact. People have observed repeatedly nuclear reactions and collected data on them, thus they now realize that the effects of nuclear physics are factual. People have observed repeatedly the effects of the development of new species over time and have collected data on these effects, eventually resulting in the knowledge that biological evolution is a fact. Because people have observed repeatedly many bodily remains or traces of dinosaurs and collected much data on them, they also know that the former existence of dinosaurs is a fact. Because the explanations for these observations are equivocal, however, science does not stop with just the gathering of facts. In science, facts and how they occur as real phenomena require interpretations, not just acceptance of their existence.

FIGURE 2.2 Scientific assessment of the skull of *Coelophysis baurii*, a Late Triassic theropod. (A) Skull of adult *Coelophysis*, showing qualitative traits (two holes in rear right side of skull, prominent eye socket, sharp teeth); Denver Museum of Science and Nature. (B) Bar graph of skull lengths ($n = 15$) for *Coelophysis bauri*, arranged in order of increasing length. Based on data from Cope (1887) and summarized by Colbert (1990).

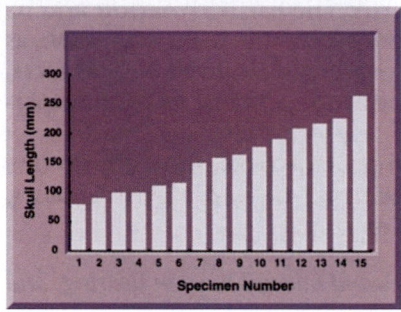

What Is a Hypothesis and How Is It Tested?

A **hypothesis** is a conditional explanation of an observation, or series of observations, that typically proposes a cause for the observations (Fig. 2.3). A hypothesis results from asking the questions:

- "What caused the observed fact?"
- "How did the observed fact occur?" and
- "Why did the event that I observed happen, and not some other phenomenon?"

Important characteristics of a hypothesis are that it must be:

1 testable;
2 falsifiable;
3 based on independently verifiable and observable factual information; and
4 used to make predictions.

In paleontology, the last of these qualities is about predicting future discoveries rather than future experiments, because all fossils and sediments that covered them occurred in the past. Paleontologists, therefore, make retrodictions, not predictions.

Testing a scientific hypothesis means that another observer can make more observations repeatedly with regard to the hypothesis. An idea that does not have

FIGURE 2.3 Example of the difference between fact and hypothesis, as well as description and interpretation. Facts: Two features occur on the surface of a rock in Big Bend National Park in western Texas, USA. Both features show semi-circular and crescentic lines emanating and expanding from a point, but one feature is larger than the other and the larger one is also concave, whereas the smaller one is convex. Hypotheses to explain the features:

1 They are completely unrelated forms of unknown origin;
2 They are odd patterns made in the rock when it fractured on this surface and their similarity in form is a result of the uniformity of the rock, which controlled the fracturing;
3 They are plant leaves that were bent upward and downward in the sediment soon after they were buried in the sediment that later formed the rock;

4 They are trace fossils made by animals that lived long ago in the sediment, where the animals were progressively feeding out from a central point but one went up and the other went down;
5 They are the outside and inside imprint of animal bodies, where the animals were similar but just had different sizes at the time they were buried;
6 They are carvings made in the rock by Native Americans, before Europeans settled in the area, that symbolize the light coming from the sun (the larger one, because it gives more light) and the moon (the smaller one because it gives less light, and also carved in opposite relief to the "sun" to symbolize the opposites of night and day).

any factual evidence supporting it is not considered a scientific hypothesis because it cannot be subjected to any testing. For example, the following statement is not a scientific hypothesis:

Some dinosaurs were invisible, weightless, and left no bodily remains or other traces of their existence.

No verifiable evidence can be gathered independently to either support or test this idea even once, let alone repeatedly. As a result, belief in the former existence of such dinosaurs would be entirely based on faith, not evidence. This is not to say that faith-based reasoning is wrong, just that it does not qualify as science.

Testing of a hypothesis, usually through collection and analysis of data, and review of reports that interpret the data, can lead to any of three possible outcomes at a given time:

1 complete rejection of the hypothesis;
2 complete acceptance of the hypothesis; or
3 modification of the hypothesis that accepts part of it and rejects part of it.

The third of these outcomes is the most common; scientists rarely come to agreement on all details of a hypothesis! Nevertheless, if the main points of a hypothesis are supported after further testing and not disproved, then it is conditionally accepted

but with the acknowledgment that it might later be proved wrong. A researcher also may have **multiple hypotheses** proposed as alternative explanations for the observations, but each of these must be tested for their veracity through the same methodology. The researcher also must be open to any new evidence that supports one of the originally rejected hypotheses, which would prompt a reinvestigation of their explanations.

The original investigators of some observed phenomena, who summarize the results of their testing and conclusions in a presentable form, typically test a hypothesis first. Then the results and conclusions are given to reviewers, who critically examine the evidence and explanations of the observations. Experts on the subject may attempt to repeat the methods and results described by the investigators, so they can compare the proposed results to their own experiences. This procedure is called **peer review**, a form of independent confirmation that is an essential part of the formal scientific process. In dinosaur studies, any potentially new discovery of dinosaur body or trace fossils is followed typically by a process where the investigators analyze their find, test results from the analyses, summarize and illustrate their find in a report, and submit that report to recognized dinosaur experts for peer review.

If a scientific report, with its hypotheses, is accepted for publication, then it is shared in topic-specific journals or at professional meetings with scientists who have similar interests. In the latter situation, papers are given as formal presentations in front of peers, either as a talk or a poster. These papers then undergo more peer review from those who view these presentations. This process means that just because a paper is accepted for publication or presentation does not mean that it is correct. It may still be disproved or modified by further critical analysis from the scientific community, sometimes many years after it was conditionally accepted.

The original manuscripts of some science books also undergo peer review. Because of this variability in procedure in comparison to most journals, the material presented in books should be examined for evidence of peer review before accepting that any hypotheses within it are scientifically based. In fact, this book underwent peer review and was considerably improved in its scientific accuracy through that method, although it still may contain some factual errors and disproved hypotheses. Fortunately for the students using this book now, it is a second edition. This means that many, although probably not all, of the mistakes from the first edition (written in 2000) were corrected and new evidence and hypotheses were added. It is also largely a **secondary source**, which means that little of the information presented here represents original dinosaur research done by the author. Likewise, many books that are considered as reliable sources of information, such as encyclopedias or textbooks, are actually at least one step removed from their original sources of information. Thus, these are more liable to error because, for example, the authors could have misinterpreted the works of others. An analogy to this situation can be illustrated by photocopying a document, then photocopying each successive photocopy; after about 20 reiterations, the words from the document may be unreadable.

Internal documents written in private corporations or in some government agencies, even if done by scientists who use scientific methods, are also not considered as peer reviewed because they are rarely shared with the rest of the scientific community for evaluation. Reports that are issued from such entities must be examined very critically. This is especially the case if they have conclusions that positively affirm the mission statement of the corporation or government agency with no additional self-critique (hence indicating a possible bias). Such distortions are a result of *a priori* reasoning, where conclusions are first accepted as correct, then "facts" are selected afterwards on the basis of how well they conform to the conclusions. Recent examples of such misuse of internal reports were those written by tobacco-company scientists, which were later submitted as evidence by tobacco companies in US courts, and revisions to a US EPA (Environmental Protection Agency) statement on global warming. These documents were not submitted to scientists who had interests differing from the companies' or government agency's goals for peer review. Similarly, some pharmaceutical firms specify that if the results of testing performed under contract by scientists from outside of the firm, such as those at universities, contradict the firm's commercial interests, the firm will reserve the right to prevent the scientists from publicly disclosing the information. Such scientific research falls under the realm of **proprietary** information (owned by the company), which means that much of the science that goes on in private corporations is not revealed to the worldwide scientific community. Some instances in which proprietary science does reach the rest of the mainstream scientific community, however, are from petroleum and mining companies that have shared their results with paleontologists or geologists, including discoveries of some important fossils.

In the past decade, the widespread use of home computers and Web pages has also revolutionized how publishing is done.

As a result, valid scientific hypotheses have been confused with ideas that have no factual bases. Formal peer review is rarely applied to the vast majority of Web pages, so a plethora of seemingly informative sites may be only expressions of the authors' imaginations. In fact, because single individuals might produce Web pages, they rarely go through an editorial process and usually lack any outside input. Peer-reviewed scientific journals, however, are now becoming more commonly adapted to the Web, and some journals, such as *Palaeontologia Electronica*, are published entirely online. Although this practice is becoming more normal, most web sites that propose supposed hypotheses should be viewed with a critical eye.

Similarly, a common mistake associated with some scientific discoveries is for researchers to go to press too quickly with their results. They will let newspapers, magazines, or web sites publish their claims, rather than going through peer-reviewed journals first. The rapidity of publication in the popular media typically causes the promotion of hypotheses about scientific discoveries well before they have been scrutinized through peer review. Furthermore, mainstream journalists are rarely expert enough in any scientific field to sufficiently perform their own peer review. This means that hypotheses seen in print, reported on television, or published on a web site may not necessarily be based on fact. This circumstance of scientific uncertainty about what is reported is an especially common problem with dinosaur finds because of the inordinate amount of media attention given to such fossils.

Two examples of dinosaur-related stories that went to the popular press before they had adequate peer review involved discovery of a dinosaur track and of a feathered theropod. In the former example, a news report in 1998 stated that a paleontologist had found dinosaur tracks in Bolivia, indicating that the dinosaur that made them was about 350 meters long. Considering that all dinosaurs described over the past 175 years of dinosaur studies were much less than 100 meters long, either the information presented to the media was incorrect or it was incorrectly reported. One hypothesis to explain this discrepancy is that the original information was garbled when the paleontologist relayed it to the media, such as the dinosaur trackway may have been 350 meters long, not the dinosaur. This indeed turned out to be the case, but an unquestioning reporter placed the erroneous information into the story anyway. In the second example, a fossil found in Lower Cretaceous rocks of China in 1999 seemed to be that of a feathered theropod with other features shared by both theropods and birds. The fossil was quickly given a species name, *Archaeoraptor liaoningensis*, and it was first seen in print in a mainstream, popular magazine that had partially sponsored the research. Later examination of the fossil revealed that it was actually a very clever forgery, having been

FIGURE 2.4
Comparison between *Archaeopteryx*, a Late Jurassic bird, and *Compsognathus*, a Late Jurassic theropod, as an approximate example of a transition between fossil forms as predicted by Darwin's hypothesis of natural selection. Reprinted by permission. From Paul (1988), *The Predatory Dinosaurs of the World*, Simon & Schuster, NY, p. 115.

composed of rocks pasted together that contained two different animals, a fossil non-avian theropod and a fossil bird. Ironically, both of the fossils were new to science at the time too, which initially gave it added credibility! Unfortunately, such mistakes can damage the credibility of a genuine find. Because of this potential pitfall, scientists are often quite reticent about reporting their preliminary findings when interviewed by mainstream journalists. Also, some journals will reject a paper if the results were previously published in a mainstream source, especially if the authors of the paper actively sought the publicity before they submitted it to the journal.

The ability to predict future observations on the basis of a hypothesis is one of the most effective and powerful ways to test its relative strength, and is central to scientific methods. An example of this predictability in paleontology was Charles Darwin's book *On the Origin of Species*, published in 1859, in which he predicted that transitional forms between major groups of organisms would be found in the fossil record. This prediction was followed two years later by the discovery of *Archaeopteryx*, a Late Jurassic fossil that shows numerous shared characteristics of dinosaurs and modern birds (Fig. 2.4). Hypotheses that do not predict observations in such a manner are incomplete (although not necessarily wrong), and consequently may not be built on a firm scientific foundation.

Two sequential steps can summarize the essence of hypothesis building: description and interpretation. The *description phase* involves the gathering of data (observations), which should be as meticulous and detailed as is humanly practical. For example, an analysis of hundreds of dinosaur bones might involve measuring and describing every feature of each individual bone, then performing statistical analyses of the quantitative data and verbal summaries of the qualitative data. Another description might require measuring and describing a dinosaur trackway that extends for 50 meters. Yet another description might be preceded by crawling on hands and knees in the hot summer sun to count the number of dinosaur eggshell fragments in a meter square. Descriptions, however, should be done with some objective in mind, such as testing hypotheses. The descriptive step requires extraordinary patience and trust in a process that has no guarantee of success.

Nearly all researchers consider the *interpretation phase* to be the most exciting part of the scientific process. In this phase, imagination is encouraged but, of course, only within the confines of what is described by the data. This is when scientists say that the spike on an iguanodontian hand was used for defense against predators (Chapter 11). This is when they explain that the missing tracks in a dinosaur trackway represents the dinosaur hopping on one foot (Chapter 14). This is when they say that the eggshell fragments in a dinosaur nest were broken originally by an egg-stealing dinosaur (Chapter 9). But this is also the phase when they might endure the critical scorn and derision of the rest of the scientific community, especially if they made major mistakes during their first step, the description. Good interpretations are nearly always preceded by good descriptions, although good descriptions are not guarantees of good interpretations.

If the first step is done well, then the second step may eventually result in an explanation that will satisfy most scientists. This is the case whether that explanation is based on original descriptions or a reinterpretation of the descriptions of others, maybe long after those original researchers have died. A hypothesis should not be made with the expectation that it will please all scientists. Although the complete dismissal of egos is unrealistic, scientists should also expect to develop a skin as thick as an ankylosaur (Chapter 12) and separate themselves personally from their work. As a scientist or thinking human being, getting used to constructive criticism and learning from it each time should result in improvement with each new attempt to answer the questions: "What is this?" or "How did this happen?"

What is a Theory?

A **theory** is a hypothesis, or set of related hypotheses, that withstands repeated testing to the point of widespread acceptance by the scientific community. Moreover, theories interrelate and overlap with one another; they do not stand alone in isolation from one another. Because they are also typically based on interrelated hypotheses, theories are still subject to further testing and are potentially falsifiable, but the likelihood of their being proved absolutely wrong is unlikely. At worst, theories are refined and better understood with time. Among the best-known theories are:

1 **gravitation**, which explains observations of the attraction of matter for matter;
2 **biological evolution**, which clarifies observations of organisms that are (or were) modified through descent;
3 **atomic theory**, which gives reasons for observed behaviors of atomic and subatomic particles, and
4 **plate tectonics**, which offers explanations for observed geologic phenomena, such as earthquakes and volcanoes that occur in definite places on the Earth.

Of these theories, plate tectonics is the youngest but, as will be explained later (Chapter 4), sufficient evidence has accumulated during the past 35 years for it to be accepted by the vast majority of geologists worldwide, and will very likely serve as a working model of global processes for the future.

Hypotheses that have not yet withstood peer review and ideas that have no supporting factual evidence are often mistakenly called theories. For example, some paleontologists might say that they have a "theory" about what types of nests were left by pachycephalosaurs (Chapter 13). Other paleontologists, however, might examine this statement and find that no pachycephalosaur eggs, nests, or even embryonic skeletons of these dinosaurs have ever been interpreted from the geologic record. Therefore, such a theory is actually a weakly supported hypothesis, having been based entirely on the factual existence of pachycephalosaurs and the probability that they laid eggs.

A theory is often distinguished by its universal applicability, whereby scientists in most countries of the world, regardless of their cultural, religious, or political backgrounds, are applying major concepts of a theory for practical reasons. Examples include the use of plate tectonic theory for earthquake prediction, evolutionary theory for antibiotic development, quantum physics for radiometric age dating (Chapter 3) and nuclear reactors, and gravitation for space exploration. Consequently, these theories are being tested every day. Continued use of a theory and its constant passing of these daily tests can only mean that it is being refined and improved through such practical applications. Thus, the admonition that an integrated concept in science, which has withstood the tests of generations of skeptical scientists worldwide, is "just a theory" exposes considerable ignorance or deliberate misrepresentation of what constitutes a theory in science. A recent example of such casual dismissals by lay people of well-documented scientific hypotheses is the criticism of global climate-change models (also known as global-warming models), which are accepted by professional climatologists in more than 100 countries yet denounced by a few non-scientists in the USA as "just a theory." This statement and others like it are wrong in two respects:

1 the global climate-change models being criticized are based on a huge body of independent observations made by scientists worldwide that nevertheless show remarkable consistency; and

2 the non-scientists are also appealing to a concept of "equal time," in that their unscientific opinions should be weighed as equally valid (or superior) to hypotheses of the mainstream scientific community.

The latter consideration is inappropriate because the scientific community only considers evidence-based reasoning as deserving of equal treatment. An analogous situation is to tell a person to consider seriously the advice of a few paleontologists about a mechanical problem with a car, rather than listen to the consensus of several thousand professional auto mechanics. The paleontologists may actually have the correct advice, but a rational person should be more assured by the cumulative experiences of the mechanics.

What Is an Opinion?

An **opinion** is an idea that is based more on how a person feels, and it may or may not be based on factual information. For example, someone might say, "I really dislike *Compsognathus*" (Fig. 2.4). When asked why, the person might say, "Because someone told me that it was a scavenger and I don't like scavengers." In this instance, what this person has expressed is an opinion. A listener has few ways of knowing what evidence or rationale supports that feeling, as well as the subsequent statement. Opinions are not necessarily incorrect and may actually coincide with factual information, but they are not derived scientifically. Thus, a flippant rejection of an evidence-based hypothesis or theory as "just an opinion" is fundamentally incorrect. The dismissal itself is an opinion because it has no expressed factual information supporting it and was not formed through evidence-based reasoning.

John Bell Hatcher, a dinosaur paleontologist (Chapter 3), expressed a similar perception about the relative value of opinions in a 1907 publication, where he wrote about the errors made by two other paleontologists regarding the identification of some ceratopsian dinosaur remains (Chapter 13):

They [the errors] are, moreover, striking examples of that axiom so often disregarded in vertebrate paleontology, namely, that one observed fact is worth any amount of expert opinion.

Hatcher's thought also relates well to the use of single or personal observations and their value in science. **Anecdotes**, which are personal-experience stories communicated by one person to another, are not considered scientific, especially if they are related as second-hand information. The use of anecdotes to support a hypothesis risks the possibility of an **individual fallacy**, which means that a single observation by one person is applied universally in a potentially incorrect way. For example, someone might say, "I have a friend who discovered a new species of dinosaur, therefore anyone can discover one." Because not all people have the right geographic location, training, skills, funding, or luck to find a dinosaur within their lifetimes, let alone a new species, such an assertion can be easily disproved. Just because a circumstance is *possible* does not mean it is *probable*, nor does it mean that it actually happened or will happen.

Argument by authority is another method that uses the views of an "expert" associated with a scientific discipline to support what may turn out to be mere opinion. In this case, an authority, such as someone who may have numerous degrees from well-known universities, might be quoted in a way that shows that person's support for a particular idea. For example:

A. J. Martin, who is a famous paleontologist at a prestigious university, said that dinosaurs are actually the descendants of extraterrestrial aliens. Therefore we must consider this possibility.

Notice that no documentation was provided showing that Martin (no relation to the author of this book) actually made this statement. Even if Martin did make the statement, its entire context must be examined to see if it was preceded by a clarifying sentence, such as "early in his career, A. J. Martin ingested large amounts of hallucinogens." Also notice that even if Martin did make this statement, it presents no supporting evidence. Finally, if evidence is associated with the statement, further investigation would determine whether the statement underwent any sort of peer review by experts in paleontology or if it was simply published in the popular press, mentioned in an e-mail message, or garnered through hearsay. As a result, Martin's status as a famous paleontologist who works at a prestigious university, or is otherwise an authority in his field, is irrelevant to the strength of his argument. The evidence and how it is presented are what really matter in science.

To illustrate the last point, someone could point to the earlier quotation from Hatcher as an "argument from authority" and speculate that it is taken out of context. A responsible researcher would address such a criticism by providing the full bibliographic reference from the peer-reviewed, scientific literature:

Hatcher, J. B. 1907. In Hatcher, J. B., Marsh, O. C., and Lull, R. S. The Ceratopsia. United States Geological Survey Monograph 49. Washington, D.C.: US Government Printing Office, 1907, 300 p.

The researcher is thus providing the original source of the information for the perusal of anyone who would like to check on the quotation and its context. This places the burden of disproof upon the critic, while simultaneously showing that the researcher has nothing to hide.

A wonderful aspect of science is that it is not an autocracy, nor is it a democracy. What a single authority states should be irrelevant unless that person has documented repeatable and testable evidence supporting that statement, regardless of how qualified that person is in a scientific field or whether that person had previously made some notable scientific discoveries. Likewise, if a popular opinion poll was taken tomorrow and it revealed that 51% of the people polled believed that humans and dinosaurs co-existed, the paleontological community would not change all of its voluminous findings that clearly contradict this view so that they conform to public opinion. For this reason alone, scientists normally would make extremely poor politicians and, not coincidentally, the vast majority of elected officials are non-scientists.

Unfortunately, an actual public opinion poll in 1998 quizzed US adult participants on their scientific knowledge and did show this same percentage (51%) of people believing in the co-existence of dinosaurs and people, despite the well-established, factually-based 65-million-year gap between them. However, a testing of the validity of the original polling methods might also falsify or otherwise modify the hypothesis presented by the pollsters. Ways to test it include asking:

1 how the question was asked;
2 how the data were collected;
3 the quality and classification of the data; and
4 whether the data were based on facts.

Opinions are unscientific enough, but a poorly-conducted gathering of a sample of opinions risks compounding the original inaccuracies to the point of meaninglessness.

What is Meant by "Proof"?

Proof is a word associated with science that is commonly misapplied by non-scientists. For example, although media reports might say that scientists have proof of the relationships between birds and dinosaurs, a reporter actually would be more accurate in saying scientists have documented yet more convincing evidence supporting the relationships between birds and dinosaurs (Chapters 8, 9, and 15). Scientific methods do not deal with absolute proof of a hypothesis or even a theory; "proof" is a completely accepted premise that is often erroneously synonymized with "truth," although the latter is closer in meaning to the previously-defined term "fact." Proof implies unchangeable conclusions in idealized situations, such as those offered in mathematical proofs of geometric relationships. In other words, mathematicians seek to prove their ideas, whereas scientists attempt to test and disprove them.

What scientific methods can do is *disprove* (falsify) hypotheses or theories. Thus, proof does not enter scientific discussion because scientists do not expect to find a perfect explanation for what they have observed. Nevertheless, they hope in the future to approach a more correct explanation than what they have now. This attitude requires typically more observations (data collection), analysis, testing, and peer review. Consequently, a scientist's job is never done because science, by its very nature, is always changing, self-correcting, and being continually refined by new discoveries, never achieving proof. Paleontology is a wonderful example of this type of change. As poor as the fossil record might seem in comparison to all of the life that has lived on the Earth during its 4.6-billion-year existence, it improves every day as yet more new fossils are found, described, and interpreted. In fact, as more fossils are found, they provide a framework whereby paleontologists become increasingly less surprised by new fossil finds.

An example of how the concept of proof can be superseded by scientific methods is seen in the practice of law. If a trial results in a guilty verdict, the jury is making this decision on the basis of asking themselves if the defendant is guilty "beyond a reasonable doubt." This ruling is typically made on the basis of the evidence presented in the trial, so it approximates a scientific methodology and may involve the testimony of expert witnesses, some of whom might be professional scientists. The now-common application of DNA testing to people convicted of crimes, however, sometimes years after they were convicted, has the potential to show with 99% probability that another person committed the crime. Thus, the new results exonerate (falsify) the jury verdict, and the prosecuting attorney's proof is rendered invalid.

So a challenge to all readers of this book, in their applications of scientific methods, is to ask:

1 which statements about dinosaurs presented in this book are based on hypotheses, and
2 which are based on opinions.

A suggested procedure for this line of inquiry is to ask the following questions:

- What factual evidence supports the statement?
- Is the statement testable?
- Is the testing repeatable and independently verifiable?
- Did the idea expressed in the statement undergo peer review in the scientific literature or other scientific forums?
- Can the statement be used to make predictions?
- Can the statement be proved false?

A cautionary note in this respect is to beware of people who claim to be scientists and say such things as "But I have *proof*!" This person is most likely *not* a scientist because most scientists are very careful, after years of experience and conditioning by their mentors and peers, to use this word sparingly in their scientific vocabulary. Much of science consists of mostly friendly argumentation prompted by curiosity that rarely ends with the final acceptance of a hypothesis, because hypotheses are, by definition, conditional. Even a well-supported hypothesis or theory should provoke more questions, rather than a single, definite answer.

Observational Methods: The Beginning of Questions

Hints on How to Observe

Before any questions in science can be formed, observations have to be made.

Observations, as mentioned earlier, can be gathered through all the senses, but in paleontology the two most important are sight and touch. A seeing-impaired paleontologist can still perform important work, as demonstrated by Geerat J. Vermeij, a blind paleontologist who has published detailed taxonomic identifications and interpretations of evolutionary and ecological relationships between fossil and modern gastropods (snails) in peer-reviewed scientific journals. Interestingly, all his original data were gathered through touching the shells of his subjects. The vast majority of paleontologists, however, rely on sight for their observations. The observation methods discussed here will emphasize that sense, while also acknowledging that ignoring senses other than sight risks a loss of much additional data.

Paleontologists learn how to find fossils through experience, in which their errors are continually corrected with scientific methods. In this respect, the old saying "practice makes perfect" should actually be "perfect practice makes perfect." A paleontologist who sees a small piece of bone in the ground, picks it up, and identifies it as a seashell, will continue to make that mistake until corrected. Correct experience in identifying fossils typically begins with looking at already identified specimens, preferably well preserved, or excellent photographs or illustrations. Then the observer should associate distinctive and memorable characteristics with each fossil. Actual specimens are preferable because the observer can see the color and feel the texture and density of the fossil, or otherwise manipulate the fossils in three dimensions; some computer-generated illustrations or animations of digital photographs can also imitate the latter action.

An observer can also draw a specimen, which is strongly recommended for learning more about a fossil. Drawing encourages careful consideration and deliberation on the defining features of a fossil. Whether the artist is a beginner or an expert, pencils are the best tools for drawing fossils. In the process of erasing and redrawing, the observer can gain new insights on the subject and correct parts of the previous sketch in the light of newly-discovered features. Adding a **scale** to the sketch, such as showing the length of 1.0 cm in comparison to your fossil, is very important for communicating its size to other people. The observer can also read a description of the fossil either before or after sketching it. Descriptions are important because their vocabulary will be learned in conjunction with a specific fossil's image. This visual and verbal record, which involves evidence gathering and testing of the evidence, can prepare an observer before going into a field situation. Some people, however, are entirely trained in these methods in the field.

Making Observations in the Field

"**In the field**" is a favorite phrase of English-speaking geologists, paleontologists, and some ecologists for referring to their outdoor work, and the same types of scientists in other countries use very similar sayings (e.g., "*en el campo*" in Spanish). Fossils come out of the earth, so many paleontologists go in the field to search for them. A typical beginning for some paleontological investigations consists of wandering through open countryside, looking at the ground. In dinosaur paleontology, this searching often takes place in deserts or other arid areas lacking appreciable vegetation, although any area containing Upper Triassic–Upper Cretaceous rocks representing terrestrial environments might be examined. This simple methodology has worked well for the past two centuries and probably will not change very much in the near future.

Knowing where to look for fossils, whether of dinosaurs or other organisms, requires some previous knowledge of where they are likely to be found, which means that one should first become familiar with the geology and geography of a prospective field area. For example, if geologists have previously documented rocks only from the Paleozoic Era in an area or they reported rocks that normally do not contain fossils, then the searching for dinosaur fossils will be fruitless. If research shows that some observations of fossils are likely, make sure that the following are taken into account before going into the field:

1 find out who owns the land,
2 get permission to search on the land from whoever owns it, especially if you plan to collect specimens, and
3 learn what is needed to make the field experience a safe and productive one (Chapter 4).

Experience of looking at fossil specimens in a classroom or in a book is no substitute for the real thing when a paleontologist goes into the field, and a field partner who will act as an independent checker of preliminary identifications is invaluable. In this respect, feedback from a field partner is extremely helpful for correcting any mistakes of identification. Identifying any organic-looking rock or feature in a rock as a fossil is a common mistake of inexperienced field practitioners, so skepticism of initial hypotheses should be the norm. For example, paleontologically

FIGURE 2.5 Field occurrence of dinosaur bone, Morrison Formation (Late Jurassic), western Colorado, USA. Notice its fragmentary nature and lack of resemblance to specimens seen in mounted displays of dinosaurs in museums.

FIGURE 2.6 Students on a field trip examining possible fossil finds in Tertiary Period rocks of central Georgia, USA. Their descriptions and hypotheses were independently tested through peer review (with each other), then presented as a single hypothesis in modified form to an expert (the author of this textbook), who conditionally accepted their hypothesis but then presented it to another expert (another geologist at the field site) who was more of an expert on the rocks in that area than the author. This geologist reconfirmed the hypothesis of the students: they had found fossil plant leaves. All of this process, from discovery to reconfirmation, took about 15 minutes.

untrained people often label large oval objects as dinosaur eggs. In such cases, both novice and expert alike should remember that clouds sometimes resemble horses or dragons. Field partners can provide instant reality checks that prevent imaginations from running wild.

Paleontologists discover quickly and early in their careers that fossils are rarely preserved as complete specimens and more likely to occur as mere fragments (Fig. 2.5). With enough correcting of observations and accounting for variations in fossil preservation (Chapter 7), however, identifications become easier. A **search pattern** is a mental image used by geologists, wildlife biologists, and ecologists, in which they scan an area with certain shapes or colors in mind, based on previous experiences, looking for matching objects. In some cases, these items correspond to what the observer is looking for. In all cases, scientific methods can be applied instantly to the observation in the form of the simple but very appropriate question "What is this?" (Fig. 2.6).

When an object is found that is identified tentatively as a fossil, a paleontologist will normally observe everything about it that comes to mind. This is the description phase, which has worked well for previous generations of scientists. The paleontologists will gather both qualitative and quantitative data by drawing, photographing, measuring dimensions, describing shapes, and noting any resemblance of a possible fossil to known objects. If the object was not lying on the ground surface, they also will especially document how it was found in the context of its host rock. All these observations will connect in some way with a hypothesis; this is the interpretation phase, an attempt to explain the data that may or may not reconcile with a hypothesis held by the observer before going into the field. If possible, this hypothesis can be tested initially in the field while the source of data is still in front of the observer. As mentioned earlier, a skeptical field partner is a big help to paleontologists in this respect. In the case of the solitary professional paleontologist or amateur collector, thorough descriptions become even more essential for communicating results to other paleontologists for evaluation. In this case, measurements (quantitative data) are among the most important descriptors for absolute comparisons and testing of a hypothesis. They can also be used later for calculating ratios, areas, volumes, or statistical tests, which can all be used in hypothesis testing.

When making these types of comparisons, the larger a sample set of measurements, the more meaningful the description. For example, if some paleontologists have multiple measurements for what they hypothesize is the same type of fossil, such as dinosaur tracks, an **average** value is useful. They can also report a **range** of values, which is the maximum number coupled with the minimum number, to give an approximation of the variability of the data. An average, also called the **mean**, is calculated through the following formula:

$$\bar{x} = \frac{\sum x}{n} \qquad (2.1)$$

where $\sum x$ is the sum of all values measured and n is the number of values. An example of how average and range can be demonstrated is by using seven measured dinosaur tracks with the following lengths: 80, 64, 78, 72, 82, 75, and 69 cm.

$$\bar{x} = \frac{(80 + 64 + 78 + 72 + 82 + 75 + 69)\ \text{cm}}{7}$$

Step 1. $= \dfrac{520\ \text{cm}}{7}$

Step 2. $= 74.3\ \text{cm}$

Based on the given data set, the range of sizes is 64–82 cm.

> *Of course, calculating an average and range is not the end of describing a set of measurements.*

Dinosaur tracks have a mean length that fits other known dinosaur tracks, and the smallest and largest lengths also conform to previously interpreted tracks, but the variation of the data is otherwise not well defined. A well-known and useful measurement for variation is **standard deviation**, which describes the spread of data around a mean. **Standard deviation** is the positive square root of another statistical measurement called **variance**. Standard deviation, which is easily calculated by popular spreadsheet programs, can be applied to a **normally distributed** sample, which is described by a bell-shaped curve. One standard deviation represents 68% of all measurements on both sides of the mean; two standard deviations represent 95% and three standard deviations represent 99% (Fig. 2.7). Many sets of data from the natural world are not normally distributed, which means that the **median** (middle value of the data set) will not be in the exact peak of the distribution, making it a **skewed distribution**. Likewise, many measurements of dinosaurs, such as femur lengths and widths, track lengths, or egg volumes, have skewed distributions, which may reflect the original life distribution or may be artifacts of the sampling and fossil preservation (Chapter 7). With our given

FIGURE 2.7 Diagrams showing how quantitative data can be summarized into histograms with curves approximating the distribution of the values. (Left) Normal distribution. (Right) Skewed distribution. The horizontal axis (abscissa) is in order of increasing value, whereas the vertical axis (ordinate) is in number of observations or data points.

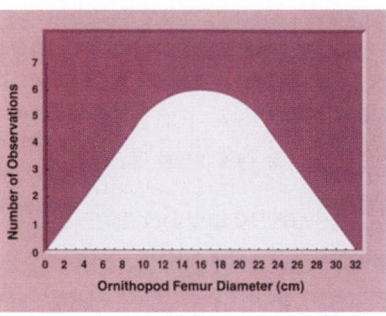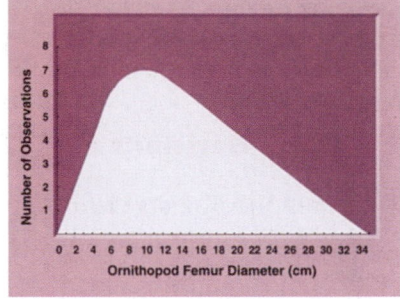

FIGURE 2.8 Sketch of a suspected small dinosaur track from the Middle Jurassic Sundance Formation of Wyoming, with measurements included and indicators of where measurements were taken on the specimen. An observer may have a different definition of "width" and "length" of a track that would be difficult to determine through only a verbal description, whereas the sketch shows clearly what was measured.

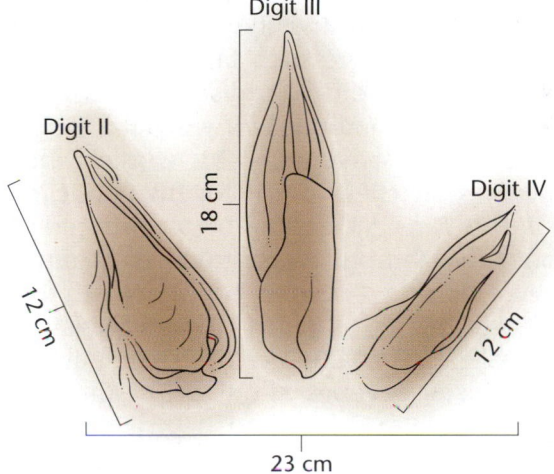

data set for the dinosaur tracks, the standard deviation would be 6.3 cm, so the sample then can be described as having a mean and standard deviation of 74.3 and 6.3 cm, respectively.

Good scientists also report units of measurement after the numbers. Otherwise, someone examining the data has no idea what parameter was measured (meters, liters, or rutabagas). In the example provided, the average length may be important because not all dinosaur tracks are as large or small as the average. Sketches of measured specimens, showing exactly what was measured, are also extremely helpful for follow-up research (Fig. 2.8). A detailed description, preferably with illustrations and careful measurements, will communicate results better and encourage further study of a paleontological find.

Most professional paleontologists have had experience with amateur paleontologists who have incredible skills at discerning fossils in places where many experts have looked before but have never found any. One way to explain the skills of these people is that they have developed a very efficient search pattern, which helps them sift through all the extraneous or otherwise distracting stimuli and instantly focus on their objective. Paleontologists who have not developed a similarly efficient search pattern and do not understand the combination of experience and talent behind it might simply refer to that person as lucky. Although some people may have an obvious knack for finding fossils, training to look for them is continually self-correcting and incorporates education about gathering both qualitative and quantitative data. This training is probably the best way to ensure the development of improved fossil-finding abilities, whether in looking for dinosaur fossils or other fossils.

Ethics and Dinosaur Studies

Ethical Decisions and Their Impact on the Science of Paleontology

Two paleontologists, who began as friends, soon became bitter enemies after they started to compete for the same fossils in the same field area. This incident was exacerbated when one of the paleontologists publicly exposed a major scientific mistake made by the other. The paleontologist in error was so deeply embarrassed that he attempted to buy, with his own personal funds, all of the journals that contained his mistake. The two rivals soon began employing spies to report on the dinosaur localities and finds made by the other's field crews. Some of their employees even destroyed dinosaur bones in the field to prevent them from being found or used by the other's employees. Each of the paleontologists took out ads in major newspapers defaming the other, compounding their enmity for one another. And, because both were such famous paleontologists, many people were eager to join their quests for dinosaur discoveries and were willing to work for small wages under difficult field conditions for years at a time. One of the paleontologists barely acknowledged the discoveries of his workers; he rarely visited them in the field, and he openly stole their results for his publications to further his fame.

Is this a recent exposé by an intrepid news crew for a tabloid television special about two famous paleontologists at renowned US universities who are exploiting their graduate students? No, it actually happened between two American dinosaur paleontologists, Otheniel Charles Marsh and Edward Drinker Cope, in the late nineteenth century (Chapter 3). Their tale of greed and considerable egos, fueled by dinosaur discoveries, is one of the most unsavory yet fascinating stories in the history of paleontology, and it provides an excellent example of how ethical problems in science are not limited to today's media. However, it also serves as a starting point for discussing the importance of **ethics** to science in general, and dinosaur studies in particular. Ethics in dinosaur studies are often directly related to the money that can be made from dinosaurs, an issue that is well publicized in the stories of the popular press today.

Ethics comprises a set of principles of conduct or behavior in human society and how that behavior affects people's relationships with one another. The principles in ethics, of what is "acceptable" or "unacceptable", are variable, depending on the norms and needs of a given society. (Think of norm and variation in principles as analogous to mean and standard deviation, respectively.) Standards are agreed upon sufficiently by a majority of people (as consensus) that certain behaviors are either considered wrong or right within a given society, and those standards may change throughout time. For example, many world societies normally consider killing other people as wrong, but the same societies allow for the variance that killing under the conditions of war or self-defense is right. In terms of how standards change with time, many cultures considered slavery to be acceptable during much of the nineteenth century, but it is illegal in most countries today.

Ethics is a necessary subject for dinosaur studies, if for no other reason than because the interest in dinosaur fossils has led to the attachment of monetary value to them, which makes them economic commodities to be exchanged or sold. History has shown that economic interests can lead to ethically problematic decisions. In fact, the sale of fossils began nearly as soon as they were first scientifically described. For example, early nineteenth-century English paleontologist Mary Anning sold her fossil finds that came from Jurassic seacliffs near Lyme Regis, England. Interestingly, the tongue-twister "she sells seashells by the seashore" was written about the commercial practices of Anning.

Even though the economic value placed on dinosaur fossils has been at the center of ethical dilemmas for paleontologists and fossil collectors since the late nineteenth century, ethics also enters into decisions that are often unique to paleontology in comparison to most other sciences, as illustrated by the following hypothetical examples:

- Does the amateur fossil collector who originally found a fossil get co-authorship on a scientific report of the fossil? Does the professional paleontologist acknowledge the collector or does the paleontologist deserve sole authorship because of his or her advanced educational background?
- What if a fossil is found on private land? Should a paleontologist be expected to pay the landowner the market value for that fossil before it can be studied?
- What recourse do graduate students have if their advisors publish their field discoveries without their consent and the advisor is listed as the first (or sometimes only) author in the resulting publication? Do the graduate students report this transgression or do they accept that the use of their works by an advisor goes with the territory of being a graduate student?
- What happens when one paleontologist performs research on certain fossils knowing full well that another researcher is already studying them? What if the first paleontologist scoops the second by submitting the results to a journal first, leading to the second researcher's work being rejected? Is this just an example of how science, like other aspects of a capitalistic and goal-oriented society, is a competitive venture?

But these are merely possible situations. As scientists, we need real, factually-based examples of ethical dilemmas. Thus, here are some actual documented instances from recent years of the ethical conflicts caused by the popularity and economic aspects of dinosaurs:

- Private fossil collectors in South Dakota uncovered a nearly complete *Tyrannosaurus rex* skeleton, only to have it seized by FBI agents acting on behalf of US government claims of ownership. The specimen, nicknamed "Sue" for its discoverer, languished in federal storage for several years until it was finally placed on the auction block at Sotheby's and sold for $8.36 million. The purchaser, the Field Museum of Natural History in Chicago, was backed by corporate sponsors Disney and McDonald's. The grand unveiling of the mounted skeleton in the Field Museum in May 2000 coincided with the opening of the Disney movie *Dinosaur* (Chapter 1), and it was broadcast live on ABC, a TV network owned by Disney. McDonald's also had a promotional tie-in of their products with the movie.
- Academic paleontologists working in western Australia in cooperation with indigenous tribes found probable stegosaur footprints. These prints are rarely reported from the geologic record (Chapter 12) and were the only ones of their kind found in Australia. Very soon after they were discovered, thieves came into the area and cut the footprints from the rock, using power tools. Because the site is considered sacred ground by the tribes, tribal spirituality was permanently damaged by this act; furthermore, the desecration and mistrust caused by the theft meant that paleontologists might not be given permission to work in the region again. The stolen footprints were never recovered and are presumably in a private collection, so they are still unknown to science.
- A large theropod was discovered on federal land in Montana and was being excavated by an academic paleontologist and his research team. While he and colleagues were temporarily away from the site, a nearby rancher and his family tried to excavate the fossil for themselves by using a backhoe.

When the federal government intervened, the rancher claimed that the government had unjustifiably taken the land from him. A family member reportedly said that selling the fossil would help to feed their family.

■ Officials of a natural history museum in Japan were proud to have gained from China a rare, Early Cretaceous bird fossil, *Confuciusornis sanctus*, only to be embarrassed later when they discovered that the fossil was probably smuggled illegally out of its country. China's 1989 law, prohibiting the unauthorized export and sale abroad of Chinese fossils, was also prompted in part by the widespread smuggling during the late 1980s of other fossils, including hundreds of Late Cretaceous dinosaur eggs (Chapter 8).

In these circumstances, the desire for economic gain and the quest for scientific knowledge became antagonistic. Private collectors and corporations competed for prize fossils while paleontologists associated either with museums or universities attempted to fulfill the ideals of scholarly study. However, as reprehensible as the theft or sale of fossils may seem to some people, in reality too few trained academic paleontologists are available to study all of the discovered fossils and find all of the undiscovered ones. This problem will not be solved soon because employment opportunities for academic paleontologists have decreased with diminished financial support for paleontology and geology. Consequently, despite the high public interest in paleontology, little economic incentive exists for people to pursue academic careers in paleontology, which results in a logical market response – fewer professional paleontologists.

Private collectors also make the point that, because there are not enough paleontologists, many fossils would weather and disappear long before they could ever be studied. In the eyes of many collectors, they are performing an important service by preserving fossils, whether they view them as "natural art" or as scientific curios. Most assuredly, some collectors, working as amateurs or for commercial gain, are careful to gather scientifically important information associated with legally found fossils and allow scientists to examine their specimens. One example of such cooperation took place in Alabama from 2000 to 2004, when amateur collectors worked together with professional paleontologists to collect, catalogue, and preserve 310-million-year-old fossil amphibian tracks from a site near Birmingham, Alabama. Furthermore, despite publicity given to illegal transactions, not all private collectors are vandals or thieves, with disreputable collectors very much in the minority. As dinosaur fossils have increased in market value, other players have been attracted to acquire them. Corporations have entered into the bidding for fossils so they can better their advertising, such as the aforementioned Sue. Interestingly, this specimen has been the source of several previously unknown and important insights on *T. rex* anatomy and behavior (Chapter 9). In these cases, the line between commercialism and science is obscured and people become understandably confused by the purposes and aims of paleontology.

To solve some of the problems associated with the collection of fossils, professional paleontologists, who must share some responsibility for public ignorance of fossils, could initiate more public outreach, perhaps through local schools, fossil-collecting clubs, or the Web. They could discuss proper collection procedures (especially what is legal or illegal) and openly discuss what is considered in their profession as right or wrong behavior. Sharp-eyed amateur collectors with well-honed search patterns have historically found some of the most important fossils (Chapter 3), at least partly because amateurs and paleontologists have cooperated with one another for a long time. In fact, paleontology and astronomy are the only two sciences in which amateurs make regular contributions with scientific importance. Paleontologists can help to continue this tradition by showing amateurs how to develop search patterns as they scan the Earth for vestiges of past life, how to record information about their finds, and how to prevent damaging fossils by the use of proper collecting methods. One of the outstanding fringe benefits of such apprenticeships is that today's amateur is potentially tomorrow's professional, as evidenced by some now-famous paleontologists who, long before they had degrees attached to their names, began their careers by wandering through fossil-laden areas (Chapter 3).

Most importantly, whether people are novices or experts at observing and identifying fossils, they can still go to rock exposures, look for fossils, and experience the joy of discovering evidence of ancient life being seen by human eyes for the first time in the history of the Earth. These private moments of enlightenment and feelings of connection with the ancient past can be more valuable than auctioned fossils, even if they do not always feed a family or help to promote a movie.

SUMMARY

Scientific methods, which are central to the study of dinosaurs, have a foundation of evidence-based reasoning that involves hypothesis forming, testing of hypotheses, peer review, and construction of theories through the interrelations of hypotheses. These methods also have many applications to decision making in the everyday, ordinary lives of non-scientists. Knowing what is or is not science is important for critical thinking, which can help in assessing whether a given argument deserves a second thought. In this respect, knowing the differences between a fact, hypothesis, theory, and opinion can allow people to make informed decisions. Hypothesis forming typically consists of two phases, description and interpretation; descriptions can contain both qualitative and quantitative information. Quantitative information can be further described through statistics, such as the mean, median, and standard deviation. Interpretations are explanations of the observed data that form the basis for a hypothesis.

Finding dinosaur body and trace fossils requires development of some observational skills, which can be cultivated through looking at illustrations of fossils, drawing actual fossil specimens, and reading descriptions of them. Sharpened observations of details in everyday life can help with establishing search patterns that can be applied to the natural world. Fieldwork is one way to test skills and learn how scientific methods can be applied in the course of an investigation, as exemplified by fossil identification. Additionally, descriptions of fossils include measurements and calculations. Numbers are used to describe aspects of dinosaur fossils or fossils of other organisms found. Measurements and calculations also can be used to communicate descriptions much more effectively to a paleontologist if a potentially important fossil is found.

Ethics is an especially important subject in dinosaur studies because of the economic value placed on some dinosaur fossils, which causes competition between collectors (either amateurs or commercial) and professional paleontologists. Understanding how these ethical situations affect human relationships in dinosaur studies provides some parallels for other ethical dilemmas that might be faced in both scientific areas and the personal matters of daily life. Such decisions can affect the well being of other people. Using a combination of scientific methods and an awareness of ethics, people thus have the tools for using knowledge combined with values and understanding the effects of their decisions on other people.

DISCUSSION QUESTIONS

1. Think of an example in your life lately where you used scientific methods to make a decision and explain it in terms of your initial observation, hypothesis, testing of your hypothesis, and peer review.

2. Which of the hypotheses presented in association with Figure 2.3 do you think was the most likely, based on the available evidence? If you do not accept any of the hypotheses, what is one that you propose? See if you can combine aspects of the different hypotheses to make a new hypothesis.

3. Ask your instructor to change some feature of your classroom before you meet for class but not to tell you what was changed, then try to identify what is different. What might help to make this an easier task for the people doing the identification?

4. Is the statement "a theory must be falsifiable" non-falsifiable, thus making it unscientific? Explain why or why not.

5. Take the commonly used term "conspiracy theory" and discuss it in the context of the definitions given for facts, observations, hypotheses, theories, and opinions. Which definitions best apply to this term, and how could a conspiracy theory be demonstrated as a valid hypothesis? How could the use of the term bias your first impression of the "theory" it describes?

6. Name some fields of study that are not normally classified as science in which you think scientific methods, as described in this chapter, are used. What evidence do you have to support your hypotheses?

7. Why do dinosaur paleontologists frequently look for fossils in desert areas, as opposed to forests or jungles?

8. Ask your instructor to let you borrow a fossil or use one that you already have, then draw it as accurately as possible but do not add a scale indicator to the illustration. When it is completed, show the illustration to someone else who is not knowledgeable about fossils, and ask this person to estimate the size of the fossil. How far off from reality was their answer?

9. While doing fieldwork, you discover what you think is a dinosaur nesting ground, composed of five circular depressions. You make the following diameter measurements: 1.2, 1.3, 1.0, 1.4, and 1.25 meters.
   a. Construct a histogram similar to the one depicted in Figure 2.1. Why is it different from Figure 2.6?
   b. What are the mean, range, and standard deviation for the sample?

10. In the discussion about ethics in dinosaur studies, which statements made by the author did you think were opinions and which were testable hypotheses? How could you disprove some of the statements?

Bibliography

Cvancara, A. M. 1990. *Sleuthing Fossils: The Art of Investigating Past Life*. New York: John Wiley and Sons.

Dobzhansky, T. 1973. Nothing in biology makes sense except in the light of evolution. *American Biology Teacher* **35**: 125–129.

Emiliani, C. 1992. *Planet Earth: Cosmology, Geology, and the Evolution of Life and Environment*. Cambridge, UK: Cambridge University Press.

Fiffer, S. 2000. Tyrannosaurus *Sue: The Extraordinary Saga of the Largest, Most Fought over T-Rex Ever Found*. New York: W. H. Freeman & Co.

Frodeman, R. L. 1996. Envisioning the outcrop. *Journal of Geological Education* **44**: 417–427.

Gibson, M. A. 1994. Teaching scientific integrity to geology majors. *Journal of Geological Education* **42**: 345–350.

Handford, M. 1997. *Where's Waldo?: The Wonder Book*. Cambridge, Massachusetts: Candlewick Press.

Hatcher, J. B., Marsh, O. C. and Lull, R. S. 1907. *The Ceratopsia*: United States Geological Survey Monograph 49.

Horner, J. R. 1997. "Afterword: What's a dinosaur worth?" *In* Horner, J. R. and Dobb, Edwin, *Dinosaur Lives*. San Diego, California: Harcourt Brace and Company. pp 233–244.

Kitcher, P. 1982. *Abusing Science: The Case Against Creationism*. Cambridge, Massachusetts: MIT Press.

Padian, K. 1988. New discoveries about dinosaurs: separating facts from the news. *Journal of Geological Education* **36**: 215–220.

Sagan, C. 1996. *The Demon-Haunted World*. New York: Random House.

Samuels, L. S. 1996. Antidotes for science phobia. *The American Biology Teacher* **58**: 455–461.

Schotchmoor, J. G., Springer, D. A., Breithaupt, B. H. and Fiorillo, A. R. 2002. *Dinosaurs: The Science Behind the Stories*. Alexandria, Virginia: American Geological Institute.

Shermer, M. 1997. *Why People Believe in Weird Things*. New York: W. H. Freeman.

Stone, M. and Couzin, J. 1998. Smuggled Chinese fossils on exhibit. *Science* **281**: 315–316.

Vermeij, G. J. 1987. *Evolution and Escalation; An Ecological History of Life*. Princeton, New Jersey: Princeton University Press.

Woodward, J. and Goodstein, D. 1996. Conduct, misconduct, and the structure of science. *American Scientist* **84**: 479–490.

Chapter 3

You have been hearing about dinosaurs ever since you were a child, but because so many new dinosaur books have been published in the past few years you assume that most dinosaur discoveries have been made in the last few decades. So when you glance through one of the latest dinosaur books, you are surprised to find that some famous dinosaurs, such as **Stegosaurus** *and* **Triceratops**, *were discovered and named in the nineteenth century. This makes you curious about other well-known dinosaurs, such as* **Tyrannosaurus** *and* **Velociraptor**.*

Considering that the popularity of dinosaurs seems so recent, what kind of people found and described dinosaur fossils, especially in the nineteenth century? When were the most famous dinosaurs discovered, and by whom? What about dinosaur tracks – when were they found and who first decided that they belonged to dinosaurs? Who found the first recognized dinosaur egg?

History of Dinosaur Studies

The Importance of Knowing the History of Dinosaur Studies

Within a human lifetime, several years may seem like an eternity, especially during our youth, and revived customs within society can take on a completely new appearance when people are experiencing them for the first time. Thus, what seems the latest fashion today actually may be a recycled trend of yesteryear. Such is the case with dinosaurs, which were a hot topic in scientific and public circles in the nineteenth and early twentieth centuries. Although there was sporadic interest in dinosaurs throughout the middle of the twentieth century, the "**Dinosaur Renaissance**" did not begin until during the past 35 years. This cyclicity can be attributed mostly to the people who studied dinosaurs, but it is also dependent on larger societal factors, such as relative public support of science or world wars. In eighteenth and nineteenth century Europe, North America, and South America, people used the beginnings of formalized scientific methods as they made new observations about dinosaur fossils and documented their finds. Some of these people dared to propose new hypotheses about such fossils, some of which, in the face of subsequent research, we ridicule today. However, some hypotheses first proposed during those times have undergone peer review by many generations of scientists and have stood the test of time with little or no modification.

In reading about the people who studied dinosaurs, we can:

1 acquire an appreciation of their legacy;
2 understand how the body of evidence about dinosaurs accumulated over the past several hundred years; and
3 learn how they contributed to what we know about dinosaurs today.

As mentioned before, dinosaur paleontologists were often influenced by social and political circumstances of their times, which affected the quality of their results and subsequent perceptions. As is still evident today, the common thread that connects the extremely diverse personalities in dinosaur studies is their sense of curiosity, whether it is manifested as contemplative inner explorations, or adventurous outer ones. Paleontology, and the people who have studied it, form one of the deepest and most colorful histories of all the sciences.

Dinosaur Studies before the "Renaissance"

Early Recognition of Dinosaur Fossils

Dinosaur fossils are found in the Mesozoic rocks of every continent, which means that dinosaur fossils are in the geographic proximity of many human populations.

Indeed, some anthropological examples suggest that dinosaur fossils influenced artwork, oral tradition, and other forms of expression well before written history.

As people of indigenous societies have always been experienced in the identification of animals, their anatomical traits, and the signs that they leave, they would certainly recognize the animal origin of dinosaur bones and other fossils, such as tracks and eggs, since well before any recorded history. For example, among Native Americans, dinosaur track motifs are evident in some of the Hopi clothing associated with a traditional snake-handling dance; the Hopi inhabit an area well known today for its Jurassic dinosaur tracksites. When Native Americans presented large bones to a nineteenth-century French-Canadian explorer, traveling through a part of Alberta that has abundant Late Cretaceous dinosaur remains, they referred to them as belonging to the "father of all buffaloes." Late nineteenth-century paleontologists reported that the Sioux tribe of the western USA had legends about dinosaur bones, explaining them as the remains of large serpents that burrowed their way into the ground to die after they had been hit by lightning. Near Mesozoic tracksites in southwestern Africa, dinosaur tracks and the animals interpreted as their makers are seen in cave paintings and were the subjects of native songs. In Brazil, early artwork was discovered that is directly associated with a Cretaceous dinosaur track. Additionally, Paleolithic or Neolithic people in Mongolia deliberately altered Cretaceous dinosaur eggshells, and may well have used them for ornamental purposes.

Probably the most intriguing potential reference to dinosaurs in folklore is the griffin, a legendary animal of central Asia. The griffin was said to have the body of a lion, a parrot-like beak, and a pair of wings, and its purpose was to protect its nests of gold. One scholar showed that the geographic range of this legend included the Gobi Desert of Mongolia, where abundant remains of *Protoceratops*, a lion-sized Late Cretaceous ceratopsian with a beak (Chapter 13), are near ancient gold mines. Of course, no wings have ever been found in association with a *Protoceratops*, indicating that, if these dinosaurs were their inspiration, the legend-makers embellished their story (a common practice even in modern societies).

Whether ancient texts actually refer to dinosaur bones is difficult to discern, as are references to dragons in European cultures that some authors have attempted to link to dinosaur fossils. The earliest known reports of "dragon bones" (in Mandarin, *long gu tou*) were written about 300 BCE from the Sichuan province of China, an area well known today for its abundant dinosaur bones. These bones were valued for their purported medicinal value, and some doctors in China still prescribe ground-up dinosaur bones as a cure for some ailments.

Early Scientific Studies of Dinosaurs: The Europeans

Prominent scientists of the fifteenth to the eighteenth centuries connected fossils to formerly-living organisms. Among them were **Leonardo da Vinci** (of *Mona Lisa* fame; 1452–1519), **Niels Stensen** of Denmark (also known as Steno, Chapter 4; 1638–87), **Robert Hooke** (1635–1703), and **Robert Plot** (1640–96) of England. Plot, a museum curator at Oxford, made the first known description and illustration of a dinosaur bone in 1677. The problem with his interpretation is that although he recognized the fossil was a bone, he speculated that it might have belonged to a modern elephant and not a large, extinct, reptile-like animal. **Richard Brookes** made another illustration of this bone in 1763 that shows it as part of a femur from the theropod *Megalosaurus* from the Middle Jurassic of present-day Cornwall (Chapter 9). Neither did Brookes know the true identity of the bone, but this lack of knowledge did not stop him from naming the specimen after its superficial resemblance to a part of human male anatomy, *Scrotum humanum* (Fig. 3.1). One French

FIGURE 3.1 Sketch of probable dinosaur bone *Megalosaurus*, described by Robert Plot in 1677 in *Natural History of Oxfordshire*.

philosopher, **Jean-Baptiste-René Robinet** (1735–1820), thought the specimen actually represented an attempt by nature to imitate human organs. Fortunately, subsequent scientific knowledge falsified this hypothesis. Nevertheless, French explorers in the eighteenth century made another anatomical analogy when they yearningly named a mountain range in the northwestern part of present-day Wyoming "Les Grande Tetons."

Thus, although a dinosaur bone had been discovered, described, and illustrated by the latter part of the eighteenth century, nobody knew that it was a dinosaur bone. Unfortunately, the original specimen is lost to science. As a result, we cannot independently verify that Plot found the first identifiable dinosaur bone. In 1728, **John Woodward** (1665–1728) of Gresham College, London, catalogued another dinosaur limb bone that was found either in the late seventeenth or early eighteenth century, but he also did not realize the identity of its former owner. Later investigations would confirm that it was indeed from a dinosaur (probably *Megalosaurus* again), which makes it the first known identifiable dinosaur bone; this specimen is currently housed at the University of Cambridge. Subsequent finds of dinosaur bones in Europe, during the late eighteenth and early nineteenth centuries, resulted in people cataloguing and giving descriptions of specimens that approached a scientific methodology. However, hypotheses were rarely offered for most of the specimens and none were considered evidence of a distinctive group of long-extinct animals. In fact, such thinking was discouraged in Europe and its colonies at that time by the strong influence of religious institutions, whose advocates held that all animals on the Earth were created at the same time and none were extinct.

Ironically, a British clergyman, the Reverend **William Buckland** (1784–1856), published the first actual scientific description of a dinosaur. Buckland made his discovery of dinosaur remains around 1815, his find consisting of several serrated, curved teeth, together with a lower jaw containing a tooth comparable to the others (Fig. 3.2). The fossils belonged to *Megalosaurus*, which seems to have been a recurring find for Europeans during the late eighteenth and early nineteenth centuries. After consultations with the renowned French anatomist, **Georges Cuvier** (1769–1832) (whose full and rather officious name was Jean Léopold Nicolas Frédéric Baron Cuvier), and the geologist (and Reverend) **William Daniel Conybeare** (1787–1857), Buckland finally read his paper before a group of scientists at the Geological Society of London in 1824. Thus, Buckland is credited with naming the first dinosaur, although **James Parkinson** (1755–1824) almost named this animal first, in 1822. Another claim to fame for Buckland was that he was an instructor to geologist Sir Charles Lyell (1797–1875), who wrote *Principles of Geology* in 1830, one

FIGURE 3.2 William Buckland (top) and remains of the first named dinosaur fossil, the lower jaw of *Megalosaurus* (bottom). From Colbert, E. H., 1984, *The Great Dinosaur Hunters and Their Discoveries*, Dover Publications, N.Y., plates 3 and 4.

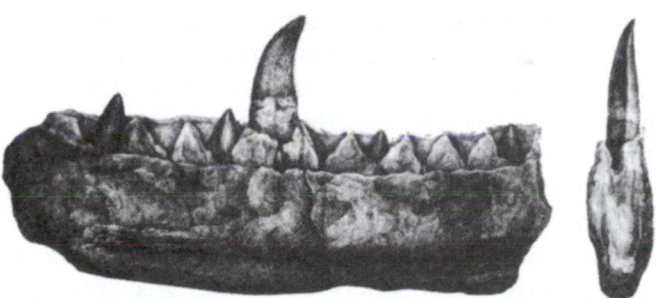

of the most influential books in the field; he also invented the term "palaeontology" (using the British spelling). As many of the geologic principles advocated by Lyell are still in use today (Chapter 4), Buckland's influence had considerable impact on modern geology.

On a more personal note, Buckland was a strange man who reveled in his oddness (which supports the idea that paleontologists really have not changed very much in the past two centuries). To say that he was eccentric is akin to saying that *Seismosaurus* (Chapter 10) was large. He apparently delighted in proving people wrong; for example, he gained some fame when he correctly identified the purported remains of Saint Rosalia at a religious shrine in Palermo, Italy, as goat bones. He kept a menagerie in his home that included jackals, which were known to eat his free-roaming guinea pigs, and a bear named Tiglath Pileser (named after an Assyrian king, 745–727 BCE). The bear was Buckland's frequent companion at academic functions and was normally clothed in a cap and gown. Buckland's interest in animals extended to consuming them, so through much experimentation he attempted to develop a system of classifying them on the basis of taste alone. Regardless of these quirks, all who knew him regarded him as brilliant and he certainly contributed much to the scientific study of dinosaurs.

FIGURE 3.3 The Mantells, Gideon Algernon (left) and Mary Ann (right), who probably were not co-discoverers of *Iguanodon*. From Psihoyos and Knoebber (1994), *Hunting Dinosaurs*, Random House, N.Y., p. 10.

A contemporary of Buckland and another important contributor to the early scientific investigations of dinosaurs was physician **Gideon Algernon Mantell** (1790–1852), also of England (Fig. 3.3). Only a year after Buckland's description of *Megalosaurus*, Mantell was the first person to name a herbivorous dinosaur and ornithopod, *Iguanodon*. According to a popular anecdote, Mantell's wife, **Mary Ann Mantell** (1796–1869); Fig. 3.3), found the teeth and bones of *Iguanodon* near the property of a patient while she accompanied her husband on a house call. However, Mary Ann Mantell was not seen to go with her husband on house calls, so Gideon Mantell was the only source of this story at first; later he claimed that he found the fossils himself. In her defense, she certainly was knowledgeable about fossils, as demonstrated by the 346 figures of fossils she prepared for a monograph published by her husband in 1822. In 1833, Gideon Mantell found fragments of an ankylosaur (Chapter 12), which he named *Hylaeosaurus*, the sauropod *Pelorosaurus*, and *Regnosaurus*, which has not been classified further because of its few remains. He was sufficiently obsessed by paleontology to fill his home with the remains of many extinct animals (as opposed to Buckland's preference for live, edible ones). This preoccupation resulted in the downfall of his medical practice, the withering of his finances, and the eventual departure of Mary Ann and their children.

Mantell's description of *Iguanodon* skeletal remains was followed by numerous discoveries of probable iguanodontian tracks in Cretaceous strata of southern Britain, leading to some of the first attempts to correlate dinosaur body fossils and trace fossils (Chapter 14). The Reverend **Edward Tagart** first presented his footprint finds in a paper to the Geological Society of London in 1846, where he attributed them to large birds. The possible reptilian origin of the tracks was proposed by 1850, but not until 1862 did **Alfred Tylor**, **T. Rupert Jones**, and **Samuel Beckles** publish separate reports on the hypothesis that these three-toed tracks came from similarly-sized three-toed feet of iguanodontians. This shared hypothesis was apparently derived independently and has not been disproved in the 130 years since; few other dinosaur tracemakers have been proposed for the tracks found in this region.

The study of dinosaurs would be quite different if not for the etymological and paleontological contributions of anatomist Sir Richard Owen (1804–92) (Fig. 3.4).

Anatomist Sir Richard Owen, the British analogue to Georges Cuvier of France, was a contemporary of Buckland and the Mantells. He was an expert on fossil reptiles to the point where

FIGURE 3.4 Sir Richard Owen, inventor of the term "dinosaur". From Psihoyos and Knoebber, 1994, *Hunting Dinosaurs*, Random House, N.Y., page 11.

many people regarded him as the authority on the subject, so his word was often unquestioned (except by Mantell), regardless of the validity of his interpretations. He is best known for his invention of the term **Dinosauria** (whose members were called dinosaurs), which he first used in 1842 in reference to the large, extinct, reptile-like animals described by Buckland and Mantell. Dinosauria is based on the Greek roots *deinos* ("terrible") and *sauros* ("reptile" or "lizard"); in Victorian England, the common usage of the word terrible connoted the awesome nature of these animals, rather than their fearsomeness, poor hygiene, or other negative attributes. As an example of how authorities are not necessarily always correct in science (Chapter 2), Owen did not include three dinosaurs known at the time in his group, Dinosauria: *Cetiosaurus*, a sauropod, *Poekilopleuron*, a theropod, and *Thecodontosaurus*, a prosauropod. Instead, he classified them as unrelated reptiles.

To his credit, Owen recognized several features that are still key to the classification of dinosaurs today (Chapter 5). Also, he was a consultant to artist **Benjamin Waterhouse Hawkins** (1807–89), who produced the first examples of dinosaur artwork (sculptures and drawings). Unfortunately, the scarcity of dinosaur material and scientific hypotheses at the time resulted in Hawkins' artistic reconstructions of dinosaurs as ponderous and heavy-set quadrupeds, thus encouraging a popular misconception that would influence future investigators until the end of the century. In 1854, Owen was the first person to describe and name a dinosaur from South Africa, the Late Triassic prosauropod *Massospondylus* (Chapter 10).

France and Germany were also sites of dinosaur fossil discoveries during the nineteenth century. A Frenchman, **A. de Caumont**, discovered bones of *Megalosaurus* in Normandy in 1828, and in 1838, **Jacques-Amand Eudes-Deslonchamps** (1794–1867) was the first person to name a dinosaur from France, the previously-mentioned *Poekilopleuron bucklandi* (named in honor of Buckland). French paleontologists were also the first to record dinosaur eggshell fragments from the fossil record (Chapter 7). **Jean-Jacques Pouech** (1814–92), a Catholic priest, gave an excellent description of eggshells that, from their size and geologic occurrence (Late Cretaceous), could only have been from dinosaur eggs. In 1869, **Phillipe Matheron** (1807–99), who had followed Pouech's work, hypothesized a connection between Pouech's eggshell fragments and the Late Cretaceous skeletal material of dinosaurs found in Provence. **Paul Gervais** (1816–79), also of France, was the first scientist to conduct detailed analyses of dinosaur eggshell fragments, the results of which he published throughout the 1870s. In Germany, one of the best-known dinosaurs of the Late Triassic, the prosauropod *Plateosaurus* (Chapter 10), was discovered and named in 1837 by **Christian Erich Hermann von Meyer** (1801–69). After this description, other specimens of this dinosaur were found frequently in southern Germany and Switzerland, and beautifully complete examples are displayed in museums throughout Germany.

Although **Charles Robert Darwin** (1809–82) of England was not directly involved with dinosaur studies, he published explanations of how fossil evidence of organisms' descent with modification correlated with his observations of living animals. The timing of his publications provoked an initial discussion of the evolutionary place of dinosaurs in the history of life (Chapter 6). Darwin was a bit shy of controversy but was defended vigorously in public by **Thomas Henry Huxley** (1825–95). Huxley expounded with much delight on the first confirmed specimen (the "London specimen") of *Archaeopteryx*, a Jurassic bird with "reptilian" (dinosaurian) features that was found in 1861 (Chapters 2 and 15). Such a fossil was excellent evidence of predicted "transitional fossils" that showed links in descent between defined, major groups of organisms (Chapter 6). Huxley and Owen often disagreed on many points of evolutionary theory. Nonetheless, Huxley contributed an important insight to Owen's original classification of the Dinosauria that was far ahead of its time. In Huxley's 1868 classification, he recognized the numerous bird-like characteristics of some dinosaurs, an evolutionary linkage that enjoys nearly total support among modern vertebrate paleontologists (Chapters 9 and 15). Before he produced his classification scheme, Huxley had named a Late Triassic dinosaur from South Africa, the prosauropod *Euskelosaurus* (Chapter 10).

While all of these contentious events were occurring, **Harry Govier Seeley** (1839–1909), of England, noticed an anatomical distinction between two major groups of dinosaurs, and his 1887 report on dinosaur hip structures is still used today for their classification. One group of dinosaurs he characterized as **Saurischia** (reptile-hipped) and the other as **Ornithischia** (bird-hipped), based on the superficial resemblance of these hip structures to modern analogues in reptiles and birds, as well as a few other skeletal traits distinctive to each group (Chapter 5). On the basis of such a distinction, Seeley argued that dinosaurs did not constitute an actual group from the same ancestral stock (**monophyletic**), but arose from separate ancestors (**polyphyletic**). This interpretation touched off a spirited debate about the origin of dinosaurs that lasted for more than 100 years. Seeley, who had grown up poor and so never gained a college degree, was wise enough to work as an assistant to the Reverend **Adam Sedgewick** (1785–1873) at Cambridge University, who along with Charles Lyell was one of the founders of modern geological methods (Chapter 4). Cambridge was the home of the Woodwardian Museum (at that time named after John Woodward, but now named after Sedgewick), which housed the extensive collections of Late Cretaceous fossil animals that Seeley studied.

The main problem faced by paleontologists during the debates over classification was that their attempts to classify and reconstruct dinosaurs were based on fragmentary skeletal remains. For at least one species of dinosaur, **Louis Antoine Marie Joseph Dollo** (1857–1931) of Belgium solved the problem of insufficient evidence with his thorough descriptions of complete skeletons of *Iguanodon* (Chapter 11). Coal miners discovered these in 1878 in Bernissart, Belgium; subsequent excavations recovered 39 individual skeletons from the site, a phenomenal number of specimens even by today's standards. As a result, through vigorous use of scientific methods and access to many skeletons, Dollo cleared up misinterpretations about *Iguanodon* that had persisted since Mantell's original description, such as the placement of its thumb as a nose spike. Most importantly, at a time when all dinosaurs were regarded as **quadrupeds** (using four legs), Dollo firmly established the **bipedal** (two-legged) nature of *Iguanodon*. Huxley proposed the same hypothesis in

1868 for a different dinosaur species found in the USA, discussed below. Dollo, in 1887 alone, published 94 peer-reviewed papers.

Europeans worked very little on African dinosaurs during the nineteenth century, although in 1896 Frenchman **Charles Depéret** (1854–1929) described bones of a previously undiscovered species of sauropod, *Titanosaurus* (Chapter 10), and a theropod, *Majungasaurus* (Chapter 9), from Madagascar. These discoveries foreshadowed the potential for later major discoveries in Madagascar nearly a century later. Another French paleontologist reported dinosaur tracks from Algeria in 1880, but little other information is available about this find. Similarly, no definite reports of dinosaur fossils came out of Australia in the nineteenth century, and it was not until 1903 that **William Hamilton Ferguson** (1861–1957) found a theropod toe bone, nicknamed the "Paterson claw," in Cretaceous rocks of Cape Paterson, Victoria.

Early Scientific Studies of Dinosaurs: The North and South Americans

Meanwhile, across the Atlantic Ocean, fossil evidence of dinosaurs was being discovered in North America in the latter part of the eighteenth and early part of the nineteenth centuries, although none of it was connected with dinosaurs at the time. The first probable dinosaur-related discovery in North America was in 1787, when anatomist **Caspar Wistar** (1761–1818) presented a bone from Cretaceous rocks of Woodbury, New Jersey to the American Philosophical Society, presided over by **Benjamin Franklin** (1706–90). **George Washington** (1732–99), who is also known for his interest in fossils, examined the same bone and mentioned it in one of his writings. Unfortunately, Wistar interpreted the bone as a large man's femur instead of recognizing it as an ornithopod metatarsal; if he had identified it correctly as reptile-like, this discovery would have preceded Buckland by 28 years. In 1802, **Pliny Moody**, a farm boy and student at Williams College, made a more definitive discovery of dinosaurs in North America when he uncovered a rock with Lower Jurassic theropod tracks while plowing his family's field in South Hadley, Massachusetts. The tracks are still in the possession of nearby Amherst College and on display there. Because they had such a close resemblance to the three-toed morphology of modern bird feet, and religion provided the primary framework for explanations of natural phenomena at the time, the footprints were attributed to "Noah's raven." **William Clark** (of Lewis and Clark fame; 1770–1838) also described a large bone in 1806 that was probably eroding from the Late Cretaceous **Hell Creek Formation** in present-day Montana; hence it was probably from a dinosaur, but Clark interpreted it as the remains of a large fish. Prosauropod remains were found in 1818 by **Solomon Ellsworth, Jr.**, in Upper Triassic deposits of the Connecticut Valley. **Nathan Smith** described these fossils in a published report in 1820, and he interpreted them as possibly human. These four examples were representative of a thankfully short-lived American tradition: the mistaken attribution of dinosaur fossils as representatives of most other recognized vertebrate groups.

*Moody's rock slabs containing the "bird" tracks were stored in a quaint building called the **Appleton Cabinet**, the first structure made for the purpose of holding dinosaur trace fossils; its refurbished version is now a dormitory for students of Amherst College.*

Nevertheless, the lack of connection between dinosaur fossils and their actual identity continued in the voluminous and otherwise groundbreaking work on dinosaur tracks by the Reverend **Edward Hitchcock** (1793–1864; Fig. 3.5). Beginning in 1836, Hitchcock's studies of tracks in Late Triassic and Early Jurassic rocks of the Connecticut Valley represented further discoveries in Moody's (and the

FIGURE 3.5 Edward Hitchcock, describer of numerous examples of Late Triassic and Early Jurassic dinosaur tracks from the Connecticut River Valley, and dinosaur tracks figured in his 1858 publication. From Amherst College Archives and Special Collections, Negative Collection, Box 1, fig. 69, and Box 3, figs 4a–5a.

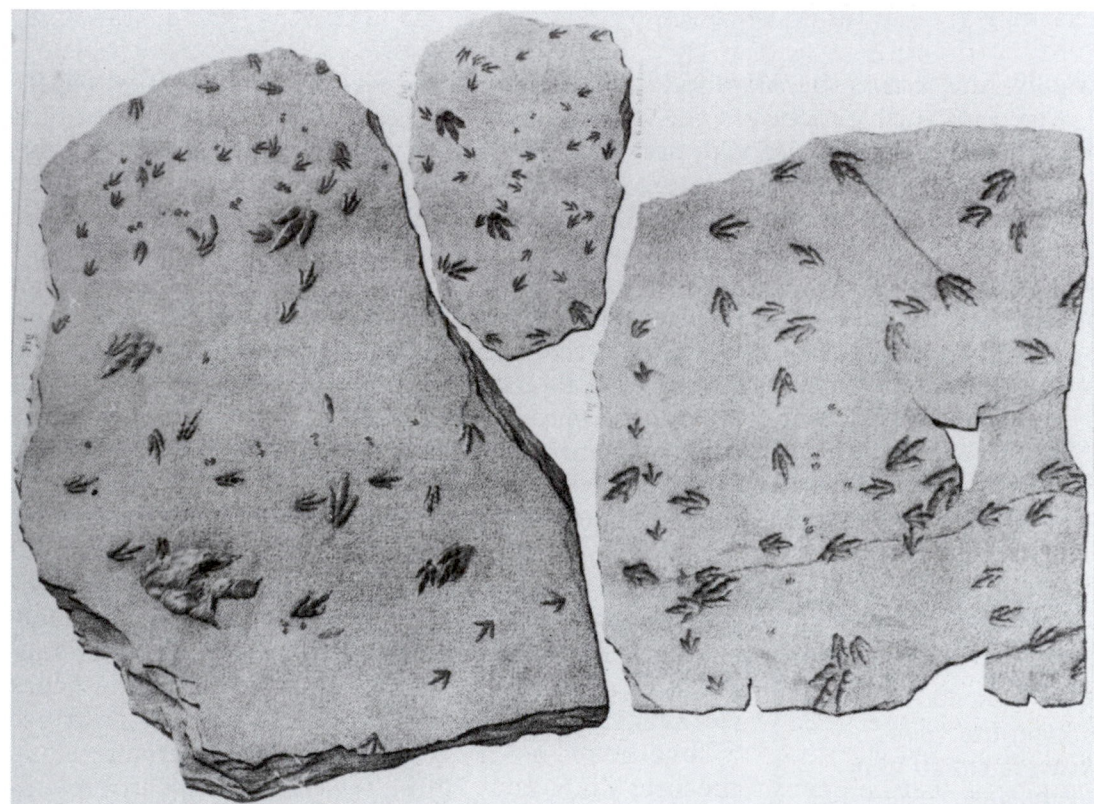

dinosaurs') old stomping grounds. Hitchcock analyzed thousands of dinosaur tracks, and tracks of dinosaur contemporaries, in his collection at Amherst College, where he was president.

Continuing the original "Noah's raven" theme, Hitchcock interpreted the numerous dinosaur tracks as originating from large, prehistoric birds, which was a perfectly reasonable hypothesis in the light of his data and then-current ideas about dinosaurs. For example, three-toed animals made many of the tracks he described and most of the trackways indicated a bipedalism that had not been yet ascribed to dinosaurs. The tracks also resemble those of flightless birds in some ways, with the notable exception of their large sizes. In an 1844 report, Hitchcock was the first person to describe probable dinosaur **coprolites** (fossilized feces, Chapter 14), which he also attributed to birds. Hitchcock's comprehensive summary of his findings, *Ichnology of New England* (1858), was the first work to prominently use the term

ichnology for the science of traces and trace fossils (Chapters 2 and 14). This classic work is still cited, not only for its extensive illustrations and descriptions of dinosaur tracks, but because it contains some of the few recorded instances of dinosaur sitting traces and tail-drag marks.

Another paleontological enthusiast in Massachusetts at the same time was **John Collins Warren** (1778–1856). Warren was a Harvard physician who also dabbled in fossils while maintaining his primary interest in anatomy; his first exposure to anatomical studies began with his father, who was the founder of Harvard Medical School. The younger Warren studied anatomy with Cuvier in Paris and later performed the first surgery with anesthesia in 1846. In 1854, Warren had the distinction of publishing not only the first photographic illustration of a dinosaur track, but also the first photograph shown in an American scientific publication. Scientific illustration, particularly for such photogenic subjects as dinosaurs, was forever changed, although photography was a new and difficult-to-use medium that would not see extensive use in dinosaur studies until later in the nineteenth century.

Despite all of this good science, no one had yet identified a dinosaur fossil from North America until the works of **Joseph Leidy** (1823–91), who initiated Americans' recurring fascination with dinosaurs. Leidy, a physician and anatomist from Philadelphia, Pennsylvania, became bored with medicine and soon turned to paleontology and other aspects of natural history. In 1856 he published a study of the dinosaur teeth found the previous year by **Ferdinand Vandiveer Hayden** in Upper Cretaceous strata of what is now Montana. Leidy named one of the dinosaurs, *Troodon*, on only the basis of these teeth (a risky scientific endeavor), which later studies revealed as one of the most interesting theropods ever found in North America (Chapter 9). It was another dinosaur, however, that would make Leidy a celebrity in the USA. **William Parker Foulke** found a Late Cretaceous dinosaur, the ornithopod (and hadrosaurid) *Hadrosaurus foulkii*, in nearby New Jersey; the dinosaur was graciously named after its discoverer. Foulke had been steered to the site by the landowner and previous discoverer of probable dinosaur bones, **John E. Hopkins**. This dinosaur was similar to and probably related to *Iguanodon*, but Leidy provided an incisive interpretation of it; on the basis of the relatively complete skeleton, he argued convincingly for the inherent bipedalism of a dinosaur. Moreover, he pointed out that, judging from its limbs, *Hadrosaurus* was likely a **facultative quadruped**, meaning that it could have walked on all fours if necessary. This hypothesis was later supported by the find of probable ornithopod tracks that reflect such behavior (Chapters 11 and 14). Furthermore, Leidy proposed a preburial history of the specimen that was probably correct. He thought that this dinosaur originally dwelled on land and its body was washed out to sea, as its remains were found in a marine deposit (Chapter 7).

In 1868, the artist Waterhouse Hawkins, who was living in the USA at the time, attempted to use the same *Hadrosaurus* specimen as a model for artistic reconstruction. Sadly, political problems and vandalism of his works-in-progress led to him being denied an exhibit of the reconstruction, which was to have been displayed in New York City's then newly-established Central Park. Consequently, the best that Hawkins could do was to make a cast of the *Hadrosaurus* skeleton, which remained on public display at the Academy of Sciences of Philadelphia for many years.

The melodramatic interactions between Cope and Marsh throughout their careers have inspired bibliographers and paleontologists alike to invoke clichés such as "bitter rivals" and "sworn mortal enemies." These two paleontologists' publicly aired hatred for one another could be the subject of an extensive psychological study

Edward Drinker Cope (1840–1897) and Othniel Charles (O.C.) Marsh (1831–99) (Fig. 3.6) (Chapter 2) produced simultaneously some of the most significant finds of dinosaurs in the world.

FIGURE 3.6 Edward Drinker Cope (left) and Othniel Charles (O.C.) Marsh (right), productive yet antagonistic contemporaries in dinosaur studies. Reprinted from *Science*, 1897 and 1889, respectively.

on megalomania. One anecdote that is incorrect but nevertheless amusing is that Marsh named coprolites after Cope. Although this book emphasizes Cope and Marsh's scientific contributions, which are unparalleled and may never be equaled, some slight digressions on their personal lives should add insight into them as both scientists and people.

Marsh and Cope had similar financial situations; both received large amounts of money from relatives and thus had few worries about earning a living, which freed their time for academic studies. Cope was the more precocious and prolific of the two, having more than 1400 scientific publications to his credit by the time he died. After he settled in Philadelphia, Cope was briefly a student of Leidy, and he associated himself with the Academy of Natural Sciences there (although Leidy would later distance himself from Cope as a result of the verbal warfare with Marsh). A peer-reviewed journal of **herpetology** (*Copeia*) was named after him in honor of his impressive contributions to the study of reptiles and amphibians. Marsh was not quite as industrious as Cope or as brilliant, but his political acumen was more finely developed, which helped him to gain much government support for his dinosaur studies. Marsh mostly worked through Yale University, where his rich uncle (**George Peabody**) had the **Yale-Peabody Museum of Natural History** built for him. He also held the title of Vertebrate Paleontologist with the newly-formed **United States Geological Survey (USGS)** for 10 years and was the president of the National Academy of Science for 12 years.

Cope and Marsh were important in the world of paleontology at the time. For example, when two schoolmasters, **Arthur Lakes** and **Oramel W. Lucas**, independently found dinosaur bones in Morrison, Colorado, and Cañon City, Colorado (respectively) in 1877, they sent news of their finds to Cope and Marsh. This started what was later called the "**Great Dinosaur Rush**," which lasted for nearly 20 years and spanned present-day Colorado, Wyoming, New Mexico, Montana, and other western states. During the ensuing frenzy of exploration and exploitation, the main producer of the numerous dinosaurs named by Cope and Marsh was the **Morrison Formation**, an Upper Jurassic rock unit that still produces many dinosaur fossils today (named after the settlement where Lakes lived). The thousands of dinosaur bones they collected were placed on railroad cars that, through the newly-built transcontinental railroad, could reach western areas that were previously inaccessible to dinosaur paleontologists.

Cope and Marsh's lasting influence is seen through their naming of so many now well-known dinosaurs, such as the thyreophoran *Stegosaurus* (Chapter 12), the sauropods *Diplodocus* and *Apatosaurus* (the latter then named *Brontosaurus*: Chapter 10), the theropods *Allosaurus* and *Ceratosaurus* (Chapter 9), the ceratopsian *Triceratops* (Chapter 13), and the ornithopod *Camptosaurus* (Chapter 11). They also attempted classification schemes, after Huxley but before Seeley, as a synthesis of the dinosaur discoveries made by them and others. Most importantly, they pointed future investigators to the areas of North America with extensive Mesozoic deposits, clearly demonstrating the potential for more dinosaur discoveries.

Marsh was apparently averse to doing most of his own fieldwork, although reportedly during one field excursion he and some assistants met with leaders of the Sioux tribe, **Red Cloud** (1822–1909), **Crazy Horse** (1842–77), and **Sitting Bull** (1831–90), to gain permission for dinosaur prospecting in their territories. Marsh kept his promise to the Sioux that he would search only for dinosaur remains rather than gold, and Sioux scouts were reportedly gratified to find only bones in the possession of Marsh's party when they left. Marsh's assistants were probably also gratified to leave with their lives intact. Cope also went infrequently into the field in the western states, but more often than Marsh and always made significant finds when he did so. During one of Cope's trips, he met with **Charles H. Sternberg** (1850–1943), and they prospected Cretaceous deposits in Montana in 1876. Sternberg later told the now-famous stories of how the two men would typically hunt for dinosaur bones by day, eat an awful late-evening meal, and go to bed. According to Sternberg, Cope would then toss and turn in the throes of nightmares that brought his Mesozoic beasts back to life, wherein they pummeled him. During this same trip, Sternberg and Cope invented a method for protecting fossil specimens for their transport back east, by boiling rice into a paste and mixing it with cloth strips that were draped around the fossils to harden. Several of Marsh's associates modified Sternberg and Cope's technique the next year by using plaster of Paris and burlap, a technique that was used to make casts for broken human bones and is still used by many dinosaur paleontologists today (Chapter 4). Sternberg undoubtedly learned much about dinosaurs during his brief apprenticeship in the field with his nocturnally-tormented mentor. His sons **George**, **Levi**, and **Charles M. Sternberg** (1885–1981) later found more dinosaur bones in Canada than any other family since.

Before the end of the century, several new workers in North America entered the fray between Cope and Marsh and made remarkable contributions to dinosaur studies. These people were **Henry Fairfield Osborn** (1857–1935), **William Berryman Scott**, **Barnum Brown** (1873–1963), **Walter Granger** (1872–1941), and **John Bell Hatcher** (1861–1904). Osborn and Scott were good friends while undergraduates at Princeton University and decided, after being inspired in class one day by their geology professor, **Arnold Guyot** (after whom seafloor volcanoes, guyots, were named), to do the 1877 equivalent of a "road trip." They hopped on a train and went to Wyoming to find fossils. During their travels by train, horse, and wagon, and in between meeting Native Americans and mountain men, they learned much about fossils in a field context. Both became friends with Cope and went to Europe to study with Huxley for a while before they became faculty at Princeton. Osborn left Princeton in 1891 to become a staff member of the now-famous **American Museum of Natural History**, where he founded the Department of Vertebrate Paleontology and later became president.

In 1897, Osborn sent an expedition to look at the Morrison Formation in Como Bluff, Wyoming, the site of much dinosaur work done by Marsh's minions (Fig. 3.7). The group included Brown and Granger, both of whom were novices at fieldwork but would become two of the most important dinosaur paleontologists of the early twentieth century. This initial foray proved, after much searching, that Como Bluff did not have the dinosaurs it used to have. They moved farther to the north

FIGURE 3.7 Barnum Brown (left) and Henry Fairfield Osborn (right) in the field at Como Bluff, Wyoming, in 1897, with a sauropod (*Diplodocus*) limb bone in the foreground and Late Jurassic Morrison Formation cropping out nearly everywhere else. Negative No. 17808, Photo. Menke. Courtesy Dept. of Library Services, American Museum of Natural History.

the next year and discovered an area where dinosaur bones littered the ground in such abundance that a local shepherd had built a cabin out of them. The site, appropriately named **Bone Cabin Quarry**, provided about 30 tons of dinosaur bones of 141 individual skeletons during that year. Seven more annual expeditions by the American Museum followed. The number of individual dinosaurs and tonnage were recorded for six seasons and these records show how such sites can become quickly depleted of dinosaur bones with continued mining (Fig. 3.8).

John Bell Hatcher, during his short life of 42 years, collected 50 ceratopsian skeletons (many with skulls) from Upper Cretaceous deposits in Wyoming, while employed by Marsh from 1889 to 1892. This feat was single-handedly the most quantitatively important contribution to the study of these wonderfully diverse dinosaurs (Chapter 13). Hatcher became so disgruntled with Marsh that he eventually left and was hired by Scott at Princeton, for whom he did more work in Colorado through the turn of the century. In his publications, Hatcher expressed some of his disgust for Cope and Marsh's occasional scientific errors. Sadly, Hatcher

FIGURE 3.8 Bar graph showing decreased productivity of dinosaur bones from the Bone Cabin Quarry in number of specimens collected versus year. Data from Colbert (1968).

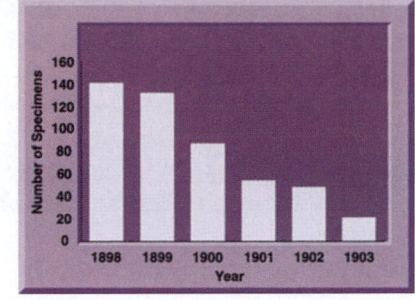

3

died of typhus while writing his classic work, *The Ceratopsia*. But fortunately for science, **Richard Swan Lull** (1867–1957), a paleontologist of some note himself (discussed below), posthumously published the manuscript in 1907.

Elsewhere in North America, **George Mercer Dawson** (1849–1901), who was the son of the important nineteenth-century geologist, **Sir William Dawson** (1820–99), discovered dinosaur bones in Saskatchewan, Canada in 1874. Like Osborn and Scott, George Dawson had studied with Huxley. Further discoveries were made in 1884 by **Joseph Burr Tyrrell** (1859–1957) in the Red Deer River valley, near Drumheller, Alberta, which was followed by other finds by one of his associates, **Lawrence M. Lambe** (1863–1919). As a member of the Canadian Geological Survey, Lambe took a boat down the Red Deer River in 1897 to document more dinosaur-bearing zones. Through the efforts of the Sternbergs, Brown, and other paleontologists, the Red Deer River area of Canada was revealed as one of the richest deposits of Late Cretaceous dinosaur bones in the world. Dawson, Tyrrell, and Lambe are also fine examples of the benefits of fieldwork for one's health. Dawson was very short and a hunchback, yet he energetically explored the wilderness areas of Saskatchewan, Alberta, and British Columbia for fossils throughout his career. Likewise, Tyrrell and Lambe began fieldwork in the late nineteenth century for health improvement; Lambe continued doing fieldwork until his death in 1919 and Tyrrell lived to the age of 98. In fact, Tyrrell exemplifies the longevity that is characteristic of many well-known, field-oriented paleontologists and geologists from the nineteenth and early twentieth centuries, who had average life spans well above typical life expectancies for their times (Table 3.1).

Only a few discoveries of dinosaurs from South America were documented as dinosaurs during the nineteenth century, although some regions later became history-making spots in dinosaur studies. The first discovery of dinosaur tracks in Columbia, South America, was by **Carl Degenhardt** in 1839, although he, like Hitchcock, thought they were bird tracks. The first discovery of dinosaur bones in the Cretaceous rocks of Patagonia, Argentina, was in 1882 by a military officer known only in historical records as **Commandante Buratovich**. Buratovich sent his finds to the renowned Argentine paleontologist **Florentino Ameghino** (1854–1911), who confirmed for the first time that Argentina had dinosaurs. Ameghino's brother, **Carlos Ameghino** (1865–1936), often assisted him by doing most of their fieldwork. **Francisco P. Moreno** (1852–1919) also found dinosaur bones in Argentina in 1891, reconfirming their presence for future workers. Another Argentine, **Santiago Roth**, began his paleontological career in the same area of Argentina soon after the Cretaceous dinosaur remains were found. In the early part of the twentieth century, Roth contributed to the dinosaur collection of the **Museo de La Plata** in Argentina, of which Moreno was the first director. These dinosaur finds were early indicators of later significant discoveries of skeletal material, eggs, nests, and tracks in Upper Triassic–Upper Cretaceous deposits of Argentina into the twenty-first century. These include some of the largest theropods and sauropods known (Chapters 9 and 11).

TABLE 3.1 Sample of lifespans of field-oriented paleontologists and geologists mentioned in the chapter who were born before 1900, names listed in chronological order from date of birth. Mean age = 76.6 ± 14.2 years, median = 78 years (n = 29, all male subjects); 34% of sampled people were older than 80 years when they died.

Name	Born	Died	Lifespan
John Collins Warren	1778	1856	78
William Buckland	1784	1856	72
Adam Sedgewick	1785	1873	88
Gideon Algernon Mantell	1790	1852	62
Edward Hitchcock	1793	1864	71
Charles Lyell	1797	1875	78
Christian Erich Hermann von Meyer	1801	1869	68
Phillipe Matheron	1807	1899	92
Jean-Jacques Pouech	1814	1892	78
Paul Gervais	1816	1879	63
Charles H. Sternberg	1850	1943	93
Henry Fairfield Osborn	1857	1935	78
Joseph Burr Tyrrell	1859	1957	98
John Bell Hatcher	1861	1904	43
Eberhard Fraas	1862	1915	53
Earl Douglass	1862	1931	69
Lawrence Lambe	1863	1919	56
Carlos Ameghino	1865	1936	71
Richard Broom	1866	1951	85
Richard Swan Lull	1867	1957	90
Ernest Stromer von Reichenbac	1871	1952	81
Walter Granger	1872	1941	69
Barnum Brown	1873	1963	90
Friedrich von Huene	1875	1969	94
Franz Nopcsa	1877	1933	56
Werner Janensch	1878	1969	91
Roy Chapman Andrews	1884	1960	76
Charles M. Sternberg	1885	1981	96
Yang Zhong-jian (C. C. Young)	1897	1979	82

Scientific Studies of Dinosaurs in the First Half of the Twentieth Century

The turn of the last century was a seamless transition for most dinosaur paleontologists, but the deaths of Cope in 1897 and Marsh in 1899 symbolized the beginning of the new era. As changes in modes of transportation and communication began to make the world a smaller place, the study of dinosaurs became more global, expanding to areas of the world outside Europe and the Americas. Dinosaur paleontologists also became more cooperative and engaged in friendly competition, a spirit that has, for the most part, continued to today. Last, fundamental connections between dinosaur body fossils and trace fossils were made that firmly established the dual importance and complementary nature of these facets for interpreting dinosaurs. But during the first half of the twentieth century, two world wars disrupted dinosaur studies; these wars not only resulted in a huge loss of human life

FIGURE 3.9 Franz Nopcsa, dinosaur paleontologist, Transylvanian nobleman, linguist, spy, and motorcycle enthusiast, shown here in Albanian costume and carrying optional field gear. From Kubacska, András Tasnáde, 1945. Verlag Ungarischen Naturwisenschaftlichen Museum, Budapest/Dover Publications.

3

but also the destruction of dinosaur skeletons. Some skeletons were sunk during submarine attacks of World War I while being transported across the Atlantic, and other skeletons in German museums were destroyed in bombing raids by Allied forces in World War II.

Of all Europeans who worked on dinosaurs in the early part of the twentieth century, **Franz (Ferenc) Baron von Felsö-Szilvás Nopcsa** (1877–1933) was the most likely contender for William Buckland's position as the most unusual dinosaur paleontologist (Fig. 3.9). The Transylvanian nobleman became a paleontologist by accident after his sister found some bones on her estate. He brought the bones to a university professor to identify, and the professor told him, "Study them yourself," which he did. Although Nopcsa had no prior training in paleontology, he subsequently published a description of the Cretaceous hadrosaur *Telmatosaurus transsylvanicus* in 1900. He then conducted more research on the dinosaurs of his home country, as well as those in England and France. Nopcsa soon broadened his scope to include large-scale concepts such as classification schemes, evolutionary relationships of dinosaurs, and integration of the (then) new idea of continental drift with dinosaur distributions (Chapters 4 and 6). Although he was not always correct, Nopcsa's thinking was original, and he may have been the first dinosaur paleontologist to look intensively for sex differences in dinosaur species, a field of study that generated much interest later (Chapters 5 and 8).

In addition to his paleontological ambitions, Nopcsa decided, after traveling through Albania and studying its cultures and dialects, that he was the most qualified person to rule it as king. He planned to accomplish this goal through various imaginative machinations, which he shared with officials of the Austria-Hungarian government. These plans included military strategies for invading Albania and generating revenue for the new nation-state through marriage with a not-then-identified daughter of an also-not-then-identified American millionaire. He figured that it would be no problem to find one after he was crowned as a king. The government declined his offer, and he never did find his hypothetical rich wife. Instead, his life took even stranger turns:

- He was involved in secret missions during World War I as a spy.
- The Romanian government seized his estates.
- He was nearly beaten to death by an angry mob of peasants.
- He was placed in charge of the Hungarian Geological Survey.
- He angrily quit the Survey.
- He took off on a 5500 km long motorcycle ride with his male Albanian secretary, who was also his lover.
- He spent all his money.
- He completed impressive works on dinosaur bone histology as related to their classification.
- He became depressed, shot his lover, and committed suicide.

On a more mundane note, the finds of western North America continued with the prodigious output of Barnum Brown in the early 1900s. In 1902, Brown discovered one of the largest (and certainly the most famous) of land carnivores, *Tyrannosaurus rex*, in the Late Cretaceous Hell Creek Formation of eastern Montana. Brown then followed Lambe by taking a barge down the Red Deer River in 1910, to explore Upper Cretaceous deposits there. For the next six years, Brown and his associates, working through the American Museum, directed by Osborn (who in 1905 named another tyrannosaurid, *Albertosaurus*, from this region: Chapter 9), collected dinosaur bones from two different geologic levels. These vertically-separated levels indicated different times in geologic history (Chapter 4), so Brown's collections contributed greatly to understanding the evolutionary sequences of dinosaurs during the Late Cretaceous. Besides tyrannosaurids, other dinosaurs documented from this area include the hadrosaur *Corythosaurus* (Chapter 11), some ceratopsians, such as *Monoclonius*, *Anchiceratops*, and *Leptoceratops* (Chapter 13), the large theropod *Centrosaurus*, and the ostrich-like theropod *Struthiomimus* (Chapter 9). In 1907, Brown was the first paleontologist to write about dinosaur **gastroliths** ("stomach stones") that dinosaurs probably used for grinding food in their digestive systems (Chapter 14). In his report, Brown noted that he found gravel associated with the skeleton of the hadrosaur *Claosaurus*, which he interpreted as gastroliths, but such hypotheses remain controversial and require much careful documentation before conditional acceptance.

The Sternbergs (Charles M. and his brothers) followed Brown's efforts. Prior to their work in Alberta, the Sternbergs made one of the most unusual and scientifically-valuable dinosaur finds of the time, when they discovered a hadrosaur in Wyoming in 1908 with skin impressions associated with the skeleton (Chapters 6). To confirm that their find was not a fluke, they later found another example nearby, which indicated similar conditions of preservation. The brothers were then employed by the Canadian Geological Survey, which was becoming tired of seeing Americans get all of the credit for dinosaurs found in Canada and then taking the Canadian dinosaurs out of the country to New York City. The Sternbergs started their work in the same Red Deer River area in 1911, using not only a barge for travel but also motorboats for occasional prospecting trips to shore. Among the Sternbergs' discoveries were specimens of the hadrosaur *Prosaurolophus* (Chapter 11), ankylosaurs (Chapter 12), the ceratopsians *Chasmosaurus* and *Styracosaurus* (Chapter 13), and a large theropod, *Gorgosaurus*, which is often viewed as synonymous with *Albertosaurus* (Chapter 9). The area prospected by Brown's group and the Sternbergs (constituting an example of the aforementioned friendly competition) is now known as **Dinosaur Provincial Park**. Within this area is Drumheller, Alberta, home of **Royal Tyrrell Museum of Palaeontology**, which contains some of Canada's finest dinosaur skeletons.

Far to the south of Alberta is another famous museum, one that is built around dinosaurs still in their entombing Upper Jurassic rock. **Dinosaur National**

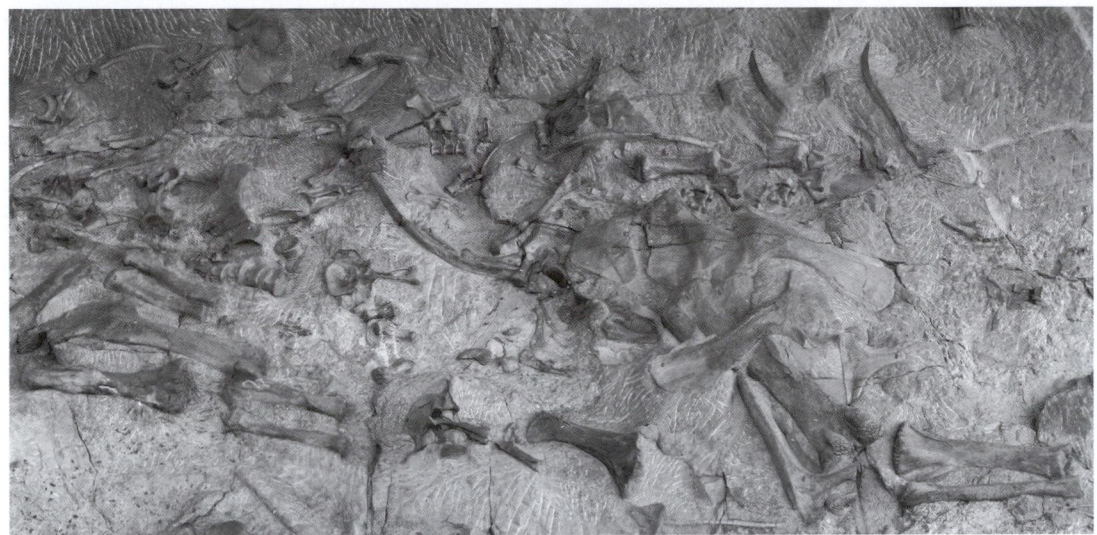

FIGURE 3.10 Outcrop of Morrison Formation (Late Jurassic) with extremely abundant dinosaur bones, discovered by Earl Douglass in 1909, Dinosaur National Monument, near Jensen, Utah.

Monument, near Jensen, Utah, contains about 1500 *in situ* dinosaur bones, which are available for public viewing today (Fig. 3.10). This happy circumstance was prompted largely by the efforts of **Earl Douglass** (1862–1931). In 1909, when Douglass discovered this site (probably the most important Late Jurassic dinosaur deposit in the world), he found what turned out to be a nearly complete specimen of *Apatosaurus* (Chapter 10). Subsequent quarrying from 1909 to 1922, sponsored by industrialist and philanthropist **Andrew Carnegie** (1835–1919) for his Carnegie Museum (now the **Carnegie Museum of Natural History**, in Pittsburgh, Pennsylvania), yielded numerous skeletal remains of *Allosaurus*, *Apatosaurus*, *Diplodocus*, *Dryptosaurus*, and *Stegosaurus*. One of the more significant specimens was a juvenile *Camarasaurus* that was nearly complete and articulated, a very unusual find (Chapters 7 and 10). Workers at this site may have been the first to use explosives, such as dynamite (invented by **Alfred Bernhard Nobel**, 1833–97, of Nobel Prize fame), for extracting dinosaur skeletons from their rocky matrix, a practice that has mercifully lessened since then. Two interesting side notes are that the trivial name in *Apatosaurus louisae* was named after Carnegie's wife, Louise, and "*Diplodocus*" was the subject of the first (but certainly not last) dinosaur-themed pub song.

Similar dinosaur quarries in the western United States, which were found and mined during the first half of the twentieth century, were the **Howe** and **Cleveland-Lloyd** Quarries. The Howe Quarry was named after rancher **Barker Howe**, who discovered dinosaur bones on his property in northwestern Wyoming and called in Barnum Brown to investigate in 1934. Brown and his assistants then uncovered a dense accumulation of Late Jurassic bones. The extreme density is confirmed by actual data. Brown mapped meticulously in approximately 1 m^2 intervals. The assemblage includes mostly sauropods, such as *Apatosaurus*, *Barosaurus*, *Camarasaurus*, and *Diplodocus* (Chapter 10). Their maps show the distribution of the bones (Fig. 3.11). Bone abundance can be calculated using the following information:

$$A = lw \tag{3.1}$$

where A is area, l is length, and w is width. The area of the Howe Quarry was 14×20 m, hence its total area was

$$A = 14 \times 20 \text{ m} = 280 \text{ m}^2 \tag{3.2}$$

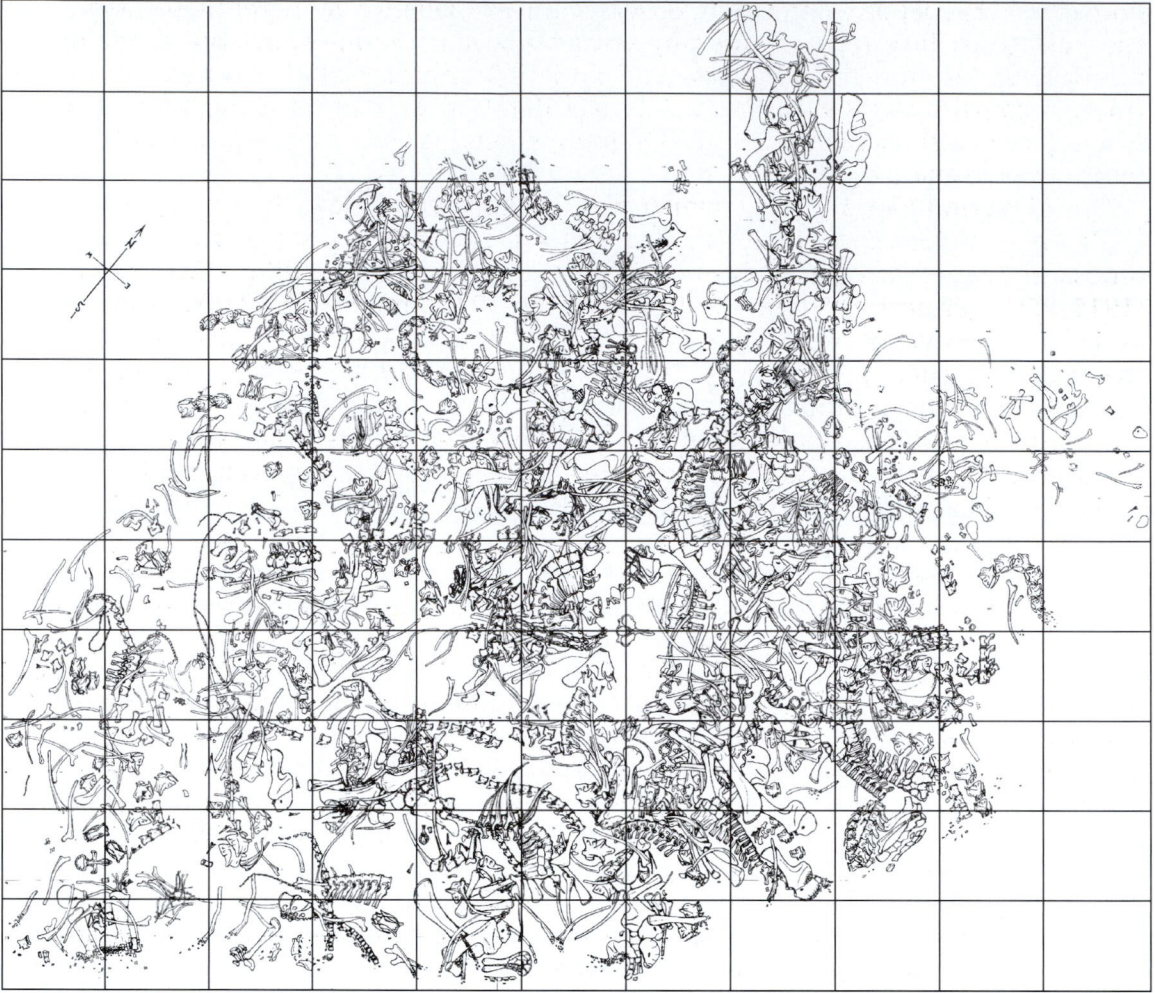

FIGURE 3.11   Map of Howe Quarry, showing horizontal distribution and concentration of dinosaur bones in the quarry area. Squares represent approximately meter squares. Neg. No. 314524. Courtesy Department of Library Services, American Museum of Natural History.

Bone abundance can be calculated by dividing the number of bones by the area:

$$B_a = N/A_t \qquad\qquad (3.3)$$

where B_a is bone density in bones per square meter, N is total number of bones, and A_t is total area. Knowing that at least 4000 bones were recovered from this area,

$$B_a = 4000 \text{ bones}/280 \text{ m}^2 = 14 \text{ bones/m}^2 \qquad\qquad (3.4)$$

Keep in mind that this number is equivalent to the mean number of bones per square meter, which does not take into account that some meter squares may not have had bones, or other squares had considerably more than 14. Additionally, looking at a bone map clearly shows how some bones transect the meter-square boundaries, leading to some restrictions about how to count bones within an area. Nevertheless, such calculations provide a measure of the relative abundance of skeletal components at a given site. Bone mapping, done at the Howe site by Brown's assistant **Roland T. Bird** (1899–1978), was the first attempt to record such information in this amount of detail. This method is now a standard procedure at any dinosaur bone deposit because it provides much evidence for hypotheses about the post-death history of a dinosaur assemblage. For example, an abundance of dinosaur bones probably indicates rapid burial of a number of dinosaurs together, through some unusual event such as a river flood or ash deposit (Chapter 7). The Howe Quarry also yielded dinosaur skin impressions, as well as a few dinosaur trace fossils, such as gastroliths and tracks.

The Cleveland-Lloyd Quarry, which was first uncovered near the small, eastern Utah town of Cleveland in 1937, also had a high bone density of Late Jurassic dinosaurs in a relatively small area. With the guidance of **William Lee Stokes** (1915–95) of Princeton University from 1937 to 1941, more than 12,000 dinosaur bones were recovered from the site, of which 60% to 70% are from *Allosaurus* and the rest from such dinosaurs as *Camarasaurus*, *Stegosaurus*, and the ornithopod *Camptosaurus*. The assemblage is unusual in its concentration of allosaur skeletons, leading to hypotheses that explain why so many meat-eaters would be in such a small area (Chapters 7 and 9). Stokes later adopted Utah as his home and spent much of his career at the University of Utah.

Roland Bird, who mapped the Howe Quarry, became more famous for his work with dinosaur tracks, particularly in eastern Texas, but also in Arizona, Colorado, and Utah. Like many early dinosaur-fossil discoverers, Bird had little formal academic training but had developed a successful search pattern for dinosaur fossils through extensive experience (Chapter 2). After finding several tracksites in Jurassic and Cretaceous rocks of some western states, he decided that these sites were too inaccessible. Subsequently, he followed up a tip and in 1939 went to Glen Rose, Texas, where he found dinosaur tracks exposed in Lower Cretaceous rocks of the Paluxy River. These tracks were made by a variety of theropods, but most importantly included undoubted sauropod tracks (some more than a meter wide), the first reported scientifically from the geologic record. One excellent paleontological point of discussion provoked by the sauropod tracks was whether they indicated that sauropods had aqueous habits (which was presumed at the time) or whether they walked on dry land (Chapters 10 and 14).

In 1909, another prolific Late Jurassic dinosaur site was investigated in an area far removed from Utah – **Tendaguru**, in present-day Tanzania. The region was a German colony at the time (German Protectorate East Africa) and in 1907 **Bernhard Sattler**, an engineer with a German mining company, discovered some large bones there. **Eberhard Fraas** (1862–1915), of the Royal Naturaliensammlung (Staatliches Museum für Naturkunde) in Stuttgart, Germany then examined these bones later that year. Fraas caught amoebic dysentery while doing fieldwork before visiting Tendaguru, but he managed to make the four- to five-day hike to the field site to confirm that the large bones were indeed from dinosaurs. While he was there, he even directed some excavation and recovery efforts. **Werner Janensch** (1878–1969) conducted later expeditions from 1909 to 1913 and was ably assisted by **Boheti bin Amrani**, a native of the region (Fig. 3.12). After World War I, the area came under British control and expeditions from 1924 to 1931 were arranged through the British Museum, again using the expert guidance of Amrani. The logistics for these forays were daunting because, unlike the American West, no railroads went into the area. There were also few automobiles and no roads, so local workers were employed to carry the dinosaur bones away on foot. Through this labor-intensive method, the local people in the employ of earlier German expeditions carried out about 225 tons of bones, the hike from the site taking four to five days. Collectively, all work done in Tendaguru resulted in the uncovering of a new species of stegosaur, *Kentrosaurus* (Chapter 12), the theropod *Elaphrosaurus* (Chapter 9), and the ornithopod *Dryosaurus* (Chapter 11), but the deposit is best known for its diverse sauropod assemblage, consisting of *Barosaurus*, *Brachiosaurus*, *Dicraeosaurus*, *Janenschia*, and *Tornieria* (Chapter 10).

FIGURE 3.12 Boheti bin Amrani at Tendaguru of what is present-day Tanzania, uncovering a sauropod rib during one of the German expeditions to the region. Dr. Bernard Krebs, Lehrstuhl für Paläontologie der Freien Universität, Berlin/Dover Publications.

In South Africa, **Richard Broom** (1866–1951), originally from Scotland and yet another physician who was much more enthused about long-dead subjects than his living patients, published papers in 1904 and 1911 on a few dinosaur finds in the Karoo basin, an area well known for its vertebrate fossils. These papers were significant because no dinosaurs had been described from South Africa since the times of Owen and Huxley, so future workers were encouraged to explore more in this area. Sure enough, **Sydney Haughton** of England and **E. C. N. van Hoepen** of South Africa soon followed Broom's works in 1915 to 1924 and expanded upon the knowledge of Late Triassic dinosaurs (such as the prosauropod *Melanorosaurus*; Chapter 10) in that region.

Also working on Late Triassic dinosaurs, especially those of southern Germany, was **Friedrich von Huene** (1875–1969). Von Huene greatly expanded the studies of the abundantly represented Late Triassic prosauropod *Plateosaurus*. Early in his long career, he reviewed critically all of the previous classifications of dinosaurs and re-affirmed in 1914, on the basis of much evidence, the dual classification system of Saurischia and Ornithischia for the dinosaurs (Chapter 5). He also described, for the first time, dinosaurs from the Upper Triassic of Brazil and did fieldwork wherever he could find Triassic rocks, which included the five continents of Europe, North America, South America, Africa, and Asia. Because of his breadth of experience with these earliest of dinosaurs, he provided much knowledge toward their evolutionary history (Chapter 6), and in 1932 published a comprehensive evaluation of the Saurischia. He also was well known for his hiking ability, and while in his 80s decided to attend a scientific meeting by walking 150 km for three days across southern Germany.

Like von Huene, Richard Lull (mentioned earlier) was an important synthesizer of knowledge about Late Triassic dinosaurs, especially those of the Connecticut Valley. Much of his work was done at the same time as his German counterpart. Lull, a former student of Osborn, was the first person to begin the arduous task of reconciling and correlating Hitchcock's dinosaur tracks with potential tracemakers, which he did by examining new discoveries of body fossil of dinosaurs in the same area. This integration of body and trace fossil evidence of dinosaurs had been attempted in other areas before, but never to the extent that Lull pursued it. His efforts resulted in a much better model of how a comprehensive approach to dinosaur fossils could enhance the quality of hypotheses about them. In 1915, Lull also reviewed all known fossil evidence (plants, insects, fish, reptiles, and dinosaurs) associated with Upper Triassic rocks of the Connecticut Valley, in an attempt to reconstruct the dinosaurs' paleoenvironments, one of the first serious studies of the paleoecology of a terrestrial ecosystem.

No history of dinosaur studies is complete without mentioning the Mongolian expeditions, first mounted by the American Museum and represented by the semi-legendary character of **Roy Chapman Andrews** (1884–1960). Chapman humbly began his career with the museum by performing janitorial duties, and eventually worked his way into the technical staff. Ironically, in light of his contributions to later discoveries, he was not primarily a dinosaur paleontologist and was mostly interested in studying mammals. Osborn, who became president of the museum in 1908, shared Andrews' interest in mammals, and Andrews persuaded him to mount a paleontological expedition to central Asia to search for the fossil ancestral remains of the most important mammals of all (to them), humans. Osborn agreed and the first Central Asiatic expedition, led by Andrews and accompanied by experienced paleontologist Walter Granger, went into the Gobi Desert of Mongolia in 1922. The trip failed in its goal to find fossils of humans, but did find the first confirmed dinosaur nests with eggs, although the identities of the egg layers were mistaken for the next 70 years (Chapter 9). French paleontologists had documented dinosaur eggs without nests in the nineteenth century. Skeletal material derived from this and successive expeditions from 1922 to 1930 included abundant specimens of the marginocephalians *Protoceratops* and *Psittacosaurus* (Chapters 7 and 13), the inappropriately-named theropod *Oviraptor* (Chapter 9), and the evil-looking *Velociraptor* and *Saurornithoides* (Chapter 9), all found in Cretaceous rocks.

Chapman has often been cited as the possible inspiration for the character Indiana Jones, a fictional **archaeologist** (someone who studies human artifacts, which is very different from a paleontologist: see Chapter 1). The producers of the Indiana Jones films have never admitted that Chapman was their source. Regardless, Chapman's expeditions and his exploits were certainly extraordinary for their time. He took advantage of his knowledge of Asian languages to work with his Chinese hosts and established his headquarters in Beijing; he navigated field crews in automobiles; and he arranged for the rendezvous of camel herds that carried gasoline and other supplies across the desolate terrain. Chapman was also an excellent marksman and was rarely photographed in the field without some type of firearm within his reach (Fig. 3.13). Bandits were bothersome in the region, and he reportedly shot some of them. Fortunately, most fieldwork in Mongolia and other areas of the world today is threatened more by bad weather or diminishing coffee supplies than hostile raiders.

In the late 1940s, Russian expeditions to Mongolia followed the American efforts through the auspices of the Russian Paleontological Institute, led by paleontologist (and famed Russian science-fiction writer) **Ivan A. Efremov** (1907–72) and herpetologist **Anatole K. Rozhdestvensky**. In these excursions they found more examples of the previously discovered Cretaceous dinosaurs of that region, as well as some important new finds, such as the ankylosaur *Pinacosaurus* (Chapter 12), hadrosaur *Saurolophus* (Chapter 11), and the large theropod *Tarbosaurus*, which is so similar to *Tyrannosaurus* that it is now considered an Asian variant of the species (Chapter 9). The continued success of the Russian expeditions ensured that more

FIGURE 3.13 Roy Chapman Andrews (right), in the Bain-Dzak area, Mongolia, with Late Cretaceous dinosaur eggs in front of him and his bandit-prevention device behind him. Negative No. 410760, Photo. Shackelford. Courtesy Department of Library Services, American Museum of Natural History.

investigators would follow; a Polish–Mongolian research group returned to the area in the 1960s, as did American Museum paleontologists in the 1990s. Renowned paleontologist **Zofia Kielan-Jaworowska** led the Polish–Mongolian expedition, which also included participants **Teresa Maryanska** and **Halszka Osmólska**, who are still considered to be Poland's leading experts on dinosaurs.

China has been one of the most productive countries for dinosaur fossils, and this status will most likely continue for many years to come. The long-known plethora of dinosaur bones in China led to initial studies, mostly by Westerners in cooperation with Chinese scientists, resulting, in 1922, in the description of *Euhelopus*, a Jurassic sauropod (Chapter 12). One of the pioneers of dinosaur paleontology was **Yang Zhong-jian** (1897–1979), known as **C. C. Young** in Western scientific literature. Yang, who officially became China's first professional vertebrate paleontologist in 1927, was also the co-discoverer of the first documented dinosaur tracks in China, which were found in Jurassic rocks of the Shanxi Province in 1929. In 1936, he led a combined Chinese and American group, which uncovered the unusually long-necked sauropod *Omeisaurus* (Chapter 10). Yang's studies in Canada, England, the USA, and Germany helped him to establish excellent contacts with Western scientists, which paved the way for exploration of the vast outcrops of Mesozoic strata in his country. Locations in China have produced one of the most productive dinosaur egg sites in the world (Chapters 2 and 8) and many species of feathered non-avian theropods and birds (Chapters 9 and 15).

Dinosaur trace fossils, other than tracks and nests, received some recognition early in the twentieth century, although some of them were not appreciated until recently. Coprolites were reported in dinosaur-bearing rocks by Hitchcock in 1844, but the first dinosaur coprolite was not interpreted until 1903 by **C.-E. Bertrand** of Belgium. His specimen came from the same Cretaceous deposit that provided Dollo with so many iguanodontian skeletons. After Barnum Brown interpreted gastroliths early in that century, von Huene in 1932 reported other gastroliths in association with bones of the Late Triassic prosauropod *Sellosaurus* (Chapter 10). In 1942, Stokes described a similar occurrence of stones found with Late Jurassic sauropod remains. **W. D. Matthew** first interpreted dinosaur toothmarks, which are often preserved in dinosaur bones, in association with a potential tracemaker in 1908. In this study, he noted that the tooth spacing of the Late Jurassic theropod *Allosaurus* matched the toothmarks on bones of *Apatosaurus*, a sauropod that lived at the same time. This approach provided an intuitive method for better determination of feeding relationships among dinosaurs (Chapters 8 and 9). Although they were always a part of dinosaur studies, dinosaur trace fossils began to gain more attention from dinosaur paleontologists in the latter half of the twentieth century, as trace fossils supplemented or, in some cases, surpassed the information derived from dinosaur body fossils.

Dinosaur Studies of the Recent Past: Beginnings of a Renaissance and a New Legacy

The Latter Half of the Twentieth Century and Globalization of Dinosaur Studies

The preceding history arbitrarily cuts off at about 1950 and is thus incomplete, but it provides a summary of dinosaur studies up to that point.

Observant readers may have noticed that one continent, Australia, has barely been mentioned, and Antarctica completely neglected. This lack of information is because Australia has become a discovery site for abundant dinosaur fossils (especially tracks) only in the past 35 years, and the first discovery of an Antarctic dinosaur was not until 1986. However, both of these continents will undoubtedly see expanded research as these finds inspire increased exploration.

Discussion of the people in dinosaur studies during the latter half of the twentieth century must be limited for several reasons. One reason is that the author of this book does not feel qualified to judge which of these people (many of whom are still active in the discipline, and perhaps reading this) deserve mention as important contributors to the long-term history of dinosaur studies. Such a pronouncement will be much easier to make in another 50 years or so, when the enduring contributions made by these investigators will be more evident. Of course, some genuinely notable discoveries already happened in the latter half of the twentieth century and beginning of the twenty-first, and those discoveries and the people associated with them will be mentioned where appropriate. Time will tell whether these contributions will make paleontological history. With that said, three paleontologists in the latter half of the twentieth century, **Edwin H. Colbert**, **John Ostrom**, and **José F. Bonaparte**, stand out for providing the most long-lasting scientific contributions from which all modern investigators in dinosaur studies will benefit.

Edwin H. Colbert (1905–2001) is best known in dinosaur paleontology for his discovery, in 1947, of a site that contained hundreds of the Late Triassic theropod *Coelophysis bauri* (Chapter 11). The site, in the **Chinle Formation** at Ghost Ranch, New Mexico, was located near the summer home of famed painter **Georgia O'Keeffe** (1887–1986), who occasionally stopped by the excavation to talk with Colbert. Late Triassic dinosaurs have always held a special interest to paleontologists, because they represent the earliest dinosaurs (Chapter 6). Thus for Colbert to document such a rich find was a major contribution to our understanding of the origin of dinosaurs and a source of detailed paleontological information about them. For example, because of their abundance, growth series and population structures for this theropod could be proposed. This is an unusual situation for any dinosaur species because of the rarity of multiple specimens of the same species. Additionally, cannibalism was first interpreted for this species based on the specimens from Ghost Ranch, where juvenile bones were thought to be inside the body cavities of an adult. This has been re-interpreted, however, because the bones of one were actually just on top of another, rather than inside of it (Chapter 9). Furthermore, the unusual occurrence of so many individuals of a single species of carnivorous dinosaur, similar to the findings in the Cleveland-Lloyd Quarry, led to some debated hypotheses regarding the pre-burial history of the assemblage, as well as implied social behavior (Chapter 6). Colbert was also well known for his excellent textbooks on vertebrate paleontology and popular books on dinosaurs, which have helped to educate aspiring vertebrate paleontologists worldwide. His fascinating works on the history of dinosaur studies also gave science enthusiasts a sense of the uniqueness of using fieldwork to search for the remains of long-dead animals. Colbert's historical works were a major source of information for this chapter, and they certainly set the standard for all future bibliographers of dinosaur paleontologists.

John Ostrom (1928–2005) is credited with sparking the Dinosaur Renaissance of the past 30 years by his detailed examination and consequent hypotheses of the Early Cretaceous theropod *Deinonychus* (Chapter 9), which he first reported in 1969. Ostrom, through convincing use of his data on *Deinonychus*, revived the idea (first proposed by Huxley and unintentionally augmented by the work of Hitchcock in the nineteenth century) that some dinosaurs were more active and bird-like in their behavior, rather than reptilian. Ostrom's interpretation was based on **functional morphology**, the study of how the form of an animal relates to its functions, an approach that had been used before in dinosaur studies but rarely so effectively. New interest thus began in studying the extent of this bird-like behavior in some dinosaurs, namely whether it was reflected by physiological indicators of endothermy (Chapter 8) or was related to evolutionary links between dinosaurs and modern birds (Chapter 15). Ostrom also made a very important discovery while examining a skeleton in a small Dutch museum. Ostrom recognized the skeleton, identified initially as a pterosaur (flying reptile), as a previously unknown specimen of *Archaeopteryx*, one of only seven ever described.

Dinosaurs from Europe and North America were studied the most during the nineteenth century and interest expanded to Africa and Asia in the first half of the twentieth century. But research in South America in the latter half of the twentieth century was prompted largely by the efforts of **José F. Bonaparte**. Of all living paleontologists, Bonaparte has named or co-named the largest number of dinosaur genera (10 as of the writing of this book), including *Argentinosaurus* (a huge sauropod; Chapter 10), *Carnotaurus* (a large, horned theropod; Chapter 9), and *Abelisaurus*, the latter a representative of a group of Cretaceous theropods unique to South America. His discoveries, primarily in his native Argentina, have shown important evolutionary relationships between dinosaurs of separate continents, especially the "southern continents" of South America, Africa, India, Australia, and Antarctica, which formed one landmass in southern latitudes, called **Gondwana**, during much of the Jurassic (Chapters 4 and 6). Bonaparte is the former student of influential American vertebrate paleontologist **Alfred Sherwood Romer** (1894–1973), and is continuing his tradition of excellence.

Other notable dinosaur paleontologists, who have already encouraged much interest in dinosaurs in the USA and abroad, include Americans **Robert T. Bakker**, **John (Jack) R. Horner**, and **Paul C. Sereno**, as well as **Martin G. Lockley**, originally from Wales but now based in the USA. Bakker, a former student of Ostrom, is best known for his role as a publicly visible cheerleader for alternative views of dinosaurs as active animals more akin to birds and mammals, as opposed to their previous stereotype as sluggish reptilians. He is also one of the best popularizers of contentious ideas about dinosaurs that have provoked much discussion and attempts at refutation; his main theses are summarized in his 1986 book *The Dinosaur Heresies*. Jack Horner, along with his now deceased friend **Bob Makela**, began his career as an amateur paleontologist and discovered dinosaur-nesting horizons of the Late Cretaceous ornithopod *Maiasaura*, the first dinosaur nests found in North America (Chapters 8 and 11). This work and further investigations changed the conception of dinosaurs from solitary and uncaring creatures to social, nurturing animals. In a relatively short time, Sereno and his research teams have chalked up a remarkable number of noteworthy dinosaur discoveries in remote areas of Argentina, Morocco, Niger, and Inner Mongolia. Included in his scientifically important contributions are the discovery and description of what are possibly the oldest known dinosaurs or dinosaur ancestors (Chapter 6), and he has otherwise made significant advances in the cladistic classification of dinosaurs. Lockley is the most recognized dinosaur ichnologist in the world, having studied and written about dinosaur tracks and their scientific pertinence in numerous peer-reviewed journal articles and books intended for public consumption. Although most of his work has been in the track-rich Mesozoic strata of the western United States, he has also studied dinosaur tracks from Argentina, Bolivia, Brazil, Portugal, Spain, central Asia, China, and Korea, thus considerably augmenting the skeletal record for dinosaurs formerly missing from many of these regions. Lockley and the other aforementioned paleontologists are especially well known for their educational outreach efforts, whether through books written for interested lay people or lectures given in public forums.

Of course, modern dinosaur paleontologists are members of an increasingly global science. A short list, for purely practical reasons of limited space, might include **Phillip J. Currie** of Canada, **Dong Zhi-Ming** of China, **Altangerel Perle** of Mongolia, **Patricia Vickers-Rich**, **Thomas Rich**, and **Tony Thulborn** of Australia, **Konstantin Mikhailov** of Russia, **Anusuya Chinsamy-Turan** of South Africa, **Fernando Novas** and **Rudolfo Coría** of Argentina, and **Armand de Ricqlès** of France. Further internationalization of dinosaur studies and inclusion of more participants from less industrialized nations should continue as Internet communications become accessible in more places and bureaucratic obstacles lessen. Stricter immigration control in the USA since 2001, however, has significantly decreased the number of foreign-born graduate students and scientists entering the USA, which may adversely affect future cooperation. Likewise, large-scale warfare in the Middle East since 2003 has hampered the participation of USA scientists in projects taking place in countries opposed to US-led war efforts.

Perspectives in the Past, Present, and Future of Dinosaur Studies

Advances in dinosaur studies in the past 25 years are exhilarating. The fast pace of these discoveries and the competition for coverage of these discoveries by the popular press ensures that the history of dinosaur studies will be continually changing, but all of this is still a direct result of the science behind such discoveries. One of the fringe benefits of the ongoing popularity in dinosaurs is that many professional paleontologists can write books for a general audience on their favorite subjects while still retaining their scientific integrity. Some of these books summarize evidence, hypotheses, and in some cases speculations about dinosaurs, whether based on information gathered in the past two centuries or just in the past few years. Knowing the history of dinosaur studies gives us a perspective as to how the science, especially in terms of its knowledge, has evolved and changed through the centuries, change that is still happening today. Scientists try to learn not just from their mistakes but from the mistakes of others, so no doubt some of the "certainties" of dinosaur paleontology today will be ridiculed (in an understanding sort of way) by future generations of paleontologists.

In contrast, some aspects of how paleontologists went about their work and made their discoveries in the past are unlikely to change. Despite the sophistication of modern technology, much of dinosaur paleontology still involves wandering through remote areas of the world, looking at the ground, and using search patterns. The human element of such explorations will also be both a source of constancy and unpredictability. As demonstrated previously, knowing about the people involved in dinosaur paleontology also helps in understanding that scientists are real people who have jealousies, fears, prejudices, greed, and occasionally nasty tempers. However, when all is said and done, they love their science. Additionally, society and politics have influenced the course of paleontological studies, and provide valuable context for certain dinosaur discoveries and their interpretations. Lastly, as we have seen, many dinosaur discoveries were made by amateurs and later described by professional paleontologists, demonstrating how paleontology is one of the few sciences where amateurs have made and continue to make important contributions.

SUMMARY

The biological origin of dinosaur fossils was probably evident to early peoples and had some influence on cultural development of some prehistoric populations. The first written reference to what may have been dinosaur fossils was in China nearly 2300 years ago, but scientific methods were not applied to these fossils until about 200 years ago. When fossils were discovered in Europe from the fifteenth to the early part of the nineteenth centuries, the voice of reason, so often associated with the rise of scientific thought in Western civilizations, rejected fossils as the remains of extinct organisms. However, these errors were eventually recognized, which demonstrates that science is a self-correcting enterprise. Early workers in England, such as William Buckland, the Mantells, and Richard Owen, were responsible for the gestation of dinosaur studies. French, German, American, Canadian, and Argentinian paleontologists investigated both dinosaur body fossils and trace fossils in the remainder of the nineteenth century. In part, the first half of the twentieth century represented a continuation of this work, but it also was marked by exploration of Asia (particularly China and Mongolia), Africa, and more of the Americas. Unfortunately, two world wars interrupted most international cooperation on dinosaur paleontology, but most relations resumed in the 1950s. The study of dinosaurs began its climb to its current exalted state when new hypotheses about dinosaurs, in the late 1960s and early 1970s, received increased publicity. The recent resurgence of new discoveries and hypotheses in the 1990s and early part of the twenty-first century ensures that the future study of dinosaurs will continue to make history. Dinosaur paleontology is now, more than ever, a global science and its evolution indicates that trend will continue in the future.

DISCUSSION QUESTIONS

1. How would you attempt to support the hypothesis that the "dragon bones" described in ancient Chinese texts refer specifically to dinosaur bones? What sort of evidence would you consider convincing?

2. The rivalry of Cope and Marsh, which was a major motivator behind their enormous number of significant dinosaur discoveries and other contributions to paleontology, brings up an interesting ethical question (Chapter 2). Did the value of their discoveries outweigh the costs of their enmity? In other words, did the ends justify the means?

3. The field party sent by Osborn to Como Bluff in 1897 was unsuccessful, but it forced Osborn to look at a different nearby locality, which resulted in the "Bone Cabin" find of thousands of dinosaur bones. What other instances in the chapter seemed to show similarly discouraging circumstances that caused the people involved to accomplish tasks that actually resulted in later success?

4. The phrase "degrees of separation" refers to how one person who has met two other people represents one degree of separation that links the two, who might never meet. Which historical examples of "degrees of separation" surprised you with regard to dinosaur paleontologists and non-paleontological figures? (For example, how many degrees of separation are there between Sitting Bull and Roy Chapman Andrews?)

5. This chapter presents data that could support the statement that a career of paleontology and geology fieldwork has known health benefits and results in a significantly increased lifespan (see Table 3.1). How would you test this statement? What are potential sources of error in these data, such as in calculations of the ages of the geologists and paleontologists?

6. After finding a bone bed in Cretaceous rocks, you set about mapping the area containing the exposed bones. The area measures 13.5 $\times$ 22.5 m and contains about 1250 bones. What is the approximate bone density of this bed (in m^2)? What are some factors that might cause variations in this average?

7. In the accounting of the history of dinosaur studies, is there any evidence of the effect of language barriers on worldwide exploration for dinosaurs in the nineteenth and early part of the twentieth century? On the basis of dissimilarities in language and culture, what areas of the world would have been the most difficult for Europeans and North Americans to arrange visits? What areas, however remote, were more conducive to investigations?

8. Identify and count how many of the nineteenth century geologists and paleontologists were clergymen. How did some of their findings conflict with the religious conventions of the time? Additionally, how many of the people mentioned from the nineteenth century were

DISCUSSION QUESTIONS Continued

    physicians? How would their medical background have helped them to become vertebrate paleontologists?

9. Written history sometimes reflects the choices of historians. How could you find more evidence of paleontologists in the nineteenth and early twentieth century than what was already mentioned in the chapter, if most mainstream books do not mention them?

10. Pick three people mentioned in the chapter that you would like to meet (they do not have to be paleontologists or geologists). What are some questions you would ask each person relating to dinosaurs and why?

Bibliography

Andrews, R. C. 1932. *The New Conquest of Central Asia: A Narrative of the Explorations of the Central Asiatic Expeditions in Mongolia and China, 1921–30*. New York: American Museum of Natural History.

Bird, R. T. 1985. *Bones for Barnum Brown: Adventures of a Dinosaur Hunter*. Fort Worth, Texas: Texas Christian University Press.

Buckland, W. 1824. Notice on the Megalosaurus, or Great Fossil Lizard of Stonesfield. *Transactions of the Geological Society of London* **1** (Series 2): 390–396.

Buffetaut, E. 2000. A forgotten episode in the history of dinosaur ichnology: Carl Degenhardt's report on the first discovery of fossil footprints in South America (Colombia, 1839). *Bulletin de la Societe Geologique de France* **171**: 137–140.

Buffetaut, E. and Le Loeuff, J. 1994. "The discovery of dinosaur eggshells in nineteenth-century France". *In* Carpenter, K., Hirsch K. F. and Horner J. R. (Eds), *Dinosaur Eggs and Babies*. Cambridge, U.K.: Cambridge University Press. pp. 31–34.

Colbert, E. H. 1984. *The great dinosaur hunters and their discoveries*. [Republication of *Men and Dinosaurs: The Search in the Field and Laboratory*, 1968, New York: E. P. Dutton.] Mineola, New York: Dover Publications Inc.

Colbert, E. H. 1995. *The Little Dinosaurs of Ghost Ranch*. New York: Columbia University Press.

Dean, D. R. 1993. Gideon Mantell and the discovery of *Iguanodon*. *Modern Geology* **18**: 209–219.

Delair, J. B. 1989. "A history of dinosaur footprint discoveries in the British Wealden". *In* Gillette D. D. and Lockley M. G. (Eds), *Dinosaur Tracks and Traces*. Cambridge, U.K.: Cambridge University Press. pp. 18–25.

Dong, Z.-M. 1988. *Dinosaurs from China*. London: British Museum of Natural History. Beijing: China Ocean Press.

Gould, S. J. 1998. An awful, terrible dinosaurian irony. *Natural History* **107**: 61–68.

Grady, W. 1993. *The dinosaur project*. Toronto, Canada: MacFarlane, Walter, & Ross.

Hitchcock, E. 1858. *Ichnology of New England: A report on the Sandstone of the Connecticut Valley, Especially of Its Fossil Footmarks*. Boston, Massachusetts: William White. Reprint, New York: Arno Press.

Holtz, T. R., Jr. 1997. "Dinosaur hunters of the southern continents". *In* Farlow James O. and Brett-Surman M. K. (Eds), *The Complete Dinosaur*. Bloomington, Indiana: Indiana University Press. pp. 43–51.

Huene, F. von. 1932. Die fossile Reptil-Ordnung Saurischia, ihre Entwicklung und Geschichte. *Monographien für Geologie und Paläontologie* (1) **4**: 1–361.

Jacobs, L. L. 1993. *Quest for the African Dinosaurs: Ancient Roots of the Modern World.* New York: Villard Books.

Jaffe, M. 2000. *Gilded Dinosaur: The Fossil War Between E. D. Cope and O. C. Marsh and the Rise of American Science.* New York: Crown Publishing.

Kielan-Jaworowska, Z. 1969. *Hunting for Dinosaurs.* Cambridge, Massachusetts: MIT Press.

Lavas, J. R. 1997. "Asian dinosaur hunters". *In* James O. Farlow James O. and Brett-Surman M. K. (Eds), The Complete Dinosaur. Bloomington, Indiana: Indiana University Press. pp 34–42.

Lull, R. L. 1915. Triassic life of the Connecticut Valley. *Connecticut State Geology and Natural History Survey Bulletin* **24**: 1–285.

Mantell, G. A. 1825. Notice on Iguanodon, a Newly Discovered Fossil Reptile, from the Sandstone of Tilgate Forest, in Sussex. *Philosophical Transactions of the Royal Society of London* **115**: 179–186.

Matheron, P. 1869. Notice sur les reptiles fossiles des dépots fluvio-lacustres crétacés du basin à lignite de Fuveau. *Mémoires de l'Académie des Sciences (belles-) Lettres et (Beaux-) Arts de Marseille* **1868–69**: 345–379.

Mayor, A. 2000. *The First Fossil Hunters.* Princeton, N.J.: Princeton University Press.

Nopcsa, F. von. 1900. Dinosaurierrete aus Siebenbürgen. Schädel von *Limnosaurus transsylvanicus* nov. gen. et spec. *Denksdchriften der kaiserlichen Akademie der Wissenschaften Wien, mathematisch-naturwissenschaftliche Classe* **68**: 555–591.

Ostrom, J. H. 1969. Osteology of *Deinonychus antirrhopus*, an unusual theropod from the Lower Cretaceous of Montana. *Peabody Museum of Natural History Bulletin* **30**: 1–165.

Owen, R. 1842. Report on British fossil reptiles. *Report of the British Association for the Advancement of Science* **11**: 60–204.

Sarjeant, W. A. S. 1997. "The earliest discoveries". *In* Farlow James O. and Brett-Surman M. K. (Eds), *The Complete Dinosaur.* Bloomington, Indiana: Indiana University Press. pp. 3–11.

Seeley, H. G. 1887. On the classification of the fossil animals commonly named Dinosauria. *Proceedings of the Royal Society of London* **43**: 165–171.

Shor, E. N. 1974. *The Fossil Feud between E. D. Cope and O. C. Marsh.* Hicksville, New York: Exposition Press.

Spalding, D. A. E. 1993. *Dinosaur Hunters: 150 Years of Extraordinary Discoveries.* Toronto, Canada: Key Porter Books.

Steinbock, R. T. 1989. "Ichnology of the Connecticut Valley: A vignette of American science in the early nineteenth century". *In* Gillette D. D. and Lockley M. G. (Eds), *Dinosaur Tracks and Traces.* Cambridge, U.K.: Cambridge University Press. pp. 27–32.

Sternberg, C. H. 1909. *The Life of a Fossil Hunter.* New York: Henry Holt and Co. [Reprint by Bloomington, Indiana: Indiana University Press, 1990]

Sues, H.-D. 1997. "European dinosaur hunters". *In* Farlow James O. and Brett-Surman M. K. (Eds), *The Complete Dinosaur.* Bloomington, Indiana: Indiana University Press. pp. 12–23.

Wallace, D. R. 1999. *The Bonehunters' Revenge: Dinosaurs, Greed, and the Greatest Scientific Feud of the Gilded Age.* New York: Houghton Mifflin Co.

Warren, J. C. 1854. *Remarks on Some Fossil Impressions in the Sandstone Rocks of Connecticut River.* Ticknor and Fields, Boston: Massachusetts.

3

Chapter

4

You are out with friends and, because you have just attended a geology class, you notice (maybe for the first time) the rocks exposed in numerous road cuts. Your instructor has already taken you on a few field trips where you learned through hands-on experience how to identify rock types and to interpret the history they represented. Consequently, you point to the different rock layers and tell your friends the order in which they were formed, what you know about their origin, which geologists had studied them, and what was happening in the world at the time these sediments were being deposited. Your friends are initially impressed, but they soon begin to doubt you and want to know how you know that certain layers were formed before other layers, how you know the approximate age of the rocks, and especially how you know that the continents were in different places to where they are today.

How did you come to your conclusions? What evidence can you cite to support your claims?

Paleontology and Geology as Sciences

4

Why are Paleontology and Geology Sciences?

Paleontologists and other geoscientists often deal with phenomena that occurred millions of years ago, which makes paleontology and geology different from most other sciences, so they are sometimes labeled as **historical sciences**. These scientists will never be able to conduct experiments on fossil subjects as living organisms or study directly the environments they lived in. This lack of direct observation of the originally living organisms and their environments has been a source of criticism of paleontology as a science, particularly from other scientists who examine modern processes. However, paleontology is similar in approach to forensic science, which uses evidence to absolve wrongly-accused suspects and implicate criminals. It is also similar to genetics, which involves indirectly observing genes through documentation of their characteristic biochemical signatures. Furthermore, astronomy also deals with objects that are extremely far removed (literally) from their observers and with events that happened in the distant past, yet few question that field's scientific basis. Paleontologists deal with facts that can actually be visited, viewed, touched, and otherwise experienced directly. In contrast, astronomers cannot derive information about Alpha Centauri, the nearest star outside of our solar system, 4.3 light years away, through the same methods. Neither is more or less scientific, just different.

Probably the best way to show that paleontology and other historical sciences are on an equal footing with other traditional scientific disciplines, such as biology, chemistry, and physics, is to review scientific methods and ask if they are still applicable to paleontology. Fields of study that normally may not be considered as sciences also involve observation, collection of data, formulation of hypotheses, analysis of data, hypothesis testing, and peer review. It is not popular opinion, but rather the proper use of scientific methods in an area of study that makes it a science. With this in mind, scientific methods are easily applicable in geologic field situations, as demonstrated throughout this book. This gathering of evidence, whether through subsequent field visits or laboratory work, requires the use of considerable knowledge, besides numerous tools. Paleontologists and geologists need to be well equipped with the right gear to collect data effectively, whether that collecting takes place in the field or the laboratory.

Tools Used in Paleontology and Geology

Many of the fundamental tools used by geologists and paleontologists today are the same ones used by their predecessors in previous centuries. Of course, the list of equipment has become more complicated, especially with the advent of easily portable electronic items that digitally record data (Table 4.1). Geophysical methods for finding fossils, by detecting their magnetic or radioactive properties, have been used with various degrees of success in recent years. With advances in portability

TABLE 4.1 List of field equipment commonly used by a geologist or paleontologist.

PERSONALLY PORTABLE GEAR USED IN THE FIELD

Item	Purpose
Notebook, pens, pencils	Recording data, making illustrations.
Hand-lens	Magnifying small features.
Camera and film	Photographic illustration, aesthetic expression.
Tape measure, ruler, scale	Measuring distances or dimensions.
Compass	Navigation, determining directions of linear features, measuring angles.
Rock hammer	Breaking rocks, catharsis.
Maps (topographic, geologic)	Integrating locations with geologic data.
Acid bottle (5% hydrochloric)	Testing for presence of $CaCO_3$.
Chisels, trowels, shovels, picks	Excavation of fossil specimens.
Global positioning system (GPS)	Determining locations, navigation.
Extra batteries	Recharging GPS and camera.
Binoculars	Viewing distant outcrops of likely fossil-bearing zones.
Calculator (solar powered)	Analyzing field measurements.
Sample bags with labels	Collecting personally portable specimens.
Newspaper	Wrapping fossil specimens for cushioning.
Toilet paper	Wrapping fossil specimens, field emergencies.
First-aid kit	Fixing oneself for more fieldwork.
Water bottle (with water)	Proper hydration for more fieldwork.
Snacks (especially carbohydrates)	Personal fuel.
Broad-brimmed hat	Keeping sun off face and rain off head, style.
Sunglasses	Eye protection, especially from sun.
Sunscreen	Skin protection (a must in desert areas).
Boots	Ankle support in rocky or mountainous terrain.
Knife	Cutting tape, burlap, or anything else that needs cutting.
Duct tape	Anything imaginable.
Backpack	Carrying all of the previously mentioned items

INSTRUMENTS USED IN THE LABORATORY

Item	Purpose
Computer	Compiling, analyzing, and storing field data; transcribing of field notes; image analysis of photographs; modeling; simulations; morphometric analysis; geographic distribution analysis; phylogenetic (evolutionary relationship) analysis; sending e-mail to colleagues.
Scanner (flatbed or slide)	Scanning photographs, maps, or other illustrations for either image analysis, publications, or sending to colleagues for review.
Binocular microscope	Examining specimens for detailed, microscopic features and measuring features.
Polarized microscope	Viewing thinly-sliced sections of rock or fossil bone (mounted on slides) under polarized light for mineralogical and textural qualities.
Scanning electron microscope	Imaging features in rocks or fossils to as little as 1 micron or using other functions to perform elemental analyses under electron bombardment.

TABLE 4.1 Continued

INSTRUMENTS USED IN THE LABORATORY	
Item	**Purpose**
Computer tomography (CT)	Scanning internal structure of fossil specimens (especially skulls and eggs), and constructing three-dimensional images of structure with computer.
Electronic caliper	Measuring fossil specimens electronically, and transmitting information to a computer.
Video camera	Capturing images of unmagnified or magnified fossils and transmitting digitized images to computer for analysis.
Mass spectrometer	Measuring atomic masses of elements and determining abundance of isotopes for radiometric age dating of rocks associated with dinosaur fossils or stable isotope analysis of bones and eggshells.

and decreasing equipment costs, these methods will no doubt be used even more in the future. Nor is the equipment used in paleontological research limited to fieldwork tools, because the study of fossils is sometimes best done by examination of specimens in museum or university collections. Likewise, a laboratory setting, possibly using the latest technological wonders accessible to a paleontologist's budget, can be used to study the specimens gathered through present-day fieldwork or uncovered in the inner bowels of a museum.

Nevertheless, fieldwork is still where many paleontologists start their investigations. It is also a less expensive method of original research for paleontologists who are not associated with institutions that have a support staff, numerous catalogued specimens in storage, and expensive equipment. The potential for making discoveries with only a minimal investment in equipment is a very distinctive feature of paleontology relative to the other sciences (Chapters 2 and 3). With this perspective, the most basic items that a geologist, paleontologist, or other interested naturalist should always take into the field are a notebook and pencils or pens to record their observations. Personal digital assistants (PDAs) are now being used more for recording data, but currently have the distinct disadvantage of limited battery life; notebooks only run out of useable pages. Secondary items for geologic investigations include a hand lens, a measuring device, a compass, maps, a camera, and a rock hammer. Optional items are sample bags with labels for recording information about collected specimens, and a small bottle of dilute acid for testing the presence of calcite or aragonite ($CaCO_3$) in a rock. A first-aid kit, water bottles and broad-brimmed hats are also recommended to treat minor injuries, prevent dehydration, and sunburn, respectively. Many areas containing well-exposed, fossil-bearing rocks are in deserts (see Fig. 1.4).

Among the most important items that an informed fossil prospector should take into the field are maps.

Maps are important for documenting the exact locations of fossils, besides being used for navigating in a fossil-bearing area and understanding the geologic context of fossils. Topographic and geologic maps are the two types most commonly used for these purposes. **Topographic maps** show the differences in elevation in a specific area as well as the area's surface features, such as forested areas, roads, and cities. Differences in elevation are represented by **contour lines**, lines of equal elevation that follow the contours of the land surface. The spacing of contour lines gives a good indication of the relative steepness of the terrain in the field area, knowledge that is useful for traversing a hilly area or avoiding steep cliffs. A wilderness excursion should never be attempted without first investigating the availability of topographic maps for the intended region.

Geologic maps typically have contour lines too, but their main value lies in how they show the outcrop patterns of rock units on the land surface. Contour lines in combination with the geology help to illustrate how the topography of an area may relate to weathering patterns of the rocks (Fig. 4.1). Perhaps most importantly,

4

Atkinson Creek Quadrangle, Colorado
1:24,000

FIGURE 4.1 Section of a geologic map, which also has contour lines of a topographic map. The contour lines show elevation changes in a landscape, and closely spaced lines indicate relatively steeper elevation changes than widely spaced lines. Rock formations are mapped on the basis of their outcrop patterns, and letter symbols on the map correspond to the age and name of the formation (i.e., Jm is the Morrison Formation, which is Jurassic). US Geological Survey Map GQ 57, 1955.

TABLE 4.2 Information conveyed by topographic maps and geologic maps.

TOPOGRAPHIC MAPS

Descriptive Information

1 Cultural features: roads, trails, pipelines, towns, streets, power lines, houses, dams, quarries, churches, schools, cemeteries, airports, mines, city, county, state, and national boundaries, hospitals, parks.
2 Natural features: streams, lakes, woodlands, mountains, glaciers, beaches, waterfalls, swamps.
3 Latitude and longitude of any point on the map.
4 Scale showing horizontal distance on the map corresponding to actual distances.
5 Elevation of the ground surface, indicated by contours and benchmarks.
6 Magnetic declination.
7 Date the map was constructed

Interpretive information

1 Shape of the land surface, which can be interpreted further through construction of profiles and block diagrams.
2 Types of landforms and how they were formed (such as glaciers, wind, surface streams, groundwater, and coastal processes).
3 Structure of the bedrock (such as folds, faults, flat-lying strata).
4 Drainage basins of surface water.

GEOLOGIC MAPS

Descriptive information

1 Topographic information. If a geologic map is drawn with superimposed topographic lines and other information, then it may provide most or all of the information that a typical topographic map would provide. However, these maps are, in some cases, difficult to read because geologic information overlaps topographic information.
2 Types and locations of bedrock units of different ages.
3 Contacts between different rock units.
4 Types and locations of surficial deposits (glacial deposits, river sediments, floodplains).
5 Types and locations of faults and folds.
6 Trends and angles of rock layers

Interpretive information

1 Rock structures (folds and faults) beneath the ground surface (can be extrapolated from surface information).
2 Rock types, both at the surface and subsurface.
3 Rock durability and how difficult the rock may be to remove for mineral deposits.
4 Origin and type of material in surficial deposits.
5 Geologic history of an area.

Adapted from Spencer (1993).

the topographic information provides visible landmarks, which help geologists to orientate themselves. Geologic maps are particularly useful in dinosaur studies because they show the probable locations of outcrops of known or suspected fossil-bearing zones. However, outside of paleontology they are important for locating mineral deposits. Overall, the information conveyed by topographic and geologic maps is extensive (Table 4.2).

All maps, no matter how simple or complex, should have several important features:

- A **scale** that shows the horizontal distances on the map, corresponding to the actual distances on the land;
- A **legend**, which is a key to all map symbols used; and
- An indicator of the direction of **geographic north** (the north pole of rotation for the Earth).

Magnetic north differs from geographic north because the Earth spins on an axis deviated slightly from where its magnetic field emanates. Consequently, a map should also show the amount of difference between the two norths (**declination**) in the area being explored. For example, a difference of 10 degrees from a compass bearing (which generally points to magnetic north) of geographic north can become a large difference over the course of a several kilometer hike to a dinosaur site.

A global positioning system (GPS) device receives signals from satellites to calculate a position on the Earth's surface with regard to latitude, longitude, and elevation.

Although this equipment list may sound old-fashioned, it is still recommended because these tools and their usefulness have been repeatedly tested by generations of geologists and paleontologists. Of course, the equipment list must be modified to meet personal preferences and needs, and should keep up with the development of new technology that makes fieldwork easier. A GPS (see box) can cross-check information on maps for the accuracy of either tool. Many GPS units are also capable of downloading geographical locality information into computers; the latter can then construct digital maps that can be compared to any previously printed maps. Digital maps can have layers of information stored in GIS (Geographic Information System) programs, such as the distribution of rock types, fossils found, and vegetation patterns. For this and other reasons, laptop computers are now part of the standard list of field equipment, and most are easily portable. Nevertheless, like personal digital assistants (PDAs), battery life is still a limiting factor in their extended use in the field. Some geologists circumvent this problem by using either solar panels or chargers connected to a field vehicle.

However inspiring the finding of a dinosaur fossil might be, the person who discovers it should first know more basic principles of geology. This can make the find more meaningful, especially in how it relates to the small picture of its local paleoenvironment and the larger picture of the ancient global system. Attempting to collect a dinosaur fossil without further knowledge of its setting is risking the loss of valuable scientific information. Making a detailed description of the geologic setting of a dinosaur fossil in the field should precede preparing it for transportation.

Basic Principles of Geology

Field Relations and Relative Age Dating

Geology is the primary scientific field associated with dinosaur studies, but people with little understanding of geology have found fossils of plants and animals, including dinosaurs (Chapter 3). However, some advanced knowledge of this subject is certainly helpful for continued success. Paleontology is the most readily recognizable subdivision of geology that applies to dinosaur studies, but other specialties, such as **sedimentology**, **stratigraphy**, and **tectonics**, are also essential to form a more complete, contextual view of dinosaurs (Table 4.3).

TABLE 4.3 Some subdisciplines of geology and their relevance to the geosciences (especially economic applications). All of these subdisciplines overlap with one another in some way, so the divisions between them are often artificial.

Subdiscipline	Definition and Application
Geochemistry	Study of properties and changes in elements in the earth: interpreting the ages of rocks, the cycling of elements in earth systems (including pollutants), and the original parameters of paleoenvironments and burial conditions of organic material (for petroleum exploration).
Geophysics	Study of the earth through principles of physics: determining presence of subsurface oil or mineral deposits.
Mineralogy	Study of minerals, such as their geological occurrence and crystallinity: examining formation of metallic ore deposits and precious stones and determining industrial applications of minerals, such as clays in ceramics and paper.
Marine geology	Study of the Earth's oceanic processes: interpreting sea-level fluctuations, earth history, and interactions of oceanic environments with other global systems.
Paleoclimatology	Study of ancient climates and their changes through time: interpreting patterns in climate change and factors affecting climate in the past, including factors that might affect modern climate.
Petrology	Study of rocks, which can be subdivided into study of igneous, metamorphic, or sedimentary rocks: interpreting conditions for formation of ore deposits.
Sedimentology	Study of sediments, including methods for their transport and deposition and post-depositional history: interpreting paleoenvironments and modern processes (formation of shorelines, rivers).
Seismology	Study of the interior of the earth and how it releases energy: evaluating and predicting earthquakes.
Stratigraphy	Study of sedimentary rocks through mapping of their vertical and horizontal extents: evaluating resources such as coal, aquifers, and oil.
Structural geology	Study of the deformation of earth materials: describing and predicting the extent and nature of folded and faulted rocks, which assists with interpreting Earth history and location of earth resources.

The basic principles of geology, which were formed through many repeated observations made by field geologists through the early nineteenth century (Chapter 3), are still used by geologists and paleontologists today and are responsible for what is seen in a geologic map. These principles are **original horizontality**, **superposition**, **lateral continuity**, **inclusions**, **cross-cutting relationships**, and **biologic succession** (Fig. 4.2). They comprise the techniques for **relative age dating** of rocks, that is, determining the relative order in which geologic phenomena occurred, without necessarily knowing the exact ages of the phenomena.

Original horizontality is the concept that **sediment**, unconsolidated material occurring at the Earth's surface, when originally deposited, settled under the influence of gravity into more or less horizontally-oriented layers. Once such layers become

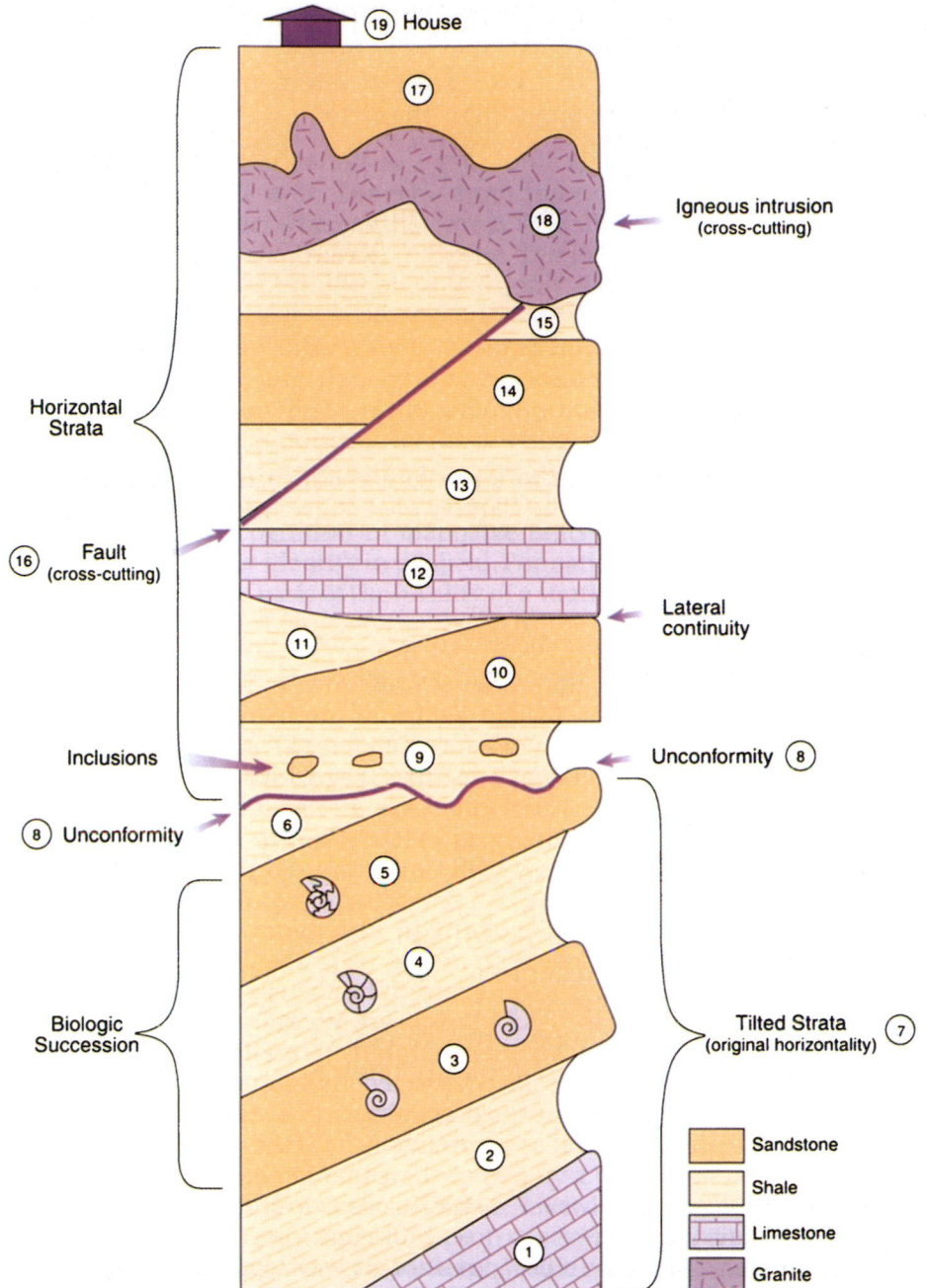

FIGURE 4.2 Idealized diagram of basic field relations of rocks that can be used to determine relative ages, using original horizontality, superposition, lateral continuity, inclusions, cross-cutting relationships, and biologic succession. Phenomena are labeled from oldest (1) to youngest (19).

consolidated by compression from overlying sediments and cementation, the sediments become **sedimentary rock**. Sedimentary rock is further classified by its **lithology**, or rock type, which is based on the composition and texture of the sediment. Layers of sedimentary rock are called **strata** (plural of **stratum**) or **beds**. A series of strata in a particular area, stacked on top of one another, may be collectively called a **stratigraphic sequence**. If the strata in a stratigraphic sequence are tilted significantly beyond a horizontal plane, then an observer knows that the tilting happened *after* deposition of all of the sediments and their formation into sedimentary rock.

Superposition is the concept that each layer of sediment deposited on top of an underlying layer is relatively younger than the latter, assuming the strata have not been tilted far beyond their original horizontality. Superposition, along with original horizontality, may seem too inherently obvious, but Nicolaus Steno (Niels Stensen) of Denmark, in the seventeenth century, was the first person to actually articulate it (Chapter 3). To visualize this principle, think of the sediments at the bottom of the sequence as being deposited first, and the sediments at the top being deposited most recently. If the rate of layering was known, the total age of the strata could be calculated. However, this is not straightforward as, in reality, there is the possibility that the rate of deposition varied through time, that the strata was overturned, or that some layers were removed throughout the history of deposition, usually by erosion.

Related to the concept of superposition is lateral continuity, which describes how sedimentary layers will continue in lateral directions until they encounter some barrier that prevents their further spread, or they otherwise run out of sediment. If a laterally continuous layer is found in widely separated places, it may represent approximately the same time of deposition, which is an example of **correlation**. Laterally adjacent layers also can succeed one another vertically, which represents how changes in environments through time can cause one environment to overlap the other.

Cross-cutting relationships and inclusions have opposite significance in relative age dating. With cross-cutting relationships, a geologic feature cutting across another feature is the younger of the two, such as a fracture that cuts through all strata in a particular geologic section (Fig. 4.3). In contrast, inclusions, which are particles of a preexisting rock incorporated into sediment, must be older than the rock including them. British geologist Charles Lyell expressed these latter two principles in the early nineteenth century (Chapter 3).

All of these principles can determine relative-age dates of rocks without the use of fossils, but the combination of these principles with rocks containing identifiable **guide fossils** adds a powerful dimension to interpreting the geologic history of an area. Guide fossils are typically body fossils (but in some cases trace fossils) that are:

1 abundant;
2 easily identifiable;
3 geographically widespread;
4 vertically restricted in their range; and
5 likely to be deposited with sediment independently of the environment in which they lived.

In terms of biologic succession, their vertical zonation, also known as **geologic range**, may demonstrate when an organism first evolved (at the bottom of the zone) and when it went extinct (at the top of the zone) in that area. When these geologic ranges for different guide fossils are noticed and recorded both locally and worldwide, they can represent different times that are identifiable whenever these fossils are found, which is another aid to correlating strata.

For example, bones of *Coelophysis*, a small theropod, have been found only in strata below those containing bones of *Allosaurus*, a larger and different theropod, and *Apatosaurus*, a large sauropod (Chapters 9 and 10). Paleontologists thus conclude that the fossils of the overlying layers succeeded the underlying ones. The strata containing the remains of *Coelophysis* are from the Late Triassic, whereas the rocks bearing *Allosaurus* and *Apatosaurus* fossils are from the Late Jurassic, which is

FIGURE 4.3 Fault cross-cutting thick stratigraphic sequence of the Santa Elana Limestone (Late Cretaceous) in Big Bend National Park, Texas. Students for scale.

based on the vertical zonation of these and other fossils. Occasionally these zonations are refined when geologists or paleontologists find fossils slightly below (older) or above (younger) the previously known ranges, which is part of the testing and subsequent improvement of the fossil record that occurs every day.

Guide fossils were applied to mapping the distribution of sedimentary rocks in parts of Europe in the late eighteenth and early nineteenth centuries, well before Charles Darwin published any of his research on biological evolution. Since then, evolution and extinction are now recognized as primary factors that controlled the vertical distribution of fossils in the geologic record (Chapter 6 and 16). **Biostratigraphy**, the use of guide fossils in mapping rocks and interpreting their ages, was largely responsible for the establishment of a worldwide, standard geologic time scale (see Fig. 1.2), with its relative time divisions, such as **eons**, **periods**, and **epochs**, each represented by distinctive fossil assemblages. The time divisions are called **time units**, whereas the rocks that represent those times are **chronostratigraphic units**: for example, the Triassic Period is a time unit and the Triassic System is a chronostratigraphic unit. The *rocks* from the early part of the Triassic are called Lower Triassic, whereas their *age* is Early Triassic (which is older than both Middle and Late Triassic). Likewise, the Jurassic is divided into Early, Middle, and Late, but the Cretaceous is divided into Early and Late. Upper and lower also refer to all rock units, not just chronostratigraphic units. Earth-resource companies use the geologic time scale and such divisions in their exploration for fossil fuels (oil, gas, and coal) and minerals, and they have been well tested since the beginning of the Industrial Revolution.

The description of strata that contain distinctive sediments and fossils, which can lead to the recognition of mappable units that a geologist can readily identify in the field, are called **formations**. Formations are given formal names by geologists on the basis of the locality of a stratigraphic section that ideally is representative of the rock types found in the formation, called the **type section**. A formation name can be the place name for the type section followed by a generic formation designation, such as the aforementioned Upper Jurassic Morrison Formation, which is named after Morrison, Colorado (Chapter 3). These names can also reflect the main rock type in the formation, such as the Upper Devonian Chattanooga Shale, which is named after the location in Chattanooga, Tennessee, and the rock type. Formations can be subdivided into **members** or grouped together into **groups**. Guide fossils can help with identifying a formation, but the easy visual identification of a formation by its lithology is more important for geologists who want to describe its distribution on a geologic map.

As stated above, strata do not always equate with time units because the same type of sediment could have been deposited at different times when an environment left sediments as it migrated laterally. For example, think of how a lateral change in the position of the shoreline to landward environments (called a **transgression**) causes the sea to cover formerly dry land and progressively deposit marine sediments farther inland through time. This type of change, perhaps caused by a rise in sea level, makes the marine sediments **time transgressive**. Likewise, a lateral change in the position of the shoreline in the direction of seaward environments (**regression**), maybe through a drop in sea level, would cause continental sediments to overlap the formerly marine sediments. However, the continental sediments are still time transgressive because they were gradually deposited through time. Geologists keep this in mind, so if different species of dinosaurs (or any other fossils) are found in the same formation in widely separated places, the organisms they represent are not always assumed to have lived at the same time.

Application of scientific methods to the basic geologic principles of relative age dating involves testing each criterion for consistency, instead of accepting that any one of them fulfills the possible geologic interpretation. For example, if inclusions that definitely come from one bed are found in the bed directly underlying it, we can conclude that the sequence is **overturned** (upside-down), and hence a hypothesis about age relations of the strata based simply on superposition could be wrong. Similarly, a fossil that seems far too young for the previously known worldwide geologic range of the strata in which it was found may have been **reworked**, that is, exhumed from its older layer and reburied, only to re-emerge in this younger strata. As a result, the hypothesis that the fossil is an inclusion and is an indicator of some sort of previous **erosion** (transport of sediment) from its original entombing bed must be tested. The hypothesis that some dinosaurs lived in the Cenozoic Era, which was based on the occurrence of dinosaur bones in the oldest Cenozoic strata, has been falsified by evidence indicating that these fossils were reworked from underlying Mesozoic strata (Chapters 7 and 16).

Surfaces of non-deposition or erosion, called **unconformities**, are very important to recognize because they signify a gap in the time record; that is, time is not represented by rocks in the area of an unconformity. In some cases, the missing geologic record consists of millions of years (Fig. 4.4). Consequently, geologists and paleontologists are often depressed by unconformities and lament the many fossils lost to the cruel processes of erosion. These surfaces can be identified with the use of the principles of relative age dating and a few other criteria.

FIGURE 4.4 Labeled unconformity near Morrison, Colorado (adjacent to Red Rocks Amphitheatre), separating the Fountain Formation (Pennsylvanian Period, about 300 Ma) from a 1.7 billion-year-old Precambrian metamorphic rock (gneiss); the surface represents a lost record of about 1400 million years.

4

So when looking at photographs and other illustrations of sedimentary rock sequences, hiking through a national park, or driving down a road where road cuts may have exposed strata, try to observe the characteristics of these sequences, attempt explanations for your observations, then test them using relative age-dating principles. Continual practice of these techniques will help lead to better interpretations of the geologic history of an area, interpretations that are based on an array of interdependent evidence.

Absolute Age Dating: How We Know When Dinosaurs Lived

When geologists confidently say that dinosaurs as a group lived for, at a minimum, 165 million years, or 165 **Ma** (Latin translation of *mega annus*), they are referring to the considerable factual evidence supporting the vast amounts of time associated with the age of the Earth. That the Earth is about 4.6 billion years old is known through **radiometric age dating**, a method used to calculate absolute ages of rocks or other materials using **radioactive elements**. Radioactive elements have known natural **decay constants**, which are unvarying rates of radioactive decay over time. Radioactive elements are unstable elements that give off matter and energy in a way that results in their eventual change of one element (the **parent element**) to another (**daughter element**). Steps in between the parent and final stable daughter element are represented by a number of elements and constitute a **decay sequence**. The general equation for radioactive decay is represented by the formula:

$$N = N_0 e^{-\lambda t} \tag{4.1}$$

where N is the number of atoms present now, N_0 is the original number of atoms from the radioactive element, e is a constant (about 2.718), λ is the decay constant, and t is time. With some algebraic re-arranging, the formula is changed to solve for t, which gives the age of the rock.

Age dates derived from such methods are also cross-checked against the many facts provided by relative age dating (Fig. 4.5). The checking and rechecking of radiometric

FIGURE 4.5 Igneous rock cross-cutting sedimentary rock (Pen Formation) in Big Bend National Park, Texas. The sedimentary rock has fossils indicating that it is Late Cretaceous, about 70 Ma. The cross-cutting igneous rock should be younger, and it is. Radiometric dates derived from it and related rocks in the region indicate that they were formed in the Tertiary Period. English professor for scale.

age dates through independently verifiable and repeatable tests, as well as the common applications of quantum physics, makes radiometric age dating scientifically certain. Likewise, decay constants of radioactive elements provide us with extremely accurate means of measuring the ages of phenomena that occurred long before human history, such as dinosaurs and their associated rocks. Rates of decay have never been observed to change in any significant way under a wide variety of laboratory conditions; hence these rates are as factually-based as the effects of gravity. For an analogy, speculating that a radioactive element may have had a random or otherwise variable decay rate is akin to saying that apples may have fallen up rather than down from trees at random times in the past.

First, some backtracking to define a few basic terms is necessary. An **element** is a substance composed of atoms that contain the same number of **protons**; this number of protons defines the **atomic number** of an element. Protons are positively-charged particles with a standard **atomic mass** of 1.0, the same mass for neutrally charged **neutrons**. Both protons and neutrons are in the **nucleus**, the center of the atom. Negatively charged **electrons**, with masses about 1/1800 of protons and neutrons, orbit the nucleus. Atomic mass is calculated simply by adding the number of protons and neutrons in an atom, ignoring the negligible mass provided by electrons. For example, some atoms of carbon have 6 protons and 6 neutrons, thus it has an atomic number of 6 and an atomic mass of 12 (written also as ^{12}C). Atoms with the same atomic number are all the same element; ^{12}C, ^{13}C, and ^{14}C have different masses as they have more or less neutrons because they have the same atomic number, and they are **isotopes** of an element. Remember that the number of protons stays the same, otherwise it becomes a different element.

A radioactive element that is actively emitting energy may be transmitting it through moving particles. Particles that are commonly emitted by radioactive elements are **alpha particles**, which are two protons and two neutrons (equivalent to the nucleus of a helium atom, which has an atomic mass of 4), and **beta particles**, which are high-speed electrons. Appropriately enough, radiation that consists of alpha particles is called **alpha radiation**, and radiation that consists of beta particles is **beta radiation**. These types of radiation result from decay of a radioactive element because when atoms lose particles they change into different elements. For example, here is part of the decay sequence for uranium-238 (^{238}U), which has an atomic number of 92 (hence having 146 neutrons):

$$^{238}U \rightarrow \ ^{234}Th \rightarrow \ ^{234}Pa \rightarrow \ ^{234}U \rightarrow \ ^{230}Th \rightarrow \ ^{226}Ra \rightarrow \ ^{222}Rn \rightarrow \ ^{218}Po$$

The alpha decays that occurred between ^{238}U and successive elements are easy to pick out: they are the ones where an atomic mass of 4 was subtracted. The beta decays are also easy to find and are where one element changed to another with no apparent loss of mass (because electrons have comparatively little mass). The change of the element happens because a positively-charged proton plus a negatively-charged electron is equal to a neutral neutron. Thus, the loss of an electron from the nucleus causes a new proton, which changes the atomic number of the element. The complete decay series for ^{238}U results in a final, stable daughter element of ^{206}Pb.

To calculate the age of a rock, a geologist has to solve for time. What are known before solving for time are the decay rate (amount of loss from the parent isotope over time) and the result of the decay (amount of daughter isotope and parent isotope). Other variables that have to be considered first are summarized by:

$$t = (1/\lambda) \ln [(d/p) + 1] \tag{4.2}$$

where t is age (time), λ is the decay constant, **ln** is the natural logarithm (which of a given number is the exponent that must be assigned to e (about 2.71) to derive that same number), d is amount of the daughter element (which is how much is measurable now), and p is the parent element (also how much is measurable now). The application of a logarithmic function is necessary because radioactive decay is **exponential**, which means that a radioactive element decays at a gradually more rapid rate with time, which contrasts it with an **arithmetic** rate, which is simply subtraction or addition with time. This mathematical distinction is important in the understanding of radiometric age dating. Because time is needed for a parent element to decay to a daughter element, a high ratio of daughter to parent (d/p) indicates a greater amount of time than a comparatively low ratio, thus the rate of increase for a daughter element is directly proportional to the rate of decrease for a parent element.

Decay constants are calculated by using a lot of math, beginning with the following equation:

$$\lambda = 0.693/t_{1/2} \tag{4.3}$$

where λ is the decay constant and $t_{1/2}$ is the half-life of the element. **Half-life** is the amount of time needed for half of the parent element to have decayed. Half-life can also be expressed by knowing that it represents half of the original radioactive element:

$$N = (1/2)N_0 \tag{4.4}$$

For a specific radioactive element that underwent decay, the decay constant represents the measured number of atoms that decay per second, compared to the number of atoms that are still in the rock sample:

$$\lambda = -(dN/dt)/N \tag{4.5}$$

where (dN/dt) is the rate of change in number of atoms (dN) in proportion to rate of the change in time (dt), and N is the number of atoms now present. The measurement of the number of alpha particles emitted by an atom per second will give an indication of the first value through a standard number of atoms, such as in one **Avogadro (Av)** of a sample, 6.02×10^{23} atoms. For example, using 1.0 Av of ^{238}U, which has a measured alpha particle emission rate of 2.96×10^6 particles per second, the decay constant is calculated as:

Step 1. $\lambda = (2.96 \times 10^6 \ \alpha/s)/6.02 \times 10^{23}$
Step 2. $= 4.92 \times 10^{-18} \ s^{-1}$

The half-life can be solved for our example of ^{238}U, because the decay constant is known:

Step 1. $t_{1/2} = 0.693/4.92 \times 10^{-18} \ s$
Step 2. $= (0.693/4.92) \times 10^{18} \ s$
Step 3. $= 1.41 \times 10^{17} \ s$

The number of seconds in a year is about 31,557,000 (give or take a thousand seconds), thus to convert the calculated number of seconds to years:

Step 4. $t_{1/2} = 1.41 \times 10^{17} \ s/3.1557 \times 10^7 \ s/y$
Step 5. $= 4.468 \times 10^9 \ y$

Therefore, the half-life for ^{238}U is nearly 4.5 billion years, which means in that time about one-half of the original amount of ^{238}U in a rock will have been lost through decay, then in another 4.5 billion years, only half of that half (25%) is left, and so on (Fig. 4.6). Once the approximate number of atoms for each element (parent and daughter) is counted from a rock sample, the ratio of one to the other is calculated, which is an indicator of the number of half-lives that have passed and then can be used to calculate the age of a rock. A device called a **mass**

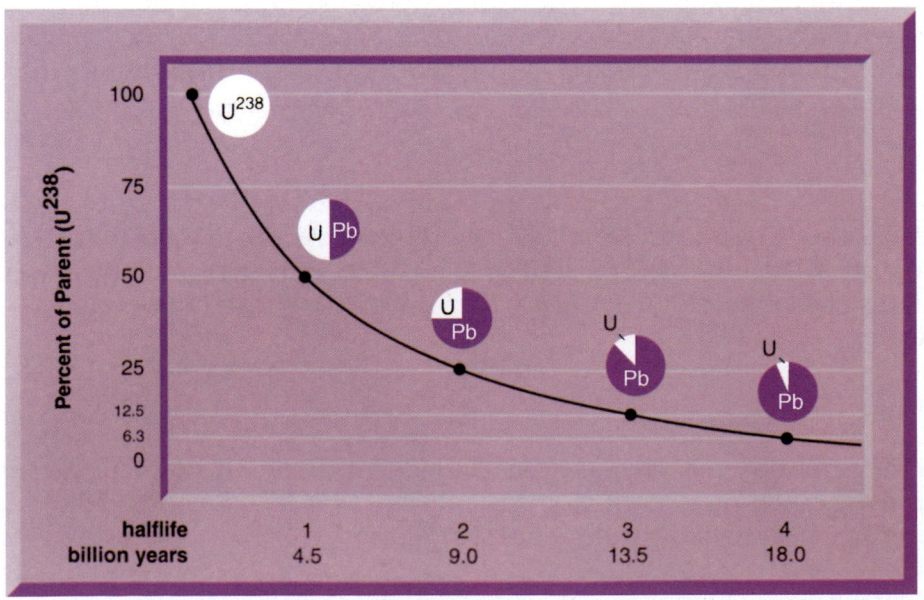

FIGURE 4.6 Exponential loss of a parent element (^{238}U) through time, showing changes in ratio with relation to the daughter element (^{206}Pb) with each half-life. Note that the plot follows a curved line, not a straight line.

TABLE 4.4 Elements used to calculate the age of a formation (Amsitoq Gneiss) in Greenland. The widely different decay constants resulted in a very narrow range (0.25 billion years, but with overlapping variation) for the calculated ages, despite the orders of magnitude differences of the half-lives for elements used.

Parent–daughter Pair	Daughter Isotope Used as Isochron	Half-life of the Parent Isotope (in billions of years)	Calculated Age for the Formation (in billions of years)
$^{87}Rb/^{87}Sr$	^{86}Sr	48.8	3.70 ± 0.14
$^{232}Th/^{208}Pb$	^{204}Pb	14.0	3.65 ± 0.08 (Th/Pb ratio)
			3.80 ± 0.12 (Pb/Pb isochron)
$^{238}U/^{206}Pb$	^{204}Pb	4.47	3.65 ± 0.05
$^{176}Lu/^{176}Hf$	^{177}Hf	35.9	3.55 ± 0.22 (Hf/Hf isochron)

Adapted from Dalrymple (1986).

spectrometer, found in the geology or chemistry departments of most research universities, counts the number of atoms of each element.

Radiometric age dating is how geoscientists know the absolute ages of some rocks. Ages derived from rocks typically are within a 1% potential error, which is estimated on the basis of probability calculations. This amount of error is reasonable, considering that the ages are often reported in millions or billions of years. If geoscientists doubt the age of a rock sample given by one isotope pair, they will cross-check with another isotope pair that has a totally different decay constant. Another cross-checking method is to involve another variable, an isotope of the daughter element, to see if the ratio between it and the daughter isotope correlates with the parent–daughter ratio. If they do, the two ratios plot as a line, which confirms that the measured sample is accurate and has not been contaminated. The method of using another isotope pair of the same element (such as $^{87}Sr/^{86}Sr$) for comparison with the parent–daughter pair is called **isochron dating**. Independently derived but identical age dates from different isotope pairs or isochrons are not coincidental; indeed, as many as five different isochrons, all with different decay constants, have shown the same overlapping age ranges for rocks in some examples (Table 4.4). This procedure works very well and, consequently, radiometric age dating, when done correctly, is a powerful and accurate tool for establishing the ages of rocks as old as or much older than those containing dinosaur fossils.

For dinosaur studies, the isotope pairs most often used are those with long enough half-lives to date material older than 65 Ma. Although radiometric dating is usually equated with **carbon dating**, carbon dating is not used in dinosaur studies. This is because the relatively short half-life of ^{14}C (5730 years) guarantees that the ratio of ^{14}C to its daughter isotope is so small after only 70,000 years (12 half-lives, meaning only 1/4096 of the original parent is left) it is undetectable to modern mass spectrometers. Isotope pairs used in dating materials from the Mesozoic Era include $^{40}K/^{40}Ar$, $^{40}Ar/^{39}Ar$, and $^{238}U/^{206}Pb$, among others. In recent years, some dinosaur paleontologists and geologists prefer ages derived from $^{40}Ar/^{39}Ar$ because of its greater precision (Table 4.5). This accuracy is a result of the calculation of an age date by directly measuring the ratio of parent to daughter element.

Unfortunately, calculation of an age for a bed actually containing dinosaur fossils, let alone the dinosaur fossil itself, is rare. This situation is because sediments will give only the age of when the sedimentary grains were originally formed, not

TABLE 4.5 Examples of isotopic age dates derived from volcanic rocks associated with dinosaur-bearing strata. (Periods are given in parentheses.)

Formation	Location	Isotopic Method	Age Dates (Ma)
Prince Creek	Alaska, USA	$^{40}K/^{40}Ar$, $^{40}Ar/^{39}Ar$	68–71 (Late Cretaceous)
El Gallo	Baja, Mexico	$^{40}Ar/^{39}Ar$	73–74 (Late Cretaceous)
Two Medicine	Montana, USA	$^{40}Ar/^{39}Ar$	74–80 (Late Cretaceous)
Judith River	Alberta, Canada	$^{40}Ar/^{39}Ar$	76 (Late Cretaceous)
Juifotang	Inner Mongolia, China	$^{40}Ar/^{39}Ar$	110 (Early Cretaceous)
Ischigualasto	Northwestern Argentina	$^{40}Ar/^{39}Ar$	228 (Late Triassic)

when they were deposited. The derived age of a sedimentary particle thus is older than the bed that encloses it (meaning that the principle of inclusions can apply equally well to individual sand grains!). The radiometric clock starts with the cooling and subsequent crystallization of a mineral from a major heating event, which fixes the radioactive elements into a small enough place (i.e., mineral grains) where geoscientists can sample them. For this reason, only rocks formed in high temperatures (greater than 250°C) at the time of their formation can yield accurate radiometric age dates. These rocks include **igneous rocks**, formed from originally molten rock, called **magma**, at temperatures of 70–1200°C; and **metamorphic rocks**, formed through pressure and heat of about 250–700°C.

Igneous rocks are preferable for age dating, especially **volcanic ash beds**, because their sediments represent the original time of formation, not the "reheating" times that are typical of metamorphic rocks or the ages of reworked sediments. Ash beds are geologically valued because they were formed by airborne ash spewed from a volcano. The ash was originally deposited in widespread, horizontal layers in a short time, which lends well to cross-checking any derived absolute ages with underlying and overlying strata. Fortunately for dinosaur studies, some dinosaur populations were proximal to volcanic areas that produced voluminous amounts of ash; these layers in combination with relative age-dating techniques have provided a good general framework for defining and understanding the timing for evolution and extinction of certain dinosaur species and groups (see Table 4.5). Igneous rocks that cross-cut Mesozoic sedimentary rocks containing dinosaur fossils, as shown in Fig. 4.5, can also be used to calculate ages, although in some instances the igneous rocks are too young for purposes of comparison to other Mesozoic dinosaur-bearing strata.

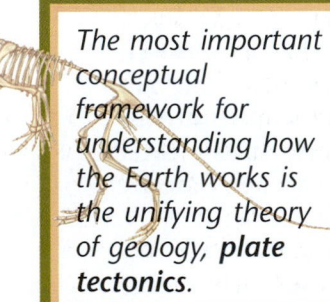

*The most important conceptual framework for understanding how the Earth works is the unifying theory of geology, **plate tectonics**.*

The Big Picture: Plate Tectonics and How it Rules the Earth

Plate tectonic theory states that the Earth is differentiated into layers with their own distinctive chemical and physical properties and that the outermost layer, the **lithosphere**, moves and interacts with an underlying layer, the **asthenosphere**. The lithosphere, the 5–120 km thick, outermost, rigid layer, is composed of the Earth's crust and the upper part of its mantle. Directly underneath it is the asthenosphere, the

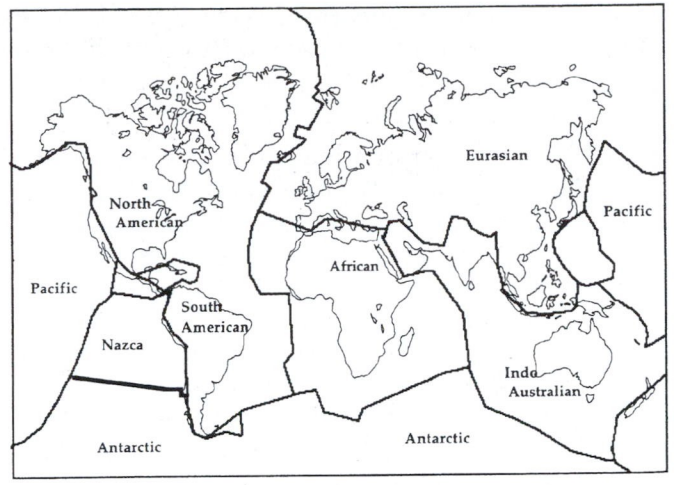

FIGURE 4.7 Map view of lithospheric plates involved in plate tectonics. Reprinted by permission from R. Cowen, *History of Life*, p. 98. (© 1995 Blackwell Science, Inc., Malden, MA.)

200–250 km thick, relatively hotter, more plastically flowing part of the upper mantle. The lithosphere is presently broken into 13 major segments (plates) that entirely cover the asthenosphere (Fig. 4.7).

Physiographic features, such as oceanic ridges and some large mountain ranges, as well as the occurrence of most earthquakes and volcanism, delineate **plate boundaries**; these are the areas where separate plates interact. Notice that plates and continents do not coincide. Continental crust is incorporated into each plate and so moves along with the plate. Large amounts of evidence support that plates in some form, similar to what we see today, have moved over the asthenosphere ever since they formed about 4.0 billion years ago, which is about the same age as the oldest known evidence of life on Earth.

Evidence for this movement of plates is multifold, some of it originating in a hypothesis called **continental drift** that was first proposed by **Alfred Wegener** (1880–1930) in 1915 and republished in 1929. In his hypothesis, the movement of the continents during the geologic past was given as an explanation for the:

- Observed "fit" of now widely separated continents.
- Presence of fossils indicating continental organisms on continents now divided by an ocean.
- Similarity of some mountain ranges and rocks on those same continents.
- Presence of warm-climate indicators in presently cold-climate places such as Antarctica.

However, Wegener was hampered by two main problems:

1 he was not trained in geology and was in fact a meteorologist, so he was not taken seriously by geologists, who subjected his work to rather vicious peer review; and
2 he proposed a seriously flawed hypothesis to explain his data, which actually justified some of the severe criticism he endured.

For example, one of his proposals for the movement of the continents was that they plowed along underlying solid rock, which led to another question that Wegener could not answer: What was the driving force for such considerable movement of continents? Moving a continent takes a considerable amount of energy and Wegener had no coherent explanation for how this happened. His observations were correct, but his explanation was wrong. Wegener died well before his hypothesis was accepted in its greatly modified form in combination with other hypotheses that stemmed from new data derived mostly in the 1960s.

The modern evidence for plate tectonics consists partially of the following:

- Submarine ridges with active volcanism and earthquakes in linear zones are coincident with the ridges, as found in the middle of the Atlantic Ocean.
- Rocks on the ocean floor with older radiometric age dates are located sequentially farther from the linear zones.
- Thicker sedimentary layers on the ocean floor are further away from these same zones.
- Older and identical ages are indicated by fossils in these same layers for the lower parts of thick sedimentary layers on either side of the zones.
- Recorded reversals in the Earth's magnetic field in igneous rocks on the ocean floor show symmetrical patterns (mirror images) on either side of the linear zones where volcanism is occurring.
- The oldest rocks on the ocean floor are only about 200 Ma (Early Jurassic), whereas those on the continents are as much as 3.9 billion years old.
- Mountain ranges on the continents are folded and faulted into defined linear zones.
- When compared to rocks on the stable interior of continents, mountain ranges with younger (more recent) radiometric ages than their igneous and metamorphic rocks have been found, which are also locations of major earthquakes and active volcanoes.
- Linear zones with no volcanism and major fault lines on the continents have violent earthquakes associated with them.
- Island chains, such as the Hawaiian Islands, that have young radiometric age dates for igneous rocks are located close to active volcanism; rocks have gradually older age dates correspondingly farther from the active volcanism.

When the evidence noted by Wegener was added to these data, the theory of plate tectonics emerged from the repeated testing of hypotheses proposed to explain all of these phenomena that originally were seen as unrelated. The most important of these hypotheses, proposed to explain the data shown in points 1 to 5 above, was **seafloor spreading**, which held that new seafloor formed at the volcanically active linear zones because of the spreading of lithospheric plates away from those zones, which allowed magma from the asthenosphere to erupt on the ocean floor. New seafloor has younger radiometric age dates (because of its close association with volcanically active zones), less oceanic sediment on it, younger fossils, and rocks that reflect the magnetic field of that time, whereas older seafloor would show opposite trends. Zones where plates move away from one another are called **plate-divergent boundaries** or **spreading centers**, which causes a tensional stress that is translated through earthquake activity. The rocks on either side of a spreading center, such as the Mid-Atlantic Ridge, can be used to calculate an average spreading rate on the basis of radiometric age dates of rocks in combination with their distance from the spreading center:

$$R = d/t \qquad\qquad\qquad (4.6)$$

where R is rate, d is distance, and t is time. An example is illustrated by calculating the average spreading rate of the North American plate for the past 155 Ma. First, a sample of ocean-floor igneous rock is taken at a known distance (1960 km) from the original spreading center (Mid-Atlantic Ridge); radiometric dates derived from $^{40}K/^{40}Ar$ yield an age of 154.5 ± 0.2 Ma. With these data, the average spreading rate (in cm/year) is calculated as:

$$R = 1960 \text{ km}/154.5 \text{ Ma} \qquad (4.7)$$

Step 1. Converting values to scientific notation for simplicity.
$$= 1.96 \times 10^3 \text{ km}/1.545 \times 10^7 \text{ y}$$
Step 2. Converting km to cm:
$$1.0 \text{ km} = 1 \times 10^5 \text{ cm/km}.$$
Step 3.
$$= 1.96 \times 10^8 \text{ cm}/1.545 \times 10^7 \text{ y}$$
$$= 1.3 \text{ cm/y}$$

Realize also that during 155 million years the average rate might have had a range, such as 0.7 to 3.1 cm/year, but these values also can be calculated as long as radiometric age dates and distances from the spreading center are known. Using this method, spreading rates (rates of plate movement) for plates have been calculated as ranging from less than 1.0 to as much as 11 cm/year.

The fact that the oldest seafloor dates from the Jurassic Period indicates that the ocean floor has been destroyed and recycled since at least the time that *Apatosaurus* and *Allosaurus* were on the continents. This brings up another integral hypothesis in plate tectonic theory, **subduction**, which is supported by points 6 through to 8. Subduction is a proposed process where, as plates move away from divergent boundaries and collide with one another at **plate-convergent boundaries**, one plate can go underneath the other. The subducted plate undergoes partial melting as it slides further into the hotter asthenosphere, which in turn forms magma for igneous rocks. The force of the collision is sufficient to generate pressures that can bend (**fold**) rocks at depth, and can also break (**fault**) rock at shallower depths. This process explains the folding of originally horizontal strata as is seen in the field, as well as fractures that cut across the strata. **Stress** (also known as pressure), the force applied to a unit area, has these associated formulas,

$$F = ma \qquad (4.8)$$

$$\sigma = F/A \qquad (4.9)$$

where **F** is force (expressed in newtons; Chapter 1), **m** is mass, **a** is acceleration (typically in m/s^2), σ is stress, and **A** is area (m^2). A moving plate has a sufficiently large mass so that its slow acceleration or velocity (the latter in m/s, rather than m/s^2) is irrelevant as it hits the other plate, which is moving as well, providing additive force applied to the area of contact between the plates. The result of this stress is **strain**, which is manifested by folds and faults. This provides an explanation for the built-up tensional energy that is periodically released through earthquakes, the pressures required for the formation of some metamorphic rocks, and the development of mountain ranges that exceed 9000 m above sea level. Such high mountain ranges occur when a thick continental lithospheric plate collides with another continental plate and neither is subducted, which explains why continents are composed primarily of rocks older than those on the seafloor. This circumstance is very fortunate for dinosaur paleontologists, as dinosaur fossils in mountainous areas would have been destroyed if they had entered a subduction zone.

Other phenomena associated with plate tectonics are **transform fault** movements and **hot spots**. Transform faults are areas where movement of the lithosphere is

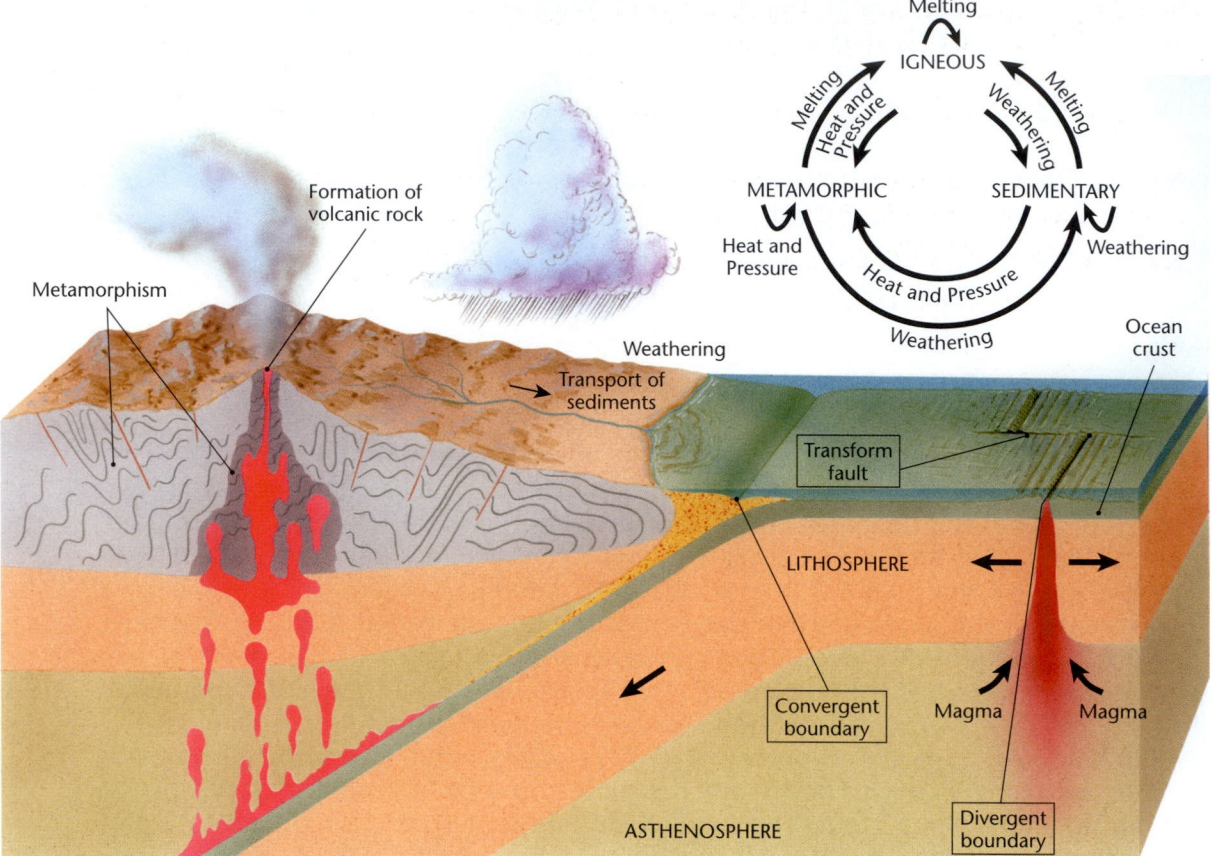

FIGURE 4.8 The rock cycle as explained through plate tectonics, showing relationship of lithosphere and asthenosphere, as well as convergent, divergent, and transform-fault boundaries.

interpreted as a result of plates moving laterally against one another without any accompanying volcanism. Evidence for this movement consists of the aforementioned stress eventually resulting in earthquakes; measurable movement along the fault plane can be defined through offset features in the landscape. Hot spots are interpreted as plumes of magma that pierce the lithosphere and rise up consistently in the same place for millions of years. Evidence for hot spots is best exemplified by strings of islands, such as the Hawaiian chain, which have active volcanism on a single island. However, other islands and undersea volcanoes (**guyots**) in the chain have volcanic rocks that show increasingly older radiometric ages farther away from the island with active volcanism. Plate tectonic theory has an elegant solution for this pattern: the hot spot stays in the same place while the plate moves over it.

This rudimentary knowledge of plate tectonic theory facilitates a better understanding of the rock cycle (Fig. 4.8). The way of the Earth is constant change and all rocks are in a state of transition, although they appear static to us during our short lifetimes. In one simplistic and linear example, elements composing sedimentary rocks become incorporated into igneous rocks through subduction and melting of the sedimentary rocks; igneous rocks become heated enough to change into metamorphic rocks; then metamorphic rocks, uplifted by plate convergence, are exposed at the surface and weathered so that their broken-down elements are cemented together into sedimentary rocks. Plate tectonics results in the following:

1 creates new seafloor and destroys old seafloor;
2 brings magma from the mantle to the surface in volcanoes, folds, and fractured rocks; and
3 provides uplifted areas that later undergo erosion and shed sediments.

It is, therefore, the source of the Earth's constant recycling that makes the planet a dynamic place for life. In fact, the Earth seems to be the only body in the solar system that shows such comprehensive evidence for plate tectonics.

Preview of the Importance of Plate Tectonics to Dinosaur Studies

Why do we need to know about plate tectonics when studying dinosaurs? Because everything on the surface of the Earth is affected by plate tectonics. It determines the location of the continents, mountain ranges, volcanoes, and earthquakes, as well as the configuration of the world's oceans. The location of the continents and their inherent geographic features affect the distribution of all land plants and animals. The entire global environment (especially climate) is influenced by the placement of the oceans relative to the continents because patterns of oceanic and atmospheric circulation are controlled by whether a continent is in an equatorial or polar position (Chapter 6). Local climates are changed by the presence of mountains, which are formed by plate tectonics. The amount of volcanism on the seafloor causes the sea level to either rise or fall, and rising sea level can cause landward environments to become more crowded for terrestrial organisms, with associated ecological stresses. Conversely, uplift of a mountain range caused by plate convergence causes land that formerly was shoreline to be more emergent, which expands continental areas for animals and plants. Volcanism caused by plate convergence places ash in the atmosphere, blocking solar radiation and cooling the Earth, which can negatively affect plant communities that animal communities depend on to live (Chapter 16). Earthquakes alter the course of rivers, change the landscape, or generate **tsunamis** (seismic sea waves) that drown many people and other living things in coastal communities. For the purposes of exploration for dinosaur fossils, some dinosaur-bearing strata were uplifted by continental collisions, and their present surface distribution in folded and faulted rocks is directly attributable to plate tectonics.

Basically, many (if not most) humans are affected by plate tectonics every day in the form of earthquakes, volcanoes, tsunamis, mountains, climates, and shorelines. All evidence from Mesozoic rocks indicates that plate tectonics was just as active a global process then. Thus, the dinosaurs lived and eventually died at the consent of plate tectonics as well, although as a group they were permitted to thrive for about 165 million years.

Recovery and Preparation of Dinosaur Fossils: How They Are Collected

From the Field to the Institution

Chapter 2 mentioned that searching for fossils can be hard work, but it did not examine the difficulties of properly collecting fossils, especially those of dinosaurs. Fortunately, many dinosaur body fossils are either from relatively small individuals or just a few parts of larger individuals. However, finding a nearly complete specimen of a large theropod (Chapter 9) or sauropod (Chapter 10) means that the logistics

(A)

(B)

(C)

(D)

for excavating and transporting the fossil from the field site are involved and expensive. Large excavations can potentially take an entire field season (usually summer) and require the services of heavy equipment for extraction and transport of the specimen. Furthermore, after transport and deposition of the fossil to a preparator, the preparation time for extracting a dinosaur skeleton may take more than a year. Dinosaur trace fossils, especially tracks (Chapter 7), typically do not require recovery, but some are taken from the field site for further study or display; entire beds containing the tracks may have to be moved thousands of kilometers.

To describe a typical recovery procedure, assume a dinosaur fossil consists of skeletal material. Upon identification of partially exposed and recovery-worthy skeletal remains, the area immediately surrounding the fossil is carefully cleaned. This action is followed by a full assessment of the horizontal and vertical extent of the skeleton, which normally involves mapping the distribution of the bones on a grid (Chapter 3). The orientation of any bones at the surface is noted; most may be flat-lying (parallel to bedding) and predictable in their extent, but others might project outside the exposed area below the surface. Erring on the side of caution is always good, even if it means carrying out too much rock for a small amount of fossil material. Of course, the person doing the recovery uses a scientific methodology: a prediction and its accompanying evidence determine the probable extent of the skeletal material. Then that person consults with any colleagues at the site to learn their estimations to seek consensus.

Any glues needed to keep bone fragments together are then applied. But choosing which glues are used should be left to a professional, as not just any glue should be applied to a 65+ Ma fossil! Excavation then begins on the area around the fossil. In some instances, the skeletal material (especially teeth or small vertebrae) may already be loose on the ground. Such material is placed into labeled sample bags after its distribution has been noted. Whether excavation is easy or difficult depends on the surrounding rock (Fig. 4.9). If the bones are in well-cemented sandstones, jackhammers or backhoes are not unreasonable tools for breaking up the rock. Some rocks, such as a mudstone or poorly-cemented sandstone (Chapter 7), can be picked away with rock hammers, trowels, shovels, or other hand tools. The excavation should then proceed around the prescribed area and to the perceived maximum depth for the fossil (maybe a little more, to be safe). Once this depth is reached, the excavation starts to cut underneath the fossil, although not far enough so that it collapses. This procedure causes it and the surrounding rock to form a pedestal. Water-soaked paper towels or toilet paper are then placed on the pedestal to form a barrier between the fossil and the final surrounding layer. At this stage dry plaster of Paris is mixed with water for dipping strips of burlap, which are placed around the towel-enveloped pedestal as a jacket. New materials that are less dense and more cost-effective than plaster of Paris, yet not sacrificing strength, have been proposed in recent years, but many dinosaur workers still

FIGURE 4.9 (*opposite*) Steps in excavation of a vertebrate fossil, in this case a partially exposed skull and other bones of a metoposaur, a large amphibian that lived at the same time (and in this case, the same region) as early dinosaurs, Chinle Formation (Late Triassic), Arizona. (A) After cleaning the area, workers estimated the extent of the fossil and dug around the defined area. (B) Digging of the rock underneath the fossil established a pedestal. (C) One worker placed wet paper towels on the top to cushion and separate the fossil from the plaster. (D) Another worker placed the plaster-soaked burlap strips for the jacket all around the pedestal. The workers then waited until the next day for the plaster to have hardened before breaking the pedestal, turning over the rock, and jacketing the underside.

FIGURE 4.10 Pelvis of *Apatosaurus* from the Morrison Formation (Late Jurassic), western Colorado, still partially encased in its protective jacket and in a preparatory lab associated with the former Museum of Western Colorado, Grand Junction, Colorado. Notice the plastic model sauropod in the background, ready to help with estimating the weight of the original animal (Chapter 1).

prefer the plaster of Paris method, which has been in use for more than 100 years (Chapter 3).

The specimen is left while the plaster hardens completely. Only then is the support under the pedestal broken so that the fossil can be turned over carefully to apply the remainder of the jacket. For later cataloguing, important information about the fossil, such as the date collected, preliminary identification, specimen number, orientation (indicated by a north arrow), and location, are written on the jacket. This also keeps the fossil from being mixed up with other, similar-looking, jacketed specimens. The snug and safe fossil is now ready for transport out of the field area, carried by people on foot (if the specimen is small enough), in land-based vehicles, or in extreme cases by helicopter.

In a preparatory laboratory, a jacketed specimen is cut open and the excavation begins anew, with the goal of liberating the fossil from its surrounding rock (Fig. 4.10). A preparator will use human energy and a variety of tools to separate the fossil from its entombing sediments. Just as in the field, the amount of time taken to extract bones from rock depends on the cementation of the rock and fragility of the fossil. Skeletal material is also commonly fragmented, requiring the preparator to handle each small piece with care so that paleontologists can re-assemble the pieces accurately later. Preparators are among the most patient and skilled people in paleontology, some operating with the precision of surgeons.

Once the dinosaur bones are prepared, they can be placed in dynamic public displays, baring their teeth or bearing their young. However, most skeletal remains of dinosaurs return to dark quarters, tucked away in storage drawers or shelves for future research. Because of its great weight, real bone is rarely mounted in a museum; supporting these hard-earned but heavy specimens and keeping them from being damaged or vandalized is an expensive technical problem. Instead, casts are made from the original bones using artificial materials, such as fiberglass. These strong,

FIGURE 4.11 *Tyrannosaurus rex* mount, which uses artificial casts of the bones and thus allows for the unusual pose of the display; Denver Museum of Science and Nature, Colorado. Author (imitating the pose in the foreground) for scale.

lightweight replications of bones are much more amenable to mounting dinosaurs in poses that may not be based on strong scientific evidence but provide for much discussion (Fig. 4.11).

Other dinosaur fossils, such as a well-outlined nest with eggs or tracks in rapidly eroding (poorly cemented) rocks, might require recovery techniques that are similar to those for skeletal material. Eggshell material presents a special problem for recovery because the fragments can be so small and numerous that their scientific value might be limited (Chapter 8). Nevertheless, the recent resurgence of studies of dinosaur eggs and juveniles has helped increase recognition of such fossils, so they are now more likely to be recovered. In contrast, most dinosaur tracks are studied in place, although casts are made occasionally with latex or some other material that does not damage the fossil. Natural casts of tracks (Chapter 14) might be loose and on their own, which encourages their collection for further study or use as display or teaching specimens. However, entire trackways are sometimes extracted, such as one from east Texas that was split into three parts: one part was taken to the American Museum of Natural History in New York, one to National Museum (Smithsonian) in Washington, and one to University of Texas in Austin. The part that went to the American Museum was then used in combination with skeletal mounts of analogous trackmakers for an imaginative linking of dinosaur body and trace fossil evidence.

SUMMARY

Many tools, physical and mental, are required for dinosaur studies, so people who are serious about learning dinosaur paleontology must know what to take into the field, both in their backpacks and brains. In their backpack, they should carry a variety of tools that allow them to find, measure, and record information about dinosaur fossils safely, efficiently, and effectively. Because dinosaurs are contained primarily in rocks, a basic knowledge of geological principles is needed to understand the local setting for their fossils and the larger-scale factors that affected their distribution. Geological principles, such as original horizontality, superposition, cross-cutting relationships, lateral continuity, and inclusions, can be applied while driving by a road cut that exposes strata, or while hiking in a wilderness area, lending a new dimension to interpreting the natural history of an area. Similarly, identification of guide fossils can immediately indicate the relative age (era, period, or epoch) of the rocks in a particular area. The combination of these low-cost observations with laboratory measurements of naturally decaying radioactive elements and their by-products gives a more complete picture of the immensity of geologic time. These observations also make allowances for demonstrating the considerable changes in dinosaurs and other fossil species through time. Knowing the mathematical and other scientifically-based reasoning behind radiometric age dating, as well as the cross-checks made through relative age dating methods, shows that the geologic time scale is based on reality and is a well-tested, accurate representation of the ages of rocks. Geologists and paleontologists use these dating methods every day, not just for finding and documenting dinosaur fossils but also especially for prospecting for the minerals and fossil fuels that make possible the lifestyle choices of industrialized nations.

With some basic knowledge of paleontological and geological principles, as well as a lot of energy, dinosaur fossils are discovered, recovered, brought back, and studied by paleontologists for the public appreciation of their inherent knowledge and beauty. The education required to find and study dinosaur fossils is well worth the long hours of studying geological principles, radiometric age dating, and plate tectonics that are necessary to form a more eclectic picture of dinosaur lives and afterlives. Finally, the existence of dinosaurs over 165 million years can be viewed against the background of the all-encompassing theory of plate tectonics, which would have affected dinosaur populations throughout most of the Mesozoic Era. Through the interactions of the lithosphere and asthenosphere, plate movement is responsible for phenomena as diverse as earthquakes, volcanism, and the occurrence of island chains. Because plate tectonics causes the movement of continents either away from one another or closer together through the course of geologic time, it is the main driving force behind the proximity of continents.

DISCUSSION QUESTIONS

1. Is one scientific field necessarily better than another? List what sciences have been mentioned so far in this book and rank them in order of what you perceive as most scientific to least scientific. What evidence do you have to justify such a list? Compare your list with other students in the class to see if they share your consensus.

2. Practice mapmaking in your everyday life, Draw a map for someone who needs directions and include the three features that should be on every map. Has someone ever given you poor verbal directions or a badly drawn map? If so, what would have helped to prevent the considerable time you spent being lost?

3. A stratigraphic sequence has a limestone bed at the base, which is overlain by a shale, which in turn is overlain by a sandstone; the entire sequence is cross-cut by a fault. For this sequence, what is the order (from oldest to youngest) of the geologic events represented? What if you find inclusions of the sandstone in the shale? Would you change your assessment, and if so, why?

4. You go to an area with dinosaur fossils and find fragments of *Coelophysis*, a theropod previously known only from the Late Triassic, in the same stratum that contains the remains of *Allosaurus*, a theropod only known from the Late Jurassic. What are at least two hypotheses to explain your observation? How can both be falsified?

5. How could a transgression occur without a change in sea level (which is caused by more water in the world's oceans)? How could a regression occur without sea-level changes?

6. What is the minimum number of alpha decays that occurred between the parent element of ^{238}U and the final stable daughter element of ^{206}Pb? How did you arrive at this number?

7. How is compound interest in savings accounts similar to radiometric age dating? Provide mathematical proofs through some examples.

8. What alternative explanations could account for the sameness of age dates derived from the five different radiometric methods given in Table 4.4? What evidence would be needed to falsify the accuracy of these age dates?

9. How can plate tectonics be responsible for the following circumstances in both relative and absolute age dating:
   a. Stratigraphic sequences that have the oldest fossils at the top and the youngest fossils at the bottom.
   b. Strata tilted into a vertical position.
   c. Volcanic ash layers that show older radiometric ages than found in underlying strata.
   d. Unconformities that show considerable angles between strata below and above the unconformity.
   e. Younger radiometric ages for metamorphic rocks than found in surrounding sedimentary rocks in a mountain range.

4

DISCUSSION QUESTIONS Continued

    f. Calculated radiometric ages indicating 125 Ma that are equidistantly 1670 km away from a mid-ocean ridge?

10. Discuss how plate tectonics relates to the following items:
    a. Smog in Los Angeles, California.
    b. Sediments in the Amazon River of South America.
    c. The low population of Tibet.
    d. Materials composing a typical automobile.
    e. A dinosaur skeleton that was under more than 1000 meters of rock now being found at the surface, available for excavation and preparation.
    f. This textbook.

Bibliography

Chapman, R. E. 1997. "Technology and the study of dinosaurs". *In* Farlow, J. O. and Brett-Surman, M. K. (Eds), *The Complete Dinosaur*. Bloomington, Indiana: Indiana University Press. pp. 112–135.

Chure, D. J. 2002. "Raising the dead: excavating dinosaurs". *In* Schotchmoor, J. G., Springer, D. A., Breithaupt, B. H. and Fiorillo, A. R. (Eds), *Dinosaurs: The Science Behind the Stories*. Alexandria, Virginia: American Geological Institute: pp 127–136.

Conrad, J. E., McKee, E. H. and Turrin, B. D. 1992. Age of tephra beds at the Ocean Point dinosaur locality, North Slope, Alaska, based on K-Ar and "SUP 40" Ar/"SUP 39" Ar analyses. *U.S. Geological Survey Bulletin*, 1990-C.

Cox, A. and Hart, R. B. 1986. *Plate Tectonics: How It Works*. Palo Alto, California: Blackwell Science.

Dalrymple, G. B. 1991. *The Age of the Earth*. Stanford, California: Stanford University Press.

Davis, D. W., Gray, J., Cumming, G. L. and Baadsgard, H. 1977. Determination of the ^{87}Rb decay constant. *Geochimica Cosmochima Acta* **41**: 1745–1749.

Dietz, R. S. and Holden, J. C. 1970. The break-up of Pangea. *Scientific American* **222** (April): 30–41.

Eberth, D. A., Russell, D. A., Braman, D. R. and Deino, A. L. 1993. The age of the dinosaur-bearing sediments at Tebch, Inner Mongolia, People's Republic of China. *Canadian Journal of Earth Sciences* **30**: 2101–2106.

Emiliani, C. 1992. *Planet Earth: Cosmology, Geology, and the Evolution of Life and Environment*. Cambridge, U.K.: Cambridge University Press.

Fichter, L. S. 1996. Tectonic rock cycles. *Journal of Geoscience Education* **44**: 134–148.

Freeman, T. 1992. *Procedures in Field Geology*. Columbia, Missouri: Friendship Publications.

Gillette, D. D. 1997. "Hunting for dinosaur bones". *In* Farlow J. O. and Brett-Surman M. K. (Eds), *The Complete Dinosaur*. Bloomington, Indiana: Indiana University Press. pp 64–77.

Goodwin, M. B. and Deino, A. L. 1989. The first radiometric ages from the Judith River Formation (Upper Cretaceous), Hill County, Montana. *Canadian Journal of Earth Science* **26**: 1384–1391.

Harland, W. B., Armstrong, R. L., Cox., A. V., Craig, L. E., Smith, A. G. and Smith, D. G. 1989. *A Geologic Time Scale 1989*. Cambridge, U.K.: Cambridge University Press.

Lucas, S. G. 2002. "It takes time: dinosaurs and the geologic time scale". *In* Schotchmoor, J. G., Springer, D. A., Breithaupt, B. H. and Fiorillo, A. R. (Eds), *Dinosaurs: The Science Behind the Stories*. Alexandria, Virginia: American Geological Institute. pp 39–44.

Moores, E. (Ed.) 1990. *Plate Tectonics*: *Readings from* Scientific American. New York: W. H. Freeman.

Renne, P. R., Fulford, M. M. and Busby-Spera, C. 1991. High resolution "SUP 40" Ar/"SUP 39" Ar chronostratigraphy of the Late Cretaceous El Gallo Formation, Baja California Del Norte, Mexico. *Geophysical Research Letters* **18**: 459–462.

Robertson, H. 1998. How to design a lightweight jacket for a dinosaur: theoretical evaluations of selected fillers mixed with plaster of Paris. *Palaios* **13**: 301–304.

Rogers, R. R., Swisher III, C. C. and Horner, J. R. 1993. "SUP 40" Ar/"SUP 39" Ar age and correlation of the nonmarine Two Medicine Formation (Upper Cretaceous), northwestern Montana, U.S.A. *Canadian Journal of Earth Sciences* **30**: 1066–1075.

Rogers, R. R., Swisher III, C. C., Sereno, P. C., Monetta, A. M., Forster, C. A. and Martinez, R. N. 1993. The Ischigualasto tetrapod assemblage (Late Triassic, Argentina) and "SUP 40" Ar/"SUP 39" Ar dating of dinosaur origins. *Science* **260**: 794–797.

Sander, P. M. and Gee, C. T. 1992. A volunteer-powered dinosaur excavation in the Upper Triassic of Switzerland. *Journal of Geological Education* **40**: 194–203.

Spencer, E. W. 1993. *Geologic Maps: A Practical Guide to the Interpretation and Preparation of Geologic Maps*. New York: Macmillan.

Thomas, R. G., Eberth, D. A., Deino, A. L. and Robinson, D. 1990. Composition, radioisotopic ages, and potential significance of an altered volcanic ash (bentonite) from the Upper Cretaceous Judith River formation, Dinosaur Provincial Park, southern Alberta, Canada. *Cretaceous Research* **11**: 125–162.

Vine, F. J. 1966. Spreading of the ocean floor: New evidence. *Science* **154**: 1405–1415.

Wegener, A. 1929. *Die Entstehung der Kontinente und Ozeane (The Evolution of the Continents and Oceans)*. Germany: Braunschweig.

Wilson, J. T. (Ed.) 1970. *Continents Adrift: Readings from Scientific American*. San Francisco: W. H. Freeman Company.

4

Chapter

5

You are watching a documentary on television about dinosaurs and you get hungry about halfway through, so you go to your refrigerator to look for some leftover chicken wings. Because you are standing there with the refrigerator door open, marveling at the antiquity of some items in front of you, you only hear the last part of a statement made by a paleontologist in the documentary. In response to a question about how birds and dinosaurs are related, she says, "Birds are dinosaurs." As a result, you look at your chicken wings with newfound admiration.

What did she mean by this statement? How would you explain it to a doubting person who did not see the program or had never heard this statement?

Dinosaur Anatomy and Classification

Refined Definition of "Dinosaur"

Chapter 1 gave a preliminary definition of a dinosaur as a reptile- or bird-like animal with an upright posture that spent most of its life on land. In this chapter we can now expand this definition. A fossil must have the following characters before it can be called a dinosaur:

- Three or more sacral vertebrae.
- Shoulder girdle with backward-facing (caudally pointing) glenoid.
- Asymmetrical manus with less than or equal to three phalanges on digit IV.
- Acetabulum with open medial wall.
- Tibia with cnemial crest.
- Astragalus with a long ascending process fitting into the anterior part of the tibia.
- Sigmoidally-shaped third metatarsal.
- Postfrontal absent.
- Humerus with long deltopectoral crest.
- Femur with ball-like head on proximal end.

This chapter dealing with dinosaur anatomy should provide a better understanding of how the following can be done:

1 identify anatomical traits essential for clearly distinguishing dinosaur fossils from other closely related forms;
2 classify dinosaurs on the basis of shared traits helping to link evolutionary relatedness of different dinosaur groups; and
3 assemble these groups through a classification based on shared anatomical attributes (cladistics) to form a picture of dinosaur evolution throughout the Mesozoic.

Such a picture is still being sketched to connect dinosaurs to their ancestors and living descendants.

Skeletons are what primarily define dinosaurs, rather than the soft parts of their anatomy or trace fossils. The classification of different dinosaurs into groups based on their anatomical traits depends on correct assessment of their body plans as revealed by their bones, necessitating a thorough knowledge of skeletal anatomy. Such methods are practical because of the geological circumstances that favor preservation of skeletal material. In contrast, dinosaur **integuments**, that is, derivatives of their skin, including feathers, are rarely found in the fossil record (Chapter 7). Notable exceptions to this generalization are foot impressions preserved as tracks (Chapter 14) and now-less-rare finds of feathered theropods from a Lower

Cretaceous deposit in China (Chapters 9 and 15). These infrequently preserved parts are also anatomical characters of dinosaurs and provide useful supplementary information for independent cross-checks of the interpretations of evolutionary lineages based on skeletal data (Chapter 6). Dinosaur body plans also indicate inferred behavior of dinosaurs based on functional morphology. These then can be compared to behavior indicated by dinosaur trace fossils (Chapter 14) or to the behavior of extant animals that serve as analogues of dinosaurs.

Anatomical Vocabulary for Dinosaur Skeletons: Which Way is Up?

Dinosaur paleontologists have their own set of terms that allows them to communicate effectively with one another regarding dinosaur anatomy. As demonstrated above, a lack of knowledge about anatomical vocabulary in particular is a barrier to understanding dinosaurs, but fortunately it can be overcome through some study and comparisons to what is already known. For example, many of the names given to bones in the human body are also applied to dinosaurs, so a few of these names should be familiar. In fact, bones that have the same name in different kinds of animals are called **homologues**. Although the bones are basically the same, they may have a different appearance because they may be evolutionarily adapted for different functions in each animal, such as arm bones associated with swimming in dolphins and with flying in bats (Chapter 6). Additionally, the orientation of most living animals, especially pets, friends, family members, or casual acquaintances, is readily observed from their anatomies. Thus, terms can be applied to different parts of animal bodies every day.

A student armed with the anatomical vocabulary for dinosaur skeletons has a starting point for learning the basic skeletal anatomy of dinosaurs (Fig. 5.1).

Another encouraging fact about dinosaur skeletal anatomy is that the number of individual bones in a complete skeleton of a 30-meter sauropod can be about the same as in a 1.5-meter tall human (slightly more than 200). The gross anatomy of dinosaurs can vary considerably from clade to clade and species to species, but a basic body plan can be a standard of comparison for each specimen encountered in the field, in a collection, or in a specimen mounted in a museum. Because of

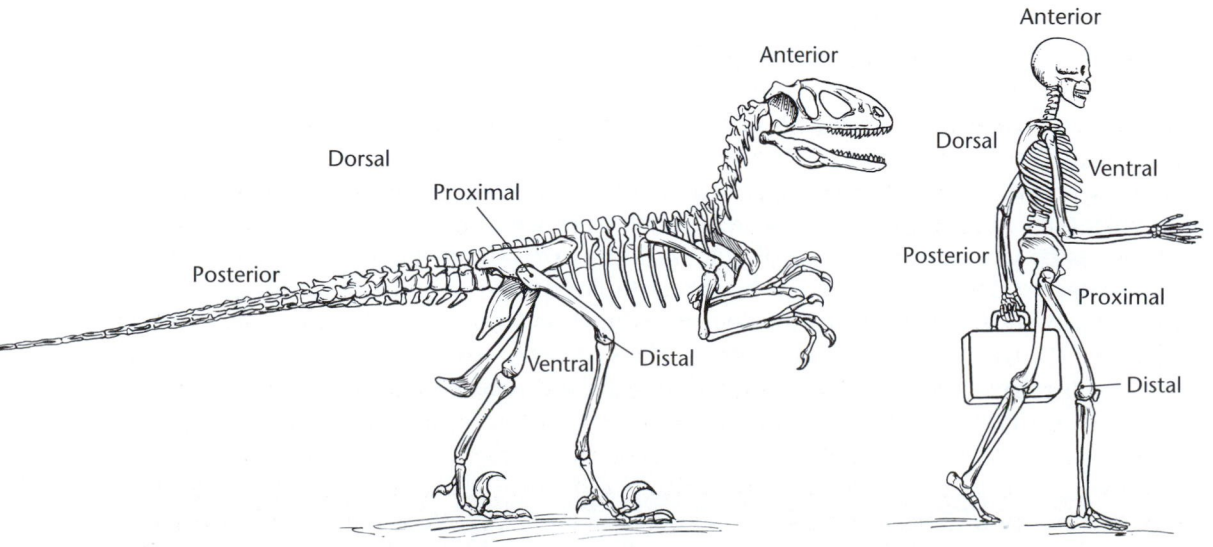

FIGURE 5.1 Orientation terminology as applied to anatomical features in vertebrates, using the skeletons of the Early Cretaceous theropod *Deinonychus antirrhopus* (left) and a modern human *Homo sapiens* (right).

variations in a species and the nature of fossilization, paleontologists usually only have pieces of what used to be entire bones. Thus, dinosaur paleontologists need to be extremely skilled anatomists who are used to working with fragmentary material in their reconstructions and restorations of ancient vertebrates. This tradition extends back to the eighteenth and nineteenth centuries, with paleontologists such as Georges Cuvier and Richard Owen (Chapter 3).

Before progressing with dinosaur anatomy, learning front, rear, sides, and other references on a dinosaur skeleton is necessary. Learning this orientation requires familiarity with some directional terms that anatomists use to describe surfaces on an animal's body or its parts. The head region of an animal is its **anterior**, whereas its rear end is the **posterior**. The same terms can be applied to any bone or place on a dinosaur skeleton that is located more toward a forward- or backward-facing region, hence the anterior part of the **tibia**, the posterior part of the **fibula**. These terms are also applied as the prefixes "ant" or "post", such as **antorbital fenestra** or **postorbital fenestra**. Another prefix used for bones that occur in front of another bone is "pre" (e.g., **premaxilla**, **predentary**). These prefixes should not be confused with the adjectives **dorsal** and **ventral**; dorsal is on the back and ventral is on the front of a standing person, but they denote top and bottom when applied to a crawling baby. For example, chordates are animals characterized by a **dorsal nerve cord**, which means that the nerve cord is located near the animal's spinal area as opposed to its ventral belly region.

The **medial** (middle) part of the body can be defined on the basis of proximity to the midline, which demonstrates the **bilateral symmetry** (left–right sameness) of a typical vertebrate body, observable from either the dorsal or ventral surface of an animal. The medial part of the body can also define its axis, of which its associated bones are the **axial** skeleton (skull, spine, hips, and tail). Any body parts extending to the sides of the body and away from the midline, such as **appendages**, are **lateral** and these bones comprise the **appendicular** skeleton. Dorsal, ventral, medial, and lateral can be applied as directional modifiers to any given skeletal part or area. For example, a hat held by its chinstrap is ventral to the jaw and lateral to the cheek, whereas the hat itself is dorsal to all other skull bones and covers the medial surface of the skull.

Appendicular parts also have their own modifiers based on how close they or their parts are to the medial part of the body. The **proximal** parts of a limb are in close proximity to the medial part of the body, but the **distal** parts of that same limb are more distant. For example, the longest leg bone in the human body, the **femur**, fits into the hip region and so is more proximal than foot bones (**metatarsals**, **tarsals**, and **phalanges**, in increasingly distal order) on the same leg.

Probably the easiest way to divide a dinosaur skeleton is to use the terms **cranial** and **postcranial**. Cranial refers to all of the bones and other features associated with the skull, which is at the anterior part of the dinosaur, whereas any bones or features toward the posterior of the skull are considered postcranial. This seemingly disparate division is justified by the complexity of some dinosaur skulls, which can contain more than 30 different bones and as many as 200 individual teeth (Chapter 9); the description of these parts details their distinctive characters. Postcranial parts can also be considered as oriented toward the **caudal** (tail) region of the dinosaur.

Axial Bones of a Dinosaur: Hips, Backbone, Tail, and Ribs

One way to start with basic dinosaur anatomy is at the hips. Traditionally, the primary distinction between the two most fundamental clades of dinosaurs, the Saurischia and the Ornithischia, is their hip structure. This division was well described by Harry Govier Seeley in the late nineteenth century and bolstered later

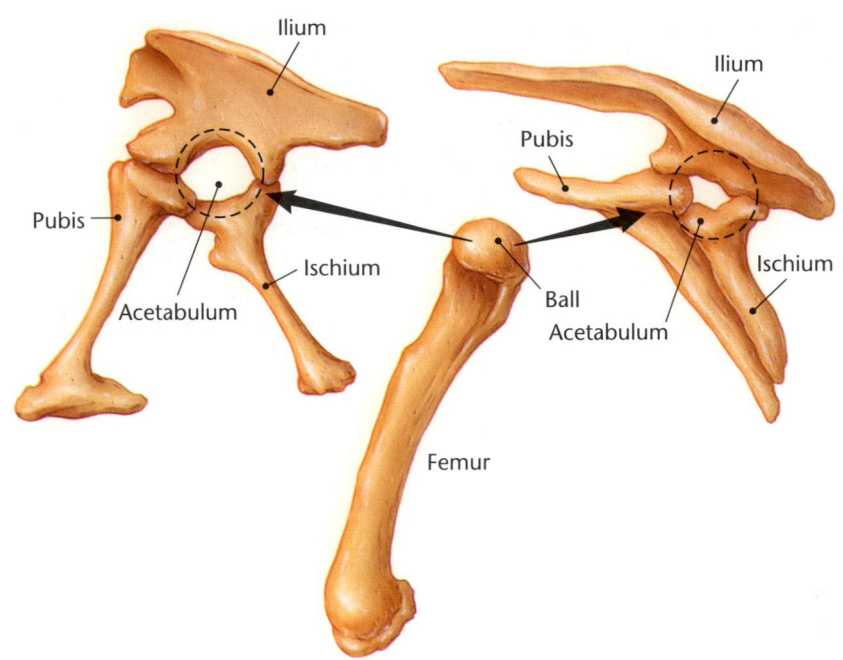

FIGURE 5.2 Left-lateral view of pelvic bones in relation to acetabulum and proximal end of femur for typical saurischian (left) and ornithischian (right) hips in dinosaurs.

in the early twentieth century by Friedrich von Huene (Chapter 3). The differentiation lies essentially in the arrangement of the hip bones (**ischium**, **ilium**, and **pubis**) in a dinosaur pelvis (Fig. 5.2). In both clades, the ilium extends laterally to both sides from the axis of the body and dorsal to the ischium and pubis. The pubis in saurischians points anteriorly (cranially), whereas in ornithischians it points posteriorly (caudally) and joins with the ischium so that it is ventral to the ischium. The ischium in both saurischians and ornithischians is posteroventral, in that it points toward the rear of the animal and is closer to its belly than its back.

In dinosaur anatomy, the distinction between saurischian and ornithischian hips is critical, because these hip structures are related to other aspects of dinosaur studies:

- History of dinosaur studies (Chapter 3)
- Dinosaur classification (Chapters 5, 9 to 13)
- Physiology and functional morphology of dinosaurs (Chapters 8 and 14)
- Evolutionary history of dinosaurs (Chapters 6 and 15).

Hip bones are found not only in land-dwelling vertebrate animals of the geologic past, but also in those existing today. As a result, they represent a recurrent structure as an adaptation to supporting movement in terrestrial environments (Chapter 10). The hip bones join laterally to form an open hole on each side called the **acetabulum**, which is articulated with the anterior end of the femur (Fig. 5.2). The fact that the medial wall of a dinosaur acetabulum is open enough that a rope can be threaded through it is one of the characters of dinosaurs that differentiate them from their ancestors. The articulation of the head of the femur with the acetabulum also contributes to another dinosaurian trait, a shelf built into the inner wall of the ilium that accommodates the ball-like projection of the femur. This adaptation is one of the main indicators that dinosaurs walked upright with their legs directly underneath their bodies, instead of crawling with their limbs sprawled out to the sides like large lizards (Chapters 1 and 14).

Vertebrae are repeated and interconnected bones that form the main axial elements in the dorsal part of a dinosaur skeleton. Vertebrae can be classified as **cervical**, **dorsal**, **sacral**, and **caudal**, in order cranially to caudally. The cervical vertebrae are associated with the neck region and the skull, the dorsal vertebrae with

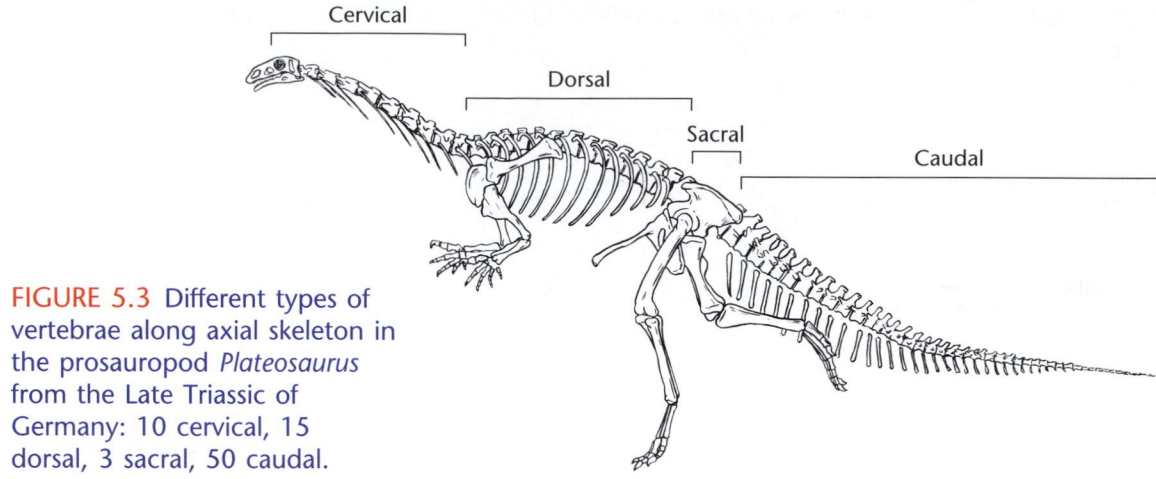

FIGURE 5.3 Different types of vertebrae along axial skeleton in the prosauropod *Plateosaurus* from the Late Triassic of Germany: 10 cervical, 15 dorsal, 3 sacral, 50 caudal.

the back, and the sacral with the hips. The caudal vertebrae form the tail; some dinosaur tails, particularly those of sauropods, were long because of a large number of vertebrae (Chapter 10). Three or more sacral vertebrae, in association with the hips of a dinosaur, comprise another defining characteristic (Fig. 5.3).

Each vertebra has a central part (appropriately named a **centrum**), an arch dorsal to the centrum (the **neural arch**), and a hollow area between the centrum and neural arch (the **nerve** or **spinal canal**) where a dinosaur's nerve cord was located. Vertebrae also have different knobs emanating from them called **processes** that articulated with one another or with **ribs** (also known as **costae**); the latter places of articulation are called **transverse processes**. Ventral and medial to the paired ribs of some dinosaurs are smaller ribs called **gastralia** that evidently lent support to dinosaur bellies. Caudal vertebrae were reinforced by tendons in some dinosaurs, such as hadrosaurs (Chapter 11) and a few theropods (Chapter 9). Some long-tailed dinosaurs, such as sauropods (Chapter 10), also had bones ventral to the vertebrae called **chevrons**. Chevrons probably protected blood vessels on the ventral part of the tail.

The Cranium

One of the easiest ways to attract the attention of dinosaur paleontologists is to announce that you have found a cranium, especially one of a sauropod (Chapter 10). Most dinosaur paleontologists want to find a dinosaur skeleton that is at least 90% complete, but skulls are relatively rare prizes that can be extremely meaningful for classification and interpretations of behavior. Because of the large number of bones in the cranium, which was attached to the anterior part of the axial skeleton through the articulation of the **occipital condyle** with the first cervical vertebra, it is one of the most complicated structures of a dinosaur.

Adjectives applied to common bones in a dinosaur skull are (in alphabetical order) the **angular, basioccipital, basisphenoid, dentary, frontal, jugal, lachrymal, maxilla, nasal, parietal, palatine, premaxilla, postorbital, prefrontal, pterygoid, quadrate, quadratojugal, surangular, squamosal,** and **vomer** (Fig. 5.4). Keep in mind that many of these bones are paired, which nearly doubles the number of bones from this list. Openings in the skull are **foramens, fenestrae** (plural of **fenestra**), and **orbits**. Foramens and fenestrae are named after their proximity to the bones surrounding them (that is, the **antorbital fenestra** and **surangular foramen**), whereas orbits refer to the former positions of a dinosaur's eyes. The size of a dinosaur's orbit gives an approximation of the original size of its eyes, which, of course, is connected to hypotheses on dinosaur vision (Chapters 9 to 13). Most of the skull

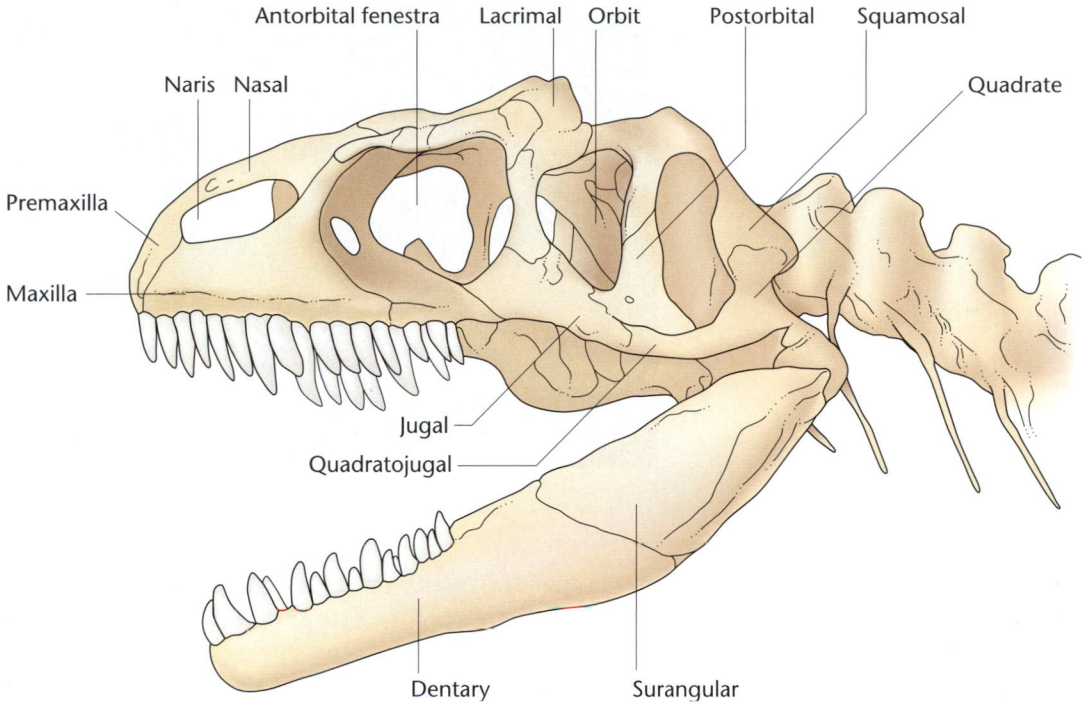

FIGURE 5.4 Cranial bones in *Allosaurus fragilis*: compare with Figure 1.6.

bones are sutured tightly to one another and have no evidence of former movement. Others, such as the bones associated with the jaws, were capable of some rather complex movements (Chapters 9 to 13). Missing from this list of cranial bones is the **postfrontal**, which was present in ancestors of dinosaurs; its absence assists in their definition.

To invoke lukewarm responses from dinosaur paleontologists, tell them that you have found only teeth. Although teeth were part of the skull, in some areas they are commonly the only body fossils that remain of some dinosaur species, so their discovery seldom adds much to a paleontological survey. However, some teeth are distinctive enough to identify the presence of species that were previously undescribed. They also are extremely important for interpreting dinosaur paleobiology as they can be a guide to dinosaur dietary preferences, especially if the teeth can be matched to toothmarks (Chapter 14).

Teeth are preferentially preserved more often than other ossified tissues because they are typically composed of dense, compact bone (Chapter 8). Additionally, dinosaur teeth were in sockets (instead of being fused to the jaws, as in lizards), which caused them to pop out occasionally. These lost teeth were then continually replaced with new ones, which tended to bias the fossil record further. A problem with this abundance of dinosaur teeth in the fossil record is that teeth, because they are relatively smaller than many dinosaur bones and are composed of more durable material, have been more subject to reworking or transportation. So their occurrence in strata may not be in the same place or originated at the same time as the original dinosaur, making them suspect as guide fossils (Chapters 4 and 7).

Appendicular Skeleton: Legs and Feet

Dinosaur limb bones, especially if they are associated with a **manus** (anterior foot, or hand) and **pes** (posterior foot), are wonderful finds, because they can provide information on how dinosaurs moved about their environments. Foot anatomy in

FIGURE 5.5 Pectoral girdle of the Late Cretaceous hadrosaur *Edmontosaurus* of North America. Denver Museum of Science and Nature, Denver, Colorado.

particular is valuable for correlating with footprint data, leading to more refined attributions of tracks to their tracemakers (Chapter 14). The bipedal posture of some dinosaurs also means that their hands were freed for other tasks, such as grasping food or potential mates. Specialized functional adaptations suggested by dinosaur limbs have also been the subject of lively discussion, such as why supposed active hunters like *Tyrannosaurus* or *Albertosaurus* had such tiny forelimbs with only two fingers in proportion to their large bodies (Chapter 9), the possible function of a "thumb spike" in *Iguanodon* (Chapter 11), and whether ceratopsians had semi-sprawling versus erect forelimbs (Chapter 13).

Forelimbs (arms) in dinosaurs were attached proximally to the main torso through the **pectoral girdle**. The pectoral girdle had as its main bones the **scapula** (shoulder blade) and **coracoid**, which interacted directly with the **clavicle** (Fig. 5.5); the latter is present in some dinosaurs such as saurischians (Chapters 9 and 10) and a few ornithischians (Chapter 13). The pectoral girdle interacted with the ribs of the chest region (**thoracic ribs**) and **sternum**, with the clavicle (if present) as an intermediary bone. **Sterna** (plural of **sternum**) have been reported from some theropods, sauropodomorphs, ornithopods, thyreophorans, ceratopsians, and birds, but they are not always present in the geologic record because some may have been cartilaginous (composed mostly of collagen: Chapter 8) and thus were not preserved. The clavicles fused in some post-Triassic theropods to form a **furcula**, equivalent to a wishbone. The place on the scapula where it articulated with the **humerus** is the **glenoid**, which pointed caudally and is yet another trait of dinosaurs (Fig. 5.6). The humerus rotated in whatever range of motion was defined by the glenoid. More importantly, in combination with the length of its forelimbs, it determined whether the dinosaur could have brought food to its mouth with these forelimbs (Chapter 9).

Another characteristic of dinosaurs is a long **deltopectoral crest** on the humerus. The way to remember this unwieldy term is to use a human body as a guide. The main shoulder and chest muscles are called a **deltoid** and **pectoral**, respectively, so any body part related to both of these would have the combined name of deltopectoral. Now think of a raised portion (crest) on the humerus that related to both the deltoid and pectoral muscles of a dinosaur and this image should illustrate the approximate position and purpose of this name for this anatomical landmark. The humerus, just as in humans, formed an elbow joint with the **radius** and **ulna**. To work out which one is which, turn your hand so that you are looking at your palm, and then turn it so that you are looking at the back of your hand. The forearm bone that moved to the medial part of your body was the radius, which

FIGURE 5.6 Scapula and its articulation with the humerus and glenoid.

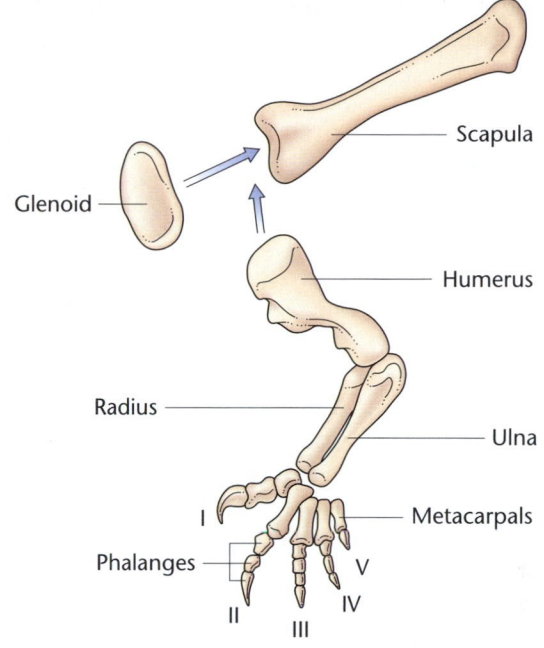

FIGURE 5.7 Phalangeal formula applied to a human hand as an example.

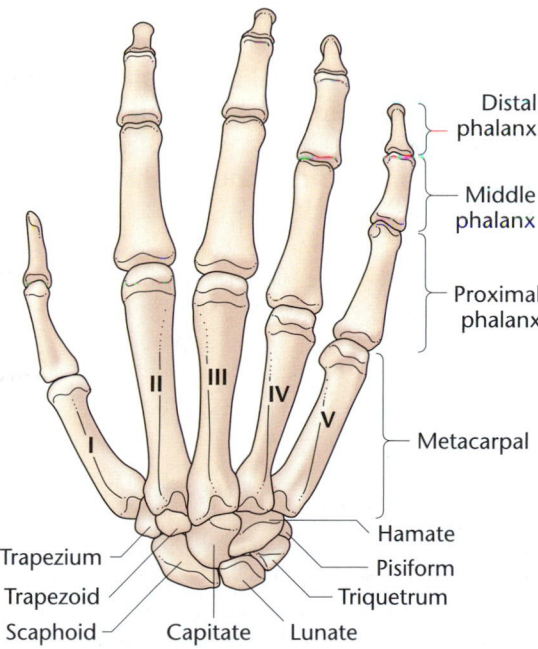

is just below (posterior to) your thumb. Distal to the radius and ulna were, in order, the **carpals** (wrist bones), **metacarpals**, phalanges (plural of **phalanx**), and **unguals** (claws or hooves); the latter two compose the digits (or fingers). The phalanges are divided from the metacarpals in a human by the location of the knuckle joints and the same is true of dinosaus.

The main difference between a dinosaur manus and a human hand is seen in the asymmetry of the dinosaur manus, which can be discerned through a **phalangeal formula** (Fig. 5.7). To demonstrate this formula, turn your hand so that you are looking at the back (dorsal surface) of it. Label your fingers from the thumb (closest to the midline of your body) to the smallest finger from I to V, then count the number of separate bones (each phalanx) in each of these labeled fingers. Test the results by comparing them with the following data:

I.   2
II. 3
III. 3
IV. 3
V.   3

The data should confirm the observation of this phalangeal formula in humans: 2-3-3-3-3. Dinosaurs have, as another trait distinguishing them from their ancestors, a manus that has such asymmetry with less than or equal to three phalanges on digit IV and less than or equal to two phalanges on digit V. In some dinosaurs, digits IV and V became modified or reduced enough in their manus that they eventually became vestigial or disappeared, which gave some dinosaurs (particularly theropods) three-fingered hands. A few theropods, such as *Tyrannosaurus*, even evolved two-fingered hands (Chapter 9). A similar circumstance happened with digits I and V in the pes of some dinosaurs, which resulted in such dinosaurs leaving four- and three-toed tracks (Chapter 7). However, other dinosaurs retained all five digits on either their manus or pes (Chapters 10 and 12), indicated in some dinosaurs' tracks as well as their skeletons (Chapter 14).

In the posterior portion of a dinosaur, hind limbs were associated with the sacral vertebrae (collectively called the **sacrum**) by the **pelvic girdle**. As mentioned earlier, the sacrum was connected with the hips (dorsal and medial to the ilium), and the ball-like, proximal end of the femur fitted into the acetabulum (Fig. 5.8A). Distal to the femur were the tibia and fibula, where the tibia, more medial than the fibula, formed the knee joint with the femur. The tibia is key to two dinosaurian traits: it has a **cnemial crest** and its distal, anterior surface fits with the ascending process

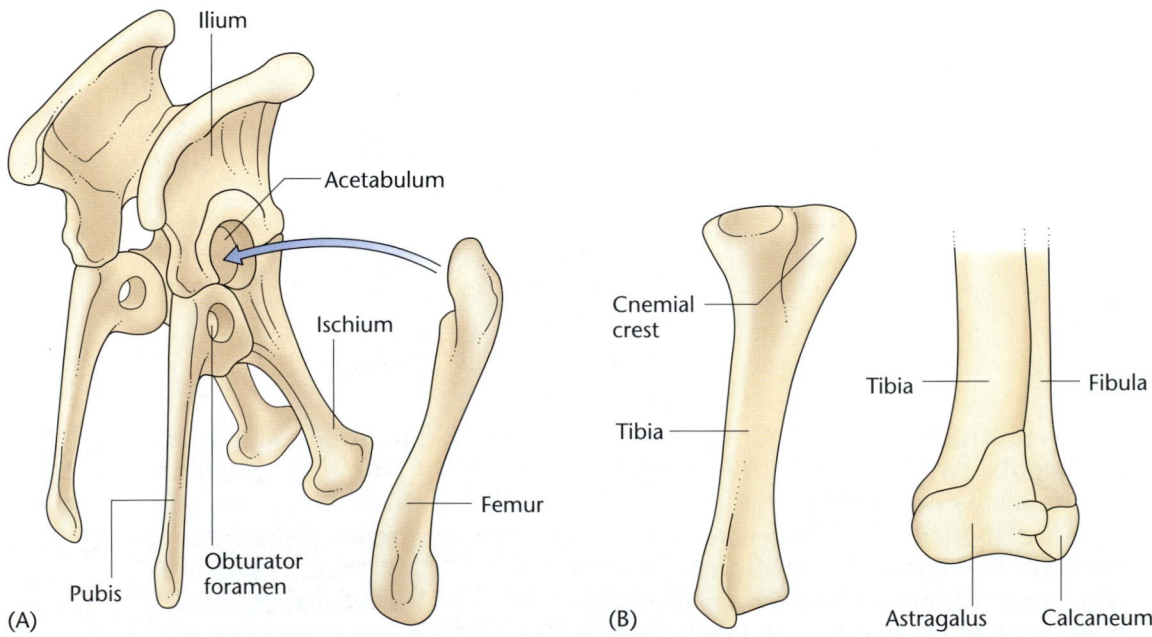

(A)

(B)

FIGURE 5.8 Characters for dinosaurs involving the appendicular skeleton. (A) Sacrum, proximal end of the femur, and the fit of the latter into the acetabulum. (B) Tibia, showing two traits of dinosaurs: cnemial crest and astragalus.

of an anklebone called the **astragalus** (Fig. 5.8B). The astragalus and another ankle-bone, the **calcaneum**, collectively formed a dinosaur's tarsals. These terms relate directly to the evolution of dinosaurs from reptilian ancestors with different arrangements of their tarsals (Chapter 6).

Distal from the tarsals and relating to a dinosaur's feet were the metatarsals, phalanges, and unguals. Although humans normally walk with their metatarsals in contact with the ground, in most cases, dinosaurs had metatarsals above the ground (Chapter 7). Such a condition, called **digitigrade**, is also observable in dogs, which have a posteriorly pointing joint between the metatarsals and ankle bones (Fig. 5.9A). A digitigrade stance can be approached in humans by standing on the balls of the feet or wearing high-heeled shoes (Fig. 5.9B) but strictly speaking is achieved only by on-point ballet dancers who stand on the tips of their toes. This stance contrasts with relaxing and standing with most of the body weight on the metatarsals (heels), which is **plantigrade**. By far, most dinosaurs were digitigrade, placing their weight on their phalanges (Fig. 5.9C), although some dinosaurs walked plantigrade under some conditions (Chapter 14).

Dinosaur Skin, Feathers, and Organs

Skin and its derivatives in modern vertebrates, such as nails, feathers, hooves, and hair, are composed of the structural protein **keratin**. Because most skin is soft tissue and has lower preservation potential than skeletal material (Chapter 7), the discovery of dinosaur body fossils that have any evidence of soft tissues are the cause of much celebration, often followed by much debate. Both older and recent finds of evidence of soft tissue clarify better what some dinosaurs looked like in life, and may indicate the placement of their internal organs, and validate the predicted presence of feathers or feather-like projections in a few species (Chapter 9 and 15). Not all dinosaur skin was soft, however, and **dermal armor**, formed originally as ossified plates set into the skin of a dinosaur, was a common accouterment to ankylosaurs (Chapter 12) and a group of sauropods called titanosaurs (Chapter 10). Other ossified dermal derivatives, **osteoderms**, include the dorsal plates or spines on stegosaurs (Chapter 12).

Skin impressions of dinosaurs typically indicate a similarity to modern reptilian skin in that they have patterns of scales (Fig. 5.10). In some cases the patterns are of equally-sized scales but others show definable patches of scales of different sizes. Stunning finds of feathered theropods from Lower Cretaceous deposits in China (Chapter 9) also demonstrate unequivocal evidence of some dinosaurs having feathers. These impressions of feathers are mostly associated with the dinosaurs' forelimbs and tails, but two theropods, *Microraptor rui* and *Cryptovolans pauli*, had feathers on both forelimbs and hind limbs (Chapter 9). Other Early Cretaceous theropods from China have, in close association with skeletal material, either carbonized fibers (probably downy feathers) or clearly-defined contour feathers (with central shafts). The presence of feathers and scales in modern birds thus lends support to hypotheses regarding the dinosaurs as ancestors of modern birds (Chapters 6 and 15). For a modern example of an animal that has both scales and different types of feathers, look at a chicken's foot for its scales and the rest of its body for down and contour feathers.

Evidence of dinosaur coloration is implied by patches of differing skin impression patterns, an interpretation that constitutes a weak hypothesis because of the lack of actually observed color differences. Banding in the aforementioned feather impressions, seen now as black and white stripes, however, is much more convincing evidence for dinosaur coloration. Because so little actual scientific data support interpretations of dinosaur colors, no multicolored or otherwise speculatively adorned dinosaurs are illustrated in this book. Evaluations of dinosaur vision, based on skeletal

5

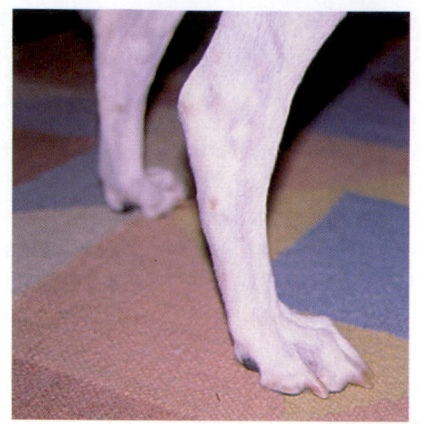

(A)

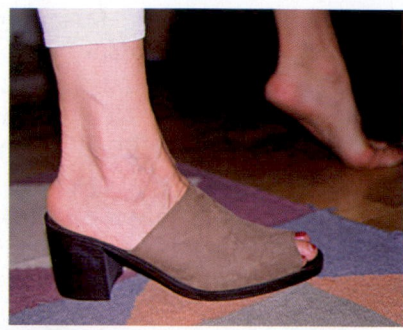

(B)

FIGURE 5.9 Lateral view of right hind limbs in digitigrade mode. (A) Modern canine. (B) Modern human. (C) Early Cretaceous theropod *Giganotosaurus* of Argentina. Fernbank Museum of Natural History, Atlanta, Georgia.

(C)

FIGURE 5.10 Hadrosaur skin impression, Late Cretaceous of North America. Mesa State Community College Museum, Tucumcari, New Mexico.

anatomy and brain endocasts, nevertheless may provide indirect information suggesting that some dinosaurs were brightly colored, possibly for species identification and attracting mates (Chapter 8). Some researchers also propose that some dinosaurs were brightly colored because this trait is exhibited by some of their modern relatives (some reptiles and birds). On the other hand, dull colorations is just as likely for herbivores, allowing them to blend more easily into Mesozoic landscapes and avoid the attention of hungry predators. This could also have been the case for predators, with a dull coloration camouflaging them from their prey.

As mentioned previously, muscle attachment sites on bones are another way to tell where some soft tissues were located, but any remnants of the muscle tissue itself are yet to be discovered in association with dinosaur bones. **Pseudomorphs** (false forms) of dinosaur muscles for the theropod *Pelecanimimus polyodon*, however, were found in Lower Cretaceous rocks of Spain. In this instance, a special, rapid postburial mineralization process (Chapter 7) mimicked the three-dimensional structure of the musculature. Musculature and other soft tissues also can be inferred for dinosaur digits through toe-pad impressions in some dinosaur tracks, giving a sense of the actual size of a dinosaur foot with all of its original skin and muscles (Chapter 14). Toe-pad impressions help to identify phalangeal formulas of dinosaur feet, which can help to narrow down the potential tracemakers of some dinosaur footprints.

Other soft parts of dinosaurs include their internal organs, of which we have much better information based on actual evidence, rather than the speculation of only about 15 years ago. For example, one of the most spectacular discoveries of preserved dinosaur soft parts was that of a small Early Cretaceous theropod found in Italy, *Scipionyx samniticus*. The specimen has a clear outline of its intestine, some muscles, and possibly a **trachea** (windpipe) and liver. Besides this direct evidence, gastroliths in a small, localized area within the rib-cage region of a dinosaur constitute indirect evidence of a former **crop** and **gizzard**, which are muscular digestive organs located anterior and posterior to the stomach, respectively, in modern birds. Gastroliths were most likely an aid to grinding up difficult-to-digest plant material in some herbivorous dinosaurs, such as sauropods (Chapters 10 and 14). At least one Early Cretaceous theropod (*Caudipteryx zoui*), presumably a meat eater, also had a concentration of small gastroliths in its abdominal region (Chapter 9). Remains of small animals within the body cavity of a dinosaur also can point toward

(A)

(B)

FIGURE 5.11 Restoration of skin and musculature for: (A) Late Triassic theropod *Coelophysis bauri* of North America: Denver Museum of Science and Nature; (B) Late Jurassic sauropod *Apatosaurus louisae* of North America: Dinosaur National Monument, Utah.

the former location of either its stomach or intestine (Chapter 9). For evidence of reproductive organs, one feathered theropod from China contains two egg-like structures within the posterior part of its abdominal region, which is evidence for an **oviduct** (birth canal).

Well-preserved skulls can provide three-dimensional approximations of dinosaur brain exteriors. This helps in determining whether room existed in parts of the skull for other tissues, such as in the nasal region, related to physiological functions (Chapter 8). Brain **endocasts**, which are casts of the braincase in a skull, reveal a minimum value of brain volume, which is measured in cubic centimeters. Endocasts also indicate the relative sizes of the different parts of the brain used for sensory perception, such as the **olfactory bulb** (related to smell), and they show the locations of nerves and blood vessels that related to a dinosaur's physiology (Chapter 8). When a skull has been crushed from compaction or damaged by further preparation, computed tomography (CT) scans render digital images that represent accurately the skull interiors that also can be rotated in three dimensions for easier study (Chapters 9 and 15).

Although still relatively sparse in comparison to skeletal anatomy, the fossil record of dinosaur soft-part anatomy is actually better than the popular perception that only bones tell the story of dinosaur bodies. The preceding examples also provide important supplemental information to help dinosaur paleontologists better "flesh out" these animals, rendering their restorations more accurate than if they only used skeletal material. Thus, using all of the information available from dinosaur fossils and comparative anatomy with extant vertebrates, restorations of dinosaur musculature result in what are probably better estimates of the overall forms of dinosaurs. Dramatic models derived from this information can show us what some dinosaur limbs and bodies may have looked like if we had the opportunity to dissect a recently dead specimen (Fig. 5.11).

Dinosaur Anatomy Related to Classification: Old and New

Linnaean Classification of Organisms

As mentioned in Chapters 1 and 3, the classification of dinosaurs has been a contentious subject for dinosaur paleontologists since the early and late nineteenth century, which illustrates the difficulty of classifying organisms (fossil or modern) in general.

Although the Linnaean classification scheme enjoyed a long history, for all practical purposes it has been replaced by cladistics (Chapter 1). One of the problems with the Linnaean system was that organisms with characters that did not fit into standard grades (phylum, class, and so on) were sometimes placed in between them, with the addition of an appropriate designated prefix. For example, superfamily includes more than one family but does not constitute an order, infraclass is within a class but not specific enough to warrant being called a superorder or order, and subphylum is not quite at the level of a phylum. Such splitting of categories presents difficulties on how to delimit them, as well as how to justify the differences between artificial subdivisions such as a superfamily and a suborder. Thus, one of the unwieldy facets of the Linnaean system is evident: where does one draw the line between its hierarchical levels?

Another criticism of the Linnaean system is that the justification for grades is made regardless of new evidence indicating the evolutionary relatedness of descendants, especially when it is applied to fossil organisms. For example, recall that dinosaurs classified in this system belonged to the following categories:

> Phylum Chordata
>   Subphylum Vertebrata
>   Class Reptilia
>   Subclass Diapsida
>   Infraclass Archosauria
>   Superorder Dinosauria
>   Order Saurischia
>   Order Ornithischia
>   Class Aves

Unfortunately, the hypothesized relatedness of theropods (in order Saurischia) and birds (in class Aves) is not indicated explicitly here. In this system, a paleontologist might want to challenge the separation of these two grades by saying that anatomical, physiological, and paleontological evidence supports the descent of birds from saurischian dinosaurs. Consequently, birds actually belong within class Reptilia and order Saurischia. Strict adherence to the Linnaean classification would not allow any appeals to new evidence that would place a formerly separate taxonomic group within the group where it shares an ancestry. Under a Linnaean classification, a reptile is a reptile, and a bird is a bird, on the basis of how they look today and how closely a fossil form might anatomically fit the appearance of a common ancestor. Such anatomical similarities are ideally determined by how closely two or more fossil forms are evolutionarily related, which the Linnaean classification approaches but does not quite achieve.

The purpose of this discussion, however, is not to belittle the Linnaean classification scheme, but to learn about it. The main justification for this is its long tradition, which led to its subsequent familiarity and widespread use in the scientific literature of the eighteenth, nineteenth, and most of the twentieth centuries. Hence, ignoring its use in biology and paleontology is akin to omitting the historical context and basis for what has constituted scientific discovery in those sciences during those times (Chapters 2 and 3). Knowledge of the Linnaean classification also helps in understanding and communicating with the biologists and paleontologists who still use it. Finally, some of the original names proposed for phyla, classes, and other Linnaean categories still form the basis of taxonomic categories applied to groups of organisms (not just dinosaurs) that share certain anatomical traits. This illustrates the difference between taxonomy (naming) and classification (sorting the names). Dinosaurs were surrounded by an abundance and diversity of organisms that resembled those of modern ecosystems. Consequently, knowing the general names based on previous classifications for life forms other than dinosaurs is also important for communicating about them. For the sake of simplicity, general groupings that correspond to conventional Linnaean classifications, such as arthropods, mollusks, and amphibians, are used throughout this text, although different names have been given in recent years to such groups on the basis of evolutionary relatedness.

This nearly complete change in taxonomic methods highlights some disadvantage of the Linnaean classification system. It does not address adequately the evolutionary relationships of organisms on the basis of shared anatomical characters in both ancestors and descendants. This disadvantage inspired the use of cladistics, which recognized the derivation of certain groups from ancestral groups on the basis of shared traits.

Phylogenetic (Cladistic) Classification

One of the major revolutions in dinosaur studies in the past 20 years has been in how they are classified, which is now accomplished through cladistics. Although evolutionary theory has been an essential part of biology and paleontology since the late nineteenth century, cladistics was not proposed in the scientific literature until 1950, and even then it did not become well known in mainstream scientific circles until 1966. This change was prompted by publication of a book in English that outlined its original concepts, which were first published in German by the entomologist **Willi Hennig**. Most dinosaur paleontologists did not begin to adopt cladistics until around 1984, although some discussion of the monophyletic versus polyphyletic nature of dinosaurs was a recurring point of debate in the 1970s and early 1980s. Since then, cladistics has become the standard classification system for dinosaur paleontologists, which probably would not have occurred so rapidly if not for the development of computer technology that analyzed quickly large data sets of anatomical characters in dinosaurs.

Cladistics is based on examination of anatomical features that can be broadly categorized as:

1 primitive (**plesiomorphies**);
2 shared and derived (**synapomorphies**);
3 new (**novelties**); and
4 convergent.

Of these, synapomorphies, which are characters shared between two or more groups of organisms and derived from earlier features, are the most important for defining clades (Chapter 1). In biological evolution, a character must be genetically inheritable and not acquired during the lifetime of an animal. Characters thus relate to an organism's **genotype**, or how its genes were expressed (Chapter 6). In contrast, a population of mice that originally had sight but lost it, then had their tails cut off, do not constitute a clade because their blindness and tail losses are acquired traits that are not inheritable. The outward physical expression of an organism, caused by a combination of environmentally caused traits and the genotype, is called its **phenotype**, which can vary considerably from the potential of the genotype (Chapter 6). In other words, these mice becoming tailless and blind were traits that were acquired, not inherited.

If a group of blind mice from a population of ancestral mice evolved a genetically inheritable lack of tails, sight, or both, then these novel traits would show up in subsequent generations (descendants). The group without tails or the blind group therefore comprise a new clade, as will any successive group that shows these synapomorphies, indicating relatedness (Fig. 5.12). Consequently, cladistics is a method used to hypothesize the phylogeny (evolutionary history) of a group of organisms, which is why it is also called a phylogenetic classification (Chapter 1). Plesiomorphies (primitive features) can also help with discerning ancestry, in that descendants may have retained a trait from far back in their evolutionary

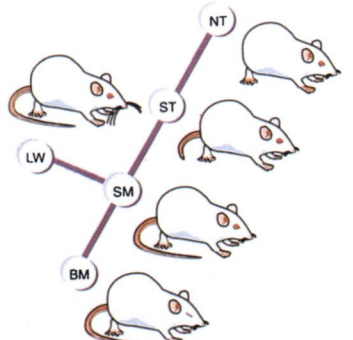

FIGURE 5.12 Hypothetical evolution of mice and how their synapomorphies (novelties) would contribute to their cladistic classification.

5

lineage: examples include the formation of teeth in embryonic chickens and pharyngeal gill slits in human embryos. A clade, because it shows the evolutionary origin for all descendants from a common ancestor, is thus monophyletic. In contrast, taxonomic groups that have multiple evolutionary origins, such as more than one clade, are polyphyletic (Chapter 3).

The result of such analyses and the consensus reached by most dinosaur paleontologists is summarized by the following clades, introduced in Chapter 1:

Chordata
  Tetrapoda
    Amniota
      Reptilia
        Diapsida
          Archosauriformes
           Archosauria
             Ornithodira
               Dinosauria
                 Saurischia
                 Ornithischia

However, the mere listing of clades, even with progressive indentations given to the list, does not explain adequately the relations between them. Their interrelationships are best illustrated through a cladogram, which shows how clades branch from one another at points called **nodes**, where a common ancestor of all subsequent clades first developed a new synapomorphy. The influence of the Linnaean classification is retained through some taxa (i.e., Diapsida, Archosauria) that were originally based on some shared characters recognized long before the invention of cladistics. Yet another aspect of Linnaean classification that still remains is the binomial nomenclature of fossil species. As a result, the embrace of cladistics by biologists and vertebrate paleontologists has not erased colorful species names such as *Triceratops horridus* (Chapter 13).

Vertebrate paleontologists, who employ cladistics, attempt to be scientifically rigorous in their approach by examining evidence for the inheritability of any character observed in fossil specimens. Each cladogram is essentially a hypothesis for a phylogeny that is tested through peer review. Typically, the least complicated hypothesis (the one requiring the fewest steps for establishing the relatedness) is regarded as the most likely, and such less complicated cladograms are said to have **parsimony**. New evidence, such as a fossil find from the field or a museum with previously undescribed or unrecognized characters, requires re-examination of previous hypotheses about the evolutionary relationships of certain clades. Hence, use of this methodology can falsify the justification for a new clade, or argue more firmly for a previously-defined clade (Chapter 15). This situation means that cladistics, in its earliest stages, can be quite volatile as new information is added, which results in hindered communication.

Outline of Main Dinosaur Groups Using Cladistics

Cladistics works as a classification system by showing how organisms with certain inherited traits have common ancestors, which makes any organism with those characters a member of a clade. Consequently, all animals that have a notochord, pharyngeal gill slits, and a dorsal nerve cord belong to Chordata. Classification of animals with these shared traits places humans in the same clade as sharks and Dinosauria. Similarly, the formation of bones in vertebrates is an ancestral trait that characterizes the clade Vertebrata. However, subsequent evolutionary innovations of vertebrates since their formation resulted in new clades through geologic time, which paleontologists try to define on the basis of character data. For example, Chordata eventually gave rise to the character of four limbs, which partially defines Tetrapoda. Tetrapoda had a clade develop, Amniota, with the evolution of egg-laying ability (Chapters 6 and 8), which then had other reptile-like clades form that eventually had clades develop from each of them. All of this diversification of amniote clades resulted ultimately in the origin of Dinosauria as a clade, most likely by the Middle Triassic or earliest part of the Late Triassic (Chapter 6). Mammalia also originated as a clade from a common amniote ancestor shared by dinosaurs and mammals.

As is typical in scientific endeavors, total agreement is still lacking about exactly what traits define Dinosauria as a clade. The consensus reached thus far is that Dinosauria is defined primarily on the basis of synapomorphies related to locomotion, such as these previously mentioned anatomical traits:

- Three or more sacral vertebrae.
- Acetabulum with open medial wall.
- Femur with ball-like head on proximal end.
- Tibia with cnemial crest.
- Astragalus with long ascending process that fits into the anterior part of the tibia.
- Sigmoidally shaped third metatarsal.

Saurischians apparently are the earliest known clade of dinosaurs, and they are certainly the most abundant dinosaurs in both the body and trace fossil record of the Late Triassic (Chapter 6). The first ornithischians occur in slightly younger strata than the first saurischians, although both groups were derived from a still-unknown archosaurian ancestor that was the node from which they diverged. These two clades had other monophyletic groups branch from them that form the presently understood ancestor–descendant relationships of dinosaurs (Fig. 5.13). Familiarization with representative species from each clade (Tables 1.1 and 5.1) will help with visualizing some of the dinosaurs associated with these clades. More detailed knowledge of anatomical differences, covered in remaining chapters, is necessary to understand the scientific basis for paleontologists showing where these clades branch into more derived clades.

Of course, human factors complicate even the most scientific of classification schemes. In some cases, the best-defined clades are biased, favoring the views of the particular dinosaur paleontologists who are most active in researching them, as well as the relative abundance of skeletal material available (or not available) for character analysis in certain clades. For example, theropods are described cladistically in more detail than any other dinosaur group (Chapter 9). This circumstance is at least partially attributable to the recurring fascination most dinosaur paleontologists have with theropods, but it is also related to the abundance of theropod material for study in nations where scientific methods have a long tradition (Chapter 3). Furthermore, the logistical problems associated with unrecoverable dinosaur remains add to the bias of their cladistic classification. For example, sauropods, most of which were disproportionately larger than the other dinosaurs, may not be defined as well in a clade because recovery of their extremely large body parts from remote field areas is difficult. The extended amount of time, money, and labor required for preparing those parts in a laboratory prevent proper study of their characters (Chapters 3 and 10). Yet another human factor to consider is experience. Because cladistics has been applied to dinosaurs for only about 20 years, not all dinosaur paleontologists have gained the expertise necessary for identifying all of the characters.

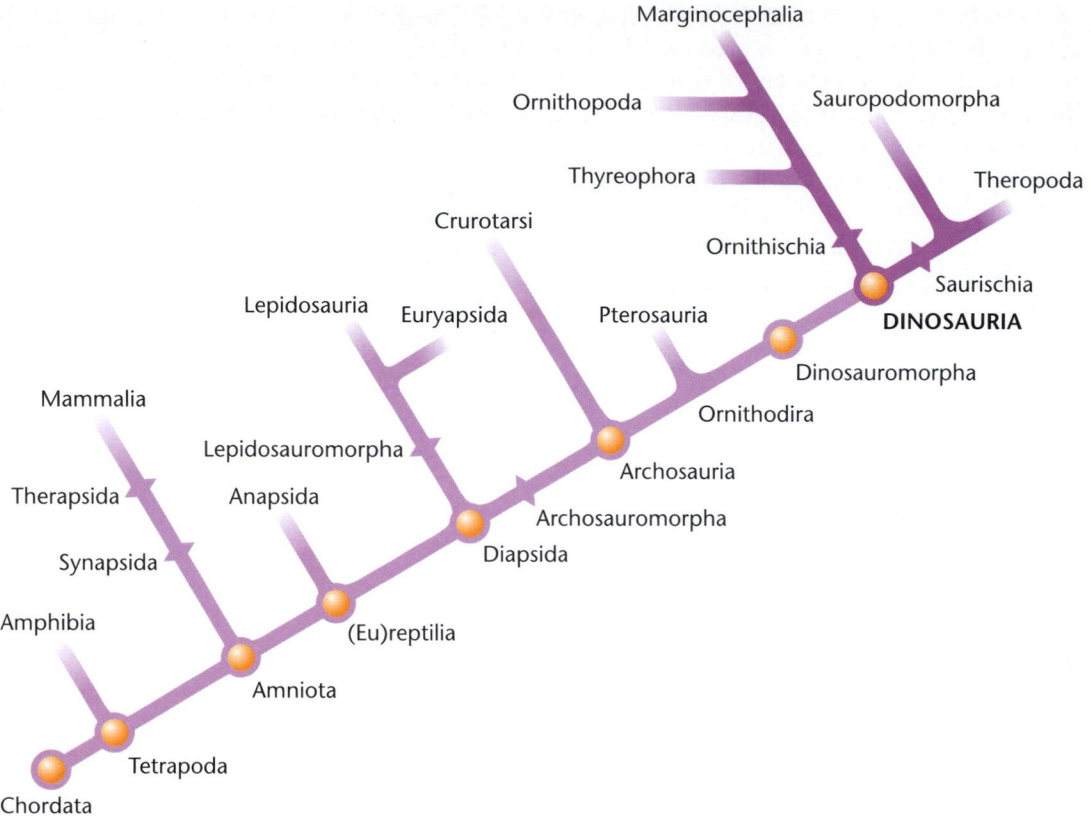

FIGURE 5.13 Currently accepted cladogram for dinosaurs, beginning with Chordata and ending with the Saurischia and Ornithischia and main monophyletic groups within these clades.

Because cladistic classifications are based on considerable amounts of descriptive data, revisions of these classifications are likely to be made with each new fossil find that either supports, refutes, or modifies an inferred node for two or more clades. Another method, independent of the fossil record, that has been used to test cladistic classifications is **molecular phylogeny**, which looks at the relatedness of certain proteins or nucleic acids (which are characters, too) of extant organisms, to extrapolate evolutionary rates of some lineages (Chapter 6). In some rare instances, fossil material yields biomolecules with sufficient nucleic acids available for comparison with living counterparts. Finds of proteins (osteocalcin), amino acids related to hemoglobin, and still-elastic tissues in dinosaur bones are encouraging in this respect, although the likelihood of nucleic acid preservation in dinosaur bones is very low (Chapter 4). This constant revision of dinosaur phylogenies results from the application of scientific methods, is evidence-based, and dissuades scientists into making unsound interpretations. For these reasons, cladistics is currently the preferred method for classification and will likely continue its prominence in dinosaur studies well into the future.

So You Want to Name a New Species of Dinosaur

Let's say that after reading only five chapters of this book an ambitious student might decide to name a new species of dinosaur. Such a task is not easy and would require exhaustive peer review and much double-checking of the scientific literature to ensure that mistakes are not made during such a quest. Several possible

TABLE 5.1 Major clades of the Dinosauria and examples of dinosaur genera classified with each clade. Realize that other clades branch from those mentioned here.

THEROPODA	
Ceratosauria	*Abelisaurus, Carnotaurus, Ceratosaurus, Coelophysis, Dilophosaurus, Elaphrosaurus, Majungasaurus, Podokesaurus*
Tetanurae	*Allosaurus, Bambiraptor, Baryonyx, Beipiaosaurus, Compsognathus, Deinocheirus, Deinonychus, Megalosaurus, Microraptor, Oviraptor, Pelecanimimus, Poekilopleuron, Protarchaeopteryx, Saurornithoides, Scipionyx, Sinornithosaurus, Sinosauropteryx, Spinosaurus, Struthiomimus, Suchomimus, Troodon, Tyrannosaurus, Velociraptor*
SAUROPODOMORPHA	
Prosauropoda	*Coloradisaurus, Euskelosaurus, Lufengosaurus, Massospondylus, Melanosaurus, Plateosaurus, Riojasaurus, Sellosaurus, Thecodontosaurus*
Sauropoda	*Apatosaurus, Argentinosaurus, Barosaurus, Brachiosaurus, Camarasaurus, Cetiosaurus, Dicraeosaurus, Diplodocus, Dystrophaeus, Euhelopus, Janenschia, Omeisaurus, Paralititan, Pelorosaurus, Seismosaurus, Titanosaurus, Tornieria*
ORNITHOPODA	
Hadrosauridae	*Claosaurus, Corythosaurus, Edmontosaurus, Hadrosaurus, Prosaurolophus, Saurolophus, Telmatosaurus*
Heterodontosauridae	*Abrictosaurus, Heterodontosaurus, Lycorhinus*
Iguanodontia	*Camptosaurus, Dryosaurus, Iguanodon, Ouranosaurus*
THYREOPHORA	
Ankylosauria	*Ankylosaurus, Hylaeosaurus, Nodosaurus, Pinacosaurus*
Stegosauria	*Huayangosaurus, Kentrosaurus, Stegosaurus, Tuojiangosaurus*
MARGINOCEPHALIA	
Ceratopsia	*Anchiceratops, Centrosaurus, Chasmosaurus, Leptoceratops, Monoclonius, Protoceratops, Psittacosaurus, Styracosaurus, Torosaurus, Triceratops*
Pachycephalosauria	*Homocephale, Pachycephalosaurus, Prenocephale, Stegoceras*

mistakes that would be sources of irritation for successive generations of paleontologists include:

- Assigning a different species name to a dinosaur that has already been given another name (called a **synonymy**).
- Assigning a species name to a dinosaur that is already being used for another organism, which may not even be a dinosaur (called a **homonym**).
- Naming a dinosaur on the basis of little material, such as one tooth, so that the definition of its characters is too vague for other workers to find it again.

Other complications of any names applied to fossil organisms, particularly at the genus and species level, stem from the recognition (or lack thereof) of **sexual dimorphism**, or different forms for different sexes. Most nineteenth- and early twentieth-century paleontologists failed to acknowledge that sexual differences could be determined from fossils. Consequently, they gave different species names to very similar (but distinctively different) dinosaurs that occurred in strata from the same place and time. Franz Nopcsa (Chapter 3) was one of the first dinosaur paleontologists to attempt a separation of then-established dinosaur species into different sexes, but he had inadequate evidence and applied it poorly to support his hypothesis. Some paleontologists have inferred plausible sex differences for the same species from dinosaur skeletons, thus justifying a reduction in the number of species names. However, more convincing fossil evidence supporting these inferences (such as male and female skeletons of the same species together in *flagrante delicto*) is, unfortunately, still lacking. Another problem is presented by **ontogenetic** variations, which are changes caused as dinosaurs grew from juveniles to adulthood. The recognition of juvenile specimens of some dinosaur species has resulted in a reduction of species names, as smaller versions of some dinosaurs are no longer assumed as species separate from their similar but larger counterparts.

With these potential pitfalls in mind, someone who names a new dinosaur species would need to pay attention to strict rules of biological nomenclature. This code has to be followed by all zoologists and paleontologists wishing to name new species of recent or fossil animals, and is summarized in a document called *The International Code of Zoological Nomenclature* (ICZN). A similar document is used for the naming of living and fossil plants among botanists and paleobotanists. As mentioned in Chapter 1, a species name should be Latinized but does not always have to contain roots from classical languages, and so can be formed from any language or acronym. Despite this latitude given in naming, names that are the most encouraged are those that are relatively easy to remember and pronounce. With dinosaurs in particular, the suffix "saurus" is used so often in genus names that any prefix used in association with it is normally the most difficult part of a dinosaur's name to remember and pronounce. Additionally, names that provide appropriate literal translations of key descriptive criteria are helpful, such as *Triceratops* ("three-horned face").

Once a species name is chosen, then a detailed description must be written of the material on which this new species is based. The material should be well illustrated through photographs, drawings, and digital images (such as the aforementioned CT scans), and the results must be submitted to a scientific journal for peer review (Chapter 2). The peer review may result in scientific acceptance of the proposed new species. It may not be accepted if the material is too fragmentary or poorly-defined, or if it closely resembles a variation of an already named species, to warrant a new name.

If the name is accepted, then the material used to describe it constitutes the **type specimen** or **holotype** of the species. The type specimen should include, as a minimum:

1 a catalogue number;
2 the name of the formation in which it was found;
3 the exact locality (geographically and stratigraphically) where it was found; and
4 the names of the people who described it.

The material is deposited in a safe place (such as a museum or university) where future workers can use it as a basis for comparison of other dinosaurs that might resemble it. A safe place for a type specimen cannot be overestimated: type specimens for *Podokesaurus*, *Spinosaurus*, and a few other dinosaur species have been lost to science because they were destroyed by fire or warfare, or were merely misplaced.

Synonymies constitute the greatest problem associated with dinosaur names. The fame associated with dinosaurs probably encourages more naming of new species than in any other fossil group, but naming may not be warranted. Another potential problem of the past, but not as prevalent now, was the publication effect: competitive paleontologists attempted to name as many new species as possible through excessive publishing (see Cope and Marsh, discussed in Chapters 2 and 3), a circumstance that artificially inflated dinosaur diversity. This problem was clarified when the co-editors of a major volume on dinosaur systematics, *The Dinosauria*, required contributing authors to meticulously review which names were valid or not, resulting in a considerable reduction of species names created before the book's publication in 1990. A subsequent edition of the book in 2004 documented how the number of species has slightly increased since then, but not at the rates seen before the more careful assessments of species names.

Paleontologists who advocate the reduction of synonymies are colloquially called "lumpers", whereas those who justify more species names are called "splitters." These attitudes represent two extremes of the spectrum, and are categorizations that can be applied to any paleontologist at different times in his or her career. If the given evidence justifies a species name, then examination of that evidence and peer review will hopefully affirm that fact. If the species name is later found as unnecessary, peer review hopefully is behind this change, too.

SUMMARY

A basic knowledge of dinosaur anatomy is required for understanding why dinosaurs are classified as dinosaurs (as opposed to reptiles, mammals, or birds) and why they are then subdivided into different groups. The anatomical distinctions that facilitate the classification of dinosaurs are most readily apparent as features preserved in their hard parts (bones and teeth). Learning about dinosaur anatomy is not difficult if the location names (e.g., anterior, posterior, dorsal, ventral), universally used in describing the anatomy of other animals, are applied to dinosaurs. Furthermore, anatomical features, primarily represented by bones, are homologous to those found in many vertebrates, so learning about these features clarifies and reinforces knowledge of general vertebrate anatomy. Bones of the axial, appendicular, and cranial skeleton of dinosaurs, as well as rare examples of soft anatomy (skin, feathers, or organs), provide paleontologists with evidence for reconstructing dinosaurs as living animals and assist in the classification of dinosaurs into different groups.

The two main classification systems used for dinosaurs and other organisms, Linnaean and cladistics, are similar in their use of categories and some taxonomic designations, but differ in key ways. Advocates of the Linnaean classification attempt to classify dinosaurs on the basis of how closely they resemble one another and place them into hierarchal groups. Cladistics can resemble a hierarchy but also emphasizes identification of inheritable traits (synapomorphies) and defines dinosaur groups on the basis of how many of these traits are shared. This method thus includes all descendants of an ancestor in a group as a clade, which is a phylogenetically-based grouping. Knowing how some clades of dinosaurs are classified, such as the Theropoda, Sauropodomorpha,

SUMMARY Continued

Ornithopoda, Thyreophora, and Marginocephalia, provides an introduction to the dinosaurs associated with those clades. Although current dinosaur paleontologists overwhelmingly prefer cladistics, learning both classification systems is advantageous for understanding the taxonomy applied by past paleontologists, particularly those who performed work before the 1980s.

The naming of a dinosaur species is relatively difficult and requires a comprehensive knowledge not only of the anatomical features of known dinosaurs but also of the literature on previously named species. The main purpose of knowing previous work on dinosaurs is to avoid synonymizing, a practice that results in too many species names, which complicate the study of dinosaurs. If a new species of dinosaur is warranted by a new discovery, the dinosaur's remains become a type specimen that is then kept as a reference for comparison with any dinosaurs discovered later that might be of the same species or differ only slightly in characters. Paleontologists can use either "lumping" or "splitting" in their identification of fossil species. Dinosaurs in particular have been subjected to both approaches, but present assessments of dinosaur species are certainly justified better than in the past.

DISCUSSION QUESTIONS

1. Determine, through discussion with your classmates and instructor, which parts of the following inanimate objects you would designate as anterior, posterior, dorsal, ventral, medial, and lateral. If you determine that some designations are not possible for the parts of some of the objects, justify why.
   a. An automobile.
   b. This textbook.
   c. A globe.
   d. An Egyptian pyramid.
   e. A pencil.
2. Why was the development of an acetabulum in a dinosaur hip important for an upright posture? What other changes in the appendicular skeleton had to happen in combination with the acetabulum to accommodate this major adaptation in vertebrate locomotion?
3. Place the following appendicular bones in order of most proximal to most distal:
   a. fibula, phalanges, femur, metatarsal, tarsal, tibia, ungual
   b. ulna, coracoid, ungual, radius, scapula, metacarpal, humerus, phalanx, carpals

DISCUSSION QUESTIONS Continued

4. Draw a manus (either left or right, labeling which it is) that corresponds to each of the following phalangeal formulas for these dinosaurs. Also, what feature revealed by the phalangeal formulas indicates that these are all dinosaurs?
   a. *Allosaurus fragilis:* 2-3-4-0-0
   b. *Apatosaurus louisae:* 2-1-1-1-1
   c. *Iguanodon bernissartensis:* 2-3-3-2-4
   d. *Triceratops horridus:* 2-3-4-3-1
   e. *Stegosaurus stenops:* 2-2-2-2-1
   f. *Pinacosaurus grangeri:* 2-3-3-3-2

5. What bones in a dinosaur skull seem to have been involved in movement of the jaw? Develop a hypothesis on how the number of teeth in the skull might have had an effect on jaw movement, which corresponds to adaptations of bones associated with this movement. How would you test this hypothesis?

6. Get together with some study partners, purchase a whole cooked chicken, and attempt to look at muscle attachments to the bones as you dissect it. Try to see how many of the bones in the skeleton are the same or comparable to those mentioned in this chapter. With this knowledge in mind, look at a live chicken or a picture of one and try to identify which areas of the chicken should contain certain bones. How did the dissection aid in your recognition of bones beneath the flesh and feathers?

7. Is a chicken a dinosaur? With your instructor, try to find as many of the characters on the chicken used for Question 6 that define Dinosauria as a clade. How do you account for any that might be missing?

8. Humans are classified under the Linnaean system as follows: Phylum Chordata, Subphylum Vertebrata, Class Mammalia, Order Primates, Family Hominidae, Genus *Homo*, species *Homo sapiens*. What information would you need to convert this classification to a phylogenetic one?

9. Mammals have seven cervical vertebrae as a character of the clade Mammalia, regardless of their body size or neck length. Thus, mice have the same number as giraffes. Explain how all dinosaurs are placed in the same clade, despite the extreme variation in the number of cervical vertebrae for different dinosaur species (that is, they range from 5 in some theropods to 19 in some sauropods).

10. Take an existing species of dinosaur, such as *Coelophysis bauri*, *Apatosaurus louisae*, *Hadrosaurus foulki*, *Stegosaurus stenops*, or *Triceratops horridus*, and predict what its immediate descendant would look like, justifying the naming of a new species. What characters would you expect to have changed in accordance with different environmental or biological factors, such as global warming or mate selection?

Bibliography

Bakker, R. T. and Galton, P. M. 1974. Dinosaur monophyly and a new class of vertebrates. *Nature* **248**: 168–172.

Czerkas, S. A. 1997. "Skin". *In* Currie, P. J. and Padian, K. (Eds), *Encyclopedia of Dinosaurs*. San Diego, California: Academic Press. pp. 669–675.

Dal Sasso, C. and Signore, M. 1998. Exceptional soft-tissue preservation in a theropod dinosaur from Italy. *Nature* **392**: 383–387.

De Queiroz, K. and Gauthier, J. 1990. Phylogeny as a central principle in taxonomy: Phylogenetic definition of taxon names. *Systematic Zoology* **39**: 307–322.

Dodson, P. 1976. Quantitative aspects of relative growth and sexual dimorphism in *Protoceratops*. *Journal of Paleontology* **50**: 929–940.

Dodson, P. 1990. Counting dinosaurs: how many different kinds were there? *Proceedings of the National Academy of Sciences* **87**: 7608–7612.

Fastovsky, D. E. and Weishampel, D. B. 2003. *The Evolution and Extinction of the Dinosaurs* (Second Edition). Cambridge, UK: Cambridge University Press.

Gaffney, G., Dingus, L. and Smith, M. 1995. Why cladistics? *Natural History* **104(6)**: 33–35.

Hennig, W. 1950. *Grundzüge einer Theorie der phylogenetischen Systematik*. Berlin, Germany: Deutscher Zentraverlag.

Hennig, W. 1966. *Phylogenetic systematics*. Urbana-Champaign, Illinois: University of Illinois Press.

Holtz, T. R., Jr. 2002. "Chasing *Tyrannosaurus* and *Deinonychus* around the tree of life: classifying dinosaurs". *In* Schotchmoor, J. G., Springer, D. A., Breithaupt B. H. and Fiorillo, A. R. (Eds), *Dinosaurs: The Science Behind the Stories*. Alexandria, Virginia: American Geological Institute. pp. 31–38.

Holtz, T. R., Jr. and Brett-Surman, M. K. 1997. "The taxonomy and systematics of the dinosaurs". *In* Farlow, J. O. and Brett-Surman, M. K. (Eds), *The Complete Dinosaur*. Bloomington, Indiana: Indiana University Press. pp. 92–106.

Linné, C. 1758. *Systema Natura per Regina Tria Naturae, Secundum Classes, Ordines, Genera, Species cum Characterisbus, Differentiis, Synonymis, Locis, Editio decima, reformata, Tomus I: Regnum Animalia*. Laurentii Salvii, Holmiae.

McGowan, C. 1998. T. Rex *to Go: Build Your Own from Chicken Bones*. New York: Harper Collins.

Padian, K. 1997. "Phylogenetic system". *In* Currie, P. J. and Padian, K. (Eds), *Encyclopedia of Dinosaurs*. San Diego, California: Academic Press. pp. 543–545.

Sereno, P. C. 1991. "Clades and grades in dinosaur systematics". *In* Carpenter, K. and Currie, P. (Eds), *Dinosaur Systematics: Approaches and Perspectives*. Cambridge, UK: Cambridge University Press. pp. 9–20.

Wu, X.-C. and Russell, A. P. 1997. "Systematics". *In* Currie, P. J. and Padian, K. (Eds), *Encyclopedia of Dinosaurs*. San Diego, California: Academic Press. pp. 704–712.

Chapter 6

While reading a book on dinosaurs, you notice that they existed in a time span of about 230 to 65 million years ago in the Mesozoic Era, and you wonder what the first dinosaurs looked like. When you find an artist's rendering of these creatures, you notice that they look similar to modern monitor lizards (such as Komodo dragons), crocodiles, or alligators. Moreover, you are surprised to find out that flying reptiles, marine reptiles, and many other dinosaur contemporaries were not actually dinosaurs. As you read about dinosaurs toward the end of the Mesozoic Era, you also see the phrase "birds are dinosaurs".

Are dinosaurs, monitor lizards, crocodiles, and alligators related to one another? If so, what common ancestors did they have? What is the basis of the phrase "birds are dinosaurs"? If this premise is acceptable to you, then how are birds related to monitor lizards and crocodiles?

Introduction to Dinosaur Evolution

6

Why Learn about Evolutionary Theory?

Because scientific theories are by definition falsifiable, scientists freely acknowledge the possibility that modern evolutionary theory is modifiable. However, the overwhelming amount of evidence supporting evolutionary theory, as well as repeated testing and modification of its numerous interconnecting hypotheses by scientists worldwide during the past 150 years, illustrate its robustness and degree of certainty. In short, no other theory in science has endured and survived as much critical peer review as biological evolution. Consequently, scientists have no rational reason to suppose that evolutionary theory is closer to being incorrect than correct, and they no more "believe" in evolution (in a faith-based sense) than they believe in gravity. Indeed, its factuality is the central pillar of support for understanding the history of life on Earth.

Evolution is both a fact and a theory: evolution and its by-products have been observed, but a theory has also been constructed to explain these observations (Chapter 2).

Dinosaurs represent excellent test subjects for, and examples of, the basic principles of evolutionary theory. The rich history of amniote evolution, which began at least 350 million years ago and continues today, can be used as a framework for understanding the roots of dinosaur evolution. Once dinosaurs had evolved into a definable group in the Mesozoic Era, their proliferation into a wide variety of forms alludes to both the genetic and environmental changes that they experienced throughout their 165-million-year history. Data relating to the genetic components of dinosaur evolution are largely incomplete but can be inferred based on their character traits, the foundation of the phylogenetic (cladistic) classification system (Chapter 5). Broad-scale environmental changes in the Mesozoic, especially those related to plate tectonics (Chapter 4) and paleoclimatology, are well documented as stages for dinosaurs changing as the world changed. Additionally, some researchers have proposed that the evolution of dinosaurs contributed to major evolutionary changes in other organisms. Such hypotheses are supported by intriguing correlations of biological trends, including the origin of the dinosaurs from amniote ancestors that became, at least partially, embedded in the geologic record and continue to be augmented by fossil discoveries made daily. Understanding the evolution of dinosaurs is thus not only important to know as a well-documented process of the past, but is pertinent in the sense that we are connected to the current by-products of dinosaurian interactions with past environments. As ecosystems changed, dinosaurs changed with them and they were active participants in those changes, as part of their role in the web of life.

Basic Concepts in Evolutionary Theory

Part I: Genetics and Natural Selection

Evolution is defined here as the change in a population between generations, where a **population** is a group of interbreeding organisms, such as a species (Chapter 5). Darwin originally summarized this process in the late nineteenth century with the phrase "descent with modification," which is still apt today, despite much revision of his hypotheses since then. A population that goes through generations, from ancestors to descendants, comprises a **lineage**. Changes that happen to an individual organism during its life do not constitute evolution, although any effects that altered organism confers on its population could have a small impact on evolution of the population. Likewise, changes that happen to an environment surrounding a population also do not represent evolution, although the effects of that environmental shift on that population could influence its evolution.

The evolution of one species into another species is called **speciation**; separateness of the two species is defined by **reproductive isolation**, whereby neither species can reproduce with the other to form offspring that also can reproduce. Mules represent an example of reproductive isolation as **hybrids**, in that they are the sterile offspring of two different species, *Equus caballus* (horses) and *Equus asinus* (donkeys). Although speciation is popularly perceived as requiring long periods of time (i.e., millions of years), fast-breeding populations under certain environmental conditions can evolve into different species within a typical human lifespan. This type of evolution has been observed repeatedly, which is one reason why evolutionary theory is a fact, not "just a theory" in the pejorative sense (Chapter 2). Examples of speciation were first documented early in the twentieth century in flowering plants, such as the evening primrose (*Oenothera lamarckiana* to *Oenothera gigas*), and were later observed with various other species of plants, as well as fruit flies, houseflies, and other insects. Moreover, pharmaceutical companies must continuously update formulas for antibiotics because strains of bacteria evolve that are resistant to these treatments. Some insect populations also evolve quickly in response to insecticides, so chemical companies must change their insecticide formulas in response to their decreased effectiveness. Consequently, evolutionary theory is not an esoteric, untested philosophy with little or no real-world applications. The reality of evolution is a social and economic concern for nations, corporations, and individuals worldwide, and practical applications of the principles of modern evolutionary theory help to solve their problems.

Why do people tend to look like their parents? The answer is mostly related to inheritance of physical traits from the parents, which is caused by the passing of **genes** from one organism to the next generation. A gene is a nucleotide sequence in a DNA molecule that provides a code for a protein or part of a protein. The location of a specific gene in a chromosome is its **locus**, and any variation of that gene at the same locus is an **allele**; a pair of genes (or alleles, if the genes vary) constitutes an organism's genotype at a locus (Chapter 5). The sum total of genes conveyed in a DNA molecule and coding for all of an organism's proteins is its **genome**, representing the genetic potential of an organism. For example, geneticists defined the human genome in the year 2000. However, the genome is not the same as the **gene pool**, which is the sum total at a given time of all genes in a population and represents different individuals.

The genotype of an organism directly relates to an organism's physical appearance and behavior, or its phenotype (Chapter 5). People who look like their parents, or in some cases behave like their parents, are simply showing their

6

phenotype. However, environmental factors acting upon the phenotype could produce a radically different physical form or behavior than anticipated from the original genotype of an organism. For example, tailless mice that acquired their physical trait through severing of their tails (Chapter 5) still have a genotype for a tailed condition, although their phenotype shows otherwise. Their offspring still have tails when they are born, regardless of the environmentally-caused features possessed by the parents. Similarly, a bipedal theropod trackway that consistently shows three digits on one foot and only two on the other foot can be concluded as representing an environmentally-induced condition (probably from an injury or other pathological cause) that was not passed on to any of the dinosaur's offspring (Chapter 14). Behavior also can be greatly influenced by environmental conditions, rather than inherited predispositions. For example, certain breeds of dogs can be bred selectively for aggressive behavior, but breeds that are ordinarily passive also can be taught to attack and be threatening.

Modifications of phenotypes encourage the argument of "nature versus nurture" (inherited characteristics versus acquired characteristics) in examining the physical appearance of an organism. Acquired characteristics cannot be inherited. This hypothesis that promoted the contrary view, articulated by French naturalist **Jean-Baptiste Lamarck** (1744–1829), was critically examined and effectively falsified by the end of the nineteenth century.

Related to the phenotype of an organism is another fundamental property of any individual in a population: it shows **adaptations** to its environment. An adaptation is a physical attribute of an organism that can help it to survive at least long enough to reproduce successfully. Accordingly, a lack of this attribute will decrease an organism's chances of surviving to reproductive age. For example, hadrosaurs had impressive rows of teeth (**dental batteries**) that were well-adapted for the processing of vegetative material, presumably for better digestion (Chapter 11). A lack of these teeth would have considerably decreased their life spans, perhaps to the extent that they would not have reached reproductive age. In evolution, adaptations such as these must be inheritable from one generation to the next. An acquired adaptation is meaningless in the change of a population over generations. An example of an acquired adaptation is the development of a suntan in a normally light-skinned person. Melanin is produced in response to an environmental stimulus (sunlight over time), but this suntan is not inherited by any successive generations coming from this individual. Likewise, a human adult's lifetime habit of dyeing ordinarily dark-brown hair to blonde is perhaps an adaptation used for social enhancement and subsequent reproductive advantage in some societies. However, this adaptation does not necessarily affect whether any offspring of a chronic colorist will also have the same conferred reproductive benefit of blonde hair.

Natural selection, a hypothesis proposed conjointly by Darwin and one of his contemporaries, Alfred Russel Wallace (1823–1913; Fig. 6.1), helped explain why populations change through time and organisms composing these populations have inheritable adaptations. This explanation was proposed with the following tenets, based on numerous observations in natural settings by Darwin and Wallace:

> *The preceding background information, especially regarding adaptations, is necessary to understand **natural selection**, the central hypothesis of modern evolutionary theory.*

- Species have variations within their populations that are inheritable.
- Species tend to overpopulate, producing more individuals than will actually survive to reproductive age.
- A struggle for existence occurs within the population, perhaps through competition over resources, habitat, or mates.

FIGURE 6.1 Main originators of the hypothesis of natural selection, Charles Darwin (left) and Alfred Russel Wallace (right). From Ridley (1996), *Evolution*, 2e, Blackwell Science, Inc., Malden, MA, pp. 9 and 10.

■ Those individuals with variations favorable for survival from this struggle (the more adaptable ones) will live to produce offspring that also have these variations, thus changing the population over time with each successive inherited variation and eventually resulting in species different from the ancestral species.

A phrase associated with natural selection is "**survival of the fittest**," which is potentially misleading because "fitness" is not necessarily related to the popularized idea that "the strong survive and the weak perish." **Fitness** in this sense actually means "better adapted" or refers to the number of offspring produced by an individual, and thus has little or nothing to do with strength. Mammals of the Mesozoic exemplify this concept, as they were physically weak and small in comparison to their dinosaurian companions but clearly were better adapted than dinosaurs for surviving the environmental changes that resulted in the extinction of the dinosaurs by the end of the Mesozoic (Chapter 16).

The tenets of natural selection have been modified since the time of Darwin and Wallace but still form the foundation of evolutionary theory. The older version of the hypothesis of natural selection is **Darwinism.** Although Darwin and Wallace knew that certain inheritable variations in organisms translated into adaptations, they did not know the source of the variations or the exact mechanism for their inheritance. Ironically, another scientist at the time, **Gregor Mendel** (1822–84), was providing the answer to this question, but his results were not widely recognized by other scientists until early in the twentieth century. Mainly through cross-breeding pea plants, Mendel discovered the basic factors underlying heredity – genes, alleles, genotypes, and phenotypes. For example, a pair of genes at a locus (comprising a genotype) is paired because each gene came from a different parent. Consequently, sexual reproduction is responsible for most of the genetic variation in

Neo-Darwinism is a modified descendant that takes into account modern genetics, the study of heredity and variations in organisms.

an organism, because one-half of its genes came from its mother and one-half from its father. This is related to the haploid nature of male and female gametes, formed by meiosis, which combine to form a diploid zygote. Dinosaurs are also presumed, with a high degree of certainty, to have reproduced sexually through male–female pairs and not through parthenogenesis (Chapter 8). This hypothesis is supported by the numerous dinosaur eggs (a few containing embryos) and nests, the sexual dimorphism interpreted from some dinosaur skeletons of the same species, and the sexual reproductive life cycles in their closest living relatives, crocodilians and birds (Chapter 8). Dinosaurs thus had a constant source of genetic variation, as with other sexually reproducing organisms.

Another discovery by Mendel was that one of a pair of genes tends to overshadow the other gene in its physical expression, which affects the phenotype of the organism, so that the **dominant** gene is expressed over the **recessive** gene. An individual with two dominant or two recessive genes at a locus has a **homozygous** condition, in contrast to one with dominant and recessive genes, which is **heterozygous**. A heterozygous condition is defined by alleles, because a pair of genes at the same locus represents variations, or alternatives, of one another. Interestingly, proportions of these dominant and recessive traits can be predicted in offspring from parents with homozygous or heterozygous conditions through probabilities. For example, the gene for brown eyes in humans is dominant over that for blue eyes, but both parents can have brown eyes and a recessive gene for blue eyes, so they will both have a heterozygous condition. The **gene frequency**, which is the frequency of each gene in relation to another gene at its locus, is 0.5 for each allele in a heterozygous condition, which corresponds to a 50% probability for each (otherwise known popularly as "50 : 50"). In contrast, a homozygous condition would have a gene frequency of 1.0 for the single gene, whether it is for a homozygous dominant or homozygous recessive.

Armed with probabilities, geneticists can make predictions about the genotypes and phenotypes of pairings. In the example of eye color, the probability for any one of their offspring to have blues eyes is 25%. Probability is calculated through assigning letters to both the dominant allele (B) and recessive allele (b) in the homozygous pairs and crossing them in a diagram used by geneticists, called a **Punnet square**:

	B	b
B	BB	Bb
b	Bb	bb

The probability of a brown-eyed, homozygous-dominant individual (BB) is 1 in 4, or 25%. The probability of a brown-eyed, heterozygous individual (Bb) is 2 in 4, or 50%. Lastly, the probability of a blue-eyed homozygous-recessive individual (bb) is 1 in 4, or 25%. Therefore, two heterozygous individuals can produce three possible genotypes, but these genotypes can differ in their expression as phenotypes. These probabilities are related as **genotype frequencies** with values between 0 and 1, such as 25% = 0.25, 50% = 0.5, and so on. Notice how the gene frequencies and genotype frequencies are different from one another.

This shuffling of genes produces variation in a population that can be predicted by calculating probabilities for successive generations, based on gene frequencies and assuming random mating with no natural selection. The expected ratio of genotype frequencies in such a case is called the **Hardy–Weinberg ratio**. The ratio was named after its originators, mathematician G. H. Hardy and physician Wilhelm Weinberg, who independently devised a formula describing it early in the twentieth century. For example, the preceding example has two alleles (B and b), which has three possible genotypes: BB, Bb, and bb. The following binomial equation describes the frequency of each genotype:

$$p^2 + 2pq + q^2 = 1 \tag{6.1}$$

where p is the gene frequency of the dominant allele and q is the gene frequency of the recessive allele. Thus, p^2 is a result of multiplying the probability of allele B by itself (BB), q^2 is the result of multiplying allele b by itself (bb), and $2pq$ is the multiplication of both probabilities (Bb), which is also multiplied by two. Because the total probability for the three genotypes is 100%, then all of the genotype frequencies must have a sum of 1.0. The Hardy-Weinberg ratio is considered as the starting point for discussion of **population genetics**, the study of factors that affect gene frequencies.

Using our example, where each heterozygous parent contributed a gene frequency of 0.5 for each allele, the Hardy-Weinberg formula predicts the genotype frequencies for the first generation of the pairing as:

Step 1. $(0.5)^2 + 2(0.5 \times 0.5) + (0.5)^2 = 1$

Step 2. $(0.25) + (0.5) + (0.25) = 1$

Step 3. $1 = 1$

which corresponds to 0.25 for BB, 0.5 for Bb, and 0.25 for bb in Step 2. Using the formula is a good way to double-check the frequencies derived from crossing them in a Punnet square. Calculated either way, the expected gene frequencies for each generation of offspring can be predicted for all possible pairings by parents with known gene frequencies (Table 6.1). The ultimate result is that observers will expect a 75% probability of the brown-eyed phenotype and a 25% probability of the blue-eyed phenotype in a large population.

However, one of the truisms of statistics is that probabilities do not always translate into certainties. One of the most important facets of evolutionary theory is that expected genotype frequencies can differ considerably from observed genotype frequencies, as represented by the anomaly of more frequent appearances of phenotypes that were not predicted from the original pairings. The primary agent responsible for changing the frequencies is natural selection, which demonstrates the intimate interaction between Mendelian genetics and environmental factors.

TABLE 6.1 All possible mating combinations for a hypothetical male–female pair, crossing for two alleles (B and b) and three genotypes (BB, Bb, bb)

Genotype Crossing	Genotype Frequencies of Offspring
BB × BB	BB = 1.0
BB × Bb	BB = 0.5; Bb = 0.5
BB × bb	Bb = 1.0
Bb × BB	BB = 0.5; Bb = 0.5
Bb × Bb	BB = 0.25; Bb = 0.5; bb = 0.25
Bb × bb	Bb = 0.5; bb = 0.5
bb × BB	Bb = 1.0
bb × Bb	Bb = 0.5; bb = 0.5
bb × bb	bb = 1.0

Modified from Ridley (1992, Table 5.1, p. 94).

How does all of this genetic theory apply to dinosaurs? Using a dinosaurian example, let us say that a male *Centrosaurus apertus* (a Late Cretaceous neoceratopsian: Chapter 13) with a homozygous dominant gene for a small nasal horn (HH) mated with a female *C. apertus* that had a homozygous recessive gene for an enlarged nasal horn (hh). The expected genotype frequency would have been 1.0 for a heterozygous condition (Hh) in all offspring of the first generation, based on the following Punnet square:

	H	H
h	Hh	Hh
h	Hh	Hh

This means that there was a 100% probability of offspring from this pairing having a phenotype of reduced nasal horns, based on the dominance of the smaller-horn allele. The second generation should have then produced the following genotype frequencies:

	H	h
H	HH	Hh
h	Hh	hh

As a result, HH = 0.25, Hh = 0.5, and hh = 0.25, meaning that the offspring have a 75% chance of having smaller nasal horns (Fig. 6.2). This represents a reduction of 25% from the previous generation; but if it is representative of the population as a whole, smaller-horned *Centrosaurus* individuals will still be more abundant than the larger-horned individuals, as predicted by the Hardy-Weinberg ratio.

Natural selection then could have gone to work, such as through the following potential scenarios:

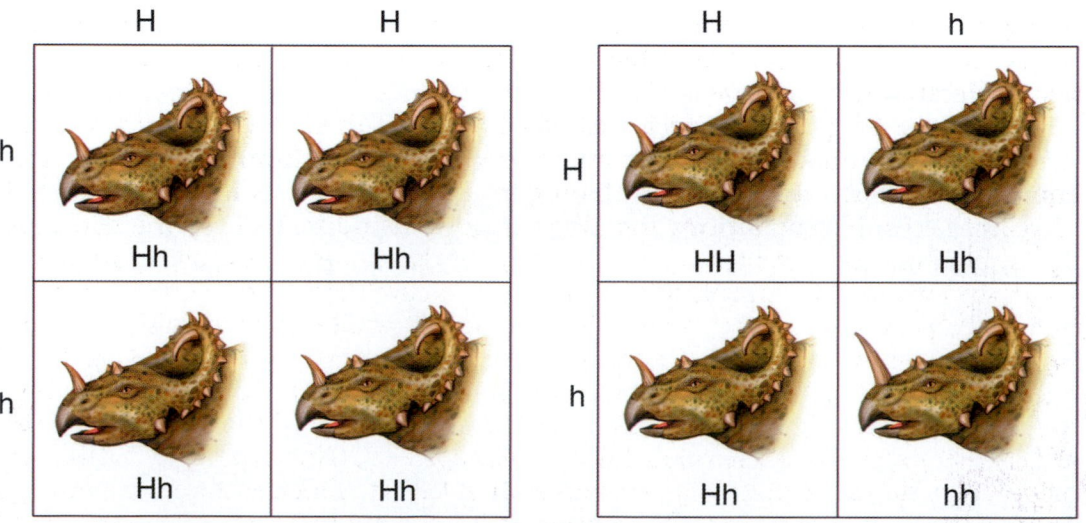

FIGURE 6.2 Hypothetical example of changes in genotype frequencies in the ceratopsian *Centrosaurus* with a dominant allele (H) for a smaller-horned phenotype. (A) First generation, with one parent homozygous dominant and the other homozygous recessive. (B) Second generation, with both parents heterozygous.

- Most other females in the breeding population refused to mate with the smaller-horned male offspring because larger horns in the males serve as better advertisements of their gender and species (a form of sexual selection: Chapter 8).
- Avians that detached parasites from the horns saw the larger horns more often because of their visual prominence, which resulted in more parasitic-borne diseases in the smaller-horned individuals and an increased mortality rate in their juveniles before they reached reproductive age.
- The smaller horn was not as effective in **intraspecific** (within species) competition as the larger-horned condition when males jousted with one another for the attentions of a potential mate. Consequently, the smaller-horned males were out-competed and did not have the opportunity to mate as often as the larger-horned ones.
- The smaller horn was a poor attribute for defense against theropod predators, which caused a higher mortality rate before the smaller-horned individuals reached reproductive age.
- All or any combination of the preceding scenarios could have occurred.

Over enough generations, the end result would have been a reduced frequency of the genotype that caused the phenotype of the smaller-horned condition in *C. apertus*. This circumstance would have happened despite the initial 100% probability from the mated homozygous individuals and the second-generation 75% probability of retaining the phenotypes from the homozygous dominant and heterozygous individuals. Assuming random mating with no natural selection, a population of 1000 *Centrosaurus* individuals should have had about 750 representatives of the smaller-horn phenotype. However, natural selection, through the offered scenarios, would have caused the reduction to a number much less than 750, possibly to zero after enough generations. The reduction of the genotype frequencies for Hh and HH, as well as a decrease in the gene frequency for gene H, was thus facilitated through natural selection that favored adaptations offered by the homozygous recessive (hh, or larger-horned) condition. Part of the natural selection also involved non-random mating, which counters random mating as an assumption of expected frequencies calculated through the Hardy-Weinberg ratio. The example also illustrates an observation in genetics that a dominant gene does not necessarily connote superiority. The word "dominant" unfortunately conveys a sort of hierarchy in genes, which is certainly not the case when a recessive gene is selected over generations.

The change of gene frequencies, added to other inheritable differences, could have caused reproductive isolation and a species different from *C. apertus* if given enough time. Such a small-scale change in gene frequencies in a population is often termed **microevolution**; the larger-scale transitions, such as the evolution of amphibians to amniotes or dinosaurs to birds (Chapter 16), are examples of **macroevolution**. Macroevolution is simply the cumulative effect of microevolution. The *Centrosaurus* example also illustrates **directional selection**, which is a consistent change in a population through time in a particular direction. Directional selections that have been hypothesized for dinosaurs include increased body size in some sauropodomorph lineages (Chapter 10) or reduction of the number of digits in the manus of some theropod lineages (Chapter 9). This type of selection was proposed by Edward Drinker Cope (Chapter 3) in the late nineteenth century through a hypothesis that became known as **Cope's Rule**, which stated that organisms showed a directional trend toward larger body size in their lineages through geologic time. Numerous exceptions have been demonstrated since Cope's time that have restricted it to a general trend observed for only some organisms; in other words, Cope based his "rule" on preliminary data from the fossil record, which has considerably improved since his time. For example, **George Gaylord Simpson** (1902–84), a paleontologist hired originally at the American Museum of Natural History by Henry Fairfield Osborn (Chapter 3), documented the numerous separate species that resulted from the ancestors of horses, a view of evolution that directly contradicted the linear progression for fossil horses proposed by O. C. Marsh (Chapter 3). Simpson's concept has since been applied to other vertebrate lineages, including dinosaurs.

Additionally, directional selection should not be allied with a concept of evolution as a linear trend. Rather, one ancestral population can have many subsequent combined directions that resulted in evolution, making a many-branched "tree" or "bush" (i.e., cladograms: Chapters 1 and 5) rather than a "ladder." Darwin illustrated this concept in a notebook from 1837, where he showed species branching outward from common ancestors in many directions (not just upward) in a "tree of life." His diagram seems amazingly prescient when compared to modern cladograms, but later scientists using cladistics were merely reinforcing Darwin's concept of descent with modification from common ancestors.

Natural selection and Mendelian genetics are currently regarded as the main contributors to changes in the gene frequencies of populations, but other sources of variation can occur through **recombination** or **mutation**. Recombination sometimes happens during meiosis through the exchange of genes between a pair of chromosomes, meaning new allele pairs that previously were unlinked can be formed in one organism's gametes before the contribution of a mate's gamete. Recombination is the basis for applications of recombinant-DNA research, also known as **bioengineering**, which has, for example, resulted in human-manufactured microbes that consume oil spills or produce insulin. Bioengineering has also created genetically altered fruits and vegetables through manipulation of genes in laboratories, and successful **cloning**, which is the production of a genetically identical organism by placing its genetic material from a diploid somatic cell into a gamete (egg). British scientists first achieved cloning of mammals in 1997 when they produced the sheep "Dolly" (1997–2003). Recombinant-DNA research is causing changes in gene frequencies much more rapidly than could be produced through either selective breeding programs or natural selection. The long-term repercussions of this work and of cloning are currently unknown and are a cause of concern among many people, including some scientists. Of course, cloning of dinosaurs has not occurred nor has its possibility been advanced anywhere except in science fiction.

Mutations constitute another source of genetic variation but differ from recombination in how they form. When a cell divides during meiosis or mitosis, its DNA is copied, but like in a photocopier or a computer printer, small errors can happen during the copying that cause the copy to be an imperfect duplicate of the original DNA. In this case, the slightly altered DNA codes different proteins. Mutations are typically caused by environmental factors, such as intense (short-wavelength) electromagnetic radiation or chemicals (often present as pollutants) called **mutagens**. Mutations have their greatest effect when expressed in gametes and many are harmful to an organism, conferring faulty information that will result in selection against the mutated trait. However, some may confer a trait that is advantageous for natural selection in the light of certain environmental factors.

Both recombination and mutation rates are measurable and can be rapid under certain conditions. Whether recombination and mutations occurred in dinosaurs is unknown, but they must be considered as likely because both are common processes in modern vertebrates. No genetic material, which would provide evidence of recombination and mutation, has been recovered yet from a dinosaur, despite some well-supported evidence of proteins (the by-products of DNA coding; Chapter 5) in a few specimens and amino acids in eggshell material (Chapter 8). Some claims of dinosaur DNA were published in peer-reviewed literature, but subsequent review has resulted in a consensus that modern DNA contaminated the analyzed samples. The reality is that dinosaur remains, with a minimum age of 65 million years, have been considerably altered from their original state (Chapter 7). This means that the direct use of dinosaur DNA for interpreting the population genetics of dinosaurs (let alone for their cloning) is very unlikely. Nevertheless, phenotypes (represented by body fossils), behaviors, paleobiogeography, and paleoenvironmental settings for dinosaurs are well documented, which provide a good framework for understanding the origin and evolution of dinosaurs.

Part II: Mechanisms for Macroevolution

As mentioned earlier, macroevolution has occurred (and is occurring) as a result of the cumulative effects of microevolution. The overwhelming evidence for this process is found in the fossil record. Fossils recognized originally for their biological origin have been placed within a relative age dating scheme (biologic succession: Chapter 4) that has been used by geologists for the past 200 years. The principle of biologic succession is simple – fossils in lower strata are older than ones in the overlying strata (superposition: Chapter 4). Consequently, those fossils with similar forms that show change through time are inferred to have evolved due to changes in their genotypes that eventually affected their phenotypes. Given the millions of years that are often represented by strata in a typical outcrop, geologists and paleontologists can, on any given day in the field, potentially view the numerous records of organisms that underwent descent with modification, and accordingly test hypotheses about biologic succession.

So-called **transitional fossils** are examples of macroevolution that are perceived as "big leaps" in evolution through what may be considered as major changes in adaptations. Examples are:

- *Pikaia*, interpreted as a primitive chordate from the Cambrian Period, represents a transition from invertebrate animals to chordates:
- *Acanthostega* of the Devonian Period is an amphibian derived from lobe-finned fish:
- *Archaeopteryx* of the Late Jurassic is a bird that evolved from dinosaurian ancestors (Chapter 15):
- *Artiocetus* is a whale from about 40 million years ago that shows clear connections to previous generations of legged, land-dwelling mammalian herbivores.

However, all organisms are in transition between generations, meaning that *all* fossils represent transitional forms or, more properly, have transitional features. Whenever a paleontologist is asked to provide an example of a transitional fossil, they can name any fossil of the millions that have been identified and would still be correct. Thus, the term "transitional fossil" (rarely used by evolutionary scientists) is often applied erroneously only to those organisms that, through their adaptations, seem to bridge a gap between habitats, such as water to land, land to water, and land to air. Such a designation consequently confuses descriptions (forms) with interpretations (functions). Using this reasoning, modern animals that could qualify as transitional fossils in the future, assuming favorable circumstances for their preservation, might include the California sea lion (*Zalophus californianus*), "flying lemurs" (such as *Cyanocephalus volans*), and emperor penguins (*Aptenodytes forsteri*), which are adapted to multiple habitats but show adaptations that favor one habitat over another.

Evolution over spans of geologic time is categorized as having occurred in two modes, **phyletic gradualism** and **punctuated equilibrium**. These modes are not diametrically opposed views, but both have natural selection and Mendelian genetics at their cores. Their difference is in the scale of evolution in its most basic sense, which is change over time. Phyletic gradualism is a hypothesis supported by evidence for small-scale, incremental changes in fossil species over long time periods, where lineages are reconstructed on the basis of morphological changes in similar fossils in a stratigraphic sequence. Darwin promoted this mode of evolution based on his knowledge of the fossil record in the mid-nineteenth century. Fossil evidence discovered since then has not yet falsified this hypothesis for some lineages. In contrast, punctuated equilibrium is characterized by long periods of no morphological changes in a fossil species, followed by rapid change. Evidence from the vertebrate fossil record that supports this hypothesis was noted by George Gaylord Simpson in the 1940s, but then paleontologists **Stephen Jay Gould** (1941–2002) and **Niles Eldredge** named and proposed it as a unified hypothesis in the early 1970s, based on fossil lineages of gastropods and trilobites, respectively. Gould, more than any other scientist of the latter half of the twentieth century, wrote extensively on punctuated equilibrium and all other aspects of evolutionary theory.

So which hypothesis does the fossil record support? The answer is both, in that some fossil lineages show slow, gradual changes and others show periods of stasis followed by rapid change. Hence, lineages should be examined on a case-by-case basis with regard to whether they are interpreted as belonging to either model or as part of a continuum in between them. Controversy exists over whether one hypothesis is more the norm for speciation, and active debate centers on the evidence supporting each. For instance, one criticism of punctuated equilibrium is that it uses its lack of evidence as actual evidence in some cases of the fossil record. Punctuated equilibrium predicts that intermediate fossil forms may not be represented in short, continuous stratigraphic intervals (corresponding to a short time span) between two distinctive fossil species. In such a case, advocates of punctuated equilibrium might propose that speciation was so rapid that most intermediate forms did not become fossilized, which is possible given that conditions must be just right to preserve some fossils (Chapter 7). Gradualists could counter that intermediate forms might still be found in other areas containing the same stratigraphic interval with more favorable conditions for preservation. The incompleteness of the fossil record, as a record of life on Earth during the past 3.8 billion years, may be an issue in this respect, but it is a record that improves every day with each fossil discovery. For example, dinosaur species have been described in ever-increasing numbers over the past few decades, filling previously perceived gaps in their lineages, especially with regard to theropod–bird connections (Chapters 9 and 15).

Regardless of the rates of change in the genotype frequencies of populations over time, the main non-genetic mechanisms that influence natural selection are environmental factors, particularly those related to biogeography. For example, members of a population can be separated geographically through a physical barrier, such as a rise in sea level that isolates an island from a mainland, a river that changes its course after a major flood, or a forest fire that divides a habitat. Separation also can be a result of migration. Members of a population may migrate thousands of kilometers away from their ancestral population, thus no longer mixing their genes with their original population. If separated populations are kept apart long enough for natural selection to cause significant changes in the genomes of each, the reproductive isolation may result in speciation. Such a hypothesis for the origin of species is called **allopatric speciation**; this type of speciation happens when the **gene flow** (the spread of genes through a population by interbreeding) is interrupted.

One version of allopatric speciation is used in the punctuated equilibrium model. When a small subpopulation at the periphery of a species' geographic range is isolated enough, it cannot reproduce with the main population. This subpopulation, because of natural-selection factors different from the parent population, will undergo rapid changes in its gene frequencies relative to the main ancestral population. The result is a new species within a small number of generations. This specific type of allopatric speciation is **peripheral-isolate speciation**. Evidence from modern biogeography that supports this mode of speciation consists of numerous plant and animal species that have small, isolated populations on islands that are morphologically distinctive (and in some cases are already reproductively isolated) from large, geographically widespread populations on mainland areas adjacent to the islands. Mountains also serve as geographic barriers between populations that began from an ancestral stock, particularly for those organisms that could not fly over them, such as non-avian dinosaurs.

Another important consideration of small populations in evolutionary theory is that they may reflect non-representative (random) samples of a larger population. Picture the following: four people randomly selected from a group of 100 people move to a small Caribbean island to start a new population, but the remaining 96 stay and mate in Kansas, in the midland part of the North American continent. The genotypes and consequent phenotypes from the mating of the four tropically placed people would not represent the group as a whole and the two resultant groups would likely differ considerably in their Hardy-Weinberg ratios after only a few generations. This divergence would happen regardless of the environmental differences between the two localities that might cause natural selection. Such a random change in the gene frequencies is an example of **genetic drift**. The dramatically rapid effect of genetic drift in small populations relative to large ones has been demonstrated in laboratory experiments with fruit flies, and may be a factor in allopatric speciation for other organisms as well. If a large number of these small populations radiate out from a central location and are isolated from one another to form species that demonstrate adaptations distinctive to each of their individual, but geographically separated, environments, then the resulting populations illustrate **adaptive radiation**. Some bird species from closely associated islands that show many "variations on a main theme" probably radiated from an ancestral species and then adapted to their respective **niches**. A niche can be envisaged as the role of an organism in an ecosystem, where it lives in a specific habitat and uses specific resources.

Probably the greatest large-scale factor now recognized as affecting the geographic distribution of populations is plate tectonic activity (Chapter 4). Although it is a much younger theory than evolution, plate tectonics has been successfully integrated with the latter to provide powerful explanations for how fossil populations became geographically isolated from one another and underwent speciation over long periods of time. The study of biogeography (or paleobiogeography) and how it relates to plate tectonics is termed **vicariance biogeography**, where the vicariance is caused by the division of a species' geographic range by movement of lithospheric plates, such as in divergence. Vicariance biogeography, as an agent for speciation, is supported by major periods of diversification in the fossil record (which presumably reflect increased speciation) that correlate with the splitting of landmasses by continental rifting, occurring at different times during the Phanerozoic Eon (Fig. 6.3). The main hypothesis for this observed higher number of identified fossil species in association with plate divergence is that ancestral populations, especially for organisms inhabiting shallow-marine and continental environments, became increasingly isolated from subpopulations as rifting continued. This separation encouraged speciation as these subpopulations, through major shifts in genotypic frequencies caused by genetic drift and natural selection, became more reproductively isolated from their ancestral populations and adapted to new environments, occupying new niches.

Vicariance biogeography is applicable to dinosaur evolution throughout the Mesozoic Era, as the continents were more or less together (forming Pangea)

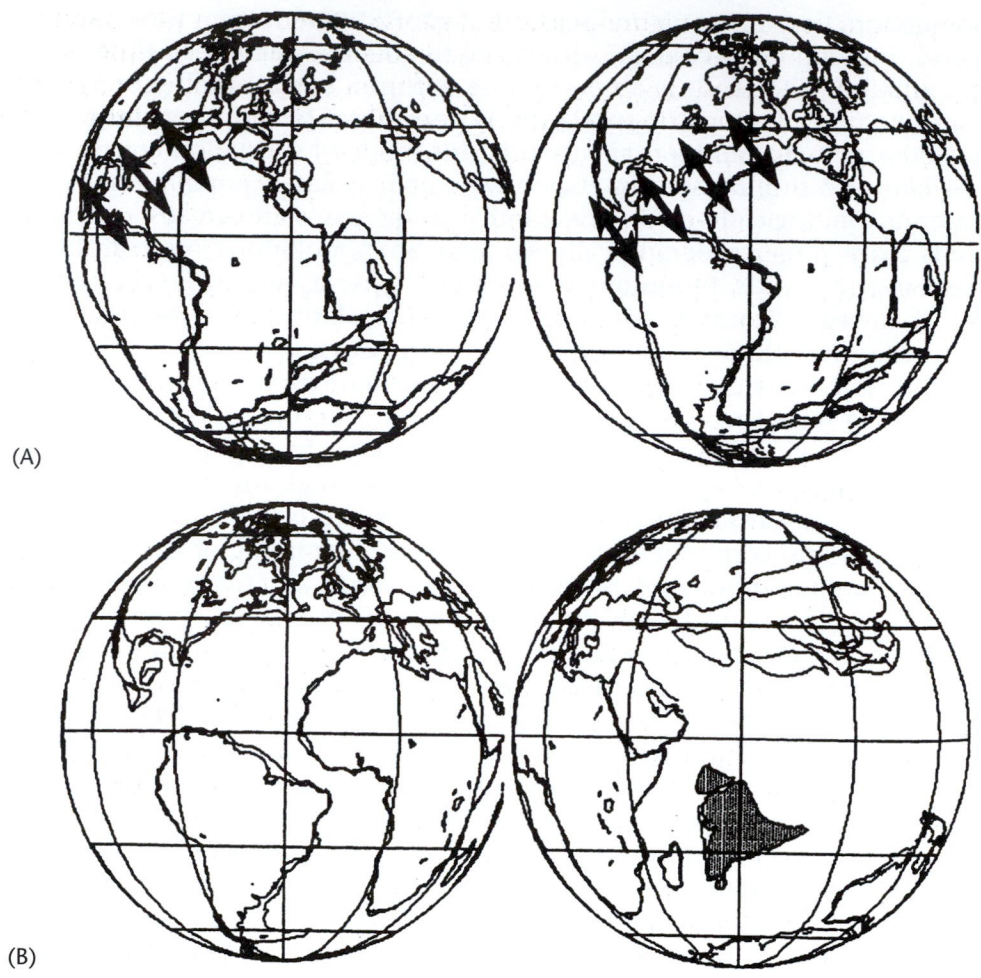

(A)

(B)

FIGURE 6.3 Continental landmasses during the Mesozoic showing how dinosaur populations became increasingly isolated through time. (A) Late Jurassic (about 140 Ma). (B) Late Cretaceous (about 80 Ma). From Cowen (1995), *History of Life*, 2e, Blackwell Science, Inc., Malden, MA, p. 82, figs. 5.13 and 5.14.

during the Late Triassic Period, when evidence for the first dinosaurs is recorded. Dinosaurs became widespread soon afterward, inhabiting every continent, except Antarctica, by the Early Jurassic (which also reflects their rapid migration rates) before significant splitting up of Pangea. However, as the continents split farther apart by seafloor spreading during the Jurassic and Cretaceous Periods, increased diversification of dinosaurs took place. Some similarities endured within species on still-connected continents, but noticeable differences appeared in those on separate continents. Thus, the most prominent barriers to gene flow and subsequent causes of reproductive isolation and allopatric speciation over time were the oceanic expanses. For terrestrially-bound dinosaurs, this circumstance meant that any of them inhabiting landmasses that later separated from Pangea then formed populations that became distinct from their ancestral populations through time. Additionally, linear mountain systems and inland seas (caused by global sea-level highs) also resulted in geographic barriers that could have been a mechanism for dinosaur speciation (Chapter 13).

However, allopatric speciation through geographic isolation is not the only hypothesis proposed for how species originated in the fossil record. Indeed, reproductively-isolated species with recent common ancestors can have overlapping geographic ranges. Those closely-related species that occur in the same region have **sympatry**, and the origin of new species from populations within these regions is possibly through **sympatric speciation**. Sympatric speciation is regarded as the result of intraspecific factors, rather than environmental factors such as climate changes or predation by other species. Sexual selection (through competition for mates) is an example of an intraspecific factor that could cause natural selection and subsequent changes in genotype frequencies in a population. This was illustrated through the hypothetical example of the less-endowed *Centrosaurus* earlier. As these differences within a species occur in the same geographic area through time, the increased genetic distance between their inheritable traits is termed **character displacement**. The role of character displacement in dinosaur evolution is poorly understood, but is hypothesized through synapomorphies (connected by cladograms) and speculations about character traits that would relate to this proposed mechanism for speciation. Examples of such characters include horns, head frills, and feathers, which might have served as sexual displays in dinosaurs or were otherwise used for intraspecific competition (Chapters 9, 11, and 13).

Natural selection and the subsequent co-evolution of two or more species that occurs as a result of their interactions are summarized by the **Red Queen hypothesis**. The Red Queen is a character in *Through the Looking Glass*, by writer and mathematician **Charles Lutwidge Dodgson** (1832–98; more popularly known by his pseudonym of **Lewis Carroll**). In the book, Alice meets the Red Queen chess piece, who appears to run across the chess board at high speed, yet never leaves her square: "Now, *here*, you see, it takes all the running *you* can do, to keep in the same place." This serves a metaphor for a co-evolutionary process in which two species of organisms continuously match one another's defenses only to maintain the *status quo*. For example, plants may evolve chemical defenses against insect herbivores, which in turn evolve resistance to the plant's chemicals, and so on. This type of equilibrium state should cause regular extinctions through time of species with two or more lineages, so the Red Queen hypothesis is scientifically testable. This hypothesis has been proposed to explain some changes in character traits of dinosaurs through time, such as in Cope's Rule, whereby prey and predatory dinosaurs became progressively larger as a result of their "arms race" interactions (Chapters 9 and 10). Additionally, increased amounts of dermal armor in ankylosaurs and apparent defensive weaponry in stegosaurs comprise other presumed evolutionary responses to pressures from theropod predation (Chapter 12). Although the preceding is a simplistic analogy with regard to modern predators and prey, this hypothesis has also been applied to changes in herbivorous dinosaur dentition and digestive systems in response to changes in vegetation types throughout the Mesozoic Era.

Finally, an important point to keep in mind with natural selection is that some species may have inheritable variations that are "pre-adapted" for a change in either the magnitude or rate of an environmental factor unprecedented in the history of a species. For instance, a large-scale volcanic eruption that deposits ash in only a few weeks over a large area of a forest may favor the reproductive survival of taller adult plants of a species, as the taller plants can still disperse their seeds above the ash layer. The shorter adult plants of the same species, completely covered by the ash, may not survive to reproduce. This chance possession of inheritable traits, favorably adapted for a selective pressure before it happened, is called **exaptation**. Exaptations also are hypothesized as features that had a neutral (non-harmful and non-beneficial) effect on an organism's adaptation that in later generations become advantageous for survival. This hypothesis for natural selection is especially applicable to explaining the survival of certain lineages of organisms after mass extinctions recorded by the geologic record. The lack of some currently undefined exaptations in dinosaurs at the end of the Cretaceous may have resulted in their demise in the face of a global catastrophe (Chapter 16).

Cladograms that hypothesize macro-evolutionary relationships of dinosaurs and other vertebrates are testable through the fossil record, and new fossil discoveries can change the cladograms. However, a test of cladograms independent of the fossil record is molecular phylogeny (Chapters 5 and 8), which compares relative differences in protein or nucleic acid sequences between extant organisms that are presumed to be descendants from common ancestors. This method is not without controversy, because molecular geneticists who look at these biochemical differences also calculate rates of change in the biomolecules, called **molecular clocks**, under the assumption that these rates do not vary over time. With these molecular-clock models, geneticists are now predicting the divergence times for major clades of organisms, figures that in some cases agree very well with the fossil record but in other cases have discrepancies of tens of millions of years. Despite these disagreements, molecular clocks provide an interesting predictive tool for paleontologists searching the geologic record.

The numerous hypotheses used here to explain how evolution happened in the geologic past, or how it is happening today, do not negate one another. How evolution occurs is not a true-or-false question but a multiple-choice one with the possible answers of "all of the above," "any of the above," or "none of the above." Based on the extensive evidence contributed, at a minimum, by the interrelated sciences of paleontology, geology, biology, and ecology, "none of the above" is the least likely correct answer and "all of the above," depending on individual circumstances, is the most likely. Of course, all hypotheses in science are subject to falsification, so completely new hypotheses about evolutionary processes that incorporate both old and new information are possible in the future.

Evolutionary Origin of Dinosaurs

Amniote Evolution and Diversification before the Dinosaurs

The origin of dinosaurs could arguably be traced back as far as the origin of life itself, which was about 3.8 billion years ago, but for the purposes of this book the evolution of amniotes is a more reasonable starting point. The development of an **amniotic** egg (one with an **amnion**, or fluid-filled sac surrounding the embryo: Fig. 6.4), from amphibian ancestors for reproduction of offspring outside of aquatic environments, is often heralded as one of the major adaptations in vertebrate evolution. Unfortunately, the first appearance of this defining characteristic of Clade Amniota, which is inferred to have happened during the Carboniferous Period, is currently unknown. No definitive fossil eggs or nests have been interpreted from rocks older than the Late Triassic. This gap in the fossil record, for such a well-established behavior, is likely an artifact of the non-mineralization of eggs before the Triassic (causing a preservation bias), or paleontologists not recognizing nest structures, or a combination of these two factors. However, similarities in the eggs

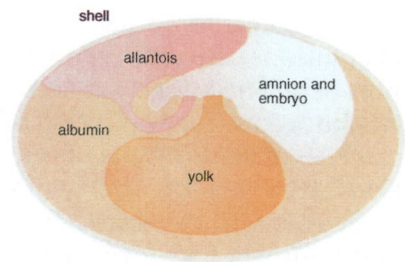

FIGURE 6.4 Components of an amniotic egg, including the eggshell, allantois, yolk sac, amnion, and embryo. Such eggs are a defining character of the clade Amniota, and by extension of dinosaurs. After Cowen (2000), *History of Life*, 3e, Blackwell Science, Inc., Malden, MA, p. 147, fig. 9.12.

of all living amniotes and their close resemblance to fossil eggs argue that this trait is a synapomorphy of amniotes, and it is currently inferred to have evolved just before the skeletal record for amniotes begins.

Despite this lack of evidence, three early species of amniotes in the fossil record are recognized from the Carboniferous Period: *Westlothiana*, *Hylonomus*, and *Paleothyris*. The interpretation of these three small vertebrates as amniotes is based on some anatomical traits distinctive from their amphibian ancestors:

- Dermal bones on the ventral surface of the skull (such as parietals, frontals, and nasals) overlying a bony braincase.
- Reduced head size relative to the overall body size and lightening of the skull.
- Highly modified pelvis consisting of a reinforced pubis and ischium.
- Astragalus and calcaneum in the ankle.

Defining whether some fossils were reptile-like amphibians or amphibian-like reptiles is problematic because of their shared features. Traits of an amniote that differ from that of an amphibian are also more numerous than those listed previously and they summarily reflect adaptations to a terrestrial lifestyle that was increasingly independent of nearby water bodies. As long as aquatic environments were abundant and widespread, amphibians probably did not undergo natural selection that would have favored inheritance of genotypes for sturdier skeletal parts adapted for moving long distances on land away from water.

> *Shared features between amphibian ancestors, such as* Acanthostega, *and amniote descendants are so close in some fossil examples from the Carboniferous that a detailed analysis by an expert anatomist is required before such fossils can be reliably placed in either category.*

So as long as aquatic environments were abundant, the buoyancy of water, which helps to relieve gravitational stresses in a vertebrate skeleton, would have negated selection for a heavier skeleton reinforced for extended periods of locomotion out of water. But with changes in environmental conditions to drier climates or the creation of niches apart from water (such as forests), exaptations or other evolutionary factors favored adaptations of pre-amniotic ancestors toward amniotes. The ability of these non-amniotic ancestors to move about freely on dry land required modifications to their skeletons that supported their weight (that is, a lighter skull, stronger hips, flexible ankles), thus natural selection may have already resulted in amphibians that were divorcing themselves from their dependency on aquatic environments.

The development of an enclosed egg among the descendants of pre-amniotic ancestors was probably the result of natural selection, as only a few eggs (from originally large numbers) had rudimentary membranes enclosing aqueous solutions and prototypes of a yolk sac and **allantois** (respiratory organ for the embryo). Only then would the embryos have survived. Another major evolutionary requirement for the development of amniotic eggs would have been internal fertilization, so sex had to have become more up-close and personal than was previously experienced by amphibians. A few examples of modern amphibians show such a reproductive mode, which means that the same inheritable behavior and anatomical attributes could have been selected in favor of increasing the chances of fertilization. Also, embryos would have been retained within the reproductive tract of the female until a sufficiently protective membrane had developed around them.

The next step in amniotic egg evolution would have been an embryo that underwent growth within the protective membrane to form a miniature version of the adult animal, in contrast to the incompletely developed and intermediate larval (tadpole) stage seen in most amphibians. Although fossil evidence for a sequence of these adaptations is lacking, the presence of many modern amphibian species that retain their eggs within their bodies for long periods of time, especially in times of drought, attests to the feasibility of this evolutionary scenario. Additionally, some modern salamanders, exemplified by **plethodontids**, lay eggs and their embryos develop completely in non-aquatic environments without any larval stage; among their preferred habitats are inside moist tree trunks or logs. Interestingly, skeletal remains of the Carboniferous amniotes *Westlothiana*, *Hylonomus*, and *Paleothyris* were all discovered within Carboniferous fossil tree trunks, so they may have occupied the same niche as modern plethodontids.

Amniota as a clade is synonymous with the older Linnaean (gradistic) classification of Class Reptilia, but with some qualifications. Under gradistic classification, reptiles, such as lizards, snakes, turtles, and crocodiles, were traditionally regarded as scaly vertebrates; most have four legs (except snakes, of course) and reproduce by laying enclosed eggs. This classification excludes mammals and birds, but cladistics recognizes shared derived characters, meaning that amniotes include all descendants from an ancestral amniote. As a result, Amniota, which includes reptiles, mammals, and birds, is a monophyletic clade. In contrast, reptiles actually comprise a **paraphyletic** group, not a clade, because it does not include all of its descendants, such as mammals and birds. The term "reptile" has been long associated only with lizards, snakes, turtles, alligators, crocodiles, pterosaurs, extinct marine reptiles (such as **euryapsids**, discussed later), and dinosaurs, among others. Nevertheless, change is a part of science (Chapter 2) and part of that change is exemplified through new classification schemes. So now most paleontologists recognize that mammals and birds are also "reptiles" in a cladistic sense.

Once amniotes had developed by the Carboniferous Period, their adaptation into numerous terrestrial ecosystems and consequent diversification was relatively rapid and impressive. The major basis for recognition of their diversification is seen in the arrangement of skull bones, specifically the presence and positions of temporal fenestrae. Amniotes can be subdivided into three major clades on this basis and other characteristics – **Anapsida**, **Synapsida**, and **Diapsida** (Fig. 6.5).

Fossil and modern turtles best represent anapsids, which lack temporal fenestrae. The first reptiles had anapsid skulls, thus turtles exhibit a plesiomorphic trait for amniotes in general. Synapsids and diapsids evidently evolved from a common

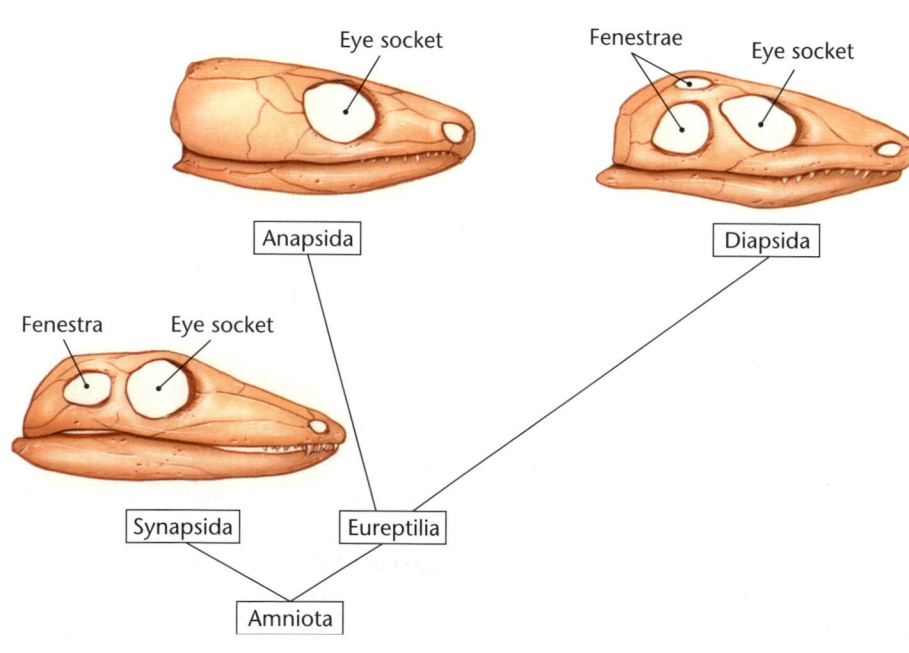

FIGURE 6.5 Three skull types, with positions of temporal fenestra outlined, characterizing the Anapsida, Synapsida, and Diapsida in the context of a cladogram, showing their hypothesized evolutionary relationships.

FIGURE 6.6
Dimetrodon, a Permian synapsid and pelycosaur that was carnivorous, but definitely was not a dinosaur. Denver Museum of Science and Nature, Denver, Colorado.

ancestor of anapsids during the latter part of the Carboniferous, but anapsids and diapsids have been placed in a single clade (Eureptilia) separate from synapsids. Some lineages of synapsids during the Permian included large herbivorous and carnivorous reptiles called **pelycosaurs**. Pelycosaurs had elongated, dorsal vertebral spines that formed sail-like structures, which along with their body size (as long as 3 meters) gave them a formidable appearance that understandably resulted in their popularized but mistaken grouping with dinosaurs (Fig. 6.6). However, synapsids also included lineages that later evolved into **therapsids**, which had some mammal-like characters, and eventually mammals. This means that pelycosaurs are actually more closely related to humans and other mammals than they are to dinosaurs. Mammals are appropriately placed in Mammalia and first show up in the fossil record, at about the same time as the first known dinosaurs, during the Late Triassic.

Diapsida is the clade most pertinent to the discussion of dinosaurs. Diapsids split into two clades, the **Lepidosauria** and **Archosauria**, a divergence of lineages that probably happened during the Permian Period. Lepidosaurs are modern lizards, which includes skinks, geckoes, iguanas, Komodo dragons, and their ancestors. A common misconception about large reptiles, such as alligators and crocodiles, is that they are closely related to lizards such as Komodo dragons, but they are

FIGURE 6.7 *Thalassomedon*, a Late Cretaceous plesiosaur, a marine reptile and an example of a euryapsid. (Euryapsids, and all marine reptiles, were not dinosaurs.) Denver Museum of Science and Nature, Denver, Colorado.

phylogenetically separate, as explained later. Snakes are also lepidosaurs because they share derived characters with lizard ancestors; they even show vestigial pelvic bones. The oldest known snakes in the geologic record are from the Early Cretaceous, thus both lizards and snakes co-existed with dinosaurs during at least part of the Mesozoic (in fact, at least one dinosaur ate a lizard: Chapter 9), and both groups were very successful in later diversification throughout the Cenozoic after the demise of the dinosaurs.

Euryapsids, mentioned previously, are also placed in Lepidosauria because of their inferred common descent from lizard-like ancestors, although they branched into a previously unexplored niche for reptiles, the seas. These diverse, abundant, and often large-bodied marine reptiles of the Mesozoic include the **ichthyosaurs**, **plesiosaurs** (Fig. 6.7), and **mosasaurs**. Among them were the first vertebrates known to have been viviparous, as shown by a few stunning fossil examples of mother ichthyosaurs with their stillborn young. These fascinating and complex reptiles, like many other vertebrates of the Mesozoic, became extinct by the end of the Cretaceous (Chapter 16). They are sometimes confused with dinosaurs because they were contemporaries and overlapped in size with some of the larger dinosaurs. However, dinosaurs were not only anatomically distinct from euryapsids, they were effectively relegated to completely different environments and niches. Probably the only interactions between these reptilian groups occurred when dinosaur carcasses floated out to sea and were scavenged by euryapsids (Chapter 7).

Archosaur Evolution and Diversification

The Archosauria is defined as having, at minimum, the following characteristics:

> *Archosauria is the clade often associated with the origin of the dinosaurs.*

- Openings anterior to the orbits (antorbital fenestrae).
- Teeth with serrations compressed laterally and none on the palate.
- Dentary fenestrae.
- Differently shaped calcaneum.
- Elongated ilium and pubis.

Some paleontologists place Archosauria within the clades Archosauromorpha and Archosauriformes, the latter originating from the former (Chapter 5). The majority of paleontologists agree upon the designation of Archosauria as a clade that had arrived by the Early Triassic, with members that evolved into lineages, both dinosaurian and otherwise. A group of fossil reptiles, known previously by paleontologists as "**thecodonts**," was once considered as synonymous with the archosaur group that gave rise to the dinosaurs, crocodilians, and birds. However, cladistic analyses show that thecodonts make up a paraphyletic grouping (such as Reptilia), hence its use as a term is now discouraged in phylogenetic classifications. However, it is commonly mentioned in older literature and represents changing ideas in science.

A likely representative fossil for a common ancestor of the archosaurs is the Early Triassic *Euparkeria* of South Africa (Fig. 6.8). *Euparkeria* was a small (about 1 meter long) but relatively long-limbed reptile that possessed antorbital fenestrae, a key feature of all archosaurs. Clades within the Archosauria, which seemingly descended from ancestors like *Euparkeria*, are the **Crurotarsi** and **Ornithodira**. Crurotarsi includes living crocodilians (alligators and crocodiles), but it encompasses many diverse fossil forms as well. An ankle where the astragalus and calcaneum form a joint between the tarsals and lower part of the limb bones characterizes this clade. Crurotarsans were well-represented during the Middle and Late Triassic by large, crocodile-like carnivorous **parasuchids** (also known as **phytosaurs**) and **rauisuchians** (Fig. 6.9), as well as the armored and herbivorous **aetosaurs**. Rauisuchians were unique among large archosaurs at the time because their forelimbs were considerably shorter than their hind limbs, which suggests that they were capable of walking on two legs. They may have been among the first such archosaurs to evolve this mode of locomotion.

Despite their abundance and success, all species of phytosaurs, rauisuchians, and aetosaurs became extinct by the end of the Triassic. However, by the Late Triassic, ornithodirans had diverged into two clades: **Pterosauria** and **Dinosauria**. Pterosaurs, the so-called "flying reptiles," were among the most famous of the

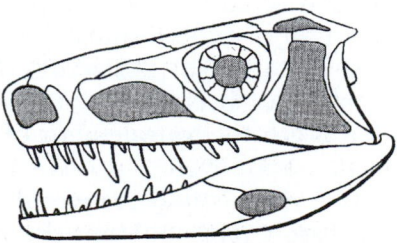

FIGURE 6.8 Skull of *Euparkeria*, a basal archosaur from the Early Triassic of South Africa, which was not a dinosaur. From Cowen (2000), *History of Life*, 3e, Blackwell Science, Inc., Malden, MA, p. 182, fig. 11.13.

6

FIGURE 6.9 Cast of *Postosuchus*, a large rauisuchian from the Late Triassic of the southwestern USA: Mesalands Dinosaur Museum, Tucumcari, New Mexico. Despite its very fierce appearance, *Postosuchus* was not a dinosaur.

terrestrial creatures during the Mesozoic, a notoriety related to the interpretation that they were the first vertebrates known to have achieved self-powered flight. Pterosaurs developed a remarkable adaptation whereby digit IV of each forelimb extended far beyond the other digits and had a membrane attached from its distal end to the torso to form a wing (Fig. 6.10). During their time on Earth, which correlated with and was nearly as long as the geologic range of the dinosaurs, pterosaurs evolved into forms as small as a sparrow to the largest animals that ever flew. For example, the Late Cretaceous pterosaur *Quetzalcoatlus* had a wingspan of about 12 meters as it soared over what is now the state of Texas. But in spite of their repute, grandeur, and chronological association with dinosaurs, the pterosaurs, like many of the other amniotic vertebrates mentioned in this chapter, are still not defined as dinosaurs, although they may have been the closest to having a common ancestor.

Dinosaur Ancestors and the Origins of Dinosaurs

Fame as a dinosaur paleontologist certainly would be justified for anyone who discovered the skeletal remains of the first dinosaurs. However, as shown by the discussion of amniote development, the phrase "first dinosaur" is in itself arguable in the light of evolutionary theory. After all, geologic ranges for fossil lineages are not always static. The possibility that fossils for ancestors of a hypothetical lineage have simply not been discovered yet leads to the concept of **ghost lineages**, meaning that a greater complement of ancestor and descendant species may still be locked away undiscovered in rocks somewhere in the world. For now, paleontologists can define the geologic range of dinosaurs as 228 to 65 Ma (Late Triassic through to the Late Cretaceous) on the basis of discovered specimens, but an understanding

FIGURE 6.10 Cast of the pterosaur *Anhanguera* from the Early Cretaceous of Argentina: Fernbank Museum of Natural History, Atlanta, Georgia. *Anhanguera*, alas, was also not a dinosaur.

of evolutionary theory allows for extrapolating a greater range represented by ghost lineages. Nonetheless, dinosaur remains discovered from Early Triassic rocks would be an extremely significant find, and similar body fossils from Permian rocks would be completely unexpected.

Tracks would be considerably less convincing evidence than skeletal remains for the first dinosaurs or their immediate ancestors, despite the valuable information potentially conveyed by such a find (Chapter 14). Even more suspect evidence would be eggs and nests, minus accompanying skeletal material (Chapter 8). Coprolites, gastroliths, and toothmarks attributable to the first dinosaurs would probably warrant the most skepticism because of the current lack of firm identity attached to such trace fossils (Chapter 14). Consequently, the origin of the first dinosaurs can only be postulated on the basis of skeletal evidence and the stratigraphic position of this evidence, although other indicators or supporting evidence of their existence is possible through trace fossils. The problem with a trace fossil approach for finding evidence of dinosaur ancestors is threefold:

1 trace fossils could have been made by tracemakers that had a similar morphology to the first dinosaurs but may have been distantly-related archosaurs;
2 the criteria for what constitutes a dinosaur in the fossil record is currently based on anatomical criteria; and
3 most dinosaur paleontologists have limited their studies to bones and have not looked for trace fossil evidence.

FIGURE 6.11 Cast of the small dinosauromorph *Marasuchus* from the Late Triassic of Argentina: Sam Noble Oklahoma Museum of Natural History, Norman, Oklahoma. *Marasuchus* is not a dinosaur, but is very, very close to being one. Length about 40 cm.

As a result, the body fossil record for dinosaur ancestors is currently considered to be the primary basis for phylogenetic reconstructions of dinosaur lineages.

Based on known lineages of archosaurs before the oldest known dinosaurs found in the geologic record and their anatomical traits, a prediction of the ancestral archosaur, the "mother of all dinosaurs," can be made. This hypothetical ancestor would have had, at a minimum, the following traits distinctive from other diapsids:

- Bipedal, with long hind limbs relative to the forelimbs.
- Four or five digits on its manus, with digits IV and V reduced in size.
- Long metatarsals and phalanges on its pes.
- Ankle with a hinge developed between the astragalus and calcaneum.
- A tibia–fibula length greater than the femur.

Of fossil finds so far, those closest to this ancestor are *Marasuchus* (Fig. 6.11), synonymous with *Lagosuchus* in some studies, and *Lagerpeton*, which are small but long-limbed reptiles occurring in the Middle Triassic strata of Argentina. *Marasuchus* and *Lagerpeton* were among the first ornithodirans, and their successors could have diverged into either pterosaur or dinosaur lineages. Additionally, small three-toed footprints documented from Early and Middle Triassic strata may be associated with ornithodiran tracemakers that preceded or were contemporaneous with the afore-mentioned species represented by body fossils.

These possible ancestral forms are succeeded in the geologic record by what are considered by many paleontologists as the earliest known dinosaurs: *Eoraptor lunensis* and *Herrerasaurus ischigualastensis* from the Ischigualasto Formation of Argentina, as well as *Staurikosaurus pricei* from the Santa Maria Formation of Brazil (Fig. 6.12). All three of these specimens are from the earliest part of the Late Triassic (Carnian Age, which was about 221 to 228 Ma); radiometric age dates of $^{40}Ar/^{39}Ar$

FIGURE 6.12 Three Late Triassic fossil archosaurs proposed as primitive dinosaurs. (A) *Eoraptor lunensis*. (B) *Herrerasaurus ischigualasto*. (C) *Staurikosaurus pricei*. Modified from Paul (1988), Sereno et al. (1993), and Sereno (1994).

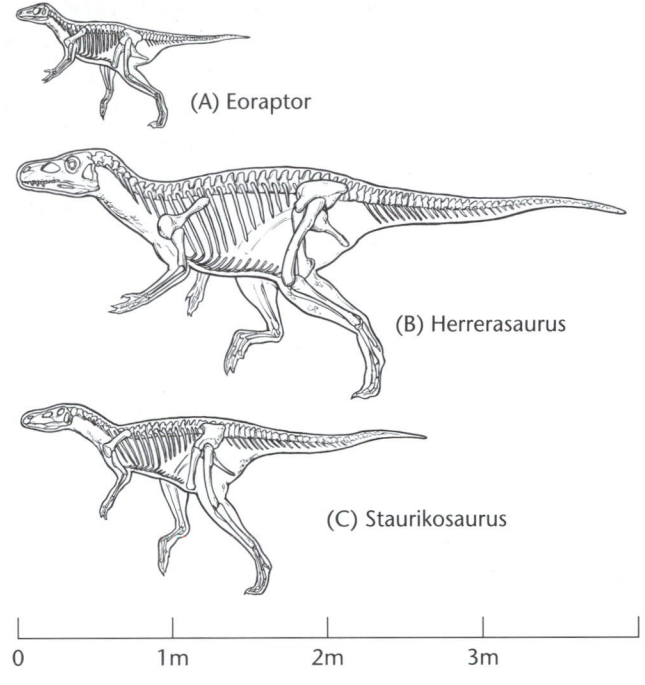

(A) Eoraptor

(B) Herrerasaurus

(C) Staurikosaurus

| 0 | 1m | 2m | 3m |

from mineral grains gave a minimum age of 227.8 ± 0.3 Ma for the Ischigualasto Formation, which contains the first two species. Of the three species, *Eoraptor* seems to have the most primitive traits, which has led to controversy over whether it actually is a dinosaur. Like many dinosaur species, it is only known from a single specimen. It also was small (about 1 meter long) in comparison to its immediate successors, as well as members of the entire clade of Dinosauria. The other two species, which collectively are represented by more than a dozen specimens, are placed within the clade Herrerasauridae, which is also occupied by the geologically slightly younger *Chindesaurus bryansmalli* of North America. All of these so-called "basal" dinosaurs are regarded as saurischians, and most paleontologists think that they are closely allied with theropods. Interestingly, the lack of agreement on their exact classification probably reflects their basal status.

The only definite ornithischian dinosaur discovered from strata of an age near the apparent "birth of the dinosaurs" is *Pisanosaurus mertii*, which is also from the Ischigualasto Formation of Argentina. Nearly contemporary with the herrerasaurs and *Pisanosaurus* in the Late Triassic was one other saurischian, the prosauropod *Azendohsaurus* from Morocco, and the ornithischian *Technosaurus* from the western USA. Only partial and fragmentary specimens represent both species; in fact, *Azendohsaurus* is only interpreted as a prosauropod on the basis of a single tooth. Hence, these dinosaurs have little to tell us about non-theropod and ornithischian evolution during the Late Triassic. The monophyletic grouping of dinosaurs, which was challenged by Harry Govier Seeley (Chapter 3) through his division of dinosaurs into the Saurischia and Ornithischia, is upheld by synapomorphies of both clades, but an immediate common ancestor for both has yet to be found. For ornithischians in particular, paleontologists have so far only found abundant and well-preserved representatives of this clade beginning in Early Jurassic strata, meaning that:

1 ornithischians were uncommon during the Late Triassic;
2 taphonomic factors prevented their preservation in Late Triassic sediments; or
3 paleontologists are looking in the wrong places for them.

This situation presents a minor evolutionary dilemma for dinosaur paleontologists: if saurischians and ornithischians were already contemporaries during the Late Triassic, when did they diverge from a common dinosaurian ancestor? The most likely answer lies in the Middle Triassic, but unfortunately the body fossil record has not been helpful in this respect. *Marasuchus* is a probable common ancestor for saurischians and ornithischians, showing characteristics (long femur, possibly open acetabulum, long metatarsals) that suggest bipedal adaptations and classification in the clade Dinosauromorpha. However, it was also proposed as a possible common ancestor for pterosaurs and dinosaurs, which would have placed it closer to the node for the clade Ornithodira. Other than *Marasuchus* and perhaps *Lagerpeton*, the other fossils are so poorly preserved or of such uncertain affinity that any declaration of a common ancestor probably would be premature.

Trace fossil data, in the form of distinctive dinosaurian tracks, provide some clues about possible dinosaur ancestors in the Middle Triassic, and some tracks found in strata of this age are similar to undoubted dinosaur tracks. Because the current anatomical evidence related to dinosaur ancestry strongly suggests that primitive dinosaurs were obligate bipeds and developed a pes with three prominent toes, they should have made tracks reflecting this bipedalism. Furthermore, they should have had a track shape that is easily distinguishable from their four-legged predecessors, as well as from other potentially bipedal archosaurs, such as rauisuchians. For example, a reduction in the number of elongated toes seen in theropod tracks from four to three (accompanying a reduction of the hallux) is predicted for the Middle to Late Triassic transition, based on evolutionary changes reflected by the skeletal record. Considering the abundance of tracks left by a living, mobile animal (maybe thousands) versus its body (one), there should be many examples of dinosaur tracks from soon after the time that they evolved (Chapter 14). However, paleontologists who are skeptical about the identity of trackmakers may first insist on correlation of footprint morphology with appendicular skeletons of known tracemakers in same-age strata. Only then might they agree that such evidence is indicative of dinosaur ancestors or dinosaurs themselves in the Middle Triassic.

Once they evolved, dinosaurs rapidly filled niches in their terrestrial environments during the Late Triassic. The extinction of large, abundant, and diverse archosaurs by the end of the Triassic coincided with the increased diversification and abundance of dinosaurs, which is reflected by their body fossils and tracks. Additionally, a prosauropod nest with eggs interpreted from the Late Triassic of Argentina indicates that dinosaurs were already reproducing in ways familiar to paleontologists who have made similar finds in Jurassic and Cretaceous rocks (Chapter 8). The change in archosaurian faunas was originally interpreted as a result of **interspecific** (between species) competition, where dinosaurian domination over other terrestrial vertebrates was ensured by their upright stance and increased speed associated with bipedalism. However, an extinction event that affected some eureptilian groups and herrarasaurids alike near the beginning of the Late Triassic (at about 225 Ma) suggests that other factors, such as environmental change, were more likely contributors to dinosaurian hegemony, which was clearly in place by the end of the Late Triassic (about 206 Ma: Chapter 16).

Possible Genetic and Environmental Causes for the Origin of Dinosaurs

For dinosaurs to have evolved from archosaur ancestors, a combination of genetic and environmental factors had to combine in just the right way to result in the fossil forms that we define as dinosaurs at least 230 million years after. Before considering what factors may have influenced the origin of dinosaurs, a review of some character traits that define dinosaurs (Chapters 1 and 5) provide a framework for how the traits reflect adaptations, which will be revisited later:

- Three or more sacral vertebrae.
- Shoulder girdle with backward-facing (caudally pointing) glenoid.
- Asymmetrical manus with less than or equal to three phalanges on digit IV.
- Acetabulum with open medial wall.
- Tibia with cnemial crest.
- Astragalus with a long ascending process that fits into the anterior part of the tibia.
- Sigmoidally shaped third metatarsal.
- Postfrontal absent.
- Humerus with long deltopectoral crest.
- Femur with ball-like head on proximal end.

Eight of these ten traits are related to modification of the appendicular skeleton that shows adaptations to bipedalism, a mode of life well-suited to nearly all subsequent theropods (Chapter 9). Discussion of the early evolution of dinosaurs should therefore focus on these adaptations, which occurred through an interaction of genetic and environmental factors.

Probably the most difficult task in figuring out dinosaur origins is evaluating the genetic factors that contributed to evolution of the characteristic traits. Nearly as difficult is discerning the environmental factors that affected a selection of these same traits. Geneticists and ecologists have problems in defining the interactions of modern populations, their genetics, and ecosystems, so why should understanding the Mesozoic be any easier? Fortunately, the skeletal record for dinosaurs and their ancestors, along with their associated geologic information, provide enough clues that a general hypothesis for the origin of dinosaurs has been proposed, tested, and refined with new information and insights.

Through cladistic analyses of Early, Middle, and Late Triassic archosaurs, the probable genetic relationships between different fossils have been well established, although cladograms are often modified with the discovery of each new fossil species or re-interpretations of previously described species. Genetic relationships between Triassic archosaurs are based on phenotypes as reflected by skeletal features interpreted as synapomorphies (Chapter 5). However, some paleontologists will acknowledge that a single specimen of a fossil species may be unrepresentative of most phenotypes in its species at that particular slice in time. Uncertainty is inevitable because some features in a body fossil may be acquired characteristics, and thus not representative of an organism's genome.

Nonetheless, the regularity and predictable occurrence of most features in a body fossil, testable through discovery of multiple specimens of a presumed species, provides a valid reason for assuming that these features are indeed reflecting inheritable traits. Such traits can be as simple as, for example, four limbs. We can safely assume that a fossil tetrapod showing four limbs does not represent a mutation inherited from an ancestor that normally had three limbs. A close examination of changes in details of the anatomy reveals what changes occurred in lineages through time, such as synapomorphies documented for typical Triassic archosaur traits – hind limbs lengthening more than fore limbs, reduction of digits IV and V, elongation of metatarsals and phalanges on the pes, etc.

Because synapomorphies are assumed in the majority of cases as representative of an archosaur's genome, morphological variations within an archosaur lineage also can be interpreted on the basis of how these features may be similar or different in time-equivalent strata. For archosaurs, this interpretation obviously depends on the sample number of specimens and the completeness of the individual specimens; hence taphonomy (Chapter 7) sets conditions on interpretations once again. Excellent examples of dinosaur species, which were abundantly preserved so that genetic variation in a population can be estimated, are provided by two Late Triassic dinosaurs, the theropod *Coelophysis* (Chapter 9) and prosauropod *Plateosaurus* (Chapter 10). In these two species, many individuals can reflect a population structure, especially if found in the same locality and deposit. Indeed, for *Coelophysis*, a proposed population structure includes juveniles to adults, although in this case too much data can mean more complications, because some of the adult variations may actually be attributable to sexual dimorphism (Chapters 8 and 9). Nonetheless, statistical descriptions of the population provide a sample of at least a few parameters of the original gene pools for *Coelophysis* and *Plateosaurus*.

As mentioned earlier, biogeography is a key facet of evolutionary theory because a close proximity of similar species is additional evidence suggesting their relatedness. The same applies to dinosaurs – because of the abundance of Late Triassic dinosaurs and their immediate ancestors in South America, the origin of dinosaurs is currently attributed to that continent, which was part of Gondwana during the Late Triassic (Chapter 4). Assuming that this was the general location for the birth of dinosaurs, the split between saurischians and ornithischians also may have happened in this area, probably about 230 Ma or slightly earlier.

Why this divergence occurred and why it was so rapid, geologically speaking, are both good questions. Because so little evidence exists for fossils showing intermediate features between dinosaur ancestors and basal dinosaurs, paleontologists hesitate to state whether this apparently rapid evolution was a result of:

1 phyletic gradualism that is simply missing parts of the lineages;
2 punctuated equilibrium; or
3 some combination of the two.

One form of natural selection invoked for dinosaur evolution, as a type of Red Queen hypothesis, is that early dinosaurs successfully competed with other archosaurs for habitats and resources throughout the Triassic, which eventually resulted in crurotarsans becoming extinct and ornithodirans (including dinosaurs) thriving by the end of the Triassic. However, some paleontologists doubt this hypothesis because re-examination of the archosaur fossil record does not show gradual inverse trends between dinosaur abundance and demises of other archosaurs. As mentioned previously, Late Triassic extinctions of archosaurs, other than dinosaurs, began before the end of the period, meaning that they may have encountered many different and changing environmental factors that selected against their survival.

Some evidence of environmental change and its effects on biota during the Late Triassic is indicated by extinctions of marine invertebrate organisms about 220 Ma, which coincided with the beginning of the breakup of Pangea (Chapter 4). Divergence of the continental masses from this supercontinent would have caused gradual changes in oceanic and atmospheric circulation patterns, which not only would have affected marine habitats but overall global climate. **Climate** is often synonomized with **weather**, but they differ considerably in their time frames. Climate is persistent long-term trends and patterns of weather, whereas weather is daily, short-term changes in atmospheric conditions. For example, if most of the years in a million-year period had low amounts of rainfall in an area, these data would allow for defining an arid climate for that area.

At any rate, climate affects evolutionary processes, in particular natural selection, and climate did indeed change during the Late Triassic. These changes are indicated by Late Triassic **evaporite deposits**, which are thick accumulations of minerals such as halite (NaCl) and gypsum ($CaSO_4 \times 2H_2O$) that form in sedimentary basins over long periods of time under predominantly arid conditions. A hypothesized effect of arid climates is that terrestrial plant communities, through natural selection, would have adapted so that drought-resistant species should have become more common. A change in plant communities meant that herbivores would have had to adapt to new food sources, and those species that could not adapt would become extinct. Likewise, carnivorous species that preyed upon the maladapted herbivores then also would have gone extinct, in the sense of an ecological domino effect (discussed in Chapter 16). Sure enough, this change in plant communities in accordance with the onset of arid climates has been observed with fossil plants from the Late Triassic, which in turn corresponds with faunal changes, justifying a cause-and-effect hypothesis.

Other than changes in climate, another possible consequence of Pangea breaking up in the Late Triassic was habitat fragmentation, which would have caused geographic isolation of dinosaur and other archosaur faunas, translating into conditions favorable for allopatric speciation and adaptive radiation that was perhaps facilitated through genetic drift. As mentioned earlier, diversification of fossil faunas seems to correspond with times of continental breakup throughout the Phanerozoic Eon, a correlation that is attributed to the formation of new habitats. Consequently, new niches also should have opened up for species that had the genetic capability to adapt. Dinosaurs certainly represented novel adaptations in archosaur lineages during the Late Triassic, which is perhaps related to their fitting into new niches caused by continental rifting and the emptying of those niches by extinct archosaurs. So rather than dinosaurs "out-competing" other archosaurs, they may have simply replaced them.

The worldwide dispersal of dinosaur faunas by the end of the Triassic, within 25 million years of their origin, is remarkable in itself, but other aspects of dinosaurs in the latter part of the Late Triassic argue for how they had already made their mark on the world. Three trends in particular are notable:

1 increased body size (corroborated by larger dinosaur tracks in same-age strata as larger dinosaurs);
2 increased number as a percent composition of terrestrial vertebrates; and
3 increased diversity with time.

Within those 25 million years, saurischians in particular increased in size, from the 1-meter long *Eoraptor* to the 11-meter long prosauropod *Riojasaurus* of Argentina. Meanwhile, dinosaurs went from about 6% of terrestrial amniote species to as much as 60%. The abundance of some dinosaur species is worth mentioning; interestingly, the dinosaur species most abundantly represented in the fossil record is the Late Triassic *Coelophysis*, but most dinosaurs (such as *Apatosaurus*, *Tyrannosaurus*, *Stegosaurus*, and *Triceratops*) lived much later in geologic time. Paleontological information suggests that *Coelophysis* was not only abundant, but had already developed social behavior, traveling together in large groups (Chapter 9). Likewise, *Plateosaurus* and other prosauropods represented a pinnacle of herbivore evolution in body size by the Late Triassic, anticipating the sauropod leviathans that would emanate from their common ancestors later in the Mesozoic (Chapter 10). The diversity of dinosaurs is indicated by the large number of species described from strata formed toward the end of the Late Triassic in comparison to the few species known from the beginning of the Late Triassic (Table 6.2).

Dinosaurs also survived one of the most well-documented mass extinctions in the geologic record, an extinction that eliminated all aetosaurs, phytosaurs, rauisuchians, and some other formerly successful archosaurs by the end of the Triassic. Compatriots of the early dinosaurs that survived this mass extinction included some euryapsids, anapsids, pterosaurs, and mammals. Various hypotheses proposed

TABLE 6.2 Dinosaur genera of the Late Triassic: (A) earlier part of the Late Triassic, Carnian Age (228–221 Ma), (B) latter part of the Late Triassic, Norian through Rhaetian Ages (221–206 Ma). Notice that some genera (*Blikanasaurus, Euskelosaurus, Melanorosaurus*) span both intervals, but these genera do not extend into the Rhaetian.

GENUS AND CLADE	PLACE OF DISCOVERY
A. Carnian Dinosaurs	
Saurischians	
Eoraptor (?Theropoda)	Argentina
Staurikosaurus (Theropoda)	Brazil
Herrerasaurus (Theropoda)	Argentina
Azendohsaurus (Sauropodomorpha)	Morocco
Blikanasaurus (Sauropodomorpha)	South Africa
Euskelosaurus (Sauropodomorpha)	South Africa
Melanorosaurus (Sauropodomorpha)	South Africa
Saturnalia (Sauropodomorpha)	Brazil
Ornithischians	
Pisanosaurus (Uncertain)	Argentina
Technosaurus (Uncertain)	Western USA
B. Norian–Rhaetian Dinosaurs	
Saurischians	
Antetonitrus (Sauropodomorpha)	South Africa
Coelophysis (Theropoda)	Western USA
Chindesaurus (Theropoda)	Western USA
Liliensternus (Theropoda)	Germany
Syntarsus (Theropoda)	Zimbabwe, western USA
Blikanasaurus (Sauropodomorpha)	South Africa
Camelotia (Sauropodomorpha)	England
Coloradisaurus (Sauropodomorpha)	Argentina
Euskelosaurus (Sauropodomorpha)	South Africa
Melanorosaurus (Sauropodomorpha)	South Africa
Mussasaurus (Sauropodomorpha)	Argentina
Plateosaurus (Sauropodomorpha)	Europe (Germany, France, Switzerland)
Riojasaurus (Sauropodomorpha)	Argentina
Sellosaurus (Sauropodomorpha)	Germany
Thecodontosaurus (Sauropodomorpha)	England
Ornithischians	
None known.	

for causes of this mass extinction (one of six indicated by the geologic record) include:

1 interspecific competition;
2 changing climates;
3 habitat fragmentation from the continued breakup of Pangea; and
4 a meteorite impact.

All of these factors are similar to those implicated in the downfall of dinosaurs at the end of the Cretaceous Period (Chapter 16).

The survival of dinosaurs through what must have been a significant change in global ecosystems in the Late Triassic indicates that dinosaurs may have had exaptations that gave them evolutionary advantages, despite whatever factors (genetic or environmental) might have eliminated other species. A similar probability of exaptations in bird lineages, which most likely evolved out of theropods during the Jurassic Period (Chapters 9 and 15), must have allowed some of them to survive the extinction at the end of the Cretaceous.

The reason for dinosaur survival through a major extinction and their subsequent worldwide dominance of terrestrial faunas is that they had the right genetic makeup for adaptations to new niches and consequent diversification in environments of the forthcoming Jurassic and Cretaceous Periods. The 140-million-year span of the Jurassic and Cretaceous, often hailed as the reign of the dinosaurs, thus followed the foundation of an already diverse and successful Late Triassic dinosaurian fauna. Dinosaurs were, and still are, by-products of an evolutionary process that continues today.

6

SUMMARY

Evolution is both a fact and a theory, in that the change in a population between generations of species has been observed, but the explanation for how this process happens is still evolving. Darwin provided the first unified explanation for the origin of species and descent with modification of organisms, although his hypothesis has changed considerably with more fossil discoveries during the past 150 years as well as the addition of Mendelian genetics, which was further elaborated through the study of population genetics. Nevertheless, the basic tenets of Darwinian theory (natural selection through inheritable variations, overpopulation, struggle for existence, and survival of the better-adapted) are still applicable to understanding how environmental and intraspecific factors change genotype frequencies and phenotypes, causing speciation. Speciation that happened over longer periods of time and caused considerable changes within lineages (macroevolution) is attributed to either phyletic gradualism or punctuated equilibrium, depending on the timing of the changes. Evidence supporting both of these hypotheses comes from the fossil record, with possible mechanisms of allopatric and sympatric speciation, adaptive radiation, and vicariance biogeography, among others. Molecular phylogeny is an independent method used more in recent years to test phylogenetic relationships established through fossil lineages, although cladistic analyses based on characteristic traits still determine hypotheses for how dinosaurs evolved.

Dinosaur evolution can be evaluated by examining the fossil record for amniotes as a whole and archosaurs in particular. The development of a cleidoic egg from amphibian ancestors probably happened during the Carboniferous Period as a result of both genetic and environmental factors that favored this mode of reproduction. Subsequent diversification

SUMMARY Continued

of amniotes resulted in the origin of anapsids, synapsids, and diapsids well before the end of the Permian; diapsids gave rise to archosaurs, which proliferated throughout the Triassic. Among the archosaur lineages was the ornithodirans, which include dinosaurs and their contemporaries, the pterosaurs. Dinosaurs probably originated during the Middle Triassic, as suggested by both body and trace fossil evidence, and their increased diversification and abundance developed rapidly within the last 25 million years of the Late Triassic, particularly for saurischians (theropods and prosauropods). Although genetic factors were certainly involved, the diversification may have been prompted by the opening of ecological niches left by other archosaurs (such as rauisuchians, phytosaurs, and aetosaurs) that went extinct toward the end of the Triassic. Additionally, the onset of arid climates and the beginning of continental rifting in the supercontinent Pangea through the same time span may have contributed to changes in the roles of plants, herbivores, and carnivores in terrestrial ecosystems. Regardless of the exact evolutionary mechanisms responsible for their ascendancy, dinosaurs had become the dominant vertebrates by the end of the Triassic and they would have a magnificent reign that would last for the next 140 million years, until the end of the Cretaceous.

DISCUSSION QUESTIONS

1. Some evolutionary biologists define a species as a "closed gene pool." Justify this description or criticize it on the basis of definitions given in the chapter for a "species" and "gene pool."
2. What is the difference between "frequency" and "probability"? How are these statistical expressions related to population genetics?
3. A hypothetical female theropod with a homozygous recessive gene for a reduced hallux mated with a male theropod with a heterozygous condition that has a normal-sized hallux as the dominant trait.
   a. What was the probability of their offspring having the phenotype of a reduced hallux?
   b. What is the probability for the next generation (the offspring of the offspring) having the phenotype of a reduced hallux?
   c. What is the assumption of the preceding probabilities? In other words, what factors could change the expected gene frequencies?
4. Out of all of your friends, think about how representative one of them might be for the phenotype of *Homo sapiens* if he or she was randomly picked as a "type specimen."

DISCUSSION QUESTIONS Continued

   a. How much variation would your friend have in his or her living appearance?

   b. Out of those observed variations, what inheritable features do you think would be evident in the fossil record that might define them as typical of your species?

   c. What acquired features do you think would be unrepresentative of their genome and thus would be a source of confusion for paleontologists of the future (whatever their species might be)? For example, do they have dyed hair, tattoos, piercings, or other modifications?

5. Explain how the evolutionary development of a cledioic egg for amniotes could have occurred through the following models:

   a. Allopatric speciation

   b. Sympatric speciation

   c. Phyletic gradualism

   d. Punctuated equilibrium

   e. Character displacement

   f. Red Queen

6. Based on the information presented in the chapter, make your own cladogram showing the ancestry of the following modern reptile ... turtles, snakes, lizards, crocodiles, and alligators. Which pair ... these five groups seems to be the most related and which ... ms the least related?

7. ...rall average height of humans has increased in the past 1000 ...ased on measurements of skeletons from that time span as ...data taken from living people. Is this increase in height an ... of directional selection (Cope's Rule)? Why or why not?

8. ...he primitive dinosaur traits of bipedalism, think about the ...g:

   a. ...ow could natural selection have caused some descendants of the first dinosaurs, such as prosauropods, to go to quadrupedalism as a mode of locomotion?

   b. What are some environmental factors that might have favored quadrupedal postures? What evidence in the geologic record would be needed to corroborate your hypotheses?

   c. What are some possible intraspecific factors that might have caused sympatric speciation in such a direction? What evidence in the geologic record would be needed to corroborate your hypotheses?

9. Of the amniotes mentioned in the chapter, which ones did you mistakenly think were dinosaurs before reading this book? How would you go about convincing someone else that these animals were not dinosaurs?

10. How could a meteorite impact have caused problems for ecosystems during the Late Triassic? List some of the effects of an impact that are unlike the more gradual changes that might have been caused by continental rifting.

Bibliography

Benton, M. J. 1983. Dinosaur success in the Triassic: A noncompetitive ecological model. *Quarterly Review of Biology* **58**: 29–55.

Benton, M. J. 1990. "Origin and interrelationships of dinosaurs". *In* Weishampel, D. B., Dodson, P. and Osmólska, H. (Eds), *The Dinosauria*. Berkeley, California: University of California Press. pp. 11–30.

Benton, M. J. 1993. Late Triassic extinctions and the origin of the dinosaurs. *Science* **260**: 769–770.

Bonaparte, J. F. 1975. Nuevos materiales de *Lagosuchus talampayensis* Romer (Thecodontia-Psuedosuchia) y su significado en el origen de los Saurischia. Chañarense inferior, Triásico medio de Argentina. *Acta Geologica Lilloana* **13**: 5–90.

Bonaparte, J. F. and Vincent, M. 1979. El hallazgo del primer nido de dinosaurios triásicos (Saurischia Prosauropoda), Triásico superior de Patagonia, Argentina. *Ameghiana* **16**:173–182.

Callaghan, C. A. 1987. Instances of observed speciation. *The American Biology Teacher* **49**: 3436.

Carroll, R. L. 1988. *Vertebrate Paleontology and Evolution*. New York: W. H. Freeman.

Chadwick, D. J. and Goode, J. (Eds). 1997. *Antibiotic Resistance: Origins, Evolution, Selection and Spread*. New York: John Wiley & Sons.

Charig, A. 1984. Competition between therapsids and archosaurs during the Triassic Period: a review and synthesis of current series. *Symposia of the Zoological Society of London* **52**: 597–628.

Colbert, E. H. 1970. A saurischian dinosaur from the Triassic of Brazil. *American Museum Novitates* **2405**: 1–9.

Colbert, E. H. and Morales, M. 1991. *Evolution of the Vertebrates: A History of the Backboned Animals through Time*. New York: John Wiley and Sons.

Cruickshank, A. R. I. and Benton, M. J. 1985. Archosaur ankles and the relationships of the thecodontian and dinosaurian reptiles. *Nature* **317**: 715–717.

Darwin, C. R. 1839. *Journal of Researches into the Geology and Natural History of the Various Countries Visited by the* H.M.S. Beagle. London: Henry Colburn.

Darwin, C. R. 1859. *On the Origin of the Species by Means of Natural Selection*. London: John Murray.

Demathieu, G. R. 1989. "Appearance of the first dinosaur tracks in the French Middle Triassic and their probable significance". *In* Gillette, D. D. and Lockley, M.. (Eds), *Dinosaur Tracks and Traces*. Cambridge, U.K.: Cambridge University Press. pp. 201–207.

El-Tabakh, M., Riccioni, R. and Schreiber, B. C. 1997. Evolution of late Triassic rift basin evaporites (Passaic Formation): Newark Basin, eastern North America. *Sedimentology* **44**: 767–790.

Ewer, R. F. 1965. The anatomy of the thecodont reptile *Euparkeria capensis* Broom. *Philosophical Transactions of Royal Society London B* **248**: 379–435.

Flynn, J. J., Whatley, R. L., Wyss, A. R., Parrish, J. M., Rakotosamimanana, B. and Simpson W. F. 1999. A Triassic fauna from Madagascar, including early dinosaurs. *Science* **286**: 763–765.

Galton, P. M. 1977. On *Staurikosaurus pricei*, an early saurischian dinosaur from Brazil, with notes on the Herrerasauridae and Poposauridae. *Paläontol. Z.* **51**: 234–245.

Galton, P. M. 1986. "Herbivorous adaptations of Late Triassic and Early Jurassic dinosaurs". *In*. Padian K. (Ed.), *The Beginning of the age of Dinosaurs*, Cambridge, U.K.: Cambridge University Press. pp. 203–221.

Graham, R. W. and Grimm, E. C. 1990. Effects of global climatic change on the patterns of terrestrial biological communities. *Trends in Ecological Evolution* **5**: 289–292.

Grant, P. R. 1991. Natural selection and Darwin's finches. *Scientific American* **265** (October): 82–87.

Hallam, A. 1985. A review of Mesozoic climates. *Journal of the Geological Society (London)* **142**: 433–445.

Holtz, T. R., Jr. 2000. "Classification and evolution of dinosaur groups". *In* Paul G. S. (Ed.), *The Scientific American Book of Dinosaurs*, St. Martin's Press. pp. 140–168.

Hunt, A. P., Lucas, S. G., Heckert, A. B., Sullivan, R. M. and Lockley, M. G. 1998. Late Triassic dinosaurs from the Western United States. *Geobios* **31**: 511–531.

King, M. J. and Benton, M. J. 1996. Dinosaurs in the Early and Middle Triassic? – The footprint evidence from Britain. *Palaeogeography, Palaeoclimatology, Palaeoecology* **122**: 213–225.

Kitching, J. W. 1979. Preliminary report on a clutch of six dinosaurian eggs from the Upper Triassic Elliot Formation, Northern Orange Free State. *Paleontographica Africana* **22**: 41–45.

Lewontin, R. C. 1986. How important is population genetics for an understanding of evolution? *American Zoologist* **26**: 811–820.

Lucas, S. G., Hunt, A. P. and Long. R. A. 1992. The oldest dinosaurs. *Naturwissenschaften* **79**: 171–172.

Martin, A. J. 2002. "Dinosaur evolution: from where did they come and where did they go?" *In* Scotchmoor, J. D., Breithaupt, B. H., Springer, D. A. and Fiorillo, A. R. (Eds), *Dinosaurs: The Science Behind the Stories*. Alexandria, Virginia: American Geological Institute. pp. 23–30.

Novas, F. E. 1997. "Herrerasauridae. *In* Currie P. J. and Padian K. (Eds) *Encyclodedia of Dinosaurs*. Academic Press.

Olsen, P. E., Shubin, N. H. and Anders, M. H. 1987. New Early Jurassic tetrapod assemblages constrain Triassic-Jurassic tetrapod extinction event. *Science* **237**: 1025–1029.

Olsen, P. E., Kent, D. V., Sues, H.-D., et al. 2002. Ascent of dinosaurs linked to an iridium anomaly at the Triassic-Jurassic boundary. *Science* **296**: 1305–1307.

Padian, K. and Angielczyk, K. D. 1999. Are there transitional forms in the fossil record? *In* Kelley, P. H., Bryan, J. R. and Hansen, T. A. (Eds), The Evolution-Creation Controversy II: Perspectives on Science, Religion, and Geological Education. *The Paleontological Society Papers* **5**: 47–82.

Padian, K. and May, C. L. 1993. The earliest dinosaurs. *New Mexico Museum Natural History Science Bulletin* **3**: 379–381.

Parrish, J. M. 1997. "Evolution of the archosauria". *In* Farlow, J. O. and Brett-Surman, M. K. (Eds), *The Complete Dinosaur*. Bloomington, Indiana: Indiana University Press. pp 191–203.

Parrish, J. T. 1993. Climate of the supercontinent Pangea. *Journal of Geology* **101**: 215–233.

Pollard, D. and Schulz, M. 1994. A model for the potential locations of Triassic evaporite basins driven by paleoclimatic GCM simulations. *Global and Planetary Change* **9**: 233–249.

Ridley, M. 2003. *Evolution* (3rd Edition). Boston, Massachusetts: Blackwell Science.

Roush, R. T. and Tabashnik, B. E. (Eds). 1990. *Pesticide Resistance in Arthropods*. New York: Chapman and Hall.

Sereno, P. 1999. The evolution of dinosaurs. *Science* **284**: 2137–2147.

Sereno, P. C. and Novas, F. E. 1992. The complete skull and skeleton of an early dinosaur. *Science* **258**: 1137–1140.

Sereno, P. C. and Arucci, A. B. 1993. Dinosaur precursors from the Middle Triassic of Argentina: *Lagerpeton chanarensis. Journal of Vertebrate Paleontology* **13**: 385–399.

Sereno, P. C., Forster, C. A., Rogers, R. R. and Monetta, A. M. 1993. Primitive dinosaur skeleton from Argentina and the early evolution of Dinosauria. *Nature* **361**: 64–66.

Simms, M. J. and Ruffell, A. H. 1990. Climatic and biotic change in the Late Triassic. *Journal of Geological Society of London* **147**: 321–327.

Valentine, J. W. and Moores, E. M. 1972. Global tectonics and the fossil record. *Journal of Geology* **80**: 167–184.

Ziegler, A. M., et al. 1993. Early Mesozoic phytogeography and climate. *Philosophical Transactions – Royal Society of London* B **341**: 297–305.

6

Chapter

7

While visiting Dinosaur National Monument in northeastern Utah, you are amazed by the numerous dinosaur bones exposed in the wall there. Among the hundreds of jumbled bones are some parts of dinosaur skeletons that are still articulated, such as the cervical vertebrae, parts of the appendicular skeleton, and skulls of sauropods. You are surprised by this display because you have heard about the "rarity" of dinosaur fossils, the incompleteness of the fossil record and how it casts doubt on evolutionary theory, and about how fossils are always preserved as scattered bits and pieces, rather than complete skeletons

How did such a concentration of bones end up in one place? How did parts of some animals stay articulated, and why are other parts scattered? Is the environment where the bones accumulated the same place as where the dinosaurs lived?

Dinosaur Taphonomy

Why Learn about Dinosaur Taphonomy?

This chapter summarizes "the Three Ds," its subject matter being death, decay, and disintegration.

Taphonomy, the study of everything that happens to an organism's body after it dies, is a fascinating science. Taphonomy (from the Greek *taphos*, burial, and *nomos*, law) was first recognized as a science for studying the post-death phenomena associated with organisms by Russian paleontologist **J. A. Efremov** in 1940, although some of its principles were discussed by Leonardo da Vinci in the fifteenth century (Chapter 3). This field has since been broadened to include the study of preservation, which requires the understanding of fossilization processes that affect trace fossils (such as tracks, nests, and feces: Chapter 14).

Taphonomy relates to the following circumstances:

*Criminal investigators (particularly in **forensics**, the applications of science to solving legal problems), wildlife biologists, and entomologists (those who study insects) use this same science, although they may not use the term taphonomy in those fields.*

- The probable cause of death.
- What other animals (including its relatives) might have scavenged on its carcass.
- How long it took before a body started to bloat and why it bloated.
- How long a body was above the ground before any part of it was buried.
- Whether parts of a body broke off and were carried away from the original death site.
- How far from the original death site a body part might have been transported.
- Under what circumstances a body part was finally buried and affected by more biological and chemicals processes while underground.
- Whether an animal's remains were noticeably changed or not during their time buried.
- Why some parts of animals become body fossils and why body fossils of some animals are rare.

Because all non-avian dinosaurs have been extinct for the past 65 million years, taphonomy is an important science for understanding how their fossils remained preserved during that amount of time. It also indicates how representative our admittedly biased sample might be of the original dinosaur population at any given time during the Mesozoic. Paleontologists acknowledge that the fossil record is an incomplete history of all species that lived and evolved on the Earth. Body fossils of terrestrial vertebrate animals in particular are quite rare, especially in comparison to marine invertebrates or trace fossils of terrestrial invertebrates.

Dinosaurs provide a good example of such rarity. Despite their fame and 165-million-year existence, approximately 80% of all described dinosaur species are only known from less than five specimens for each species; about 50% are known from only one specimen. The only major group of dinosaurs that is well represented in terms of diversity and number of individuals representing each species are the ceratopsians (Chapter 13). Knowledge that modern vertebrate animals, in particular, show an abundance of herbivores relative to carnivores also raises questions about the circumstances that might have preserved hundreds of carnivorous dinosaurs in one deposit with no herbivores whatsoever (Chapter 9). Similarly, we might ask why other deposits contain an estimated population of thousands of herbivorous dinosaurs, apparently of the same species, whereas the only other body fossils representing another dinosaur species are errant theropod teeth (Chapter 11).

Within this type of study are the added benefits of:

1 gaining a better understanding of organisms and ecosystems associated with dinosaurs (especially in terms of nutrient cycling);
2 learning about the sedimentary environments where dinosaur bodies were buried; and
3 discerning the biogeochemical conditions of sediments enclosing dinosaur body parts.

Determining the cause of death for individual dinosaurs also provides clues toward understanding natural selection in the context of a dinosaur's ecosystem, which relates to dinosaur evolution (Chapter 6). It also helps illuminate more specific aspects of behavior, such as injuries or deaths that might be attributable to intraspecific or interspecific interactions.

Dinosaur trace fossils also are allied with taphonomy because they were subject to some of the same principles of body fossil preservation. The single largest advantage that dinosaurs' trace fossils have over their body fossils is that most trace fossils represent *in situ* (in place) fossils. Many dinosaur body fossils were probably moved from the original site of death of the animal, but most dinosaur trace fossils, such as tracks (Chapter 14), record the former presence of dinosaurs exactly where they are found. Furthermore, certain dinosaur trace fossils are more abundant than other dinosaur fossils in some areas, which may be a function of their preferential preservation in rocks that normally do not preserve dinosaur body fossils. This abundance also may be a result of one dinosaur making many trace fossils during its lifetime.

Surprisingly, dinosaur taphonomy receives relatively little mention in most books about dinosaurs. This chapter will simply point out how the study of taphonomy as applied to dinosaurs can provide many educational benefits. Taphonomy provides a way to discern the long history of what happened to a dinosaur between the day it died and the day we see its remains for the first time, many millions of years later. It makes for an interesting story.

Possible Causes of Injury, Poor Health, and Death in Dinosaurs

Typical, popular depictions of dinosaur deaths are melodramatic and limited to two scenarios: mortal combat with other dinosaurs or a direct hit from an asteroid. The latter scenario has been added only recently, dating from 1980 (Chapter 16). Dinosaur deaths, depicted in fiction or other re-enactments, usually focus on dinosaur-versus-dinosaur conflicts, except where anachronistic humans are included to finish them off in some creative way (Chapter 1). But has anyone ever seen a depiction of a dinosaur dying from eating a toxic plant? Or being stung to death by Mesozoic bees and wasps? Or tripping over a log and falling off a cliff?

Or expiring from old age? Once the mind is opened to the realm of possibilities for death, and these are compared to recorded information on actual modes of death for vertebrates, the most likely causes rarely match preconceived notions.

Application of scientific methods can soon show that a hypothesis for a proposed cause of death can be tested, at least partially, for its probability through observations made in modern, natural settings. For example, **predation**, the hunting of a live animal (**prey**) by another (**predator**) for the purposes of killing and eating that animal, has been overstated in its importance in the deaths of terrestrial vertebrates. A study of modern hyena predation in Africa reported that only 1% to 2% of all animals in a prey population actually died as a result of hyenas hunting them. The same application of actual data to causes of human deaths can also yield surprising results. Despite the fear many people have of dying from bites by poisonous snakes, factual information shows that many more people die each year from allergic reactions to bee-stings (hence the previous allusion to dinosaurs dying in the same manner). Similarly, despite much fear in the USA of terrorist attacks, the actual risks of injury or death from this possible source of harm are still far lower than the odds of dying from accidental electric shock, drowning in a swimming pool, or being maimed or killed in a car accident, especially for people living outside of major cities.

For individual dinosaurs, here are some possible causes of death and the evidence supporting these causes:

- **Suffocation** by ash and gases from volcanic eruptions, getting stuck in a muddy watering hole, or rapid burial by mass movements of sediment.
  Evidence: Large accumulations of mostly one species of dinosaur, such as the ornithopod *Hypacrosaurus*, theropod *Allosaurus*, and neoceratopsian *Psittacosaurus*, have been found in near-complete condition in volcanic ash or muddy deposits, respectively; dinosaurs seemingly "frozen" in life-like positions occur in sandy deposits, such as a specimen of the theropod *Oviraptor* sitting on its nest, or specimens of the theropod *Velociraptor* and ceratopsian *Protoceratops* intertwined with one another.
- **Intraspecific competition**, where dinosaurs fought with individuals of the same species for food, mates, or territory.
  Evidence: Toothmarks or other bone injuries, seen in some tyrannosaurids and ceratopsians, suggest they were caused by the same species of dinosaur.
- **Predation**, where a dinosaur died from injuries inflicted by a predator.
  Evidence: Remains of a dinosaur have been found in the abdominal area of another animal; toothmarks or dislodged teeth in bones inflicted by a different species of dinosaur (such as tyrannosaurid toothmarks in the ceratopsian *Triceratops* or *Edmontosaurus*) or by other potential predators, such as crocodiles (Fig. 7.1).
- **Drowning**, either in a river or an open body of water such as a lake or delta.
  Evidence: Mass accumulations of dinosaurs of the same species (the theropod *Coelophysis* or some ornithopods, such as hadrosaurs, and some sauropods) or nearly complete individual dinosaurs of differing species are preserved in a river or deltaic deposit.
- **Pathogenic conditions**, ailments caused by diseases (such as bacterial or viral infections), imbalanced nutrition, or other environmental stresses.
  Evidence: Diseases are indicated by bone overgrowths in *Allosaurus* and *Triceratops*, bone cancer in *Allosaurus*, or arthritis (gout) and fungal infections in *Tyrannosaurus*.
- **Injuries** (of uncertain causes).
  Evidence: Stress fractures are seen in phalanges of ceratopsians and sauropods, fused caudal vertebrae occur in ankylosaurs, and some theropod

FIGURE 7.1 Crocodile toothmarks in hadrosaur dorsal vertebra, Aguja Formation, Late Cretaceous, west Texas. Whether these trace fossils represent predation or scavenging of the hadrosaur is currently unknown. Photograph by Stephen W. Henderson.

CENTIMETER

7

trackways show limping or missing digits in individual tracks. There is some limited evidence of bone infections associated with probable healed fractures in a variety of dinosaurs.

■ **Dehydration**, dying from lack of water, either because of distance from a water body or drought conditions.
Evidence: Ceratopsian remains have been found in close proximity to an interpreted water hole that apparently diminished until it could not sustain the dinosaurs.

■ **Embryonic death** (while still in the egg), possibly from the egg having been drowned.
Evidence: Small bones or nearly complete skeletons of sauropods are within some dinosaur eggs, where the eggs are buried in contemporaneous floodplain sediments.

The study of sickness, injuries, and other abnormalities in the health of ancient organisms is called **paleopathology**, *and this science as applied to dinosaurs has provided some results that go against the stereotype of a brutal "struggle for existence" often envisaged for them.*

Massive, **monospecific** (single-species) bone beds, in some cases containing parts from hundreds or thousands of individuals of the same species of dinosaurs, probably represent the near-simultaneous death of a population. A more catastrophic condition was required for their death, accumulation, and preservation than would be needed for any individual dinosaur, as explained later. In the cases of individual dinosaurs found in the fossil record, most of the evidence for their deaths is gathered by looking for signs that a dinosaur was having problems with its normal, everyday functioning while it was alive.

From examination of dinosaur bones, trackways, and eggs for defects or abnormalities, we know that dinosaurs were apparently healthy animals, although a few individuals definitely had problems. Theropods in particular seemed prone to injuries, as demonstrated by the numerous examples of their bones showing healed fractures and trackways that show signs of limping (Chapter 14).

No evidence is documented yet for the following proposed causes of dinosaur death, other than similar causes of death seen in some vertebrate populations today. So they remain possible but otherwise weakly supported hypotheses for any given individual dinosaur or dinosaur assemblage found in the fossil record:

- **Hypothermia** – freezing to death.
- **Hyperthermia** – overheating.
- **Egg predation** (especially by mammals) – embryos killed by animals that eat eggs before they are hatched.
- **Starvation** – dying from lack of food.
- **Toxicity** – poisoning from eating the wrong plant or eating an animal with a toxic defense.
- **Fires** – burning to death or suffocating from fumes associated with either forest or brush fires. (Although dinosaur bones cooked in a Cretaceous forest fire are documented, this might have happened after they died.)
- **Plagues** – widespread diseases, either viral or bacterial, that wipe out significant numbers of a population.
- **Clumsiness** – tripping at high speeds while running or tumbling down steep slopes.

Another item to add to this list that has yet to be proposed for dinosaurs through demonstrable evidence is **parasitism**, a form of **symbiosis** (different species living together) where one species of organism lives at the expense of a host organism. Parasites can be either **endoparasites** (living inside a host) or **ectoparasites** (on the outside of the host). Modern endoparasites are extremely varied, although commonly known examples include one-celled organisms such as amoebas. For example, one species of amoeba probably caused the dysentery suffered by Eberhard Fraas in 1907 on his way to examine dinosaurs in Tendaguru, as mentioned in Chapter 3. Multicellular animals that commonly live an endoparasitic lifestyle include some nematodes (roundworms) and tapeworms. Ectoparasites include certain arthropods, such as lice and ticks, and annelid worms such as leeches. Although parasites have a vested interest in not causing the immediate death of their host, they can cause weakened conditions that easily hasten the demise of a normally strong, adult animal. Known forms of endoparasitism only affect soft tissue and the parasites themselves are composed of soft tissues, thus evidence of them and their effects on ancient organisms probably had low preservation potential. Indeed, no evidence of endoparasites in non-avian dinosaurs is known, although one example of a Mesozoic (Early Cretaceous) ectoparasite was reported in association with a fossilized bird feather.

The preceding listing of dinosaur maladies and mishaps does not include possible causes for **mass extinctions**. Mass extinctions, which involve the near-simultaneous extinctions of many different species, require either a combination of numerous factors that result in those species dying out or an overwhelming catastrophe that favors the survival of only the few species that could successfully reproduce after such an event. Taphonomy is also important in interpretation of mass extinctions, a subject that is covered in more detail in Chapter 16.

Sedimentary Environments and Dinosaur Preservation

The preservation of dinosaur fossils invariably required burial in sediments, as shown by the near-exclusive occurrence of them in sedimentary rocks. A sedimentary rock

TABLE 7.1 Common sedimentary rocks, classified on the basis of composition and texture.

Clastic Sedimentary Rocks: formed through consolidation of previously broken rock material, mostly composed of silicate minerals.

Shale. Mostly mud-sized particles (clay and silt), but shows parting of layers (fissility).
Mudstone. Mostly mud-sized particles, no preferred parting of layers (sometimes synonymized with shale).
Siltstone. Silt-sized particles.
Sandstone. Sand-sized particles.
Graywacke. Sandstone with an appreciable amount (>15%) of mud.
Conglomerate. Sandstone containing abundant well-rounded particles of greater than 2-mm diameter.
Breccia. Sandstone containing abundant angular particles of greater than 2-mm diameter.

Chemical Sedimentary Rocks: formed through chemical precipitation of previously dissolved rock material, mostly composed of calcium carbonate, oxides, sulfates, and chlorides.

Limestone. Wide variety of rocks of different textures composed of $CaCO_3$ (normally calcite but younger examples with aragonite).
Dolomite. Composed primarily of mineral dolomite ($Ca,Mg(CO_3)_2$).
Chert. Microcrystalline quartz precipitate with some H_2O incorporated in the structure.
Hematite. Composed primarily of mineral hematite (Fe_2O_3), typically replaces reduced iron minerals.
Gypsum. Composed primarily of mineral gypsum ($CaSO_4 \times 2H_2O$).
Halite. Composed primarily of mineral halite ($NaCl$), also known as "rock salt."
Coal. Composed of fossil plant material with sufficient organic carbon (as hydrocarbons) to be combustible.

is held together by cement that was chemically precipitated, or by a fine-grained matrix of broken rock material. Depending on the rate of **lithification**, the processes involved in the formation of a rock from sediment, this consolidation can occur within a few years or over the course of millions of years. Sedimentary rocks are varied and are classified on the basis of their **composition** (types of minerals) and the **texture** (size, shape, and arrangement) of the sediment and associated cement or matrix (Table 7.1). The two other major categories of rocks, igneous and metamorphic (Chapter 4), are of peripheral interest to dinosaur studies but nevertheless are important for interpreting the geologic history of the Mesozoic Era. Rare reports of dinosaur tracks, preserved in some former lava flows, are the only known instances of either rock type containing dinosaur fossils; accordingly, dinosaur paleontologists restrict their search image to sedimentary rocks.

Sedimentary rocks comprise about 90% of all exposed rock at the Earth's surface but not every **sedimentary environment** (places where sediments accumulated) was conducive to the burial and subsequent preservation of dinosaur fossils. Dinosaurs were land-dwelling animals, which is partially corroborated by their most frequent association with sedimentary rocks that have all of the characteristics of **continental** environments, such as **fluvial** (river), **lacustrine** (lake), and **deltaic** (delta) environments. A few dinosaur body fossils have been found in rocks formed in marine environments, for reasons examined later in this chapter. Nonetheless, the majority

189

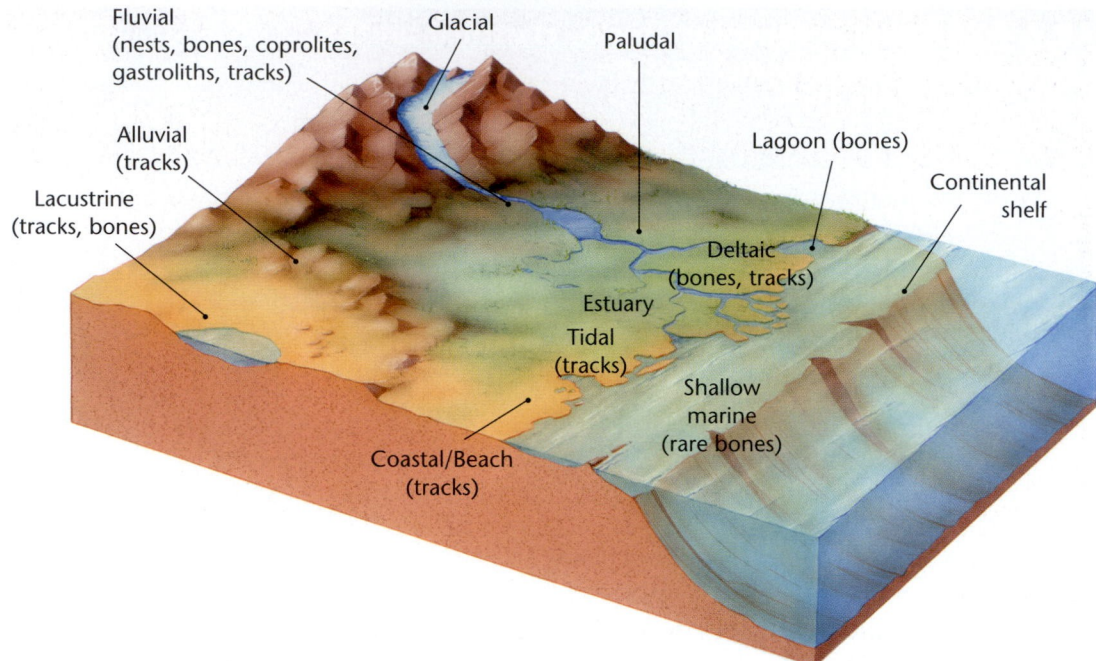

FIGURE 7.2 Common sedimentary environments, showing which ones had relatively low to high preservation potential for dinosaur body and trace fossils. Environments include glacial, alluvial, lacustrine, fluvial, paludal, estuarine, deltaic, coastal, tidal, and shallow marine.

of dinosaur body and trace fossils occur in rocks composed of sediments deposited in areas around and in lakes, rivers, deltas, **estuaries** (coastal embayments where fresh and salt water mix), and marine shorelines (Fig. 7.2). Another prerequisite for preservation of dinosaur fossils is that the sedimentary environments had to have existed during the Mesozoic Era, specifically from the Late Triassic through to the Cretaceous Periods.

The proper interpretation of sedimentary environments that preserved dinosaur fossils from the Mesozoic depends on a thorough description of their **facies**, which is all the characteristics imparted by an environment to its sediment at approximately the same time. Facies can be analyzed on the basis of lithologic characteristics (**lithofacies**), body fossils (**biofacies**), and trace fossils (**ichnofacies**). This type of description and the subsequent interpretation of a sedimentary environment comprise **facies analysis**, which is similar to paleoenvironmental analysis (Chapter 1). Geologists will refer to some rocks as fluvial facies, deltaic facies, or shallow-marine facies, based on a holistic and thorough examination of the many environmental parameters that left their clues in the sediment (Table 7.2). This eclectic approach hints at the breadth of geologic training that is required of a dinosaur paleontologist (Chapter 4); overlooking a single environmental parameter can considerably change the interpretation of the original environment that entombed the dinosaur fossils.

For example, parts of the Late Triassic Chinle Formation in the western USA contain abundant remains of fossil forests (as found in Petrified Forest National Park of Arizona), body fossils of numerous reptiles, a few small theropods, large amphibians, fish, and a few mollusks. All of these fossils are preserved in various muddy and sandy sediments, but no body fossils of insects or most other arthropods were reported. Previous workers did not look for, or otherwise recognize, many of the trace fossils made by invertebrates in the same sediments that contained the body fossils. Consequently, reconstructions of Chinle environments rarely included any invertebrates, showing a bias in interpretations on the basis of lithofacies and

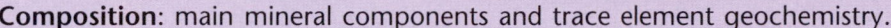

TABLE 7.2 Parameters in a sediment deposit or sedimentary rock at the outcrop scale that lend to facies analysis. Keep in mind the two-step method for scientific hypothesis formation: *Description* – What is it?, and *Interpretive significance* – What is it used for?

Composition: main mineral components and trace element geochemistry.

Texture: size, shape, and arrangement of sediment grains or cement.
Sedimentary structures: features in a rock formed by either biological, physical, or chemical processes.
Geometry of sedimentary body: lenticular (lens-like), tabular.
Physical structures: bedding, cross-bedding, ripples, dunes, graded bedding.
Biogenic structures: burrows, borings, tracks, trails, nests.
Chemical structures: nodules, stalactites, stalagmites, styolites.
Body fossils: microfossils (algae, protozoans), plants, invertebrates, vertebrates.

biofacies. Beginning in the 1990s, finds of numerous trace fossils attributed to insects (termites, ants, bees), crayfish, horseshoe crabs, and mollusks (clams, snails), as well as vertebrate tracks and nests (some made by phytosaurs: Chapter 6), have added an ichnofacies component. As a result, the previously incomplete picture of paleo-environments in the Chinle has been expanded to include fluvial and lacustrine environments in association with forest ecosystems. The ichnofacies thus helped to enrich our understanding of the life history of the theropods and other vertebrates that lived in those same environments.

Dinosaurs used to traverse sedimentary environments, leaving tracks, making nests, dropping feces, and munching on foliage (or each other). *In situ* evidence for all of these various life activities is restricted to a narrow range of facies, which are fluvial, deltaic, or close to a marine shoreline, such as estuaries, tidal flats, or beaches. After the dinosaurs died, their bodies often remained in the same places as where they had lived or they were buried very close to where they died. Alternatively, entire bodies and the sum of their parts were carried away from their life and death sites before being buried for the first time. Evidence indicates that some dinosaur bones were buried, exhumed, transported, and buried again. Here is where the two major challenges of dinosaur taphonomy are encountered:

1 how far away is a dinosaur body fossil with respect to its original life site (which may not coincide with its death site, let alone its final resting spot)? and
2 how separated in time is a dinosaur body fossil from the strata that represent the time it was alive, as opposed to the strata that contain it?

Postmortem Processes: Pre-burial

Definition of Biostratinomy

After a dinosaur died, its body may have undergone a complicated series of post-death (also known as **postmortem**) processes that involved both biological and physical factors, perhaps well before any part of it became buried. **Biostratinomy** is the study of what happens to an organism between its death and its final burial. In most cases, dinosaurs would have died before their remains became buried, although a hypothesis was proposed for live burial of some dinosaurs in Late

Cretaceous deposits of Mongolia (Chapters 9 and 13). Another possibility to consider for some dinosaur body parts found in the fossil record is that a dinosaur may have lost a part before it died, and that part ended up in the fossil record but the rest of the dinosaur did not. As mentioned earlier, some dinosaur tracks show a missing toe on one foot, indicating that the dinosaur was still walking after it had its digit-depriving mishap (Chapter 14). The best example proposed for body-part loss among dinosaurs while they were still alive is their probable shedding of teeth, as seen with certain theropods (Chapter 9). However, finding convincing evidence supporting the hypothesis that a particular dinosaur body part other than a tooth was detached from a dinosaur long before it died would be extremely difficult. Nevertheless, this possibility serves as an alternative hypothesis to explain the anomalous occurrence of an expendable dinosaur body part in a deposit lacking any other sign of its owner.

Although the following sections are divided into biological and physical processes, any of the processes from each category could have happened simultaneously. For example, a dinosaur caught in a flooded stream would have been transported while it drowned, and biological processes (such as scavenging) probably would have affected the dead body while it was still moving. In other words, a spectrum of interacting events may have led to the preservation of any given dinosaur body fossil, which should be considered when interpreting all of the available evidence associated with the fossil.

Biological Processes: Decay and Scavenging

The decomposition of a dead body is largely a biological process, called **necrolysis**. This discussion will feature a fictional case study that illustrates the common scenario of biological processes associated with dinosaur taphonomy.

One day during the Mesozoic, a 3-metric-ton ornithopod died from a horrible, nasty, and painful viral disease. After much aimless staggering, the ornithopod collapsed on to the floodplain of an inland river. Soon after it died, its corpse underwent **rigor mortis** (the stiffening of musculature in the body) within the first 10 hours or so of death, followed by a relaxation of the muscles. It initially landed on either its right or left lateral surface, with its limbs stretched away from its body and its head turned toward its back as the stronger muscles on the dorsal surface of the neck contracted. The same process affected the tail, and its dorsal muscles pulled the tail toward the back of the body, with the neck and tail forming an arc.

At the same time, the **anaerobic bacteria** (which live in the absence of oxygen) in the gut of the animal and the **aerobic bacteria** (which need oxygen for respiration) from the environment outside of the body broke down the proteins and other organic compounds in the corpse to obtain their food. These simple, one-celled organisms were responsible for as much as 90% of the dinosaur's initial decomposition. Their consumption of the corpse gave off gaseous metabolic by-products that bloated the body and gave it a noticeable malodorous scent, a process called **putrefaction**. The gases released as waste during the aerobic decomposition of the organic matter (**oxidation**) were CO_2 and H_2O; this process recycled much organic carbon back into the ecosystem (an important phase in nutrient cycling). Anaerobic decomposition proceeded through **fermentation**, exemplified by the following equation:

$$C_6H_{12}O_6 \text{ (glucose)} \rightarrow 2CO_2 + 2C_2H_5OH \text{ (alcohol)} + Energy \qquad (7.1)$$

where glucose (a sugar) was a simple biomolecule broken down by the bacteria. Thus, CO_2 was the main gas produced, although methane (CH_4) is another potential by-product of some anaerobic bacteria.

The first effluvia of these gases attracted the attention of scavenging animals, which regarded the body as a potential food source. Insects were probably the first to arrive: flies laid eggs, which soon hatched into feeding larvae, and carrion beetles began stripping away some of the outer flesh. The flies normally laid their eggs in open parts of the body, such as the nostrils, eyes, anus, or any obvious wounds. Maggots may also have already been present in any open wounds on the dinosaur's body before its death. The carrion beetles were numerous and relentless; they worked on the corpse for weeks and eventually left their distinctive, pitted gnawing marks in any exposed bone. Ants were also industrious, outnumbering the beetles as they carried away millions of pieces of the body to their nests on a nearby, vegetated riverbank.

The emission of greater volumes of the fetid gases from the corpse also alerted vertebrate scavengers to a potential meal, such as theropods, pterosaurs, small mammals, and, if this was during the Late Jurassic or Cretaceous, birds. The river flooded just enough to cover part of this area while the body was still on the floodplain, so crayfish brought in by the floodwaters joined in the feast. If the conditions were relatively hot and humid on the floodplain, then the metabolic activity of the bacteria accelerated and the corpse's bloating was more rapid than normal. This continued for nearly a week before the body exploded (aided by the numerous punctures left by scavengers) and then deflated. The result was a flattened profile to what had originally been a voluminous piece of putrid flesh.

Within six to eight weeks, all soft parts were stripped from the bones, and the bones themselves were attacked by yet more beetles and any other animals that were interested in obtaining some calcium and phosphorus in their diets (Chapter 8). More delicate parts of bones, such as the epiphyses (wide ends) of femurs and other limb bones, were especially susceptible to being broken by animals with strong jaws; some cranial bones might have been wholly consumed. The bones, once they were exposed to sunlight, also had their organic content depleted more rapidly than before, bleaching them white and making them more brittle. Other dinosaurs strolling through the area stepped on the now-weakened bones, pulverizing or otherwise fracturing them. This increased their surface area and made them more susceptible to dissolution by natural acids and further bacterial or fungal decay.

All of the preceding processes depended on temperature and humidity; for example, faster rates of decay and insect scavenging are associated with increased temperature. Assuming summertime conditions and high amounts of rainfall, within eight to ten weeks there might have been little trace of the several-ton animal, except for maybe a little more vegetation growing in its resting spot as a result of the natural fertilizer provided by its body. Unless the corpse, in whatever state, was buried by the sediments of a river flood during the preceding time frame, or parts of it were carried away by floodwaters or scavengers and subsequently buried, this ornithopod would not have made it into the fossil record in any shape or form.

Evidence supporting the preceding scenario is based on the following data:

- The rates and processes of decomposition and scavenging of large modern mammals (such as elephants) or other vertebrates with body sizes comparable to some dinosaurs (small ones, too) have been observed, some of which died in environments similar to those interpreted for containing dinosaur remains (Fig. 7.3).
- Studies from **entomology** (the study of insects) have documented and calculated the life cycles of fly species associated with laying eggs in animal bodies; carrion-feeding beetles and other scavenging insects have also been studied for their effects on corpses.
- The types and amounts of metabolic by-products from both anaerobic and aerobic bacteria involved in the decomposition of organic material have been

FIGURE 7.3 Taphonomic information derivable from an opossum (*Didelphis marsupialis*) on Sapelo Island, Georgia. Opossum was observed dead in the road at 8:00 a.m., July 31, 2004, and seemed freshly killed at the time; hypothesized cause of death was from being struck by an automobile. At 4:00 p.m. that same day, seven black vultures (*Coragyps atratus*) were seen around the body; their tracks and a drag mark ending with the body confirmed that it had been moved about 3 meters from the spot where it was originally sighted. Body had been almost completely eviscerated; internal organs and musculature were more than 90% gone. Tire tracks and crushed bones indicated that several vehicles had run over the body both before and after scavenging. Flies were on the body and a noticeable odor was present, the latter probably as a by-product of aerobic bacteria. Temperature was about 30°C, with nearly 100% humidity.

measured, as have the rates of this type of decomposition in association with temperature and other climatic factors.

■ Paleontological information from both body fossils and trace fossils in Mesozoic rocks indicate the presence of probable scavengers in continental ecosystems, such as flies, crayfish, beetles, and ants (Fig. 7.4).

■ Toothmarks in bone made by other dinosaurs, that are in certain patterns or areas of the bones, suggest scavenging behavior rather than predation, as do teeth left behind by theropods in monospecific beds of herbivorous dinosaurs (Chapter 9).

■ Crushed bone found in areas of dinosaur bone accumulation suggests trampling by large, heavy animals (that is, other dinosaurs: Chapter 14). The same process has also been documented for damage of modern elephant bones by other elephants.

■ Relatively large olfactory bulbs derived from brain endocasts of some carnivorous dinosaurs suggest an enhanced ability to smell. This adaptation

FIGURE 7.4 Pockmarks in sauropod bone from the Late Jurassic of Utah, interpreted as trace fossils formed through beetles gnawing on exposed bones.

7

could have been used to detect a decaying carcass over long distances, and is seen in modern vultures with their scavenging lifestyle (Chapter 9).

Currently, no definitive evidence supports fly, ant, crayfish, mammal, or bird scavenging of dinosaur corpses as in the above scenario, but these animals were contemporaneous with dinosaurs and some of their species are observed scavenging today. Although this is a weakly supported hypothesis, such evidence may be found if paleontologists look for traces of such activity, either in bones or sediment associated with bones. Similarly, no evidence supports the scavenging behavior of pterosaurs on dinosaurs, although some pterosaurs are interpreted as carrion feeders on the basis of functional morphology. In contrast, a documented example of a dinosaur feeding on a pterosaur has been interpreted on the basis of matching teeth with distinctive toothmarks found on pterosaur bones (Chapters 8 and 9).

Modifications to the outlined scenario can be made easily by placing the dinosaur body in different environments and times and varying the behaviors of the scavengers. For example, if the body had been washed into a river channel instead of lying on a floodplain, crocodiles and fresh-water fish might have fed on it. If it had fallen near a marine shoreline, crabs, which are common coastal scavengers today, would have been more likely consumers of the corpse instead of crayfish, which live exclusively in fresh-water fluvial and lacustrine environments.

If waves at a shoreline washed up enough to pull a dinosaur body out to sea, its gas-filled corpse would have rafted on the open sea. Likewise, dinosaurs that dwelled in lowland environments, such as coastal swamps, deltas, or estuarine marshes (a life habit shown by numerous dinosaur tracks in such environments: Chapter 14), could have had their bodies washed into nearby channels that emptied into a seaway. Once at sea, a dinosaur corpse would have provided a rich food source for sharks, smaller fish, and marine reptiles, which would have nibbled on its appendages hanging down into the water. This hypothesis, often termed the **"bloat-and-float" hypothesis** by vertebrate taphonomists, provides an explanation for the rare occurrences of certain dinosaur bones in shallow marine deposits. Joseph Leidy first proposed it in 1858 to explain *Hadrosaurus* remains in a Cretaceous marine deposit (Chapter 3). Later analyses of dinosaur bones in Middle Jurassic marine deposits of England indicated that they were displaced a minimum of 80 km from the nearest continental environment, which meant that there was plenty of time for marine scavengers to enjoy some dinosaurian snacks. In the cases of marine-deposited dinosaur remains, mainly the distal parts of the animals are preserved (those parts that would have been dangling in the water) and some of these bones exhibit the distinctive toothmarks (or teeth) of their shark or marine crocodile contemporaries.

Whether scavenging happened on the land or at sea, it may mean that body parts of a dinosaur could have been carried far away from the original site of death.

Physical Processes: Water and Wind

One important concept to keep in mind is that a dinosaur body (or any of its body parts in a fluvial environment) was subject to the same physics as other sedimentary particles of similar size, shape, or density.

Physical processes probably caused the most dramatic changes in locale for a dinosaur's body, moving it from where it died to where it was buried, as illustrated by the bloat-and-float hypothesis. Dinosaur remains buried near the same spot where they died are **autochthonous**, whereas those that traveled a significant distance from their death site are **allochthonous**. This section discusses allochthonous remains.

By water or wind, sediments are moved through **traction** (dragged along a surface) or in **suspension** (lifted into the fluid medium above a surface). An intermediate form of movement, where a particle "jumps" intermittently, is called **saltation**. The combination of sediment moved through traction and saltation constitutes the **bedload**, and suspended particles are the **suspended load** of a stream (Fig. 7.5). Sediments moved by flowing water show these respective behaviors depending on the **competence** of the flow, which is a

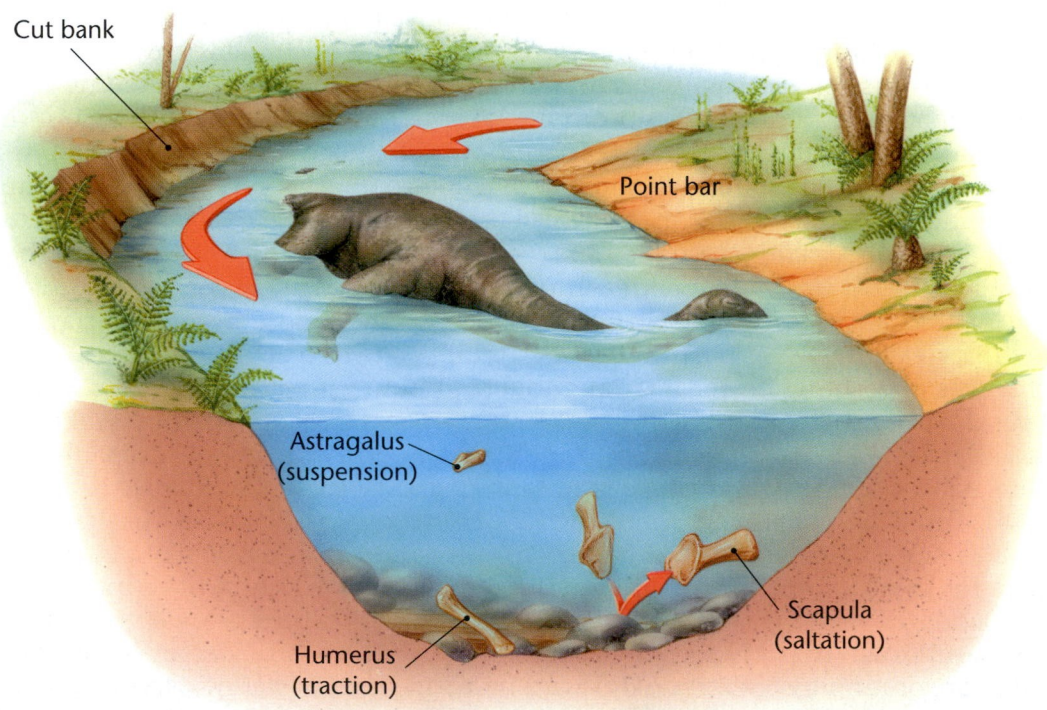

FIGURE 7.5 Dinosaur bones demonstrating their behavior as sedimentary particles. Suspension – astralagus; Saltation – scapula; Traction – humerus. Floating sauropod (with an apparent density of less than 1.0 g/cm³) for scale.

measure of the maximum-size particle it can move. Competence in turn is determined by **discharge**, the volume of water passing a certain point during a certain amount of time, measured through the following formula:

$$Q = AV \tag{7.2}$$

where Q is the discharge, measured in m^3/s; A is the cross-sectional area of a stream in m^2; and V is the average velocity of the stream in m/s. For example, if the cross-sectional area at a specific place in a stream during the Jurassic was measured as 16.5 m^2 and the velocity of the stream at that same place was 4.3 m/s, then the discharge was

Step 1. $Q = (16.5 \text{ m}^2)(4.3 \text{ m/s})$
Step 2. $Q = 71 \text{ m}^3/\text{s}$

In our example, the water (which has mass) was moving, which means that it had **momentum**, expressed by this formula:

$$M = mV \tag{7.3}$$

where M is momentum and m is mass. This means that our same example, with its 71 m^3 of water converted to kilograms (remember that 1 cm^3 of water = 1 g, hence 1000 cm^3 = 1000 g; see Table 1.2) and an assumed velocity of 1.0 m/s, had a momentum of

Step 1. $M = (71 \text{ m}^3)(1.0 \text{ m/s})$
               (Converting m^3 to kg)
Step 2. $M = (71 \text{ m}^3 \times 1000 \text{ kg/m}^3)(1.0 \text{ m/s})$
Step 3. $M = 71{,}000 \text{ kg} \times \text{m/s}$

These calculations pertain to how dinosaurs or parts of them were moved, because such movement would have taken momentum. For example, a dinosaur such as our unfortunate example in the previous section, which weighed about 3000 kg and died on an emergent floodplain, would have required a flow with a momentum greater than 3000 kg × m/s to move it 1.0 meter in 1.0 second. Even higher momentums would have been needed to move the same mass any appreciable distance from the death site of the dinosaur within a given time.

What evidence can we use as the basis for calculations of how fast a stream was flowing during the Jurassic, and how can we work out its cross-sectional area? Although exact values cannot be calculated, as with modern streams, the approximate speed and cross-sectional area of a Mesozoic stream can be calculated from indirect evidence preserved in the lithofacies. For calculating speed, numerous experiments conducted with modern sediments of different sizes and known densities have shown that certain velocities are needed to lift and otherwise transport sediments with these parameters. This information can then be applied to dinosaur parts. Additionally, distinctive **bedforms** are made in water by flows of a certain velocity as applied to sediments of certain grain sizes. Bedforms, such as ripple marks or dunes, are sediment bodies with a definable geometry. If distinctive bedforms are associated with dinosaur bones, they give paleontologists an estimate of what flow existed when the dinosaur parts were buried. This information may not tell us exactly how fast the water was flowing immediately before the parts were buried. However, in conjunction with the sizes and shapes of the dinosaur bones, bedforms can at least provide some limits for working out the competence of the flow, and

FIGURE 7.6 Sandstone bed formed from filling of a river channel and containing dinosaur bones near Morrison, Colorado.

thus the discharge and momentum. A bedform also can be a filled fluvial channel, which can give an approximation of the cross-sectional area of a stream (Fig. 7.6).

Dinosaur-dependent factors that affected movement of its body or body parts include the shapes and sizes of the parts being moved, as well as whether the dinosaur was recently dead (pre-putrefaction), dead for several weeks (during putrefaction), or mostly or entirely stripped of flesh (post-putrefaction), all of which affected its density. Density, in turn, is related to the **buoyancy** of a dinosaur's body or body parts once it was in a water body. Buoyancy, which is the lift force generated by a fluid (whether it is moving or not), is caused by pressure exerted by a liquid on all sides of an immersed object. This pressure makes an object weigh less while maintaining the same mass that it has in air. This principle, first articulated by Greek scientist **Archimedes** (of "Eureka!" fame, *c.* 287–212 BCE), states that a floating or immersed body in a fluid will experience a lift force equal to the weight of the displaced fluid. For example, the immersion experiment conducted on dinosaur models in Chapter 1 showed a displacement of volumes for each model; these models had lift forces exerted on them equivalent to the weight of whatever water volume was displaced. Using a human body that weighs 64.8 kg as an example (see Eqn 1.2, Chapter 1), if it displaces 72,000 cm^3 of water, the weight in water is

$$Wi = 64.8 \text{ kg} - (72,000 \text{ cm}^3 \times 1.0 \text{ cm}^3/\text{g}) \qquad (7.4)$$
$$= 64,800 \text{ g} - 72,000 \text{ g}$$
$$= -7200 \text{ g}$$
$$= -7.2 \text{ kg}$$

where *Wi* is immersed weight. A negative number in the final value means that a person should float when immersed in water, because they weigh less than the mass of the water displaced. This is why objects less dense than water float: in the preceding example, the density is 0.9 g/cm^3 (64,800 g/72,000 cm^3), whereas fresh water has a standard density of 1.0 g/cm^3. Conversely, an object with a density greater than 1.0 g/cm^3 will sink, unless the water is made denser, which changes the buoyant force. For example, ocean water, which is **saline** (containing a relatively high concentration of dissolved elements) in comparison to fresh water, has more buoyant properties than fresh water because it is denser. The Great Salt Lake in Utah or the Dead Sea in Israel are more saline than ocean water, so these are accordingly more buoyant than ocean water. Buoyancy allows steel ships weighing thousands of tons to float. When those ships strike icebergs, the rapid influx of water subsequently increases their density, causing them to sink. Likewise, buoyancy is what caused some multi-metric-ton dinosaur carcasses to float, depending on the respective densities of the dinosaurs (or their parts) and the bodies of water carrying them.

The density of a typical vertebrate land animal may vary from about 0.7 to 1.1 g/cm^3, meaning that some recently dead animals can float. However, their flotation might be similar to icebergs in that most of their bodies will be below the surface of the water. The implication here is that a recently dead dinosaur body would have dragged along the bottom of a river that was shallower than half of the dinosaur's body width (limbs included) because of its limited ability to float. Therefore, the dinosaur body would have behaved much like a sedimentary particle undergoing traction or saltation. In contrast, a bloated (post-putrefaction) carcass that retained large volumes of internal gases may have been much less than 1.0 g/cm^3 (Fig. 7.5). As a result, more of the animal would have been above a stream surface in suspension, which would have provided the potential for it to move greater distances from its original site of death. This possibility allows for largely complete dinosaur bodies to have moved considerable distances but later accumulated in stagnant areas. For example, carcass flotation has been invoked as the cause of monospecific beds of remains for the theropod *Coelophysis* (Chapter 9) in Late Triassic fluvial deposits of New Mexico.

Yet another factor in dinosaur buoyancy is whether the dinosaur had feathers or not (Chapters 5, 9, and 15). Modern birds are relatively more buoyant than other animals because of air trapped between their feathers, as well as their less dense skeletal structures. Consequently, some birds can float for long distances in a water body until their feathers become waterlogged, which can finally cause sinking. Similar scenarios have been proposed for fossil birds, beginning in the Late Jurassic (Chapter 15). The occurrence of relatively small, feathered theropods in Cretaceous lake deposits is also suggestive of such floating and sinking to lake bottoms (Chapter 9).

Once they were floating, numerous dinosaur bodies in a stream also may have created "dinosaur jams," similar to log jams where felled trees clog a stream. Such jams may have contributed to massive deposits of nearly entire dinosaur skeletons, providing an explanation for the numerous, well-preserved sauropods (Chapter 10) and theropods (Chapter 9) seen in the Late Jurassic Morrison Formation at Dinosaur National Monument in Utah, as well as mixed assemblages of dinosaur remains in the Late Cretaceous Judith River Formation of Montana.

Figuring out the preceding scenarios for whole dinosaur bodies is relatively easy in comparison to another possibility, the movement of detached body parts with flesh or bone material. Just to simplify these possibilities, bone (without flesh) can be used as an example. Dinosaur bone was composed originally and primarily of the mineral **dahllite** (Chapters 5 and 8), which has a density of 3.15–3.2 g/cm^3. This guarantees that this mineral by itself, even in ocean water, would have sunk. However, skeletal material is not composed entirely of solid mineral material: average densities for teeth, **compact** bone, and **cancellous** (porous) bone (Chapter 8) in modern vertebrates average about 2.0, 1.7, and 1.1 g/cm^3, respectively. Consequently, movement of teeth and compact bone would have been more likely as bedload; such dragging would have caused visible pits and fractures in most exposed bone. Such telltale marks from the physical transport of dinosaur bones have indeed been interpreted among the bones in high-energy facies. This information provides evidence for whether dinosaur remains were reworked into deposits much younger than the time when a dinosaur was alive. However, if the bones had any flesh remaining, these parts might have been cushioned from the abrasive effects of stream transport, thus the absence of fractures is not necessarily diagnostic of an autochthonous fossil.

Most modern examples of bone are cancellous, which with included organic matter are less dense than solid dahllite; loss of the organic material results in more open spaces and correspondingly less density. Dinosaur bones were similar in this respect and some, such as those of theropods, were lightly built and noticeably less dense than those of other dinosaurs (Chapters 8 and 9). Different hard parts on the same individual could also have had different densities, such as the bone composing the parietals of pachycephalosaurs (Chapter 13), or the teeth of any toothed dinosaur versus their limb bones.

Size and shape of a body or bone are also important factors in transport. Well-rounded tarsals of dinosaurs, for example, were more likely to roll along a stream bottom than their femurs or tibias. Of course, smaller bones were more susceptible to transport, with all other factors in the bones and stream being equal. Nevertheless, shape is probably more important to consider than size, because equal density of a large or small body translates into equal buoyancy regardless of size. Shape can be measured by looking at the ratio of an object's surface area to its volume, which is expressed through the simple relation of

$$S = A/V \tag{7.5}$$

where S is shape, A is surface area, and V is volume. Using the example of a sphere, surface area is calculated by the following equation:

$$A = 4\pi r^2 \tag{7.6}$$

and volume for that same sphere is

$$V = 4/3\pi r^3 \tag{7.7}$$

Using a typical orange (before peeling or squeezing) with a diameter of 10 cm (radius of 5 cm) as an example, its surface area to volume ratio can be calculated through the following procedure:

Step 1. $A = 4 \, (3.1416)(5)^2$
Step 2. $= 314.2 \, cm^2$
Step 3. $V = 4/3 \, (3.1416)(5)^3$
Step 4. $= 523.6 \, cm^3$
Step 5. $S = 314.2 \, cm^2/523.6 \, cm^3$
Step 6. $= 6.0 \, cm^{-1}$

This ratio is actually the smallest that can be derived for any sedimentary particle; any particle shape deviating from a perfect sphere will result in a larger number.

The important application of this measurement to stream transport is that spherical particles are less likely to be lifted by a current than long or flat particles. In the same way, Frisbees™ (which have a high A/V) can stay airborne longer than baseballs (low A/V). Thus, long, flat particles are lifted more easily than spherical particles because of an important principle first formulated by Swiss mathematician **Daniel Bernoulli** (1700–82). Bernoulli was a primary contributor to **hydrodynamics**, the physics of water flow, which is an important science to taphonomists interested in estimating transport of bodies in water. Bernoulli discovered that a moving fluid (either water or air) caused less pressure on an object than stagnant fluid, this lower pressure providing a lift force to the object affected by the flow. This principle is exemplified by wings on aircraft, which are designed so that the pressure caused by air moving rapidly over them is less on top, causing an aircraft to lift off the ground.

Of all of the bones mentioned in Chapter 5, none are spherical, which means that all dinosaur bones had higher A/V ratios than a sphere. Bones with the largest ratios were those that were long, flat, or both, such as some cranial bones (parietals, frontals), the femur, humerus, tibia, and scapula. Notice that a typical ilium is shaped more like an aeroplane wing than, say, a cervical vertebra. So an ilium was more likely to be lifted in a stream and transported far away from the original death site of a dinosaur than its semispherical parts.

Consequently, the densities, sizes, and shapes of bones varied enough that all of these factors have to be taken into account when looking at a final assemblage of dinosaur bones in a deposit. In fact, some taphonomists were industrious enough to experiment with various bones of modern vertebrates, calculating A/V ratios and proportion of compact to cancellous bone (which affects density) to categorize bones on the basis of how easily they could be transported by water. These data provide a hypothetical model to test when encountering dinosaur bones in the field and assessing their possible transport (Table 7.3).

Most of the preceding discussion on transport of dinosaur bodies was based on water as a medium, but wind was also a possible (albeit less probable) agent of transport. The physics of air and its movement is **aerodynamics**, an essential science for people who design and fly aircraft, but one that can also be applied to any effects of air movement on any objects. For example, modern hurricanes and tornadoes have carried large, multi-ton objects for considerable distances. Living animals also have been transported hundreds or thousands of meters away from their original environment. An example of the lift forces generated by some tornadoes is illustrated by the instance of a home freezer, which probably weighed about 200 kg, that was moved 2 km by a tornado in Mississippi in 1975, and a 70 metric-ton railroad car, which weighed more than most adult sauropods (Chapter 10), that was also moved a measurable distance by a tornado.

Storms have been interpreted in the geologic record on the basis of the distinctive deposits that they leave in marine and coastal sediments. Such storm deposits, called **tempestites**, are common in strata formed in shallow-marine environments from the Mesozoic, so dinosaurs certainly experienced violent storms. However, no one has ever provided evidence for transport of dinosaur bodies by wind, hence this is only an idea, not a hypothesis. Of course, observations of the impact of modern hurricanes, as well as interpreted Mesozoic tempestites, could lend themselves to the hypothesis that similar inland flooding occurred from the massive amounts of precipitation and coastal storm surges that accompanied Mesozoic hurricanes or other storms. These phenomena would have increased the amount of stream discharges and correspondingly increased the likelihood of dinosaurs either drowning or having their otherwise-dead bodies washed into water bodies and later buried.

How would a paleontologist look for clues of postmortem transport (or lack of it) once dinosaur bones are found in a Mesozoic deposit? One clue already

7

TABLE 7.3 Bones grouped according to their approximate susceptibility to movement on the basis of their *A/V* ratio and densities. Some overlap between groups can occur, meaning that bones listed in one group may belong to either the one before or following where they are placed here.

Group One: High *A/V* and cancellous/compact ratios (low density); moved easily by flows with low momentum.
    Costae (ribs)
    Vertebrae
    Sternum
    Scapula
    Phalanges
    Ulna

Group Two: Low *A/V* ratio and intermediate cancellous/compact ratio; moved by flows with intermediate momentum.
    Femur
    Tibia
    Humerus
    Ischium
    Ilium
    Pubis
    Radius

Group Three: Low *A/V* and cancellous/compact ratios; moved by flows with high momentum.
    Mandible
    Skull bones (parietals, frontals, etc.)
    Teeth

Modified from data by Voorhies (1969).

mentioned is evidence of abrasion on dinosaur bones that were transported through traction or saltation. Another clue may be gained by looking for a concentration of bones at the base of a filled channel structure, which indicates that less dense sediments were removed by flowing water. However, probably the best evidence for answering this question is gained through looking at the orientation of the bones, especially bones with appreciable lengths.

Water or air currents with sufficient momentum to move objects will tend to orientate long objects parallel to the predominant direction of the original flow, whether the flow was from water or air. Maps of bone quarries, such as the Cleveland-Lloyd or Howe quarries (Chapter 3), are extremely important in this respect because they show the distribution of skeletal parts in at least two dimensions. In such maps, long skeletal elements can be counted and plotted with vectors on a **rose diagram**. A rose diagram is so-called because a variety of directions indicated by the data can cause it to resemble a flower with petals. These diagrams can immediately show whether the bones have a preferred orientation, which may be a result of a unidirectional current flow (the simplest hypothesis). A rose diagram showing much scatter and no clearly definable, preferred direction to the bones may be a result of many different current flows, no currents, or redistribution of the bones by either biological or physical processes after they were initially deposited (Fig. 7.7). Indeed, some dinosaur remains show well-defined orientations that, in association with diagnostic physical sedimentary structures, suggest current activity. In contrast, other

FIGURE 7.7 Hypothetical rose diagrams of bone orientations from one flow direction (left) and variable flow directions (right) with femur used to point in direction of flow.

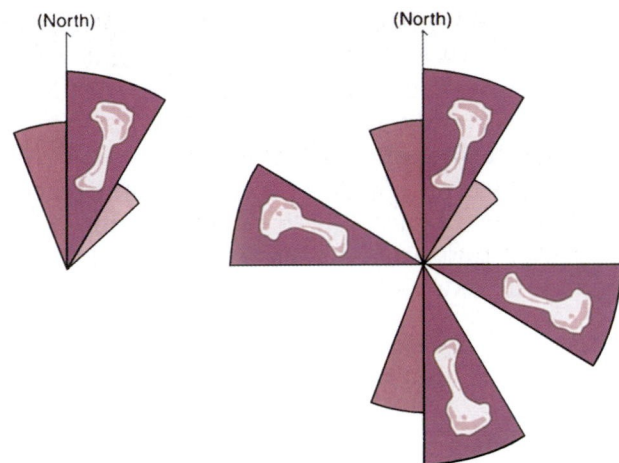

(North)   (North)

accumulations have random distributions that show no evidence of a current. The latter type of deposit has been hypothesized as a product of gradual accumulation of bones in watering holes or quicksand, which has been suggested for the unusually large assemblage of *Allosaurus* remains in the Cleveland-Lloyd Quarry of Utah (Chapters 3 and 9). Adding another dimension (vertical) to the bone distributions can test this hypothesis. If some long bones, such as tibias and femurs, have vertical orientations, this may constitute evidence of trampling by other dinosaurs in the area that caused the bones to go from a horizontal to a more vertical position

Maps of bone distributions can give other important clues about the sedimentary situation responsible for a bone accumulation. For example, when the Howe Quarry was mapped, with its abundant bones contributed by many dinosaurs of several different species, a semi-circular pattern to the deposit became evident upon excavation (Fig. 3.11). Based on this map view, one explanation for this distribution is that the deposit represents a **crevasse splay**, which is a sediment body formed outside the levee of a stream from a narrow breach of the levee (Fig. 7.8). In this scenario, a river would have been carrying a variety of dinosaur remains, then the flow broke through the levee of river and the remains were deposited in the temporary channel and small "delta" that resulted from the breach.

In summary, hypotheses applied to interpreting the transport of dinosaur remains are testable through examination of at least the following information:

- Describing dinosaur bone beds where hundreds of a single species of dinosaur have accumulated and comparing such monospecific assemblages with characteristics of modern occurrences of vertebrates (such as mammals or birds) and their modes and distances of transport.

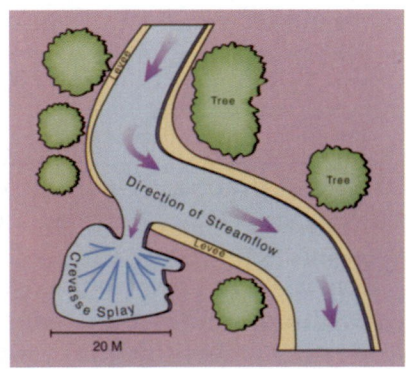

FIGURE 7.8 Map view of crevasse splay laterally adjacent to stream channel (compare with Fig. 3.11).

- Experimenting with different rates of flow and sedimentary particles (including teeth and bones) of different densities, shapes, and sizes, then comparing these results to the calculated values for dinosaur bodies or body parts.
- Looking for evidence of abrasion (or lack of) in dinosaur skeletal material and looking for similar evidence from experimentally treated modern bones.
- Carefully examining the sedimentary characteristics (lithofacies and ichnofacies) of the rock enclosing the dinosaur bones for evidence that argues for high flow rates, such as well-defined ripple bedding, large particle sizes, concentrations of bone material in the bases of channel fills, or a lack of trace fossils.
- Mapping the distribution and orientation of bones in a bone bed to detect whether the bones show a preferred or random direction, which may indicate the former presence or absence of water currents.
- Comparing the known geologic ranges of other fossil species or the radiometric age dates of any associated volcanic ash deposits (Chapter 4) with the geologic ranges of certain dinosaur remains to see if they seem too "young." If the fossils occur in deposits stratigraphically above where they normally occur in deposits worldwide, they were likely reworked.

Hopefully, the most important point gained from the preceding section is that when a nearly complete dinosaur skeleton is found, a paleontologist should never assume that this skeletal integrity represents an autochthonous specimen. In some cases this might be a reasonably accurate assessment, but in others the dinosaur might be kilometers away from where it met its demise.

Postmortem Processes: Accumulation, Burial, and Post-burial

Accumulation and Burial

An accumulation of dinosaur bones in the geologic record may indicate the following:

- A lack of movement sufficient to have carried the original bodies or bones from the site of the dinosaurs' death (an autochthonous assemblage).
- The cessation or insufficient momentum of a flow that carried the bodies or bones to that spot.
- A reworking of bones, possibly from several pre-existing bone beds, into a concentrated deposit.

Dinosaur paleontologists should be very interested in a mass accumulation of dinosaur bones determined as autochthonous. This is because such an assemblage presumably contains a representative sample of dinosaurs that lived in the same environment and died at approximately the same time, which provides direct information regarding dinosaur communities. Almost complete skeletons of dinosaurs, also of great interest to paleontologists, present another interesting problem in taphonomy. Did the bodies stay exposed above a surface for a while, subject to biological processes of decay and disintegration (yet still staying mostly together)? Or were they buried immediately after death, thus escaping scavengers or physical weathering?

Massive bone beds, such as those of the Howe Quarry, Cleveland-Lloyd Quarry, and Dinosaur National Monument area, certainly have been the source of much important paleontological information on dinosaurs. However, such beds also have taphonomic significance, providing data about the paleoenvironmental context of the dinosaurs. Additionally, their associated geological settings can be used for predicting other dinosaur-bearing zones in the fossil record. Perhaps the most germane question is whether the dinosaur remains are autochthonous, and thus resemble a **biocoenosis** (life assemblage), or are allochthonous and comprise a **thanatocoenosis** (death assemblage), which has major implications in reconstructing the paleoecology of a dinosaur site.

Co-occurrences of potential predators with prey in a small area, such as several *Deinonychus* (a rather fearsome theropod: Chapter 9) and an individual *Tenontosaurus* (a ornithopod: Chapter 11) found in an Early Cretaceous deposit of Montana, are intriguing to some paleontologists. They have speculated that these associated remains represent actual interactions between the species and thus reflect biocoenoses. Nevertheless, extreme caution must be practiced to consider all of the taphonomic variables before proposing any dramatic interpretations on the basis of associated remains alone. After all, each individual animal in an assemblage may have lived far apart from one another in space, time, or both, and all of them may have had their own separate pre-burial and burial histories. More convincing examples of individual dinosaurs that likely interacted with one another are provided by those entangled in dynamic poses in the same deposit, as discussed below.

Autochthonous remains of dinosaurs, whether found as a few individuals or as mass accumulations, are interpreted as a result of sediment traps, such as the Late Jurassic theropod *Allosaurus* of the Cleveland-Lloyd Quarry and the abundant remains of the Late Triassic prosauropod *Plateosaurus* in Germany. Drought-related mortality, perhaps associated with a watering hole, is another possibility for autochthonous assemblages; a Late Cretaceous example, composed primarily of the hadrosaur *Parasaurolopus* (Chapter 11) and ceratopsian *Styracosaurus* (Chapter 13), was proposed as being a result of this process. Fissure accumulations, where dinosaurs either fell into depressions or their bones accumulated in them, provide yet another way to form autochthonous dinosaur fauna. This hypothesis was proposed for the beautifully preserved Bernissart *Iguanodon* specimens, first described by Louis Dollo in the nineteenth century, from the Early Cretaceous of Belgium (Chapter 3). A large, monospecific assemblage of the Late Cretaceous hadrosaur *Hypacrosaurus* in Montana is interpreted as a population that died from a volcanic ashfall. As a result, it is considered as autochthonous, too. More recently, 34 specimens of the neoceratopsian *Psittacosaurus* from the Early Cretaceous of China were discovered close together in a volcanic ash deposit. The pristine and complete nature of this monospecific group of dinosaurs also argues for their burial in the same place where they died (Chapter 13).

Autochthonous assemblages are typically characterized by low-energy facies, little evidence of abrasion, perhaps some evidence of subaerial weathering, and low diversity. Although similar characteristics have been described for the previously-mentioned monospecific beds of *Coelophysis*, evidence for high-energy facies and lack of abrasion of those bones also suggests that bodies were transported whole or nearly whole as a group. A lack of scavenging traces also indicates that bodies were probably buried quickly, meaning that this criterion can be applied to either autochthonous or allochthonous assemblages, depending on their burial rate.

Mass accumulations of dinosaur remains can be the product of any number of possibilities, but what all of these scenarios have in common is that the dinosaur remains eventually had to be buried, probably within four to six weeks after death. Longer periods of time before burial would show evidence of the onslaught of biological and chemical processes that otherwise would have affected these body parts. Consequently, rapid burial in sediments that were conducive to preserving the dinosaur body parts was almost a prerequisite for fossilization, whether the dinosaur was already dead and was later buried, or died while becoming buried.

7

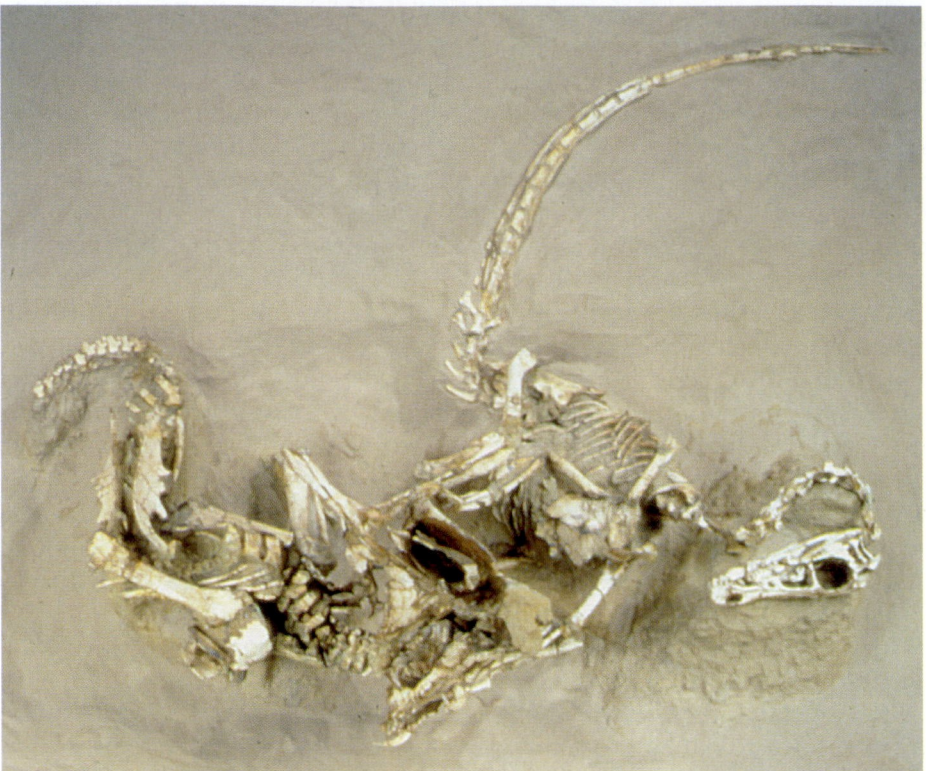

FIGURE 7.9 *Protoceratops* (left) and *Velociraptor* (right) in close proximity to one another (nicknamed "The Fighting Dinosaurs") from the Late Cretaceous of the Gobi Desert, Mongolia. These specimens likely represent rapid burial and autochtonous fossils. Transparency No. 18973, Courtesy Department of Library Sciences, American Museum of Natural History.

Rapid burial of dinosaur remains has been mostly attributed to deposition by river floods, but sandstorms have also been proposed as a mechanism. The unusually well-preserved Late Cretaceous ceratopsians, ankylosaurs, and theropods of the Gobi region in Mongolia are often cited as examples of such rapid burial (Chapters 9, 12, and 13).

With reference to the Gobi, the most famous example of individual dinosaurs that were likely interacting with one another when they were buried comes from the Late Cretaceous. In this instance, the complete skeletons of the theropod *Velociraptor* and ceratopsian *Protoceratops* were found with the *Velociraptor* lying on its right lateral surface with a fore limb in the mouth region and a hind limb in the abdominal region of the *Protoceratops*. The latter dinosaur is oriented with its dorsal side up (in normal "life" position). This extraordinary co-occurrence, found by one of the Polish expeditions (Chapter 3), has been nicknamed "The Fighting Dinosaurs" and indeed such an interpretation is quite reasonable in the light of the evidence (Fig. 7.9). Another dinosaur that seemed to have been stuck in time, frozen in a life position from the Late Cretaceous, is a specimen of *Oviraptor* that shows a body posture consistent with that of a brooding female over a nest of eggs. Preservation of this posture implies that the dinosaur was likely buried while protecting its nest. Just to show that this find, as startling as it might seem, is probably not unusual in terms of the behavior it implies, a similar specimen of *Oviraptor* situated above a nest of eggs was also found in strata of the same age in Mongolia. Moreover, most of the skeleton of a small theropod from the Late

Cretaceous of Montana, *Troodon*, was found on a clutch of eggs that contained embryonic remains of the same species (Chapter 9).

The taphonomic problem with such apparent "snapshots" from the Mesozoic is that they show little to no signs of struggle from whatever burial process preserved the dinosaurs. In other words, if the dinosaurs in "life position" were buried alive, then the burial was so rapid that they died before they could react to being buried or they were unable to move once buried. The "fighting dinosaurs," the brooding *Oviraptor*, and other numerous, beautifully preserved Late Cretaceous dinosaurs of the Gobi all show that burial was instantaneous and death must have been immediately before or simultaneous with burial. The traditional explanation for the extraordinary preservation of the Gobi dinosaurs is that they were buried rapidly by fierce sandstorms, which still regularly pound the Gobi Desert today. Perhaps these modern dramatic changes in weather influenced such interpretations. Beginning with paleontologists of the American Museum expeditions of the 1920s, they would have found such storms a useful (and harrowing) modern analogue (Chapter 3). However, these interpretations incorporated little detailed sedimentological evidence that convincingly demonstrated sandstorms during the Late Cretaceous caused rapid burial of dinosaurs.

A recently proposed alternative hypothesis invokes the sudden deposition of wet sand (associated with rain storms) onto the hapless animals within alluvial fans. An equivalent volume of wet sand is much heavier than dry sand, which would have limited the movement of any animal trapped in it and allowed for a minimum of struggling before they succumbed to suffocation. Other evidence arguing against sandstorm burial is the lack of physical structures and sediments that typify wind-blown (**eolian**) deposits in the thick deposits containing some of the best dinosaur remains. Furthermore, many of the dinosaur tracks and invertebrate trace fossils that have been found in what are interpreted as eolian deposits indicate some breaks in sedimentation. These facies also lack abundant dinosaur body fossils. Consequently, the new hypothesis is that stable dunes existed first, where dinosaurs lived in fruitful abundance during times of moist climate, and these animals were buried later by alluvial fans caused by rainstorms. Although other facies in Late Cretaceous deposits of Mongolia clearly reflect wind deposition, probably from sandstorms, the rocks of undisputed eolian origin do not contain abundant skeletal remains, let alone the complete skeletons found in other facies.

Another way for individual dinosaurs to have accumulated and been preserved temporarily before burial would have been through deposition at the bottom of a lake, swamp, lagoon, or marsh. In this scenario, burial could have been slow in comparison to burial rates in other environments. Modern examples of these aquatic environments have more oxygenated water toward the air–water interface but less oxygenated (anaerobic) water toward the sediment–water interface. The stagnant nature of these environments allows little circulation of oxygenated water toward their bottoms. The anaerobic conditions prevent many animals from living at the bottom, preventing scavenging and aerobic bacterial decay. Accordingly, this results in a high degree of fossil articulation and, in some cases, soft-tissue preservation. For example, the small, well-preserved Lower Cretaceous theropods of China and Italy (Chapters 5 and 9) come from lacustrine and marsh deposits, respectively.

Even if dinosaur bodies or body parts were rapidly buried or buried slowly under anaerobic conditions, they had to be buried deeply enough that scavengers could not dig them up or they were not susceptible to being uncovered by rapidly moving water or wind. No doubt many dinosaurs died in the many catastrophic ways described previously in this chapter, but they did not ultimately make it into museums unless they reached the next taphonomic step, **diagenesis**.

7

Post-Burial Processes: Diagenesis and How Bones Stayed Preserved

Diagenesis typically involves biological, chemical, and physical processes in a sediment or rock that are capable of changing an organism's remains to the point of completely erasing any record that the organism's body ever existed (although the same organism might have left trace fossils: Chapter 14).

Once buried deeply enough to prevent their immediate exhumation, dinosaur remains were potentially subjected to many chemical and physical processes between their time of burial and being discovered. Diagenetic processes that favor the preservation of bodies or traces left by organisms are often synonymized with the term **fossilization**. However, the totality of taphonomic processes affect whether a fossil becomes preserved, not just the post-burial ones. Hence "fossilization" is better applied as a synonym for taphonomic processes in general.

A dinosaur body buried in sediment radically changed the character of the sediment immediately surrounding it. For example, any organic matter in the body would have been affected by anaerobic bacteria, which produce specific gases (such as CO_2) as by-products of their metabolism. Such activity potentially changed the geochemistry of the entombing sediment. To a lesser extent, aerobic bacteria also might have contributed to the biological activity, depending on whether a dinosaur body was in contact with oxygenated water. Of course, a dead body representing "worm food" is also one of the most popularly known taphonomic processes, and undoubtedly dinosaurs (especially the large ones) provided many contented worms or insect larvae with full meals as they consumed any soft tissue in a buried body.

Besides biological factors, the geochemical conditions of **groundwater** (any water occurring below the Earth's surface) are responsible for major changes imparted on buried bodies or body parts, perhaps in combination with the microbial processes. For example, relative acidity is a major factor in the preservation of bones because bones are easily dissolved in acidic groundwater. Acidity is measured by **pH**, the negative logarithm of the hydrogen ion concentration in a liquid. This measurement has a scale of 1 to 14, with 1 the most acidic (having the highest concentration of hydrogen ions in solution) and 14 the most basic (lowest concentration). For example, if a soft drink has a hydrogen ion concentration of $1 \times 10^{-2.5}$ per unit of water, then the negative logarithm of that concentration is 2.5, hence its pH = 2.5. This pH is orders of magnitude more acidic than milk, which has a pH of about 8.0 (1×10^{-8} per unit of water). The presence of a high concentration of hydrogen ions breaks down bonds in some minerals (including dahllite), although some soft tissues are preserved by the same acidic waters. Pure water, containing no dissolved elements, has a neutral pH of 7.0, but almost all natural waters (including groundwater) are slightly acidic and thus capable of dissolving bones. As a result, groundwater at the time of burial for a dinosaur should have had a relatively low acidity (high pH) and all groundwater afterward should likewise have had a pH of about 5.0 or above to ensure that the bones remained intact.

A consequence of groundwater containing dissolved elements is that some of these elements can enter abundant pores of bones and precipitate as mineral combinations, thereby filling the spaces that were originally occupied by organic material in the bones when the dinosaurs were alive. This filling of pores (**permineralization**) by minerals such as quartz, calcite, pyrite, or apatite was extremely important for preservation of dinosaur bones because they added a solidity that resisted compaction from what might have eventually become kilometers-thick piles of overlying sediment.

Additional physical stresses on dinosaur bones could include tectonic processes. Convergent plate, divergent plate, or transform fault zones (Chapter 3) may have squeezed the rocks and wiped out more than one potential prize specimen, millions of years before hominids had evolved to the point of recognizing dinosaurs as formerly living creatures.

Elements that normally occur in low concentrations in the Earth's crust (**trace elements**) also can be concentrated through permineralization in bones. For example, dinosaur bones of the Late Jurassic Morrison Formation in Utah, examined for their geochemical content, contain barium, cesium, chromium, fluorine, lanthanum, lead, manganese, nickel, rubidium, strontium, thorium, uranium, and vanadium. Consequently, permineralization also caused fossilized dinosaur bones to take on elements that make them "hot". Uranium preferentially accumulated in bones because of its affinity to bonding with a dahllite mineral structure. As a result, many dinosaur bones are radioactive and give off alpha and beta radiation in particular (Chapter 4). Uranium occurs naturally as two different isotopes, ^{235}U and ^{238}U; ^{238}U comprises 99.3% of all uranium and ^{235}U the remaining 0.7%. Both isotopes are soluble in water and are easily carried in groundwater under aerobic conditions, but will precipitate in combination with oxygen in the mineral **uranitite** (uranium oxide) under anaerobic conditions.

Another mode of fossil preservation that differs from permineralization but might interrelate with it is **replacement**, in which the dahllite in the bone was replaced by a different mineral, yet the original bone structure is still recognizable for its original structure. Apparently this mode of preservation in vertebrate bones is relatively uncommon in comparison with permineralization. Except for substitutions between OH^-, and F^- with one another and CO_3 with PO_4 in the dahllite structure (forming variations on it), the mineral part of bone stays more or less the same as when a dinosaur was alive.

Mineralization, whether it occurs as permineralization or replacement, can occur rapidly, anywhere from a few hours to a few years. This is contrary to the popular conception of fossilization as an extremely long, gradual process. Studies of relatively young bones, such as those from large mammals from only 10,000 years ago, reveal that a lack of permineralization resulted in poor preservation of the bones, in direct contrast to well-preserved and durable bones with their pores filled by mineral matter. Dinosaur fossils with preservation of soft tissue exemplify the best evidence for fast rates of mineralization. This mode of preservation required anaerobic conditions and probable bacterial mediation of chemical reactions that fixed certain elements to make a recognizable facsimile of the original organic structure. Indeed, some fossil fish from the Early Cretaceous Santana Formation of Brazil exhibit finely-detailed musculature and gills that were replaced by phosphatic minerals, a process that had to have happened within a few hours. This is not to say that all preservation processes were rapid, but some of them were, which is the subject of the next section.

Preservation of Dinosaur Skin Impressions and Soft-Part Anatomy

Some of the most memorable dinosaur finds in the history of dinosaur studies are of those that either have skin impressions, such as hadrosaur specimens (*Edmontosaurus*) discovered by the Sternbergs in Late Cretaceous strata of Alberta (Chapter 3), or otherwise show evidence of soft parts. However, not all dinosaur skin was soft, as evidenced by a few sauropods and most thyreophorans (Chapters 10 and 12). Skin impressions were typically preserved as fossils through **mummification**, a loss of water from a body or body part that caused some shriveling but otherwise preserved much of the soft tissue for a long enough period to have made an impression. In this case, the impression was literally that: a buried dinosaur body underwent dehydration and was surrounded by sediment that pressed around the outside of the body and made an **external mold**, which included an impression of its skin (Fig. 7.10). If the body decayed but the external

7

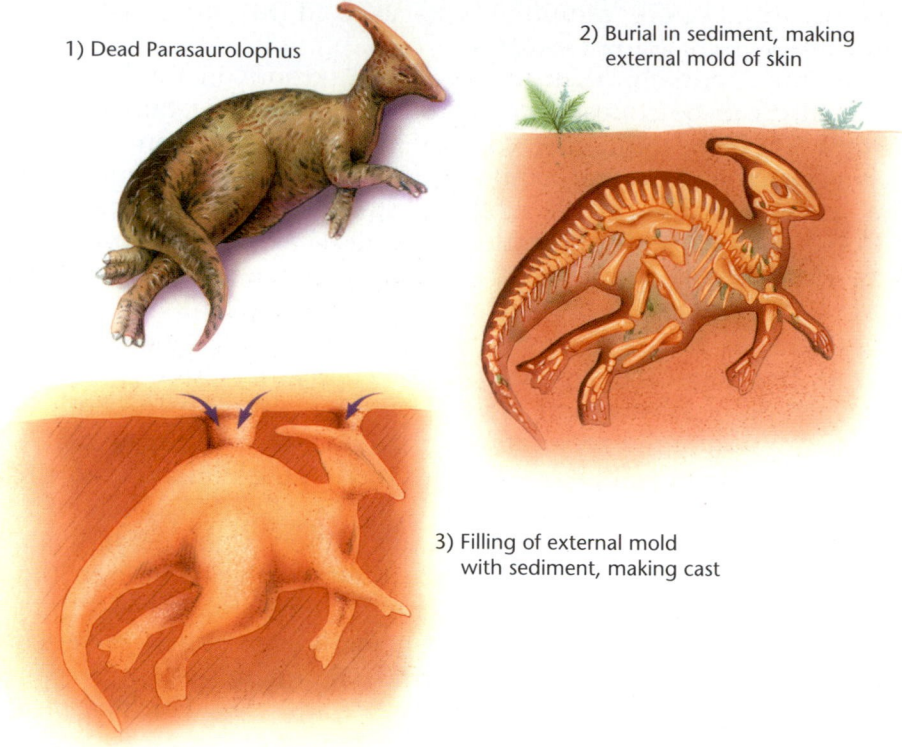

1) Dead Parasaurolophus

2) Burial in sediment, making external mold of skin

3) Filling of external mold with sediment, making cast

FIGURE 7.10 How an external mold and cast could have been made of a dinosaur body, preserving skin impressions (such as seen in Fig. 5.10).

mold remained, and sediment later filled the cavity left by the body and solidified, then the resulting fossil is a **cast**, which is preservationally analogous to a cranial endocast (Chapter 5). Skin impressions are currently known for ornithopods, sauropods, ceratopsians, ankylosaurs, and theropods. More will surely be found as researchers become more vigilant for their possible presence in association with dinosaur bones.

As mentioned before, anaerobic conditions are also conducive to preservation of soft parts, and recent finds of theropods with feather impressions and internal organs are evidence of a low rate of organic degradation that enabled sufficient preservation of at least the outlines of these anatomical features. **Carbonization** is the preservation of soft parts by a loss of **volatiles**, which are elements such as hydrogen, oxygen, and nitrogen that are normally gaseous at near-surface temperatures, their loss resulting in a carbon film being left behind. This mode of preservation has helped paleontologists discern at least outlines for soft tissues in fossil plants, invertebrates, and vertebrates alike. If diagenesis happened rapidly enough, so that phosphatic minerals replaced the original organic material, a pseudomorph of soft tissues (such as muscles) can be preserved (Chapter 5). This type of preservation gives more of a three-dimensional character to the parts than mere carbonization.

Finally, recognition of biomolecules and flexible soft tissue in dinosaur bones (Chapter 8) is an example of how dinosaur paleontologists should still look for remnants of organic remains in dinosaurs. Assuming preservation in anaerobic conditions and "sealing" of biomolecules from the ravages of diagenetic processes, dinosaurs may yet yield more amino acids, proteins, and possibly nucleic acids that provide further information related to their evolutionary relationships (Chapter 6). Most recently, flexible soft tissues recovered from the center of a *Tyrannosaurus* femur show cell nuclei and other remarkable cellular features, suggesting that such preservation may be more common than previously thought.

SUMMARY

Because most dinosaur species are based on so few described specimens, an understanding of how these specimens became preserved (or why their relatives were not preserved) provides a perspective on the fossil record for dinosaurs that acknowledges its scientific weaknesses in places but its strengths in other areas. Taphonomy provides considerable scientific information about dinosaur paleobiology, and contributes to integrated analyses of the characteristic lithofacies, biofacies, and ichnofacies of sedimentary environments. Taphonomy also examines the biological, chemical, and physical processes that occur from the time a dinosaur died to when it was buried (its necrolysis and biostratinomy), as well as what happened to its remains after it was buried (diagenesis).

Bacterial decay, through both anaerobic and aerobic bacteria, is a largely unseen process that accounts for most cycling of nutrients from dead animals in modern environments. If this same process had not occurred in the Mesozoic, numerous dinosaur carcasses would still impede daily life today. Scavenging, a feeding behavior exhibited by most humans, is also better appreciated through knowledge of all of the potential scavengers that fed on dinosaurs, including insects, crustaceans, reptiles, mammals, birds, and other dinosaurs. Taphonomy places dinosaurs in the context of their original ecosystems or more clearly defines their potential interactions. Understanding how dinosaur bodies or parts of their bodies might have been transported far away from their death sites by stream or ocean currents (as allochthonous deposits) can help define where dinosaurs lived or did not live. Dinosaurs that were apparently buried in place (autochthonous) provide the most desirable situation for working out the paleoecology of dinosaurs.

Knowing how bodies or parts could have been transported and how they were deposited not only provides more information for predicting dinosaur-bearing zones but also gives a good education in the hydrodynamics of water flow in association with sediment parameters (density, size, shape, etc.) and, to a lesser extent, aerodynamics. The dinosaur remains that are seen today were, in most cases, either buried rapidly or placed in an anaerobic environment before being buried. Most often, they were preserved through permineralization. However, replacement, formation of external molds and casts, carbonization, or a combination of these diagenetic processes also may have preserved their bodily remains. Thus, study of diagenesis requires extensive knowledge of groundwater and its characteristics (such as pH and dissolved elements), and how it could interact biogeochemically with dinosaur remains.

DISCUSSION QUESTIONS

1. Think of what animals you have seen recently that had been killed by cars. Were any kinds of animals more commonly killed than others? Why would some animals be more susceptible to being killed?

2. Of the dead animals you've seen, how long would you estimate some of them had been dead? What are some methods for testing your initial hypothesis? What did you learn from this chapter that has changed how you previously would have made such an evaluation?

3. Law-enforcement agencies often use principles of taphonomy to solve crimes. What are some other applications that you can think of that are already in use or could be used?

4. Which of the following modern assemblages of bodies or traces would you term as autochthonous, allochthonous, or a combination of the two, and why?
   a. A pile of leaves raked from a yard.
   b. Dog tracks preserved in a cement sidewalk.
   c. Animal parts in a butcher shop.
   d. Piles of hair from different individuals swept just outside the door of a hair-dressing salon.
   e. Peanut husks on a bar-room floor.

5. While working as a volunteer on a dinosaur excavation, you find a small (less than 1-meter long) theropod skeleton that is nearly complete, with beetle borings on the bones of its right lateral surface. It is in the bottom of an ancient fluvial channel deposit. Based on this information, develop at least two possible scenarios for the biostratinomic history of the theropod. What possible new information, found at the site or in the skeleton, would falsify each of your hypotheses?

6. An assemblage of dinosaur bones found in an overlying layer at the same site shows numerous bones of the appendicular and axial skeletons (no skulls) of many individuals representing the same species of hadrosaurid. The bones are arranged in such a way that a rose diagram reveals two preferred directions (northeast and southwest) for the remains. The assemblage is mixed with some large theropod (tyrannosaurid) teeth; otherwise no theropod remains are present. Develop at least two possible scenarios for the biostratinomic history of the hadrosaurids. What possible new information, found at the site or in the skeleton, would falsify each of your hypotheses, especially related to the lithofacies (not described here)?

7. What are the discharge and momentum for streams with the following cross-sectional areas and velocities? (For momentum, assume a velocity of 1.0 m/s.)
   a. 22.3 m^2, 1.2 m/s
   b. 76.4 m^2, 0.7 m/s
   c. 55.1 m^2, 2.3 m/s

DISCUSSION QUESTIONS Continued

8. Use the data from Chapter 1, Question 10, to calculate the immersed weight in seawater (with a density of 1.02 g/cm³) of the following dinosaurs: *Camarasaurus*, *Allosaurus*, *Pachycephalosaurus*, and *Brachiosaurus*.

9. The area and volume of a cylinder are $A = 2\pi rh + 2\pi r^2$ and $V = h(\pi r^2)$, respectively, where h is the height (or length) of the cylinder. Assuming that the following bones approximate a cylinder in their shapes, what are their A/V ratios?
   a. Femur of Late Triassic prosauropod *Plateosaurus*; 57 cm long, 11 cm wide.
   b. Humerus of Early Jurassic theropod *Syntarsus*; 9 cm long, 1.2 cm wide.
   c. Humerus of Late Jurassic sauropod *Brachiosaurus*; 176 cm long, 39 cm wide.
   d. Metacarpal of Early Cretaceous ornithopod *Hypsilophodon*; 8.3 cm long, 1.1 cm wide.
   e. Tooth (including root) of Late Cretaceous ankylosaur *Edmontonia*; 3.8 cm long, 0.8 cm wide.

   Once you have calculated these values, rank them in order of least to most susceptible to movement by a flow. What variable dimensions of each bone might change your calculations (especially for the assumption of a cylinder for bone diameter)?

10. Using the calculated volume for the *Brachiosaurus* humerus in the preceding question, assume that 35% of the bone volume was occupied by pores. Using a density of 3.15 g/cm³ for dahllite, calculate the "dry" weight of the bone (what it would have weighed after taking out the original organic material). After that, calculate how much it would weigh after that 35% porosity was permineralized with quartz, which has a density of 2.65 g/cm³. (Then you might appreciate better why museum mounts of sauropods rarely use original bone!) Discuss the probability of a permineralized bone being transported in a stream, versus a fresh bone.

Bibliography

Allison, P. A. and Briggs, D. E. G. (Eds). 1991. *Taphonomy: Releasing the Data Locked in the Fossil Record. Topics in Geobiology.* New York: Plenum Press.

Behrensmeyer, A. K. 1991. Terrestrial vertebrate accumulations. *In* Allison P. A. and Briggs D. E. G. (Eds), *Taphonomy: Releasing the Data Locked in the Fossil Record.* New York: Plenum Press. pp 291–335.

Behrensmeyer, A. K. 1978. Taphonomic and ecologic information from bone weathering. *Paleobiology* **4**: 150–162.

Behrensmeyer, A. K. 1988. Vertebrate preservation in fluvial channels. *Palaeogeography, Palaeoclimatology, Palaeoecology* **63**: 183–199.

Behrensmeyer, A. K. and Hill, A. P. (Eds). 1980. *Fossils in the Making: Vertebrate Taphonomy and Paleoecology.* Chicago: University of Chicago Press.

Boggs, S., Jr. 1995. *Principles of Sedimentology and Stratigraphy* (2nd Edition). Englewood Cliffs, New Jersey: Prentice Hall.

Briggs, D. E. G. 1995. Experimental taphonomy. *Palaios* **10**: 539–550.

Chen, P., Dong, Z. and Zhen, S. 1998. An exceptionally well-preserved theropod dinosaur from the Yixian Formation of China. *Nature* **391**: 147–152.

Cook, E. 1995. Taphonomy of two non-marine Lower Cretaceous bone accumulations from southeastern England. *Palaeogeography, Palaeoclimatology, Palaeoecology* **116**: 263–270.

Currie, P. J. and P. Dodson. 1984. Mass death of a herd of ceratopsian dinosaurs. *In* Reif W. E. and Westphal F. (Eds), *Third Symposium on Mesozoic Terrestrial Ecosystems*. Tübingen, Germany: Attempto Verlag. pp 61–66.

Davis, P. G. and Briggs, D. E. G. 1998. The impact of decay and disarticulation on the preservation of fossil birds. *Palaios* **13**: 3–13.

Dodson, P. 1990. Counting dinosaurs: how many kinds were there? *Proceedings of the National Academy of Sciences* **87**: 7608–7612.

Dong, Z. and Currie, P. J. 1996. On the discovery of an oviraptorid skeleton on a nest of eggs at Bayan Mandahu, Inner Mongolia, People's Republic of China. *Canadian Journal of Earth Sciences (Journal Canadien des Sciences de la Terre)* **33**: 631–636.

Donovan, S. K. (Ed.). 1991. *The Processes of Fossilization*. New York: Columbia University Press.

Efremov, J. A. 1940. Taphonomy, a new branch of paleontology. *Pan-American Geologist* **74**: 81–93.

Fastovsky, D. E., Badamgarav, D., Ishimoto, H., Watabe, M. and Weishampel, D. B. 1997. The paleoenvironments of Tugrikin-Shireh (Gobi Desert, Mongolia) and aspects of the taphonomy and paleoecology of *Protoceratops* (Dinosauria: Ornithischia). *Palaios* **12**: 59–70.

Fiorillo, A. R. 1991a. Prey bone utilization by predatory dinosaurs. *Palaeogeography, Palaeoclimatology, Palaeoecology* **88**: 157–166.

Fiorillo, A. R. 1997. "Taphonomy". *In* Currie, P. J. and Padian, K. (Eds), *Encyclopedia of Dinosaurs*. San Diego, California: Academic Press. pp 713–716.

Hill, A. 1980. "Early post-mortem damage to the remains of some contemporary East Africa mammals". *In* Behrensmeyer, A. K. and Hill, A. P. 1980. *Fossils in the Making: Vertebrate Taphonomy and Paleoecology*. Chicago: University of Chicago Press. pp 131–152.

Hubert, J. F., Panish, P. T., Chure, D. J. and Prostak, K. S. 1996. Chemistry, microstructure, petrology, and diagenetic model of Jurassic dinosaur bones, Dinosaur National Monument, Utah. *Journal of Sedimentary Research A: Sedimentary Petrology and Processes* **66**: 531–547.

Knutson, R. M. 1987. *Flattened Fauna: A Field Guide to Common Animals of Roads, Streets, and Highways*. Berkeley, California: Tenspeed Press.

Kruuk, H. 1972. *The Spotted Hyena: A Study of Predation and Social Behavior*. Chicago: University of Chicago Press.

Loope, D. B., Dingus, L., Swisher, C. C., III. and Minjin, C. 1998. Life and death in a Late Cretaceous dune field, Nemegt basin, Mongolia. *Geology* **26**: 27–30.

Lyman, R. L. 1994. *Vertebrate taphonomy*. Cambridge Manuals in Archaeology Series, Cambridge, UK: Cambridge University Press.

Machel, H. G. 1996. Roadkill as teaching aids in historical geology and paleontology. *Journal of Geoscience Education* **44**: 270–276.

Martill, D. M. 1991. Organically preserved dinosaur skin; taphonomic and biological implications. *Modern Geology* **16**: 61–68.

Martill, D. M. 1991. "Bones as stones: the contributions of vertebrate remains to the lithologic record". *In* Donovan, S. K. (Ed.), *The Processes of Fossilization*. New York: Columbia University Press. pp 270–292.

Norell, M. A., Clark, J. M., Chiappe, L. M. and Dashzeveg, D. 1995. A nesting dinosaur. *Nature* **378**: 774–776.

Roberts, E. M. and Rogers, R. R. 1997. Insect modification of dinosaur bones from the Upper Cretaceous of Madagascar. *Journal of Vertebrate Paleontology* **17**, Supplement 71.

Rogers, R. R. 1990. Taphonomy of three monospecific dinosaur bone beds in the Upper Cretaceous Two Medicine Formation, northwestern Montana: evidence of mass mortality related to episodic drought. *Palaios* **5**: 394–413.

Schäefer, W. 1962. *Aktuo-Paläontologie, nach Studien in der Nordsee*. Frankfurt, Germany: Verlag w. Kramer.

Schwartz, H. L. and Gillette, D. D. 1994. Geology and taphonomy of the *Coelophysis* quarry, Upper Triassic Chinle Formation, Ghost Ranch, New Mexico. *Journal of Paleontology* **68**: 1118–1130.

Schwimmer, D. R., Williams, G. D., Dobie, J. L. and Siesser, W. G. 1993. Late Cretaceous dinosaurs from the Blufftown Formation in western Georgia and eastern Alabama. *Journal of Paleontology* **67**: 288–296.

Schwimmer, D. R., Stewart, J. D. and Williams, and Dent, G. 1997. Scavenging by sharks of the genus *Squalicorax* in the Late Cretaceous of North America. *Palaios* **12**: 71–83.

Shipman, P. 1975. Implications of drought for vertebrate fossil assemblages. *Nature* **257**: 667–668.

Shipman, P. 1981. *Life History of a Fossil: An Introduction to Taphonomy and Paleoecology*. Cambridge, Massachusetts: Harvard University Press.

Varricchio, D. J. 1995. Taphonomy of Jack's Birthday Site, a diverse dinosaur bonebed from the Upper Cretaceous Two Medicine Formation of Montana. *Palaeogeography, Palaeoclimatology, Palaeoecology* **114**: 297–323.

Varricchio, D. J. and Horner, J. R. 1993. Hadrosaurid and lambeosaurid bone beds from the Upper Cretaceous Two Medicine Formation of Montana: taphonomic and biologic implications. *Canadian Journal of Earth Sciences* **30**: 997–1006.

Voorhies, M. R. 1969. Taphonomy and population dynamics of the early Pliocene vertebrate fauna, Know County, Nebraska. *Contributions to Geology* Special Paper No. 1.

Walker, R. G. and James, N. P. (Eds). 1992. *Facies Models: Response to Sea Level Change*. St. John's, Newfoundland: Geological Association of Canada, Memorial University of Newfoundland.

Weigelt, J. 1927. *Rezente wirbeltierleichen und ihre paläobiologische Bedeuntung*. Leipzig, Germany: Verlag von Max Weg. [Translated by Judith Schaefer and re-published as "Recent Vertebrate Carcasses and the Paleobiological Implications," 1989, Chicago: University of Chicago Press].

Williams, C. T. 1987. "Alteration of chemical composition of fossil bones by soil processes and ground water". *In* Grupe, G., et al. (Eds), *Trace Elements in Environmental History*. Heidelberg: Springer-Verlag. pp 27–40.

Wood, J. M., Thomas, R. G. and Visser, J. 1988. Fluvial processes and vertebrate taphonomy: the Upper Cretaceous Judith River Formation, south-central Dinosaur Provincial Park, Alberta, Canada. *Palaeogeography, Palaeoclimatology, Palaeoecology* **66**: 127–143.

7

Chapter 8

You have two pets, a large snake (a python) and a cat. Although both are carnivores, you notice a large difference in their feeding habits. You feed the snake a live mouse once every two weeks, which it swallows whole; this morsel satisfies it completely until the next time (two weeks later) you drop another mouse into its terrarium. In contrast, the cat eats dry food at least three times a day, and demands (rather insistently) to have wet food in both the morning and early evening. Much to your horror, he also tries to kill small birds that alight in your backyard to eat seed from a bird feeder. Compounding this problem, other birds are constantly on the bird feeder, knocking loose seeds that attract the ground-grazing birds. Squirrels also constitute a nuisance with their once-daily raiding of the bird feeder; their ingenuity in circumventing your barriers to the feeder is both admirable and frustrating, and provokes wonder about how much energy they spend on working out ways to get the food.

Why are there such differences in feeding habits in these animals? Why does the python, the largest of the animals, require so little food to sustain itself? How is it that the birds, which are the smallest, are feeding over the course of an entire day? What prompts the squirrels to come by only once (sometimes twice) a day, investing so much effort in working out how to get the food? Why does your cat require so much protein every day?

Dinosaur Physiology

8

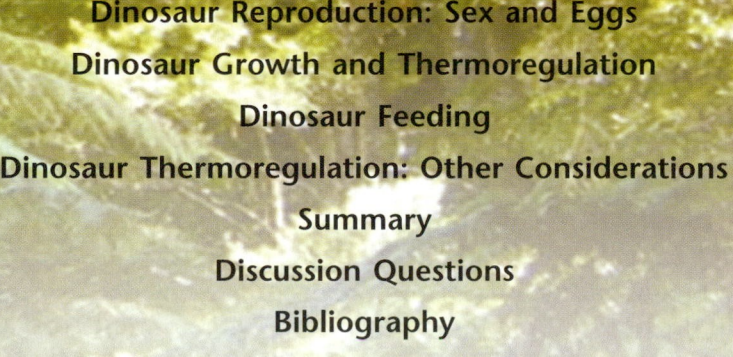

Why Learn About Dinosaur Physiology?

Physiology is how animals convert and transfer matter and energy in their daily lives. **Metabolism** is all the processes that they use to accomplish these tasks. Animals take in, use, and send out air, water, food and other nutrients that are used in ways determined by the cells of their bodies, which in turn were determined by genes, but also are affected by the surrounding environment. Because dinosaurs were living animals, undoubtedly they also converted matter and energy in accordance with their bodies and surrounding environments. The scientific questions that usually deal with these basic presumptions start with the word "How?":

■ How did dinosaurs reproduce, and how do their eggs provide information about the physiology of both mothers and offspring?
■ How did dinosaurs form bones and teeth, the most likely parts of a dinosaur to be preserved in the fossil record?
■ How quickly did dinosaurs grow, after they hatched from eggs?
■ How did dinosaur bones record clues about their metabolism, such as whether they generated their own body heat or depended on the outside environment for warmth?
■ How do other data besides bones and eggs, such as paleoenvironmental setting and trace fossils, provide insights into dinosaur physiology?

All of these pertinent questions, asked about dinosaur physiology for more than 150 years, have still not all been satisfactorily answered.

Dinosaur Reproduction: Sex and Eggs

Why Learn about Dinosaur Reproduction and Eggs?

Paleontologists have known about dinosaur eggs since early in the second half of the nineteenth century, although verifiable nests were not described and interpreted until the 1920s (Chapter 3). Despite this long history, the study of dinosaur eggs was neglected in favor of skeletal evidence and dinosaur tracks. Fortunately, the relatively lower status of these dinosaur fossils has improved in the past 20 years. This renewed study, originating from more than 200 egg and nest sites documented worldwide, includes the identification of dinosaur embryos and presumed parents associated with egg and juvenile remains. The most well-known dinosaur egg sites, some with embryos, are in Montana, France, China, Mongolia, and, most recently, Argentina.

Dinosaur eggs provide important evidence on how dinosaurs reproduced and, by extension, how egg laying and the life of an embryo in an egg reflect dinosaur physiology. Moreover, dinosaur nests lend insights into how dinosaurs ensured that their broods hatched and sometimes whether they cared for their young, which also relates to their physiology and behavioral relatedness to modern archosaurs. Dinosaur eggs are used increasingly as additional evidence for working out evolutionary relationships of different dinosaur clades. Indeed, eggs are now being classified through cladistics in an attempt to reconcile egg types with probable egg layers.

> *Like all birds, most reptiles, and a few mammals, dinosaurs reproduced through eggs formed within the body of the female and laid on land to hatch, a trait called* **oviparous**.

The oviparous trait associated with dinosaurs contrasts with giving live birth (**viviparous**), which is typical of most mammals and occurs in a few reptiles. In continental ecosystems, the formation of enclosed eggs, which probably happened slightly more than 300 million years ago, was an extraordinary evolutionary development in the history of vertebrates (Chapter 6). Unlike fish and amphibians, reptiles can migrate within continental interiors without depending on nearby water bodies. Thus, the architecture of fossil eggs provides clues as to how dinosaurs adapted to a variety of terrestrial ecosystems.

Of biogeochemical significance is the fact that some dinosaur eggshells contain amino acids, which may tell us about the dinosaurian production of those biomolecules in the absence of other evidence from their skeletal remains. Eggshells also contain carbon and oxygen isotopes, as well as some trace elements. These chemical clues provide information about the possible dietary preferences of the mother dinosaurs. These elements also may reflect environmental conditions in both the pre- and post-burial history of an egg.

Dinosaur Sex

A necessary antecedent of dinosaur eggs was dinosaur sexual activity. The certainty of dinosaur sex as a prelude to their laying eggs is supported by the numerous observations of how mating in all egg-laying vertebrates is a necessary precursor to egg development. Fertilization of an egg without the help of a male, known as **parthenogenesis**, is common in nearly every major invertebrate clade but is known in only a few vertebrates (some amphibians and lizards). As a result, this process probably does not apply to dinosaurs.

Dinosaur sexuality has been the subject of much debate on the basis of little scientific evidence. In fact, paleontologists are still uncertain about which dinosaurs in the fossil record represent male and female specimens (Chapters 5 and 6). A few researchers have recently promoted the case for sexual dimorphism in what are anatomically very similar dinosaur species from the same stratigraphic intervals. For example, some paleontologists have proposed that more robust forms of some dinosaurs (such as *Tyrannosaurus*) are females, based on a similar disparity of size seen in the sexes of modern large reptiles. Larger size of female reptiles, which is the opposite expectation for sexual dimorphism in most mammals, corresponds to a capacity to hold many eggs in reptilian body cavities.

Before mating occurs in modern vertebrates, males and females both use methods of sexual attraction in which they undergo **sexual selection**, or the choosing of their mates on the basis of preferred traits. When done by enough individuals within a population, this process ultimately affects the evolutionary history of a species by causing genetic change of that population over time (Chapter 6). Some birds that have colorful or prominent plumage in one gender, such as peacocks, provide examples of sexual selection, in which they cause a visual stimulus for a potential mate. In this respect, the elaborate and prominent head shields and horns of ceratopsians (Chapter 13) and skull crests of hadrosaurs (Chapter 11) have been proposed as possible display structures. In modern animals, displays are sometimes accompanied by other sensory signals, such as mating calls. Male crocodilians, for example, will bellow for attention and are sometimes answered by vocalizations from a nearby female. In at least some hadrosaurs, vocalization structures, also associated with cranial crests, have been postulated (Chapter 11).

In some instances visual and auditory stimuli are not the only cues to mating. For example, some animals use pheromones, which are complex biomolecules emitted by an organism into a water body or air for the purpose of causing a response in another individual of the same species. Pheromones can elicit numerous responses in organisms, such as triggering silent alarms, providing a trail for others to follow (seen readily in ants), or signaling aggregation, but they are often recognized for their sexually attractive qualities. Whether dinosaurs used pheromones or not is unknown. Many modern reptiles use olfactory sensations for mate attraction, but pheromones have not yet been detected in birds. Consequently, paleontologists have little basis for inferring this physiological function in dinosaurs through their hypothesized closest living relatives.

In many cases, once a female vertebrate shows receptivity to a male, mating will occur quickly relative to time spent in attraction and courtship. In fish and amphibians, the male's sperm is simply deposited in the water near a female rather than through more proximal association. Amniotes presumably developed sex organs that worked more effectively for getting gametes together through direct bodily contact. After all, internal fertilization was necessary before an egg could be developed internally. In some reptiles (crocodilians and turtles) the delivery of sperm into a female's oviduct, which also functions as the "birth canal" for egg laying, is sometimes facilitated by the insertion into the oviduct of a male's penis. Snakes and lizards have similar but smaller structures called **hemipenes**. Most modern bird species lack a penis, although some flightless birds and ducks, geese, and swans do possess such organs. Birds without penises mate through close contact of their **cloacae**, which are openings that double in function as outlets for **gametes** (sex cells, such as sperm and eggs) and excretion of bodily wastes in males and females, as well as egg-laying in females.

Because the preservation potential for non-mineralized dinosaur tissues is so low (Chapter 7), no direct evidence of dinosaur reproductive organs is known. Nevertheless, the arrangement of bones in the pelvic region of some dinosaurs may relate to muscles associated with penile organs in males. Furthermore, at least one example of a theropod was found with probable eggs in its body cavity (Chapter 9). This helped to corroborate the presence of an oviduct and (of course) its gender. Moreover, the statistically-defined pairing of eggs in the nest of another theropod strongly suggests that it laid the eggs through paired oviducts (Chapter 9), a trait seen in some modern crocodilians.

Dinosaur mating was probably achieved through the male approaching the female's caudal region and positioning its posteroventral anatomy in close association with the female's posterior. Such positioning was relatively straightforward for bipedal dinosaurs, but probably necessitated a temporarily bipedal posture for quadrupedal dinosaurs. A similar posture is demonstrated by modern elephants, which are the heaviest land animals available for such a comparison (Fig. 8.1). The rather lengthy tails of some dinosaurs, such as sauropods, seemingly would have been impediments to mating. Unfortunately, no trackways or other trace fossils have yet provided evidence of dinosaur mating habits. Likewise, no dinosaur skeletons of the same species have been found in what might be construed as a compromising position, even after applying much imagination to them. Documentation of trace fossils indicating dinosaur mating in particular is an area yearning for further, in-depth research, considering the amount of sedimentary disturbance that must

FIGURE 8.1 African elephants (*Loxodonta africanus*) mating, providing a model for mating positions of some large quadrupedal dinosaurs. The male is located posterodorsally with respect to the female. W. M. Colbeck/OSF/Animals Animals.

have been caused by the mating of some of these animals, which together comprised tens of metric tons.

Fertilization in modern vertebrates is the result of the uniting of gametes from a male and a female. The gamete from each parent contains half the number (**haploid**) of **chromosomes** (genetic material) of a typical body (**somatic**) cell from each parent, thus both parents together create a **diploid** cell. The formation of gametes by each parent is accomplished through a splitting of a diploid cell into haploid cells, a process called **meiosis**. The division of body cells into more body cells (diploid to diploid) is **mitosis**. A fertilized egg (**zygote**) subsequently results in mitotic division of cells through **cleavage**.

8

Dinosaur Eggs

For reptiles and most of their descendants, which includes dinosaurs and birds, an egg is an enclosed yet porous mineralized or organic structure that contained or contains an **amnion** (a fluid-filled sac) surrounding a developing embryo (Fig. 6.4).

An egg serves as a form of protection for an embryo that also keeps its nutrients in a restricted space while allowing the inflow of oxygen and exit of waste products (such as carbon dioxide) from the egg environment through its pores.

Dinosaur eggs and all other fossil eggs are body fossils. Although some paleontologists used to classify them as trace fossils, their explicit physiological function makes eggs distinctive from dinosaur trace fossils such as tracks, nests, toothmarks, coprolites, or gastroliths (Chapter 14). An eggshell secreted by a mother dinosaur was an integral, connected part of a developing embryo. This makes it an extra body part, analogous to the exoskeleton of an invertebrate, that was essential for survival of that embryo. The occasional inclusion of embryonic dinosaur remains within an egg provides a complete picture of an egg as a body fossil (Chapters 9 to 11).

In a modern reptilian or avian egg, the amnion forms around the embryo soon after cleavage. The enclosed fluid of the amnion suspends and thus protects the embryo from concussions or desiccation. The mother then secretes eggshell around the amnion, further protecting the embryo, and a second sac (**allantois**) develops between the eggshell and amnion. The allantois serves as a respiratory organ for the embryo, bringing in oxygen and giving off carbon dioxide. This type of egg is a **cleidoic** egg, which means that it provides a food supply (through the **yolk sac**) and a membrane for respiration, temperature maintenance, and waste disposal.

In oviparous animals, the egg is retained in the mother's body until a sufficiently protective layer for the developing embryo is secreted, which normally involves some **biomineralization** (formation of minerals by an organism). It is then laid outside of the mother's body for further development. Subsequent growth of the embryo within the enclosed environment of an egg for weeks afterward is made possible through the large yolk sac in the egg, which provides food. The formation of cartilage and bones happens during this time within the egg, in which the eggshell supplies calcium. Microscopic pits in the inner surface of an eggshell show where the embryo absorbed the calcium; such pitting has been described in some dinosaur eggs. Development of all other organs and muscles that are needed for an animal to hatch and move after hatching also occurs within the egg.

Pathological conditions brought on by environmental stresses, such as dehydration, are reflected by eggshell abnormalities. For example, a multilayered eggshell is a symptom of stress suffered by a mother. This condition develops when a mother retains an egg for a longer period of time than normal, such as during environmental stress, such as a drought. The physiological response of the mother is to form another membrane and shell layer on the previously complete egg. A few dinosaur eggs also show evidence of this paleopathologic condition (Chapter 7), which illustrates some of the reproductive problems faced by dinosaurian mothers, but physiological adaptations had already evolved in archosaurs.

The mineral material composing an eggshell is $CaCO_3$, either in the form of aragonite, found in turtles, or calcite, found in eggs of other reptiles, birds, and dinosaurs. Some organic materials, such as amino acids, also form in eggs and are documented from dinosaur eggshells. Eggs composed mostly of organic material are sometimes described as "leathery," a common descriptor for eggs from modern sea turtles and a few other reptiles, such as some crocodilians and lizards. However, modern bird eggs are noticeably calcified. Excellent examples of this mineralization are seen in chicken or ostrich eggs.

Dinosaur eggs are preserved in a variety of shapes. Some are nearly spherical, whereas others are ellipsoidal or semiconical (Fig. 8.2). Sizes range from a few centimeters to slightly more than 30 cm long, with comparable widths depending on the degree of egg sphericity. Approximate egg volume can be calculated by using formulas appropriate for the shape of the egg. For example, the volume of a spherical egg can be calculated simply by using the formula for the volume of a sphere (see Eqn 7.7). In contrast, an egg shape that deviates from a sphere, in that one or more of its axes may be unequal, is termed an **ellipsoid** (a body where all plane sections are either circles or ellipses).

> *A typical ellipsoid describing most eggs is **a prolate spheroid**, resembling a sphere that is elongated ("stretched") in a single axis.*

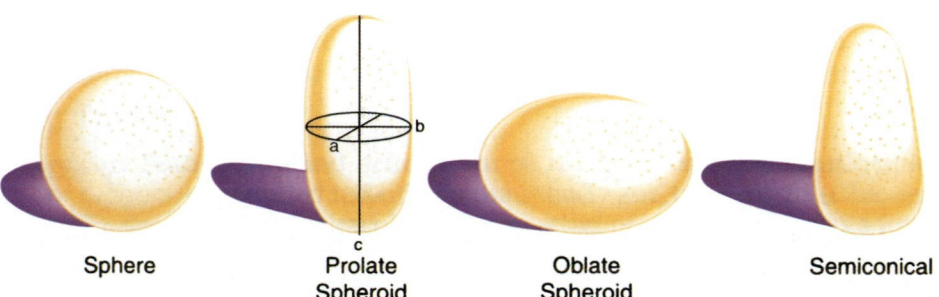

| Sphere | Prolate Spheroid | Oblate Spheroid | Semiconical |

FIGURE 8.2 Shapes, sizes, and dimensions of dinosaur eggs. From left to right, sphere, prolate spheroid, oblate spheroid, semiconical; axes and measurements associated with prolate spheroid.

The volume of a prolate spheroid is:

$$V = 4/3(\pi a^2 c) \tag{8.1}$$

where the a and b axes are equal and c is the long axis. Some dinosaur eggs were prolate spheroids, but others were **oblate spheroids**, which are spheres shortened ("squashed") in the c axis in relation to the a and b axes. The volume is still calculated with the same formula in Eqn 8.1, assuming that axis $a = b$.

For example, a dinosaur egg that is 13 cm long and 10 cm wide in its other two dimensions (hence it is a prolate spheroid) has a volume of

$$V = 4/3\pi \times (5 \text{ cm})^2 \times 6.5 \text{ cm} \tag{8.2}$$

Step 1. $V = (4.2)(162.5 \text{ cm}^3)$
Step 2. $V = 682 \text{ cm}^3$

For perspective, this volume is more than ten times that of a chicken egg (also a prolate spheroid), which is typically 6 cm long and 4.5 cm wide:

$$V = 4/3\pi \times (2.3 \text{ cm})^2 \times 3 \text{ cm} \tag{8.3}$$

Step 1. $V = (4.2)(15.9 \text{ cm}^3)$
Step 2. $V = 66.8 \text{ cm}^3$

This calculated difference means that one dinosaur egg could have been substituted for more than half a dozen chicken eggs in an omelet.

One of the qualitative characteristics described in dinosaur eggs is surface texture, which is evident in some eggs as a slight bumpiness or microrelief. The regularity of the texture is better defined with higher magnifications, revealing the distinctive shell microstructures. These microstructures relate to the functional morphology of an egg as determined through biomineralization, which began as either aragonite or calcite crystals that grew outward and perpendicular to a shell membrane surface (Fig. 8.3). The inner shell membrane, which is organic, is also called the **eisospherite layer**, whereas the crystalline exterior is the **exospherite layer**. In modern eggs, crystals are intimately interlocked with organic material throughout growth to form a lightweight but strong and flexible structure. Eggshells are also porous, with pores developing perpendicular to the shell membrane and more or less parallel to the crystals. These pores allow gas exchange, facilitated by the allantois, between the developing embryo and its outside environment.

8

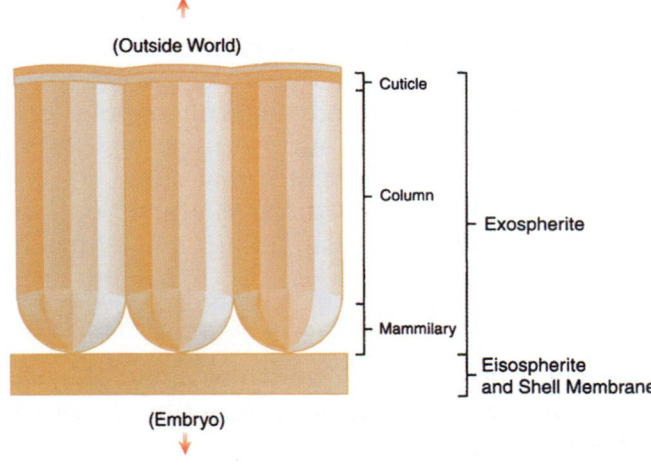

FIGURE 8.3 Cross-section of eggshell microstructure, showing (from inside to outside) eisospherite layer (with shell membrane) and exospherite layer (with mammillary layer, column layer, and cuticle layer).

The functional morphology of dinosaur eggshells is important paleontologically for interpreting the egg-laying, physiology, and nesting behaviors of dinosaurs. For example, a high number of large-diameter pores translated to a greater ease of conductance of gases between the embryo and the outside world. The calculated values of conductance from porosities in dinosaur eggshells are 8 to 16 times those of avian eggshells, which is different enough to postulate an adapted purpose for this function in dinosaur eggs. Most avian eggs are laid in an open environment, where they are not covered by either sediment or vegetation. In contrast, the high conductance values of dinosaur eggs strongly suggest that they were placed in environments with low O_2 and high amounts of CO_2 and H_2O. These presumptions imply that the dinosaur eggs were probably buried in some medium, such as in a nest structure. Furthermore, crystals in an eggshell may manifest externally as surface ornamentations, forming lengthwise ridges. These ridges probably represent microscopic channels for gas exchange along the long axis of a vertically-oriented egg.

An assemblage of eggs in close association with one another in the fossil record is often regarded as part of a **clutch**, meaning that these eggs represent one egg-laying episode. Dinosaur clutches apparently ranged from two to perhaps as many as 35 eggs, which is comparable to numbers laid by modern birds. This range is in contrast to the large numbers produced by crocodilians or other reptiles; for example, sea turtle egg clutches can contain more than 100 eggs. Furthermore, for modern reptiles, variation in clutch size is positively correlated with body size, hence larger individuals will lay more eggs. Whether a similar relationship existed for some dinosaurs is relatively unexplored, although better correlation of certain eggs with egglayers of known body size could lead to an answer.

Many modern reptiles abandon their clutches after laying them, leaving the juveniles to fend for themselves. However, some crocodilians stay close by their nests and will help to excavate any buried eggs as the juveniles in them begin to hatch. Most birds also remain near their clutches and will typically assume a seated position over the eggs for protection and insulation, a behavior called **brooding**. Brooding has been interpreted on the basis of dinosaur skeletons found sitting on egg clutches. These finds include the theropods *Oviraptor* and *Troodon* (Chapter 9) and also can be inferred from the close association of juvenile and adult remains of the same dinosaur species (*Maiasaura* and *Psittacosaurus*: Chapters 11 and 13, respectively). Interestingly, because brooding is more typical of birds than reptilians, this behavior is often explained as a trait that physiologically relates dinosaurs to birds (Chapter 15).

A dinosaur egg assemblage probably represents eggs laid by a single mother of the same dinosaur species. However, modern cuckoo birds (*Cuculus canorus*) and a few other species of birds will lay their eggs in the nest of a different species. Thus all of the eggs in a nest should not always be assumed as coming from the same parent. So far, only one possible piece of evidence would suggest that some dinosaurs engaged in cuckoo-like behavior: a nest from the Late Cretaceous of Mongolia contained the embryonic remains of two different dinosaurs, *Oviraptor* and possibly *Velociraptor*. However, taphonomic factors, such as selective transportation and deposition of hydrodynamically similar eggs from different nests, cannot be discounted (Chapter 7). Another hypothesis for explaining the juxtaposition of different eggs in the same nest is that a nest site could have been constructed by one of the two species. In such cases, a nest could have been re-used later by another species while it still contained previous unhatched eggs.

Dinosaur Egg Biogeochemistry and Physiology

Dinosaur eggs contain a wealth of useful biogeochemical information pertinent to how dinosaurs took elements into their bodies and used them. Relevant chemical constituents of eggshells include:

- Calcite, which provides information about the degree of mineralization of the egg as well as the effects of diagenesis (Chapter 7).
- **Stable isotopes** for oxygen, which can indicate the temperature of the original environment inhabited by the egglayer; and for carbon, which can reflect dietary choices by the egglayer.
- Trace elements, which may give clues about paleopathology, paleoenvironmental conditions, and diagenesis (Chapter 7).
- Amino acids and proteins, whose presence and proportions can be compared to those in modern eggshells, which in turn can be used for interpreting the relatedness of dinosaurs to extant animals, perhaps indicating a similar physiology.

Calcite was first recognized as a constituent in dinosaur eggs in 1923, and in these and other fossils it is normally re-crystallized. Original, non-recrystallized calcite is diagnosed by preservation of fine-scale structures in the eggshell and open pores. The opposite characteristics indicate at least one episode of recrystallization. A paleontologist or geochemist considering an elemental analysis of dinosaur eggs always tries to determine the degree of recrystallization first, because a recrystallized eggshell may contain chemicals that were more recently acquired.

*Once a dinosaur eggshell is determined as relatively unaltered, a geochemical analysis may begin with calculations of **stable isotope ratios**.*

Stable isotopes differ from radioactive isotopes (Chapter 4) as they do not undergo radioactive decay. Oxygen isotopes, ^{16}O and ^{18}O, are the most commonly used stable isotopes. Their ubiquitous presence in eggshell material as a component of $CaCO_3$ makes them a natural choice for study. The principle behind using them is that changes in the ratio of ^{18}O to ^{16}O can be a direct consequence of temperature changes in waters of ancient environments. Because ^{18}O is a heavier isotope than ^{16}O, it tends to remain in water undergoing evaporation. Hence, higher temperatures cause higher rates of evaporation. This results in a depletion of ^{18}O in the evaporated water and an $^{18}O/^{16}O$ ratio lower than that in the original water source. Conversely, glaciation of an area would cause an enrichment of ^{16}O in the ice. This condition causes a higher ratio than in tropical water. The fractional change in the amount of stable isotopes, where one becomes depleted while the other is enriched, is called **fractionation**. The value derived from the ratio of ^{18}O to ^{16}O is called the **$\delta^{18}O$ value**.

Organisms use water to make shells or bones. Consequently, the $\delta^{18}O$ value of an organism's unaltered bones or eggshell should reflect the temperature of the water used in the precipitation of the calcite or dahllite. This method has been used, with calcite from fossil unicellular organisms and some mollusks, to successfully interpret paleotemperatures. Unfortunately, its use with dinosaur eggshells is still controversial because of the high likelihood that most eggshells have been recrystallized. These eggs would have $\delta^{18}O$ values that represent the temperature of the water used in the recrystallization process during diagenesis, rather than the temperatures of water used by a dinosaurian mother. However, once more unaltered dinosaur eggshells are discovered, the method will have important paleoecological applications. Oxygen isotopes have also been used to interpret body-temperature regulation in dinosaurs, as discussed later. Nevertheless, this is not without dispute, again because of the same effects of diagenesis.

Oxygen isotopes provide information about paleotemperatures, but carbon isotopes are instructive about what dinosaurs ate or drank. With carbon isotopes, ^{12}C is enriched relative to ^{13}C through photosynthesis. As a result, plants have lower $\delta^{13}C$ values than dahllite in its mineral form. Modern herbivores can affect the carbon isotope ratio by eating plant material, which is enriched in ^{12}C; this manifests as a low $\delta^{13}C$ value in the herbivores' bones. The same principle has been extended to the hypothesis that dinosaur eggshells, which contain carbon (again through $CaCO_3$), should indicate whether a mother dinosaur ate plants. The ratios that have been calculated from plants from different environments can even indicate whether the plants came from coastal marine swamps or river floodplains. Yet again, the reality of diagenesis is invoked in skeptical peer reviews of carbon isotope data derived from dinosaur eggs. Nevertheless, such data may eventually provide meaningful clues to dinosaur consumption habits in the absence of other body or trace fossil evidence.

Trace elements, elements that occur in organisms in minute quantities (unlike carbon, hydrogen, oxygen, nitrogen, calcium, and phosphorus), have been consistently detected in dinosaur eggshells. This adds another geochemical parameter whose significance is not yet understood by paleontologists. Excess or deficient amounts of trace elements are linked with pathological conditions in many organisms. For example, in humans a lack of iron causes anemia or an excess of lead causes neurological damage. Anomalous trace elements in dinosaur eggshell material, particularly in those approaching the Cretaceous–Tertiary boundary, are cited as evidence for environmental problems that caused dinosaurs to lay unhealthy eggs (Chapter 16). For example, large amounts of trace elements are linked with problems in the formation of shell proteins. Some Late Cretaceous dinosaur eggshells contain relatively high concentrations of lead, vanadium, and zinc, among other elements. This led some paleontologists to link these concentrations with paleopathological conditions. Again, the effects of diagenesis constitute a potential falsifier of this hypothesis; the large concentrations of trace elements introduced during diagenesis in some dinosaur bones (Chapter 7) support the alternative.

Amino acids and proteins were first discovered in dinosaur eggshells in 1968, much to the surprise of paleontologists who did not expect their preservation. More surprising is that the quantity and quality of the amino acids closely resemble types and proportions in eggshells of some modern amniotes. In all, 17 different amino acids have been detected in dinosaur eggs and their relative percentages track very closely with chicken eggs and the "average protein." Paleontologists are still uncertain which proteins were in dinosaur eggs because the amino acids, which are more long-lived in the geologic record than proteins, only provide an approximation of their protein parents. However, modern avian eggs contain the proteins carbonic anhydrase, chonchiolin, and collagen, the latter also found in bone material, as discussed in the next section. Thus far, the only negative factor to consider is potential contamination from bacteria, fungi, insects, or other organisms from either the original egg environment or modern sources. Nevertheless, such research is promising for future revelations about dinosaur biochemistry, which, of course, relates to their physiology.

Dinosaur Growth and Thermoregulation

Biomineralization and Biochemistry of Bones and Other Hard Parts

As might be expected, the most likely body fossil of a dinosaur encountered in either a field situation or a museum is **ossified** (formed into bone) material, such as bones, teeth, and some connective tissue (tendons). The primary reason for this bias in the fossil record is because mineralized tissue is more easily preserved than soft tissue (Chapter 7). Modern bones, teeth, and ossified tendons, called "hard parts" by paleontologists, are composed of a combination of mineral and organic matter. In geology, a **mineral** is a solid, naturally occurring, crystalline substance with a definite chemical composition; organic matter is not necessarily solid or crystalline.

Dahllite is the specific mineral that composes hard parts in **chordates** (Chapter 5). Although not all chordates have hard parts, those that do are classified as **vertebrates**. According to presently-known fossil evidence, dahllite has contributed to the hard parts of vertebrates since the Cambrian Period, thus it is a probable primitive trait. Dahllite has the chemical formula (a slight variation from the formula for the mineral **apatite**) of $Ca_{10}(PO_4)_6(OH, CO_3, F)_2$. This formula shows why advertisements urging the consumption of milk claim that they cause healthy bones, as dairy products are calcium-enriched relative to most other foods consumed by humans. Likewise, fluorine is added to drinking water and toothpaste because of its essential role in the proper bonding of dahllite in teeth. Because of this mineral composition, teeth are relatively harder (less prone to wear) than hard parts formed by comparable minerals in other organisms, such as calcite or aragonite. The exact composition of dahllite or apatite depends on substitution of other elements in the crystalline framework of the mineral, represented in the last part of the formula by the elements in parentheses. For example, in **fluorapatite**, F^- substitutes for OH^-; in other forms of apatite, PO_4 can be substituted with CO_3^-. In some instances Ca^{2+} can be replaced by trace elements such as Sr^{2+}, U^{4+}, and Th^{4+}. In essence, apatite is a "garbage" mineral that unselectively accepts an admixture of the preceding elements (or others) into its structure.

Because these minerals allow for much substitution of elements, the original mineral composition of vertebrate hard parts may not have altered through time, especially considering that fluorapatite is more stable than dahllite. Consequently, most dinosaur bones are chemically different from their original state. This is not surprising, considering that the youngest of them are 65 million years old. What is surprising is that the changes may be minor – some mineral material in bone may be nearly the same as when the dinosaur originally died. Compaction by overlying sediments and other events that happen after burial, such as filling of voids, also may change dinosaur bones from their original form (Chapter 7). However, enough bone retains its original shape (or can be reliably reconstructed) so that dinosaur paleontologists can still describe detailed aspects of a dinosaur's anatomy.

Biomineralization is not limited to vertebrates, as many invertebrate animals and some one-celled organisms secrete their own shells or skeletons. Unlike vertebrates, which have their skeletons on the inside of their bodies, most invertebrates wear their skeletons on the outside of their bodies (**exoskeletons**), although a few have **endoskeletons**. The origin and evolution of biomineralization in the geologic past was a happy event for future paleontologists, and many of them would be unemployed (or at least gelatinous) if it had not happened. The vast majority of body fossils that are found, described, and interpreted by paleontologists have biomineralized hard parts, a circumstance for these fossils that increased their likelihood of preservation (Chapter 7).

But before biomineralization can occur, certain elements must be fixed through the interactions of biomolecules. Among these biomolecules are **nucleic acids**, such as **deoxyribonucleic acid, DNA**, and **ribonucleic acid, RNA; lipids**, which are fatty molecules; **carbohydrates**, also known as **sugars**; and **proteins**. No undisputed evidence of preserved nucleic acids, lipids, or carbohydrates of dinosaurs has been documented. Indeed the conventional wisdom has been that very few or none of the original biomolecules of dinosaur bones have been preserved. Many biomolecules degrade quickly, even on a human time scale. This is why any report of their discovery in dinosaur bones is greeted with skepticism. However, recently reported results of detailed analyses, carried out on dinosaur bones during the past 20 years, have provided some evidence for the presence of proteins and other soft tissues. Because these contributed to the development of bones, connective tissues, and muscles in dinosaurs, the processes that developed these body parts should directly reflect dinosaur physiology.

8

Proteins are composed of amino acids and are organic compounds that form the basis for much of the soft tissue in animals and help to facilitate biochemical reactions. Very simply, organisms cannot live without amino acids. They can produce some amino acids, but others are acquired through food. Amino acids must have a **carboxyl** group (COOH) and **amino** group (NH_2), in which the carboxyl performs as an **acid**, which is a substance enriched in hydrogen ions. On the other hand, the amino acid acts as a **base**, enriched in hydroxyl anions. The general structural formula for an amino acid is:

$$R—\underset{\underset{\displaystyle H—N—H}{|}}{\overset{\overset{\displaystyle H}{|}}{C}}—COOH$$

where R could be any chains of organic compounds that attach to amino acids, giving them distinctive formulas and names. Only 20 different amino acids, but in a myriad combinations, compose the millions of different proteins in any given animal body.

Because unaltered dinosaur soft tissues are typically not preserved, paleontologists must look to bones or eggshells for evidence of the dinosaurs' original amino acids and proteins. The proteins specifically involved in bone production are **albumin**, **collagen**, **hemoglobin**, and **osteocalcin**. Albumin and hemoglobin are proteins associated with blood circulating through a bone and help with facilitating biochemical reactions in an organism. Collagen and osteocalcin are **structural proteins**, meaning that they lend physical support to an organism's body and are typically fibrous.

Bone formation in modern vertebrates begins through structural proteins with the laying down of collagen fibers in a parallel, linear arrangement. This forms part of a flexible, non-ossified connective tissue called **cartilage**, which grows through cells called **chondrocytes**. The formation of cartilage is followed by biomineralization of dahllite in spaces between the collagen fibers. This process is performed by **osteoblasts**, which are bone cells partly composed of osteocalcin that later become **osteocytes**. Osteocytes ensure that the bone continues to function as living tissue. Crystallization takes place in both the center and the outer, middle part of the cartilaginous structure and replaces the cartilage with mineralized tissue. It also eventually opens spaces for **marrow canals**, which carry blood to the bones. The further development of mineralized tissue extends to either end of the originally cartilaginous structure and results in outward bone growth, from the development of an embryo through to adulthood.

Growth of bone eventually results in two types of bone, **cancellous** and **compact**. Cancellous bone has a "spongy" (porous) texture, where the pores (**lacunae**) are occupied by osteocytes during the life of the animal. Tiny canals called **cannaliculi** interconnect these lacunae. Cancellous bones have a density based on only 10% to 30% of the original biomineralized matter, which made for light yet strong bones in dinosaurs and other vertebrates. In contrast, compact bone was originally about 95% biomineralized matter and accordingly is more durable. In some cases, this type of bone is preferentially preserved among the body parts that were physically reworked after death of a dinosaur (Chapter 7). Within a single bone, the amount of cancellous versus compact bone can vary. For example, a limb bone has mostly compact bone on the **diaphyses** (shafts) whereas the **epiphyses** (wider ends) have more cancellous bone. A thin outer layer of compact bone protects this cancellous bone (Fig. 8.4).

FIGURE 8.4 Differences in proportions of cancellous and compact bone in the diaphyses and epiphyses (respectively) of a limb bone.

Bone growth generally stops with adulthood of **endotherms**, which are animals that produce and maintain their own internal body heat as a consequence of their metabolism. Modern mammals and birds are endotherms that show cessation of bone growth with adulthood. However, bone growth continues throughout the entire life of **ectotherms**, animals that rely mostly on the temperature of their surrounding environment to provide heat for metabolism. For example, modern reptiles are ectotherms that can become continually larger with increased age. In other words, the largest crocodiles are usually the oldest crocodiles. All of the evidence gained from dinosaur bones indicates a process of initial (embryonic) formation for the bones similar to the general model outlined here for vertebrates, which means that this is the currently accepted hypothesis for dinosaur bone growth. However, the interpretation of endothermy versus ectothermy in dinosaurs, as indicated by their bones, is still controversial and has provoked much contentious debate. Nonetheless, this subject, discussed in detail later in this chapter, is better understood than only 20 years ago.

Some dinosaur bones show the presence of collagen and osteocalcin, as well as indirect indications of amino acids that composed hemoglobin in the red blood cells of dinosaurs. However, collagen could be present as a contaminant in dinosaur bones because some invertebrates also produce it. In other words, its detection could give a false indication of its previous presence in a dinosaur. Collagen could have been introduced into dinosaur bones through scavenging insects soon after the dinosaur died (Chapter 7) or by any modern insects that were incorporated inadvertently into an analyzed bone sample. On the other hand, osteocalcin is only produced in vertebrates. Hence its presence in dinosaur bones is nearly indisputable for representing the original protein of a dinosaur. Additionally, the former presence of collagen is indirectly indicated by the parallel arrangement of fluorapatite crystals observed in most dinosaur bones. This texture is visible as a "grain" (like wood) that typically is parallel to the length of some dinosaur bones (Fig. 8.4).

Skeletal Allometry

Allometry (defined in Chapter 1) is the most appropriate approach to studying different sizes of the same bones from the same species of dinosaurs. Differently sized bones from the same species should reflect stages of growth during the lifetime of an individual, also known as its **ontogeny**. Accordingly, such bones ideally show a **growth series**, which is a sequence of sizes that reflect increased growth of an animal with time. As mentioned in previous chapters, a few dinosaur species are so well represented in the geologic record that their bones show a growth series from juvenile to adult. Dinosaur species that show such a continuum include the Late Triassic theropod *Coelophysis bauri*, the Late Jurassic theropod *Allosaurus fragilis*, and the Late Cretaceous ornithopod *Hypacrosaurus stebingeri* (Chapters 9 and 11), among others. Growth series are most visually compelling when made by simply placing the same bone from a species in order of size, from smallest to largest. However, few paleontologists have enough bones of one species in the same place to put them all together in order. As a result, they are often content to deal with careful measurements, which can then combine data from bones at widely-separated institutions. Limb bones are the most commonly measured for such studies, although skulls can be used if abundant and well preserved enough (e.g., *Coelophysis*: see Fig. 2.1), and length is a typical, simple parameter. However, this approach was replaced by measuring multiple parameters of bones and then computing ratios (e.g., length:width) or using **multivariate analysis**, a statistical method that tests for whether certain parameters correlate or not. One of the parameters used more frequently with the advent of computer-based image analysis is the picking of "anatomical landmarks" for measurement. Some of these features are the same points on a skull with relation to an orbit, nares, or other well-defined features.

The one major scientific shortcoming of assembling growth series from dinosaur bones is that these series rarely tell paleontologists anything about the rates of growth that contributed to the series. Consequently, paleontologists have to cut cross-sections of the bones to look at certain features associated with the original biomineralization processes used by dinosaurs, and how tissue functions contributed to these features. Do dinosaur bones show evidence of endothermy, ectothermy, both, or neither? The answer, confusingly, is "yes."

Bone Histology and Biogeochemistry

Histology is the study of tissues: how they are formed and how they function. Many tissues have been mentioned in this chapter and others: bones, teeth, cartilage, muscles, blood, organs, and skin. Because the most commonly preserved former tissues of dinosaurs are bone, whenever a paleontologist is discussing dinosaur histology they are referring to bone histology, with only a few exceptions. But of course, such a discussion necessarily should be accompanied by considering how those bones may have been changed since the dinosaur was alive.

Many dinosaur bones recorded growth lines, which are of two types: **annuli** and **lines of arrested growth** (also known by their acronym **LAGs**). Annuli are layers of bone fibers that were formed parallel to one another and do not show vascularization. These are detectable because of their contrast with surrounding vascularized (**fibrolamellar**) bone (Fig. 8.5). LAGs are similar to annuli but are not as thick and correspondingly have fewer bone fibers. Based on studies of modern vertebrates, growth lines in general are a result of some temporary slowing (represented by annuli) or stoppage (represented by LAGs) of bone growth. The latter can correlate approximately

FIGURE 8.5 Annuli (unvascularized area) and lines of arrested growth (LAGs) in ornithopod limb bone.

with a lack of growth each year. As a result, a minimum age can be estimated in any given individual dinosaur that has observable growth lines in its bones.

LAGs recorded in dinosaur bones represent periods of interrupted growth, which can be attributed to yearly cycles in growth that suggest ectothermy. However, LAGs also are present in some mammal bones. Subsequent research shows that a number of environmental factors, such as prolonged droughts or cold winters, can cause these features. Additionally, some dinosaurs have LAGs but others lack them, which means that they cannot be used as a universal indicator of thermoregulation. For those dinosaurs that have LAGs, and using the assumption that they represent annual growth lines, growth rates have been calculated for some dinosaur genera (Chapter 11). High growth rates should reflect endothermy, whereas slow growth rates are characteristic of ectothermy. The growth rates calculated for some dinosaurs are faster than those known for crocodiles, but slower than those of birds. Another interesting feature common to compact bone in dinosaurs is that it does seem highly vascularized. This feature is common in endotherms, but also has been seen in the bones of some endotherms.

Dinosaur Feeding

Evidence for Dietary Preferences: What Did Dinosaurs Eat?

Five main lines of evidence can be used to determine either generally or specifically what dinosaurs preferred to eat:

1 types of teeth;
2 toothmarks inflicted by those teeth;
3 stomach contents;
4 gastroliths; and
5 coprolites.

Other forms of evidence can supplement these, such as overall functionally morphology, tracks, or body fossils of potential food items in the same-age strata. Nonetheless, the preceding forms of evidence are what will be covered here, although toothmarks, gastroliths, and coprolites will be discussed in more detail as trace fossils in Chapter 14.

Dinosaur Teeth: Herbivorous and Carnivorous

Composed primarily of dense, compact dahllite but smaller than most bones, teeth are sometimes the only part of an animal preserved in the fossil record.

People with cavity-bearing teeth pulled from their jaws or who have seen X-ray photographs of their teeth have the opportunity to observe a few or most of the tooth parts and their orientations. Teeth have two primary layers, tough **enamel** on the outside and softer **dentine** on the inside. The part of the tooth exposed above the gumline of an animal is its **crown**, whereas the part within the socket is the **root**. In many cases, the root composed the majority (as much as 75%) of a dinosaur tooth. The side of a tooth closest to the tongue is its **lingual** side, whereas the outer side is **labial**. Small protuberances, in some cases found on the **apical** (top) parts of the crowns, are **denticles**. Narrow places on teeth that form blades or ridges, typically on their anterior and posterior sides, are **carinae**.

Another detailed feature in teeth of some modern and ancient animals, but absent in humans, are **serrations**. These are square, triangular, oval, or rectangular denticles on carinae separated by narrow indentations (**cellae**) along a tooth surface. Serrations considerably increase the surface area of a tooth and can be described as coarse or fine, depending on the relative size of the denticles. Coarse serrations are useful for roughly cutting soft material, such as woody tissue. Fine serrations are more efficient at ripping through both soft and hard tissue, such as flesh and bone. Serrations also serve to temporarily hold the object being cut, especially if the serrations are angled with respect to the long-axis orientation of the tooth. The cutting motion is accomplished by pulling the serrated surface across the object. A motion in the opposite direction would encounter more resistance because of the angled serrations gripping the cut object. A smooth (non-serrated) blade applies its shearing force to a much smaller area and thus causes more stress, which cuts efficiently with either a pulling or pushing motion, but does not grip the cut object. The latter principle shows why some swords, such as those made in medieval Japan (*katana* or *wakazashi*) were designed with thin, smooth blades instead of serrations. In contrast, steak knives typically are designed with serrations so that they can grip a steak as it is being cut mainly with a pulling motion.

Dentition is the sum of a dinosaur's teeth in its jawbones, which consist of the dentary, maxillary, and in some cases the premaxillary. Dinosaur teeth were individually grown in sockets of the jawbones, which differs from teeth fused to those bones, seen in modern lizards. This socketing allowed for lost teeth to be replaced by teeth below the roots. The closing of a dinosaur mouth, so that the teeth from the upper and lower parts of the jaw came together, is called **occlusion**; where teeth come together is the **occlusal surface**. If the occlusion causes teeth from the maxillary or premaxillary to cover the teeth of the dentary and predentary, an **overbite** results, whereas the opposite condition is an **underbite**. Most dinosaurs had a noticeable overbite, especially theropods (Chapter 9). Occlusion could have varied as a result of any tooth injuries that broke or damaged a tooth.

In most dinosaurs, dentition was composed of individual teeth that were similarly shaped, with tooth shapes including teardrop-shaped, peg-like (cylindrical), conical, or bladed with diamond- or D-shaped cross-sections (Fig. 8.6). However, even if teeth were similarly shaped, the sizes of the teeth could have varied within the same jaw of a dinosaur. **Heterodont** dentition shows a variety of tooth shapes

FIGURE 8.6 General tooth shapes typically associated with certain clades of dinosaurs (but with some exceptions as noted in the text). (A) Leaf-like (prosauropods, ornithopods, thyreophorans). (B) Peg-like (sauropods). (C) Conical (some theropods).

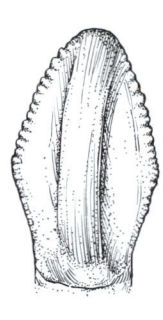  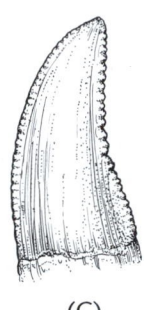

(A) Leaf-like   (B) Peg-like   (C) Conical

in a dinosaur's jaws, as is seen in the *Eoraptor* from the Late Triassic of Argentina (Chapter 6). This dentition is also a namesake feature of one clade of ornithopods, the Heterodontosauridae (Chapter 11). Another variation of teeth in some ornithopods and ceratopsians were **dental batteries**, in which the teeth in the cheek region were fused together to form compound teeth. The most complex dental batteries known are those of hadrosaurs, which had more than a hundred teeth in each battery (Chapter 11). Some ceratopsians (Chapter 13) had less numerous teeth but similar batteries. One analogy that may help with visualizing how the arrangement of teeth in a dental battery worked is to think of them as bricks in a wall. Individually they would have had little supportive strength for grinding food, but when cemented together they composed a formidable tool for increasing the surface area to grind tough plants. Dental batteries apparently developed just prior to the middle of the Cretaceous Period. Thus, they may indicate some evolutionary response to changing vegetation patterns during the Mesozoic Era.

A secondary characteristic that dinosaur teeth acquired during life were signs of wear. In most instances these wear marks were inflicted by a dinosaur's own teeth with normal occlusion during biting or chewing. These wear patterns provide valuable evidence of jaw mechanics for dinosaurs, indicating how certain dinosaur species chewed. In other cases, the teeth may have been damaged after encountering resistance from whatever object a dinosaur was biting. Worn and damaged teeth, as documented from the geologic record, bear testimony of dinosaur food preferences. For example, some paleontologists have proposed that a few carnivorous dinosaur species must have been scavengers because their teeth show little evidence of wear, suggesting that they may have fed only on soft, rotting flesh. In contrast, high incidences of worn and broken teeth may represent attempts at biting live, struggling prey or fracturing primarily compact (versus cancellous) bones. Although tooth hardness with respect to the bitten object is a factor, the role of jaw strength in tooth wear cannot be underestimated. For perspective, the biting force measured from some humans indicates that they could bite through a steel bar if their teeth were composed of diamond instead of dahllite. Indeed, some dinosaur toothmarks provide indirect evidence of their bite strength, clarifying why dinosaurs show numerous replacement teeth in their jaws. They should have lost their teeth on a frequent basis in correspondence with food choices and bite forces.

Herbivorous dinosaurs had teeth that were functional for grasping, tearing, shearing, and grinding plant material. The wide variation of tooth shapes in different herbivorous dinosaurs is an indirect indicator of the plant diversity consumed by these dinosaurs. In modern herbivorous reptiles, a model for comparison to presumed herbivorous dinosaurs, teeth are:

1 comparably sized;
2 teardrop-shaped;
3 closely associated, and
4 possess coarse serrations.

Some of the earliest interpretations of herbivorous dinosaurs, such as prosauropods, were based on these criteria (Chapter 10). Although some researchers thought the rough serrations were evidence for carnivory in prosauropods, the currently accepted hypothesis is that they functioned to cut through woody material. This adaptation would have been specifically used for separating plant material from trunks, stems, or roots.

Tooth wear in herbivorous dinosaurs with dental batteries resulted in a secondary occlusal surface that formed wide areas for grinding. Modern horse teeth and the teeth of other large herbivores show a similar adaptation. This became especially pronounced in the ancestors of these animals with the evolution and proliferation of tough plants such as grasses during the Cenozoic. Although grasses were not available for dinosaurs during the Mesozoic, numerous other plants identified through paleobotanical studies provide evidence for difficult-to-process plant material that influenced dinosaur adaptations. A different adaptation of dental batteries is presented by those dinosaurs, such as some neoceratopsians (Chapter 13), with narrow occlusal surfaces that could not have been used for grinding. In these cases, tooth wear was limited, and the batteries probably served as support for shearing through plant material. This was also a likely adaptation for toothless dinosaurs.

Some restorations of dinosaurs eating plants show them with full, bulging cheeks. These restorations are not based on speculation: some dinosaurs, such as ornithopods (Chapter 11), likely had cheeks that allowed them to temporarily store plant material in their mouths for thorough chewing. Clues to the former presence of cheeks are teeth that are more medial (inset) with respect to the jaw exterior, as well as "shelves" on opposite labial sides of the maxillary. Assuming that flesh covered the area between the upper and lower jaws in dinosaurs with these features, the area between the flesh and the labial sides of the teeth likely functioned as cheeks. No other hypothesis has been proposed for the described evidence, which is also found in some thyreophorans (Chapters 12 and 13).

Peg-like teeth, typically associated with relatively small skulls, are a hallmark of sauropods. When paleontologists first tried to interpret sauropod feeding habits, these teeth were the source of some mystery (Chapter 10). How could animals as large as sauropods have such small heads with only widely spaced, mostly cylindrical teeth available for chewing? One hypothesis is that these dinosaurs used their teeth mainly for pulling plant material into their mouths, followed by swallowing without appreciable chewing. The initial action (separating leafy material from branches) can be mimicked with a rake; if applied to a tree, the rake's widely separated and narrow tongs will pull whole leaves off but leave most major branches intact. Through this analogy, a mechanism for sauropod feeding is more easily visualized. Vegetation thus cropped and swallowed whole was then ground in either a crop or gizzard with the aid of gastroliths and digested with the help of anaerobic bacteria (discussed below).

Carnivorous dinosaur teeth are normally serrated and curved posteriorly; they resemble those of archosaur ancestors, although many variations evolved on that theme throughout the Mesozoic Era. The curved, serrated teeth of carnivores are called **ziphodont**; for carnivorous dinosaurs, these teeth represent a **plesiomorphic** (primitive) ancestral condition that is seen in presumed dinosaur ancestral species, such as *Herrerasaurus* and *Eoraptor* (Chapter 6). Interestingly, the ziphodont teeth in *Eoraptor* are only present in the maxilla; the dentary teeth are more leaf-like, resembling those of prosauropods. Ziphodont teeth in dinosaurs are restricted to the clade Theropoda, of which most members are interpreted as carnivorous.

Carnivorous dinosaur teeth were adapted for grasping and cutting through flesh, as well as crushing or punching through bone. Because the vast majority of teeth in any given theropod jaw have a conical shape, occlusal surfaces should have been minimal for chewing. This means that these carnivores probably tore off chunks and swallowed, rather than chewing thoroughly. Besides processing meat and bones, another probable function of theropod teeth was as killing implements. However, the death of one dinosaur that was directly caused by another dinosaur has not yet been convincingly documented from the geologic record. If some theropods did kill other dinosaurs, using their teeth as their primary weaponry (a reasonable hypothesis for some theropods), then a gripping strength in the jaws sufficient to hold a struggling prey was necessary. Regardless of whether some theropods killed with their teeth, the tooth morphology and jaw dynamics of carnivores are radically different from those of plant eaters. Furthermore, in some cases they show overt adaptations for the application of much greater forces than would have been necessary for the consumption of stationary, passive plant material.

Teeth in modern carnivorous fish, amphibians, reptiles, and mammals are typically elongate, conical, recurved, and bladed, with few exceptions. Although tooth shapes can be similar in these carnivores and most theropods, tooth sizes may vary within the same jaw. These different tooth sizes in theropods probably helped to cause variable levels for cutting surfaces. Some theropods, most notably spinosaurs and therizinosaurs (Chapter 9), have numerous small, similarly-sized teeth in their jaws. These, along with other anatomical information, have led to the hypothesis that some species were **piscivorous** (fish eaters), as discussed later. Some paleontologists have even suggested that therizinosaurs were herbivores, which would have made them among the few known herbivorous theropods.

As discussed before, fine serrations on carinae of both the posterior and anterior parts of teeth are also a characteristic of carnivore teeth in modern animals. The superior cutting and gripping ability that serrated teeth impart is evidently a successful and recurrent adaptation in carnivores. Such serrations have a dual purpose in some animals: modern Komodo dragons (*Varanus komodoensis*) of Indonesia cannot only slash flesh with their teeth, but the cellae of the serrations retain fibers of flesh that over time rot in the mouth and produce a bacterial or **septic** culture. Consequently, an originally wounding bite from a Komodo dragon can later become fatal as a nasty, debilitating infection develops in the prey animal and causes death within several days. In this respect, the similar serrations in tyrannosaurid teeth have been compared to the functional utility of Komodo dragon teeth, which led to multiple hypotheses of how these dinosaurs gained their meals (Chapter 9).

Some paleontologists have speculated that a few dinosaurs with dentition that does not clearly fit models of strict herbivores or carnivores were **omnivores** (eating plants and meat) or **insectivores** (insects only). An egg-eating diet was also proposed originally for some toothless, beaked theropods such as the *Oviraptor*, but subsequent data have cast doubt on this idea (Chapter 9). An insectivorous diet that would have provided an abundant source of protein was possible for juvenile and otherwise small theropods. In the case of juveniles, the eating of insects may have happened before they graduated to consuming vertebrate flesh. This behavioral transition is also observed in some modern crocodilians. However, no stomach content or coprolite evidence has yet indicated that dinosaurs were insectivores, omnivores, or had any other specialized feeding. Admittedly, these feeding patterns are also difficult to assess in modern animals without directly observing either their feeding or feces.

Nevertheless, incontrovertible evidence of dinosaur meat-eating preferences is provided by rare instances of dinosaur teeth, identifiable to a species, filling their toothmarks in the bone of another animal. One example is a *Saurornitholestes* tooth in a pterosaur bone, and others are of *Tyrannosaurus* teeth in the fibula and rib of the ornithopods *Hypacrosaurus* and *Edmontosaurus*, respectively. Toothmarks lacking their *in situ* tracemaking teeth are also convincing pieces of evidence pointing toward a carnivorous dinosaur's feeding habits. This is especially the case if a cast is made of the toothmark to mimic the original morphology of the tooth.

Toothmarks as Indicators of Diet

A **toothmark** is an impression left by the bite of an animal with teeth, regardless of what was being bitten. Toothmarks are trace fossils, whereas the medium they bit into are body fossils. Dinosaur toothmarks, first described and interpreted in 1908 (Chapter 3), have been so far only reported from bones, and no dinosaur toothmarks in fossil plant material are currently known. For those dinosaurs that fed on other animals, some left distinctive toothmarks on bones, which clearly indicates their feeding habits. Whether these pieces of evidence are representative of feeding preferences for some dinosaurs is inconclusive and a firmer understanding depends on the discovery of more toothmarks or supplementary clues from stomach contents or coprolites. The behavioral significance of toothmarks is discussed at length in Chapter 14, where they will be explained to identify what dinosaurs ate.

Toothmarks have been attributed to specific dinosaurs on the basis of their close resemblance to known tooth anatomy (especially denticles on serrations) and spacing. Some reported examples from the Late Cretaceous include *Troodon* toothmarks in ceratopsian bones, *Saurornitholestes* toothmarks in bones of an ornithomimid and *Edmontosaurus*, and *Tyrannosaurus* toothmarks in neoceratopsian, hadrosaur, and *Saurornitholestes* bones. Of these examples, a direct correlation between "dinner" and "diner" species through toothmark evidence is of *Tyrannosaurus* toothmarks in *Edmontosaurus* and *Triceratops* bones (Chapters 9, 10, and 13). Similar toothmarks attributed to a closely-related species of tyrannosaurids, *Albertosaurus*, have also been interpreted from *Edmontosaurus* bones. This means that this species was a possibly popular choice on theropod menus, having been consumed by at least three species of them. However, such toothmarks do not necessarily mean that these theropods preyed upon and killed any of the eaten dinosaurs. After all, the specimens may have already been dead when they were munched. In contrast, toothmarks from theropods that show post-wound healing have been reported for at least two specimens of the oft-victimized *Edmontosaurus*, which indicates a successful escape for a preyed-upon dinosaur, or a failed hunt for the predator.

Stomach Contents: Halfway Through

Actual remains of plants or animals in the abdominal region of a dinosaur seemingly represent unambiguous evidence supporting hypotheses about what dinosaurs ate. However, considering that fossilization of any dinosaur part was a rare event (Chapter 7), finding a specimen preserved with parts or all of a recent meal in the location of its former innards is always a surprise. The rare reports of dinosaur stomach remains provide a glimpse of a dinosaur's last meal that, despite being a sample of one, can be compared to other evidence of feeding behavior for that given species. Of course, taphonomy is an all-too-important consideration when discussing what composes these stomach remains: animals with mineralized tissues are much more likely to have been preserved than those with soft tissues or plants. Indeed, stomach remains from dinosaurs are mostly vertebrates. The consumption of insects or other invertebrates by dinosaurs is unknown, and the study of plant material associated with herbivorous dinosaur body fossils is a study in frustration.

Although the fossil record for herbivorous dinosaurs' stomach remains is poorly documented, a review of digestion in herbivores is warranted. Because plant material in many cases has low nutritional yield for large volumes, much plant material will reside in an alimentary canal for a relatively long time. However, a long residence does not necessarily mean that digestion is inefficient; on the contrary, digestive efficiency has evolved to a high degree in modern herbivores. But considering that herbivorous dinosaurs were the largest land animals that ever lived, they should have evolved comparable or superior digestive efficiency.

> *Digestion in dinosaurs was most likely facilitated by a series of organs specialized for the task, along with a little help from anaerobic bacteria.*

Decomposition of plant material with the assistance of anaerobic bacteria in the alimentary canals of terrestrial animals probably developed early among plant-eating reptiles. Modern herbivores that exemplify this process include **ruminants**, mammals that have large, multi-chambered stomachs (**rumens**) that physically mash plant material into a compacted mass called a **bolus** (sometimes called a **cud**). The bolus is then regurgitated into the mouth for more chewing, then swallowed again. Bacteria within the digestive tracts of herbivores also chemically break down some of the organic compounds in plants that otherwise cannot be digested. For example, the decomposition of cellulose, a common organic compound in plants, first produces sugars through fermentation (Chapter 7), then the formation of acetic, propionic, butyric, and formic acids, followed by amino acids, vitamins, CO_2, and CH_4. In the last stage of decomposition, bacteria reduce CO_2 to form CH_4. After fermentation, partially digested material is mixed with microbial cells from the bacteria, where it passes into the rest of the gastrointestinal tract. This constitutes the main source of protein and vitamins for a ruminant.

A ruminant would quickly die from malnutrition if it did not have a symbiotic relationship with its gut bacteria. Modern carnivorous animals have similar requirements for digestion. As a result, paleontologists assume that decomposition aided by anaerobic bacteria was also the case for carnivorous dinosaurs. Modern carnivorous analogues to theropod digestion, such as crocodilians and birds, have a stomach divided into a **proventriculus**, which produces enzymes for chemical breakdown and precedes a muscular gizzard that further aids digestion.

Modern methanogenic bacteria, such as *Methanobacterium thermautotrophicum*, have a cumulative effect of producing methane on a globally measurable scale, if present in enough herbivores that digest large amounts of plant material. Of the current global methane budget, about 80% is related to methane produced by bacteria, of which many are hosted by the guts of domesticated cattle. Termites also are significant contributors to the global methane budget (in fact, more so than cattle) and they also host bacterial colonies that assist their digestion of wood. During the Mesozoic, large herbivorous dinosaurs, along with termites and other wood-digesting organisms, were probably the purveyors of voluminous gaseous emissions that would have saturated the atmosphere of that time.

The rarity of plant material as stomach contents may actually be an artifact of preparation methods. The matrix entombing an herbivorous dinosaur may have contained disseminated plant fragments that were formerly in the gut of the animal. Thus far, only one specimen of an herbivorous dinosaur (*Edmontosaurus*) was reported with stomach remains consisting of plant material. Described in 1922, this specimen's abdominal area contained seeds, twigs, and needles from a species of conifer. Unfortunately, the plant material was taken out of the specimen during its preparation, and a thorough investigation of its taphonomy was not undertaken. As a result, an alternative hypothesis is that this plant material may not represent actual stomach remains, but rather fragmented debris washed into an open cavity of the hadrosaur's dead body.

8

Among the meat eaters, four theropods provide specific examples of interspecific predator–prey relations in dinosaurs. In the first, a *Compsognathus* specimen from the Late Jurassic Solnhofen Limestone of Germany contained a complete specimen of *Bavarisaurus*, a type of lizard. In this case, the lizard remains are enclosed by the costae of the *Compsognathus* specimen. This circumstance supports the notion that the lizard was consumed and was not merely separate remains deposited in the same location. The second example is a specimen of *Sinosauropteryx*, a feathered theropod from the Early Cretaceous of China, which also contains its last meal, an unidentified small mammal that is only present as a single dentary. This find constitutes the only evidence of any dinosaur eating a mammal, despite fiction depicting numerous such scenarios (Chapter 1). The third case is of the Early Cretaceous spinosaurid *Baryonyx*, which contained acid-etched scales from the fish *Lepidotes* and pieces of *Iguanodon*. Interestingly, the functional morphology of *Baryonyx* was originally interpreted as that of a possible fish eater, so the later discovery of fish remains in its skeleton was an excellent example of a predictive hypothesis. A similar case for stomach contents was proposed for a tyrannosaurid, *Daspletosaurus*, from the Late Cretaceous Two Medicine Formation of Montana. In this example, an acid-etched vertebra from a juvenile hadrosaur was found in association with the partial remains of a *Daspletosaurus* specimen. The acid etching was hypothesized as having resulted from partial digestion of a hadrosaur (or at least part of one) in the gut of the *Daspletosaurus*.

Former stomach remains, where a dinosaur regurgitated its meal, are also a possibility in the geologic record, although their preservation potential was probably very low. Regurgitation is a common reaction of an animal to ingested toxic substances or overeating. For example, most alkaloids, compounds commonly found in plants, induce vomiting in humans and many other mammals if ingested in small quantities. In fact, these become fatal poisons in larger amounts. Secondary causes of regurgitation are disease, dehydration, or other bodily ailments. However, not all regurgitation indicates poor health. Some predatory birds, such as owls and eagles, will ingest whole rodents, then regurgitate a pellet composed of their bones and hair after they have digested the prey's muscle tissue. Although these pellets are often confused with feces, they exit from the mouth instead of the other end. Criteria for interpreting fossil **regurgitants**, or the products of regurgitation, would include poorly-sorted masses of broken plant or animal material restricted to small areas with no evidence of sedimentary sorting. Only one possible dinosaur regurgitant, found in Early Cretaceous deposits of Mongolia and composed of turtle and dinosaur bone fragments, has been interpreted through such criteria.

Gastroliths (from the Greek gastro, stomach and lithos, stone), first mentioned in Chapter 3, are stones used primarily to help in the mechanical breakdown of food within a digestive tract. A colloquial term for gastroliths is "gizzard stones".

Gastroliths: Mostly for Herbivores

Some modern birds will swallow mineral grains several millimeters in diameter, which then reside in their gizzards and aid in the digestion of food by helping to grind tough material. Because birds do not have teeth, they need this mechanism to break down their food. The muscular action of the gizzard and the grinding caused by the mineral material helps to increase the surface area of the food for easier digestibility.

Gastroliths were first described and interpreted from a Late Cretaceous hadrosaur (*Claosaurus*), by Barnum Brown early in the twentieth century. Friedrich von Huene later found them in association with bones of the Late Triassic prosauropod *Sellosaurus*, and William Lee Stokes described some in association with Late Jurassic sauropod remains. Surprisingly, they have been studied very little since then, perhaps because of the level of skepticism they have received from many paleontologists.

Evidence for gastroliths in dinosaurs consists of the numerous polished stones associated with dinosaur body fossils. The most oft-cited examples are found within the thoracic cavity region, ventral to the cervical and dorsal vertebrae and anterior to the sacral vertebrae.

Dinosaurs with well-supported evidence for gastroliths include some sauropodomorphs (both prosauropods and sauropods: Chapter 10), a nodosaurid (Chapter 12), psittacosaurids (Chapter 13), and a few theropods (Chapter 9). Brown's hadrosaur example of gastroliths has since been regarded as unconvincing, which means that gastroliths are undocumented for ornithopods. Because gastroliths are normally associated with herbivores and the theropod specimens are seemingly carnivores, the presence of gastroliths in their gut regions is a subject of controversy (Chapter 9). Gastroliths in psittacosaurids are also enigmatic because these dinosaurs had well-developed dental batteries that should have easily ground up their roughage, seemingly negating the need for gastroliths.

Coprolites: The End Products of Digestion

Coprolites are trace fossils (Chapter 14) that are rarely matched with their trace-makers with any form of reliability. Nonetheless, a few notable exceptions provide insights into their value in working out food choices for some dinosaurs. Both fossil plants, such as conifers, and bones have been discovered as ground-up material in localized masses attributed to dinosaurs. For example, coprolites filled with ground-up conifer material occur with bones, eggs, and nests of the Late Cretaceous hadrosaur *Maiasaura* in Montana. On the carnivorous side, evidence for consumption of a juvenile hadrosaur and chewing in tyrannosaurids can be interpreted from a large (44-cm long) cylindrical Late Cretaceous coprolite from Alberta. The coprolite is attributed to a tyrannosaurid because of:

1 its unusually large size;
2 the inclusion of numerous small bone fragments of the hadrosaur; and
3 its occurrence in strata known to contain body fossils of tyrannosaurids, such as *Albertosaurus* and *Tyrannosaurus*.

However, because occlusal surfaces were so narrow for tyrannosaurids and most other theropods, chewing would not have been very efficient. Hence an alternative hypothesis is that the bone fragments represent nipping of the bone material by the anterior teeth as it pulled meat off the juvenile hadrosaurid. Another large (64-cm long) coprolite, also discovered in the Late Cretaceous strata of Alberta, contained three-dimensional impressions of muscle tissue and finely-ground bone. Also attributed to a tyrannosaurid, this coprolite indicates a brief digestive period for the tracemaker, rapid phosphatization, and burial of the fecal mass. Otherwise, the muscle tissue would not have been preserved. These coprolites and their implications with regard to tyrannosaurid behavior are further discussed in Chapter 9.

Diet and Physiology: How Much Did a Dinosaur Need to Eat?

"You are what you eat" is a commonly applied phrase that relates the general health or disposition of a person to what they eat. In the case of dinosaurs and considerations of their physiological needs, the question might be better asked as "You are how much you eat." As a general rule, ectotherms, kilogram for kilogram, will require less food than endotherms, but even some endotherms need more or less food than others of their thermoregulatory type. As different foods have varying caloric or other nutritional values, determination of whether a dinosaur was an ectotherm, endotherm, or somewhere in between directly related to the quantity

8

and quality of what they ate. Consequently, this section will examine how physiology is interrelated with diet.

A common misconception about ectothermy is that it is somehow inferior to endothermy. Ectothermy and endothermy are simply different ways for animals to make a living. For example, an ectotherm's dependency on its surrounding environment for maintaining its body temperature means that it is less dependent on seeking food than an endotherm. An endotherm is largely independent of its outside environment for thermoregulation (except in cases of hypothermia or hyperthermia: Chapter 6). The trade-off is that an endotherm must eat relatively more food to maintain its internal body temperature. This means that a crocodile can go much longer between meals in comparison to a lion of the same mass in the same environment. Maintenance of a constant body temperature qualifies an animal as a **homeotherm**, no matter whether it accomplishes this feat through ectothermy or endothermy. In contrast, a **poikilotherm** is an animal whose temperature either stays constant or varies in direct accordance with the temperature of its environment. This term was used as a synonym for ectotherm, and indeed there is much overlap between animals that are ectothermic and poikilothermic, but enough exceptions have caused physiologists to make a distinction between the two.

Food has energy content, resulting in **kilocalories**, the heat energy that is used for bodily functions such as aerobic metabolism. A kilocalorie (abbreviated as **kcal**) is the amount of energy needed to heat 1.0 kg of H_2O 1°C (from 15.5° to 16.5°C as a standard). Kilocalories are still used as units in studies of diet, but the internationally accepted standard unit for the study of energy flow is the **kilojoule (kJ)**. A **joule** (J) is a measure of energy equivalent to 1 newton (N) of force applied over 1 meter distance, or

$$J = N \times m \tag{8.4}$$

A kJ is 1000 joules, which is equivalent in heat energy to 0.24 kcal, or 4.2 kJ = 1.0 kcal.

Different foods have different energy values, exemplified by the following approximations of kilojoules per gram in each potential food:

- Wood = 17 kJ/g
- Shoots and leaves = 21–23 kJ/g
- Vegetation (average) = 21 kJ/g
- Muscle = 21–25 kJ/g
- Fat = 38 kJ/g

These estimations of food-energy values may not have changed appreciably throughout geologic time and can be safely assumed as similar during the Mesozoic. Take, for example, an ornithopod (Chapter 11) that ate woody tissue. To maintain its energy levels, it would have needed to eat a greater amount of food than an equivalent-sized stegosaur (Chapter 12) that ate shoots and leaves. Likewise, a large theropod that ate lean meat, such as other dinosaurs, would have needed more food than an equivalent-sized theropod that ate fatty fish.

However, these relationships based on the different energy contents of food also assume that the animals being compared have the same modes of thermoregulation. Endothermic animals simply require more kilojoules than ectotherms. A delightful illustration of this difference in caloric needs was provided recently by two paleontologists who calculated how many 68-kg lawyers (where 1.0 lawyer = 4.3×10^8 kJ) a 4.5-metric ton *Tyrannosaurus rex* (Chapter 9) would have needed to eat in a year, depending on whether it was ectothermic or endothermic. Their calculations revealed that an endothermic *Tyrannosaurus* of this size would have required 292 lawyers/year, whereas an ectothermic one would have only needed 73 lawyers/year. In other words, an endothermic *Tyrannosaurus* would have been four times more effective at stemming frivolous litigation than an ectothermic one.

On a broader scale, energy flow in an ecosystem, as well as the thermoregulation of the animals inhabiting the ecosystem, can affect the proportion of carnivores versus herbivores. This means that an ecosystem tells much about the physiology of its animals. If all of the predators are endothermic in a particular ecosystem, their caloric needs could decimate a prey population quickly unless the latter had large numbers. Preliminary studies of modern ecosystems with predator–prey relationships of endotherms, specifically, mammals, show that predators compose a much smaller proportion (less than 1%) of the total predator–prey biomass than in ecosystems dominated by ectotherms (about 25%). This information can be applied as a predictive model to dinosaur populations, where the ratio of theropods to all other dinosaurs in a contemporaneous deposit is calculated. The few studies that employed this approach in a comprehensive manner, using a census of dinosaurs identified as "predator" and "prey," found small proportions of theropods versus other dinosaurs (about 1 to 30, or 3% to 5% theropods), thus more closely resembling the endothermic model.

However, the predator–prey model has some of the following problems:

- The model is based on one group of endotherms (mammals) and not birds, which are also endotherms. Considering that most paleontologists consider birds as dinosaurs, an examination of the latter might be more instructive for comparison.
- Not all food is alike. The model assumes that the predators may have been equal-opportunity hunters instead of selective.
- The interpretation of an endothermic physiology can only be made about the theropods, not the herbivorous dinosaurs.
- Taphonomic bias rears its ugly but reality-inducing head once again (Chapter 7). Not all of the predator and prey animals are preserved in the assemblage, especially if the animals were small.

The last of these caveats can be tested independently through a track census, which looks at the proportion of tracks attributable to theropods versus all other dinosaurs, then normalizes the data for biomass. Footprint length gives an estimate of the size of the track-making dinosaur (Chapter 14), which is used to calculate a biomass for each track-making dinosaur. These biomasses are then totaled for theropods and other dinosaurs, represented on a track horizon to derive a predator/prey ratio. Using such methods, dinosaur ichnologists calculated predator/prey ratios from a Late Jurassic tracksite of North America of about 1 : 7 (about 15% theropods). Interestingly, this ratio is intermediate compared to those calculated for predator/prey ratios in endothermic and ectothermic populations (Fig. 8.7). The advantage of this independent measure is that, in most instances, tracks from a given horizon are more likely to be contemporaneous samples of dinosaur populations than a bone bed. The disadvantages are that this analysis gives more of an assessment of the physiology of the predators, not the prey, and that tracks hold their own distinctive biases (Chapter 14).

A similar method that would compare the ratio of herbivorous dinosaur biomass to plant biomass has yet to be carried out. This approach operates on the similar assumption that endothermic herbivores would have had much greater food needs than ectothermic herbivores in a given terrestrial ecosystem. As a result, the biomass of large herbivorous dinosaurs should have been limited by the biomass and caloric quality of the available plants. Some paleontologists have surmised on

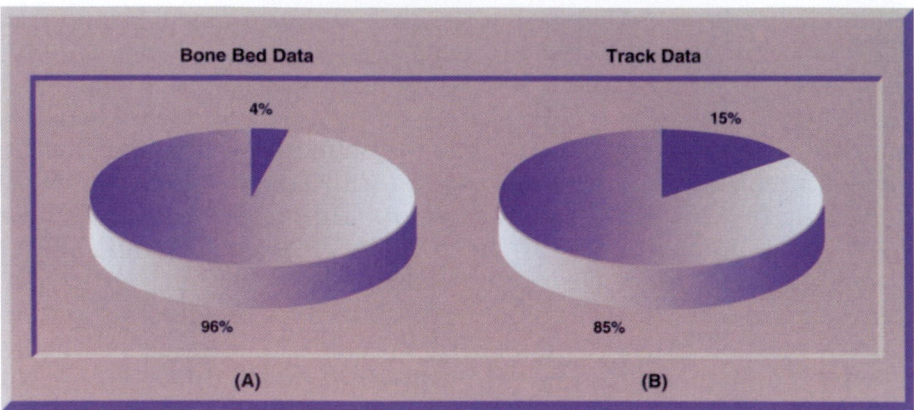

FIGURE 8.7 Estimates of dinosaur biomass in two different Mesozoic deposits as determined by predator/prey ratios and using different sources of data. (A) Bone data (based on a Late Creteceous deposit). (B) Track data, normalized for biomass (based on a Late Jurassic tracksite). Data from Lockley (1990).

such ecological reasoning that very large, 20+ metric-ton herbivores, represented by some sauropods, such as Late Jurassic diplodocids and Cretaceous titanosaurids (Chapter 10), must have ingested large amounts of low-quality (woody) plants. This would have happened regardless of whether they were endothermic or ectothermic. In short, difficult-to-find food of high quality (those with high kJ/g) was sacrificed for the sake of immediately accessible quantity. Accordingly, low-quality plant foods would have needed extensive residence time in the gastrointestinal tract for fermentation by symbiotic anaerobic bacteria. Thus, digestion required a longer and larger gut, which in turn necessitated a larger animal. Gastroliths presumably aided in this digestion, whether they were present in a muscular crop, gizzard, or both (Chapters 5 and 14). Unfortunately, these pieces of evidence, along with the small amount of coprolite data linked to herbivorous dinosaurs' dietary choices (Chapter 14), do not provide adequate answers to questions about their thermoregulation. Consequently, other sources of information from a wide range of choices must be examined.

Dinosaur Thermoregulation: Other Considerations

The controversy over whether dinosaurs were endothermic, ectothermic, or some sort of physiology that did not qualify as either (or varied on a species-to-species basis) has caused the death of many trees because of the large number of papers written on the subject during the last 30 years. Of all topics in dinosaur studies, the popularity of the dinosaur thermoregulation discussion is rivaled only by the enduring debates over theropod ancestry and origin of flight in birds, and the causes of dinosaur extinctions at the end of the Cretaceous. One of the fortunate outcomes of this continuing research is that the question of which mode of thermoregulation dominated among dinosaurs has provoked a multi-faceted approach to answering it. One of the less fortunate outcomes is a common dilemma in science: more questions are generated by the research than are answered.

The other clues used to test for endothermy and ectothermy in dinosaurs include the following, most of which involve comparison to living terrestrial vertebrates (including birds):

- Neurophysiology, where the **encephalization quotient** (EQ) is calculated and brain complexity is described (Chapter 11).
- Geochemistry of bone as related to oxygen-isotope ratios.
- Social behavior (degree of brooding or nurturing, gregariousness, intra-specific competition).
- Cranial anatomy related to respiration.
- Overall posture and locomotion.
- Body size (including the lengths of some body parts).
- Soft-part anatomy (organs and feathers).
- Paleobiogeographic distribution.
- Phylogenetic "closeness" to endothermic or ectothermic animals.

None of these approaches has resulted in 100% agreement among researchers. However, the evidence suggests strongly that some, but not necessarily all, dinosaurs were likely endothermic but showed this thermoregulatory mode while growing up. Figuring out which thermoregulatory mode was the case after adulthood is less certain. With that warning in mind, arguments and counterarguments for each facet of thermoregulation are presented briefly here. Readers should look for points of consensus, then peruse the voluminous literature on this subject for the details.

Neurophysiology

When EQs are plotted for dinosaurs against a 1.0 standard for modern crocodiles, the values range from extremely low (0.2) to substantially higher (6.5). The less "brainy" dinosaurs include sauropods, thyreophorans, and ceratopsians, which all rank below 1.0. Ornithopods and theropods show higher values, the highest being from coelurosaurs (especially troodontids), a clade that presumably includes the ancestors of birds. In fact, the EQs at the higher end of the scale for non-avian theropods overlap with some modern flightless birds (Chapter 15). The reasoning behind using EQs as a measure of thermoregulation is that modern endothermic organisms have high EQs. However, this does not necessarily mean that a cause-and-effect relationship can be inferred between EQ and endothermy. Also, less than 5% of dinosaur genera have had their EQs calculated, so the data set should be regarded as preliminary. Independent neurophysiological evidence related to EQ, such as brain complexity, also reveals that dinosaur brains are rather simple, comparable in morphology to those of modern reptiles. Nevertheless, a few (such as *Tyrannosaurus*) show enlarged olfactory bulbs, which suggests a superior sense of smell but denotes nothing about thermoregulation.

Bone Geochemistry: Oxygen Isotopes

Endothermy and ectothermy can cause oxygen-isotope signatures in different parts of an animal's body. Therefore, studying the oxygen-isotope ratio from different parts of the body should indicate which thermoregulation mode the animal used. For example, an endothermic animal should maintain a very similar body temperature throughout its entire body (homeothermy). This means that its phalanges would reflect the same temperature as its vertebrae, despite the distal location of the fingers relative to the spine. As explained earlier, oxygen-isotope ratios ($^{16}O/^{18}O$) change in direct correspondence to temperature. Consequently, the ratio of these isotopes from bones in different parts of an animal's body should either differ considerably, correlating with an ectothermic physiology, or coincide, indicating endothermy. Dinosaur bones, like dinosaur eggs, contain oxygen isotopes, meaning that their isotopic ratios can be calculated. For example, ratios calculated for the theropod

8

Tyrannosaurus, ornithopod *Hypacrosaurus*, and ceratopsian *Montanoceratops* indicate that the ratios varied little, equivalent to ±2°C. These preliminary results indicate at least homeothermy for those dinosaurs tested, with the exception of one nodosaurid that showed variations equivalent to ±11°C.

Two criticisms of this method are:

1 homeothermy implies higher metabolic rates than ectothermy, but still does not indicate endothermy on the level of birds or mammals; and
2 the ratios may be nowhere near the original ratios formed when the animal was alive, and may be a function of water chemistry from post-burial processes (Chapter 6).

This method of analysis is still relatively new and being tested, but holds some promise for future studies.

Social Behavior

Some of the social interactions interpreted for dinosaurs are also common behaviors in modern birds and mammals. These behaviors include:

1 nurturing of young by some hadrosaurs (Chapter 11);
2 brooding of a nest by an oviraptorid (Fig. 8.8);
3 herding or other large group movements by theropods, sauropods, ornithopods, and ceratopsians; and
4 intraspecific competition in theropods, ornithopods, thyreophorans, and ceratopsians.

In contrast, these behaviors are rare to absent in most modern terrestrial ectothermic vertebrates. Of course, present behaviors that are correlated with thermoregulation may not necessarily apply to Mesozoic animals, and past behaviors are based on inferences. The majority of this evidence, however, is persuasive of behavioral qualities that may have depended on endothermy. A brooding *Oviraptor* on its nest is particularly striking: why would a dinosaur sit on a nest unless it was keeping its eggs at the same temperature as the parent? Along those same lines, the outstretching of arms around a nest implies protection, which would have been provided by arms that had a greater surface area caused by feathers, another endothermic trait.

Cranial Anatomy Related to Respiration

Endothermic vertebrates have large spaces within their nasal cavities to accommodate folded bony or cartilaginous structures, which were often lined with mucous membranes called **respiratory turbinates**. These structures are essential to endotherms because they help to conserve the water and heat associated with the near-constant breathing that endotherms use for their more active metabolism. Turbinates ensure that as much as 60% of the water moisture that is inhaled is absorbed ("reclaimed") before being exhaled, otherwise its absence would quickly result in dehydration. Body-heat losses and energy demands would also accumulate if turbinates were not in place, the equivalent of leaving a window open on a cold winter day. Because of the essential function of turbinates in conserving water and heat, nearly all mammals and birds possess them and they can be used as undoubted indicators of endothermy in modern taxa. Searches for these structures (or at least enough room for them) in a few dinosaur skulls have so far not shown

FIGURE 8.8 Evidence of brooding and association of embryo with the theropod *Oviraptor*. (A) Skeleton of adult *Oviraptor* (missing its skull) on egg clutch. (B) *Oviraptor* embryo recovered from an egg that was previously interpreted as belonging to the ceratopsian *Protoceratops*. Both specimens recovered from Late Cretaceous strata in Mongolia. Transparencies 5789 (5) and K17685 (Photo. Mick Ellison). Courtesy of the Library, American Museum of Natural History.

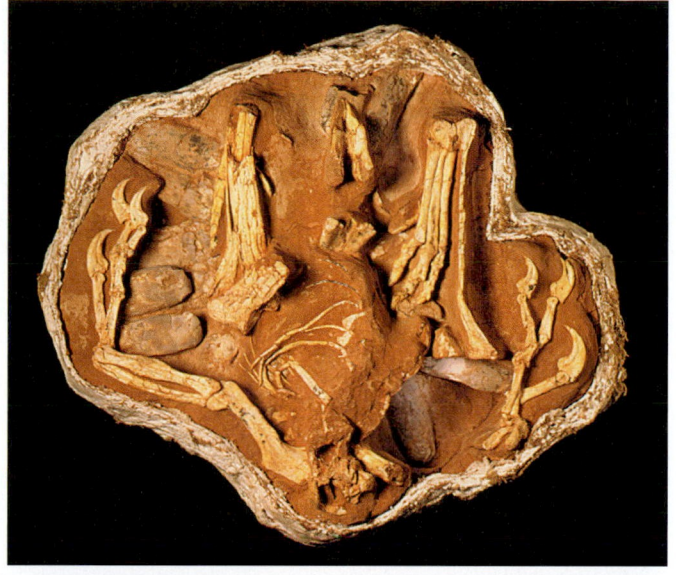

(A)

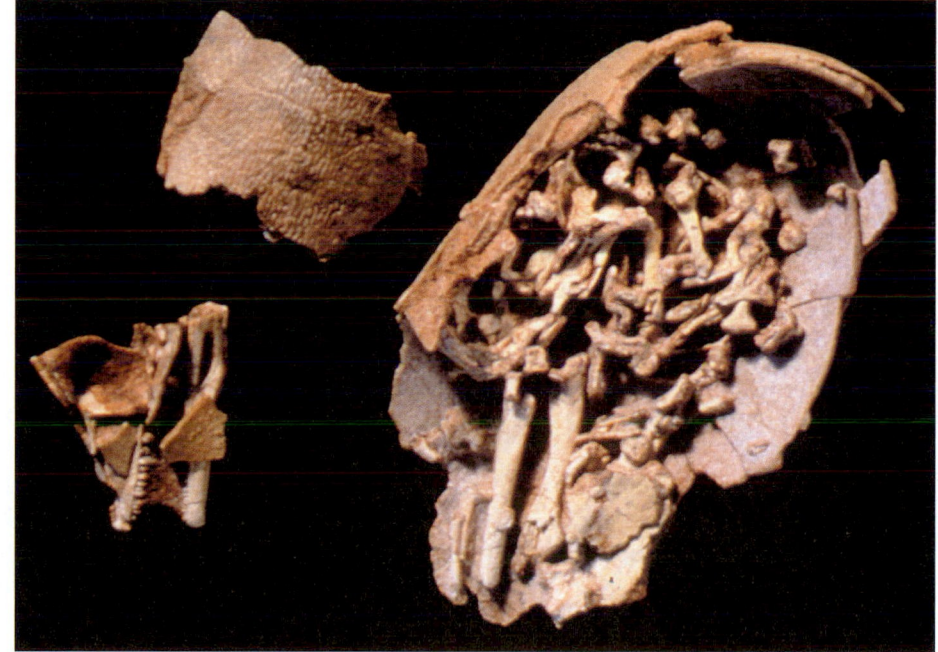

(B)

any evidence for turbinates. Thus, this constitutes evidence favoring ectothermy in those dinosaurs examined (*Ornithomimus*, *Tyrannosaurus*, and *Hypacrosaurus*).

One of the objections to this hypothesis is that turbinates may have evolved later as a more derived trait associated with endothermy, but was still preceded by the evolution of endothermy in general. In this sense, the absence of a single trait does not necessarily negate all other evidence that is in agreement. Another criticism is that turbinate-like structures may have been present in dinosaurs but they were composed entirely of soft tissue and not preserved, or otherwise are not yet recognizable. Furthermore, similar to the situation posed by EQs of dinosaurs, only a small number of dinosaur genera have been examined for the absence or presence of cranial space for respiratory turbinates. As a result, any emphatic statement that all dinosaurs lacked these structures is premature. Lastly, endothermic dinosaurs lacking such structures could have compensated behaviorally, such as by drinking more water. Nevertheless, respiratory turbinates provide yet another example of why thermoregulation has been a source of vigorous debate.

Overall Posture and Locomotion

Upright posture, a defining trait of dinosaurs (Chapters 1 and 5), is linked with endothermy because it is a characteristic of all extant mammals and birds. Upright posture is not only discernible from dinosaur skeletons, but is also apparent in dinosaur trackways, which normally show narrow straddles associated with diagonal walking patterns (Chapter 14). Upright posture also implies increased mobility and energy needs in comparison to a sprawling one. For example, adaptations to obligate bipedality in ancestral dinosaurs resulted in faster and possibly more sustained movement (Chapter 6). Numerous theropod trackways also indicate that they were active animals. Activity levels are presumably related to metabolism and food requirements. Overall, these locomotor capabilities are suggestive of endothermy.

The problems with these generalizations are that:

1 just because endothermy is correlated with upright posture in extant animals, does not necessarily mean that it also applies to extinct ones;
2 in warmer climates, ectotherms can also be quite active, and many of the Mesozoic environments inhabited by dinosaurs were mild, especially during times of global warming; and
3 skeletons and trackways of a few dinosaurs, such as those of ceratopsians, suggest more of a semi-erect posture.

Using this criterion, ceratopsians could have been either ectothermic, if upright posture is a 100% sure indicator of endothermy, or endothermic, if upright posture is a fallible indicator of endothermy. Perhaps the answer lies somewhere in between.

Body Size (Including Body Parts)

In the larger dinosaurs, such as some sauropods (Chapter 10), sheer size may have been a factor in maintaining body heat. An animal that keeps its body temperature the same, because of its mass retaining heat, is called an **inertial homeotherm**. This is regardless of whether they were endotherms or ectotherms. In fact, a sauropod with endothermy equivalent to that of a mammal would have had a problem with getting rid of heat, rather than with retaining it. One of the possible avenues for dissipating excess heat is through body parts, as seen in modern elephants with large ears (such as *Loxodonta*). This mechanism has been postulated for stegosaurs with dorsal plates (Chapter 12). Although large external ears or other flaps of skin are unknown in sauropods, the extremely long necks and tails, as well as four lengthy limbs, of some species could have also served as heat vents.

Neck lengths, however, pose a dilemma for blood pressure. If sauropods held their necks vertically or otherwise assumed a vertical posture, they would have required very high blood pressures to pump oxygenated blood to their brains (Chapter 10). High blood pressure is correlated with endothermy in modern endotherms. Accordingly, the estimates of the blood pressures needed to pump blood from the distance of a dinosaur heart to its brain can be calculated. When the two sets of figures are compared, most of the dinosaurs that were considered, with the exception of ceratopsians, show blood pressures that overlap or exceed those of modern endotherms. Still, such calculations made for sauropods assume a vertical posture for their necks. In contrast, some anatomical data now suggest horizontal postures were more common, so these animals would not have required such high blood pressures. Also, correlation is not necessarily causation, in that endothermy may not have necessarily caused high blood pressure.

Soft-Part Anatomy: Organs and Feathers

Although soft tissues are more rarely preserved in dinosaurs than their bones and teeth, when found they can provide valuable insights into dinosaur physiology. For example, feathers in dinosaurs have important implications for physiology, as do organs such as hearts and lungs. In birds, feathers have multiple purposes, such as display, flight, and insulation. In *Archaeopteryx*, the feathers were developed well enough to have served all three purposes (Chapter 15). On the other hand, feathers seen on flightless theropods – such as those on the arms and tail of *Caudipteryx* and other small theropods, as well as the downy dorsal fringe seen on *Sinosauropteryx* (Chapter 9) and possibly one specimen of the ceratopsid *Psittacosaurus* (Chapter 13) – may not have been sufficient for insulation purposes. Nevertheless, they may have made the dinosaurs look attractive to others of their species. Some paleontologists have proposed that juvenile theropods or other juvenile dinosaurs were born with downy coats that they later shed, but this idea lacks evidence. The only skin impressions known of embryonic dinosaurs are those of some Early Cretaceous titanosaurids, which show only scales (Chapter 10). In fact, the vast majority of dinosaur skin impressions show no signs of feathers, although this is a bias that also applies to most soft tissues possessed by dinosaurs.

An exception to this generalization was the recent discovery of a mineralized mass in the thoracic cavity of a specimen of the Late Cretaceous ornithopod *Thescelosaurus*, which was originally interpreted as a fossil heart. Both endotherm hearts and crocodile hearts have four chambers, but endotherm hearts differ in having single aortas and completely divided ventricles, whereas crocodile hearts have two aortas and only partial division between ventricles. CT scans of the structure in *Thescelosaurus* showed a morphology resembling four chambers and a single aorta, which matches modern endotherms. This does not necessarily mean that *Thescelosaurus* was endothermic? Since the publication of the peer-reviewed paper on this structure, it underwent even more peer review that resulted in fewer scientists accepting it as a "fossil heart." However, if any dinosaur is found with a definite heart that shows four chambers and a single aorta, it would only mean that it had high blood pressure. Again, this is a trait correlated with endothermy, but not necessarily dependent on it.

Paleobiogeographic Distribution

Dinosaur body and trace fossils have been recovered from both Arctic and Antarctic latitudes, and all latitudes between. Although dinosaurs are often depicted as living in tropical or low-latitude desert environments, a significant number of them lived in temperate or even polar environments. In fact, any dinosaur fossils from Alaska, northern Canada, Sweden, Siberia (of Russia), Australia, New Zealand, Antarctica, or the southern part of South America are designated as evidence of so-called **polar dinosaurs**. Modern ectotherms are much narrower in their biogeographic ranges, clustering in both numbers and diversity in low latitudes and the equator. Even when taking into account the plate tectonic movements that have moved some areas to polar localities of today (Chapter 4), an appreciable number of dinosaurs still lived in 60 degrees or higher latitudes, or within 30 degrees of either the geographic North or South Poles.

Such a paleobiogeographic range suggests endothermy, especially if any dinosaurs stayed in these regions during the six months of darkness associated with winters in these regions. However, considering that warm global temperatures were typical of much of the Mesozoic, especially the Cretaceous Period, even the biogeographic ranges of ectotherms should have been wider than today. Nevertheless, some paleontologists have argued that these dinosaurs still exceeded expected ranges for Mesozoic ectotherms. Other evidence relates to the possibility that dinosaurs migrated with the seasons, a behavioral trait seen in modern endotherms such as birds, many of whom breed in the northern hemisphere during the summer and then fly south for the winter (Chapter 15). Whether polar dinosaurs were year-round residents or "snowbirds" is still not known, but anatomical evidence that supplements the year-round hypothesis comes from the large orbits (eye sockets) of a few species. These traits certainly correspond to large eyes, which may represent adaptations to seeing better under low-light conditions.

Phylogenetic Closeness to Endothermic or Ectothermic Animals

According to phylogenetic analyses, birds and crocodilians are the closest living relatives to dinosaurs, but crocodilians are ectothermic and birds are endothermic. According to present paleontological knowledge, the common ancestor of crocodilians and dinosaurs probably lived in the Early Triassic. In contrast, the common ancestor of dinosaurs and birds probably lived in the Middle to Late Jurassic. The common ancestry of crocodiles and dinosaurs was originally used as evidence of ectothermy in dinosaurs. Now, however, the common ancestry of dinosaurs and birds is used as evidence for endothermy in dinosaurs. Furthermore, pterosaurs (flying reptiles: Chapter 6) are now regarded as endotherms, and they had a common ancestor with dinosaurs. Some paleontologists default to the more recent evolution of birds as evidence of a minimum time when endothermy emerged in archosaur lineages. Nevertheless, this still does not answer exactly when endothermy developed. Endothermy also was independently developed in mammalian lineages throughout the Mesozoic, and it most likely began in mammals well before birds.

The only clear conclusion from using a phylogenetic argument for the timing of endothermy is that it developed in some archosaur lineages sometime in the Mesozoic and most likely in some small theropods during the Jurassic. Nonetheless, this conclusion still does not entirely pertain to other dinosaurs, even other saurischians. As a result, the mystery of dinosaur thermoregulation is not yet firmly answered by either "big picture" or "small picture" perspectives, but the answers have been good in provoking further research.

SUMMARY

Paleontologists who study dinosaur physiology examine how dinosaurs converted and transferred matter and energy in their daily lives, which is especially related to their metabolism. This is a topic of much interest to paleontologists because of what it reveals about dinosaur biology and evolutionary history. Much of the controversy surrounding dinosaur physiology centers on whether they were endothermic ("warm blooded"), ectothermic ("cold blooded"), or some other type of physiology that was in between. Lines of evidence used in interpreting dinosaur physiology are mostly associated with reproduction, growth, and feeding. Reproductive behavior, types of eggs, clutch sizes, and the biogeochemistry of eggs all lend clues to dinosaur physiology, particularly for female dinosaurs. Studies of bone mineralization and dinosaur bone histology provide information about their rates of growth. These indicate rapid growth rates consistent with endothermy in a few dinosaur species, although it is unknown whether most or all dinosaurs had the same sort of physiology. Dinosaur nutritional needs, adaptations to feeding (such as teeth and jaws), and how they fed were intimately related to dinosaur physiology. Dinosaurs consisted of both herbivores and carnivores, although evidence from toothmarks, stomach contents, gastroliths, and coprolites can sometimes indicate exactly what a dinosaur ate or which of them were eaten. Other lines of evidence related to dinosaur physiology include EQ (encephalization quotient), geochemistry of bone, social behavior, cranial anatomy related to respiration, overall posture and locomotion, body size, soft-part anatomy, paleobiogeographic distribution, and phylogenetic relatedness to endothermic or ectothermic animals. Collectively, these data point toward dinosaurs as a physiologically varied group of animals that cannot be easily categorized.

DISCUSSION QUESTIONS

1. Describe or sketch a hypothetical trace fossil of dinosaur mating for a
   a. bipedal dinosaur species, and
   b. quadrupedal dinosaur species.
   How would a paleontologist be able to distinguish such behavior from the vast majority of trackways that merely show locomotion?
2. Acquire one dozen chicken eggs from a grocery store and measure their lengths and diameters.
   a. What shape would you characterize for the eggs?
   b. Examine the eggshell using either a handlens or a binocular (dissecting) microscope. If pores are visible, draw a square centimeter onto

DISCUSSION QUESTIONS Continued

the surface of the egg and count the number of pores within that square centimeter. Repeat the experiment for two other eggs and compare the results for all three samples. How similar are they?

c. Using the appropriate formula, calculate the volume for a single egg, then calculate individual volumes for the next 11 eggs. What statistical measurements can you use to describe the "typical" egg volume and its variations?

d. What is the cumulative (total) volume of all one dozen eggs?

e. How does this cumulative volume compare to a Late Cretaceous sauropod egg from Argentina with axis diameters of $a = 13$ cm, $b = 13$ cm, and $c = 15$ cm?

f. Test the chemical composition of an egg by immersing it in vinegar. What elements are being released from the egg by dissolving the eggshell material? Would this affect its biogeochemistry in comparison to an egg that did not have its eggshell dissolved?

3. What is an explanation for site fidelity of dinosaur nests, when dinosaurs were clearly capable of moving to a variety of environments for their egglaying?

4. How would the evolution of teeth and bones composed of calcite or aragonite, instead of dahllite, have changed the lifestyles of some vertebrates? Give some specific examples using either fossil or present-day animals.

5. Why is relatively more cancellous bone located in the epiphyses of limb bones, instead of the greater amount of compact bone found in the diaphyses? What would be the functional advantage of this unequal distribution?

6. What evidence would help to better support the hypothesis that some theropods preferentially consumed sauropod skulls? If such behavior was reasonably supported, then offer an explanation why this body part was preferred as a result of some physiological need. What nutrition would brains provide versus other organs?

7. Iguanodontians lacked teeth on their premaxillaries, yet the rest of their teeth are clearly adapted for tearing and chewing plant material. Assuming an iguanodontian pulled plant material into its mouth using teeth from the dentary but none on the premaxillary, what might their toothmarks on the plants have looked like, especially in comparison to herbivores that have incisors for shearing on both upper and lower jawbones? What modern animals have a similar adaptation to feeding and what do their toothmarks on plants look like?

8. Give a taphonomically based alternative hypothesis for how a supposed prey animal could have been placed in the thoracic region of a dinosaur without it having been actually consumed or constituting an offspring.

9. If someone asked you what is meant by "hot-blooded" and "cold-blooded", how would you explain this, using real, modern examples? How would you then apply this explanation to dinosaurs that have been extinct for more than 65 million years?

10. What is an alternative explanation for "polar dinosaurs" other than endothermy? Provide evidence that would support your hypothesis, whether it would come from body fossils, trace fossils, or analogues with modern animals.

Bibliography

Abler, W. L. 1999. The teeth of tyrannosaurs. *Scientific American* **281**: 50–51.

Bakker, R. T. 1972. Anatomical and ecological evidence of endothermy in dinosaurs. *Nature* **238**: 81–85.

Barrick, R. E. 1998. Isotope paleobiology of the vertebrates: ecology, physiology, and diagenesis, in isotope paleobiology and paleoecology *The Paleontological Society Papers* **4**: 101–137.

Barrick, R. E. and Showers, W. J. 1994. Thermophysiology of *Tyrannosaurus rex*: evidence from oxygen isotopes. *Science* **265**: 222–224.

Barrick, R. E., Showers, W. J. and Fischer, A. G. 1996. Comparison of thermoregulation of four ornithischian dinosaurs and a varanid lizard from the Cretaceous Two Medicine Formation: Evidence from oxygen isotopes. *Palaios* **11**: 295–305.

Bennett, A. F. and Ruben, J. A. 1979. Endothermy and activity in vertebrates. *Science* **206**: 49–654.

Bertrand, C. E. 1903. Les Coprolithes de Bernissart. I. partie: Les Coprolithes qui ont été attribués aux Iguanodons. *Memoires du Musee royal d'histoire naturelle de Belgique* **1(4):** 1–154.

Brown, B. 1907. Gastroliths. *Science* **25(636)**: 392.

Buckland, W. 1835. On the discovery of coprolites, or fossil faeces, in the Lias at Lyme Regis, and in other formations. *Transactions of the Geological Society of London Series 2* **3(1):** 223–236.

Campbell, H. W. 1972. Ecological or phylogenetic interpretation of crocodilian nesting habits. *Nature* **238**: 404–405.

Carpenter, K. 1999. *Eggs, Nests, and Baby Dinosaurs: A Look at Dinosaur Reproduction (Life of the Past)*. Bloomington, Indiana: Indiana University Press.

Carpenter, K., Hirsch K. F. and Horner, J. R. (Eds). 1994. *Dinosaur Eggs and Babies*. Cambridge, UK: Cambridge University Press.

Chiappe, L. M., Coria, R. A., Dingus, L., Jackson, F., Chinsamy, A. and Fox, M. 1998. Sauropod dinosaur embryos from the Late Cretaceous of Patagonia. *Nature* **396**: 258–261.

Chin, K. 1997. "What did dinosaurs eat? Coprolites and other direct evidence of dinosaur diets". *In* Farlow, J. O. and Brett-Surman, M. K. (Eds), *The Complete Dinosaur*. Bloomington, Indiana: Indiana University Press. pp. 371–382.

Chin, K. and Gill, B. D. 1996. Dinosaurs, dung beetles, and conifers: participants in a Cretaceous food web. *Palaios* **11**: 280–285.

Chin, K., Tokaryk, T. T., Erickson, G. M. and Calk, L. C. 1998. A king-sized theropod coprolite. *Nature* **393**: 680–682.

Constantine, A., Chinsamy, A., Vickers-Rich, P. and Rich, T. H. 1998. Periglacial environments and polar dinosaurs. *South African Journal of Science* **94**: 137–141.

Currie, P. J. and Jacobsen, A. R. 1995. An azhdarchid pterosaur eaten by a velociraptorine theropod. *Canadian Journal of Earth Sciences* **32**: 922–925.

DeNiro, M. J. and Epstein, S. 1978. Influence of diet and the distribution of carbon isotopes in animals. *Geochimica Cosmochimica Acta* **42**: 495.

Erickson, G. M. 1997. "Tooth replacement patterns". *In* Currie, P. J. and Padian, K. (Eds), *Encyclopedia of Dinosaurs*. New York: Academic Press. pp. 739–740.

Farlow, J. O. 1987. Speculations about the diet and digestive physiology of herbivorous dinosaurs. *Paleobiology* **13(1):** 60–73.

Farlow, F. O., Thompson, C. V. and Rosner, D. E. 1976. Plates of the dinosaur *Stegosaurus*: forced convection heat loss fins? *Science* **192**: 1123–1124.

Fiorillo, A. R. and Weishampel, D. B. 1997. "Tooth wear". *In* Currie, P. J. and Padian, K. (Eds), *Encyclopedia of Dinosaurs*. New York: Academic Press. pp. 743–745.

Fisher, P. E., Russell, D. A., Stoskopf, M. K., Barrick, R. E., Hammer, M., Kuzmitz, A. A. 2000. Cardiovascular evidence for an intermediate or higher metabolic rate in an ornithischian Dinosaur. *Science* **288**: 503–505.

Folinsbee, R. E., Fritz, P., Krouse, H. R., and Robblu, A. R. 1970. Carbon and oxygen-18 in dinosaur, crocodile, and bird eggshells indicate environmental conditions. *Science* **168**: 1553.

Francillon-Vieillot, H. et al. 1990. "Microstructure and mineralization of vertebrate skeletal tissues". *In* Carter J. (Ed.), *Skeletal Biomineralization*. New York: Van Nostrand-Reinhold. pp. 471–530.

Hopson, J. A. 1977. Relative brain size and behaviour in archosaurian reptiles. *Annual Revie of Ecological Sysems.* **8**: 429–448.

Horner, J. R. 2000. Dinosaur reproduction and parenting. *Annual Review of Earth and Planetary Sciences* **28**: 19–45.

Horner, J. R. and Makela, R. 1979. Nest of juveniles provides evidence of family structure among dinosaurs. *Nature* **282**: 296–298.

Horner, J. R., de Ricqlès, A. and Padian, K. 1999. Variation in dinosaur skeletochronology indicators: Implications for age assessment and physiology. *Paleobiology* **25**: 295–304.

Jerison, H. J. 1969. Brain evolution and dinosaur brains. *American Naturalist* **103**: 575–588.

Johnston, P. A. 1979. Growth rings in dinosaur teeth. *Nature* **278**: 635–636.

Nagy, K. A. 1987. Field metabolic rate and food requirement scaling in mammals and birds. *Ecological Monographs* **57**: 111–128.

Norell, M. A., Clark, J. M., Chiappe, L. M. and Dashzeveg, D. 1995. A nesting dinosaur. *Nature* **378**: 774–776.

Norell, M. A., Clark, J. M., Dashzeveg, D., Barsbold, R., Chiappe, L. M., Davidson, A. R., McKenna, M. C., Perle, A. and Novacek, M. J. 1994. A theropod dinosaur embryo and the affinities of the Flaming Cliffs dinosaur eggs. *Science* **266, 5186**: 779–782.

O'Connor, M. P. and Dodson, P. 1999. Biophysical constraints on the thermal ecology of dinosaurs. *Paleobiology* **25**: 341–368.

Ostrom, J. H. 1969. Osteology of *Deinonychus antirrhopus*, an unusual theropod from the Lower Cretaceous of Montana. *Peabody Museum of Natural History Bulletin* **30**: 1–165.

Padian, K., de Ricqles, A. J. and Horner, J. R. 2001. Dinosaurian growth rates and bird origins. *Nature* **412**: 405–408.

Paladino, F. V., Spotila, J. R. and Dodson, P. 1997. "A blueprints for giants: Modeling the physiology of large dinosaurs". *In* Farlow, J. O. and Brett-Surman, M. K. (Eds), *The Complete Dinosaur*, Bloomington, Indiana: Indiana University Press. pp. 491–504.

Parrish, J. M. et al. 1987. Late Cretaceous vertebrate fossils from the north slope of Alaska and implications for dinosaur ecology. *Palaios* **2**: 377–389.

Paul, G. S. 1988. *Predatory Dinosaurs of the World*. New York: Simon & Schuster.

Reid, R. E. 1997. "Dinosaurian physiology: the case for intermediate dinosaurs". *In* Farlow, J. O. and Brett-Surman, M. K. (Eds), *The Complete Dinosaur*, Bloomington, Indiana: Indiana University Press. pp. 449–473.

Ricqles, A. J. de. 1974. Evolution of endothermy: histological evidence. *Evolutionary Theory* **1**: 51–80.

Ricqles, A. J. de. 1983. Cyclical growth in the long limb bones of a sauropod dinosaur. *Acta Palaeontologica Polonica* **28**: 225–232.

Ruben, J. A., Jones, T. D., Currie, P. J., Horner, J. R., Espe, G., III, Hillenius, W. J., Geist, N. R. and Leitch, A. 1996. The metabolic status of some Late Cretaceous dinosaurs. *Science* **273**: 1204–1207.

Ruben, J. A. et al. 1999. Pulmonary function and metabolic physiology of theropod dinosaurs. *Science* **283**: 514–516.

Ruben, J. A., Jones, T. D., Signore, M., Dal Sasso, C., Geist, N. R. and Hillenius, W. J. 1999. Pulmonary function and metabolic physiology of theropod dinosaurs. *Science* **283**: 514–516.

Sander, P. M. 1997. "Teeth and jaws". *In* Currie, P. J. and Padian, K. (Eds), *Encyclopedia of Dinosaurs*. New York: Academic Press. pp. 717–725.

Sarkar, A., Bhattacharya, S. K. and Mohabey, D. M. 1991. Stable-isotope analyses of dinosaur eggshells: paleoenvironmental implications. *Geology* **19**: 1068–1071.

Sasso, C. D. and Signore, M. 1998. Exceptional soft-tissue preservation in a theropod dinosaur from Italy. *Nature* **392**: 383–387.

Schmidt-Nielsen, K. 1972. *How Animals Work*. Cambridge. UK: Cambridge University Press.

Thomas, R. D. K. and Olson, E. C. (Eds). 1980. *A Cold Look at Warm-blooded Dinosaurs*. Boulder, Colorado: Westview Press.

Varricchio, D. J. 1999. Gut contents for a Cretaceous tyrannosaur: implications for theropod digestive tracts. *Journal of Vertebrate Paleontology 19*, **3**: 82A.

Varricchio, D. J., Jackson, F., Borlowski, J. and Horner, J. R. 1997. Nest and egg clutches for the theropod dinosaurs *Troodon formosus* and evolution of avian reproductive traits. *Nature* **385**: 247–250.

Weaver, J. C. 1983. The improbable endotherm: the energetics of the sauropod dinosaur *Brachiosaurus*. *Paleobiology* **9**: 173–182.

8

Chapter 9

Of the dinosaurs depicted in fiction, the ones you most often encounter are theropods. Theropods are inevitably portrayed as active, voracious, and vicious predators that ruthlessly pursued and killed their prey. Many movies show theropods running in packs, which suggests that they had greater activity levels and energy requirements than normal modern reptiles and behaved more like mammals or birds.

What fossil evidence supports that any, let alone most, theropods were predators? In contrast, which of the theropods could have been scavengers? Were some possibly omnivores or even herbivores? Finally, what fossil evidence answers questions about their activity levels and social behavior, and how do these compare to modern mammals or birds?

Theropoda

9

Why Study Theropods?

Of all dinosaur clades, Theropoda is the most intensively studied and discussed by paleontologists. Based on their frequent appearances on the covers of prestigious science journals and mentions in news reports of the past 20 years or so, theropods remain the most famous of dinosaurs. This fame is at least partially related to an enduring, perhaps vicarious fascination with large predatory animals in general, such as sharks, crocodiles, Komodo dragons, large cats, and grizzly bears. Theropods fit well in this group of meat-eaters but with one exception – some of their members were the largest carnivores that ever lived in terrestrial environments; a few species may have approached 8 metric tons in weight and 15 meters in length. Consequently, an informal competition for discovery of the largest land-dwelling carnivore focuses exclusively on members of the Theropoda, a sideline issue for paleontologists for more than 100 years. This contest began with the announcement by Barnum Brown in 1902 of the first specimen of *Tyrannosaurus rex* (Chapter 3) and continued more recently with *Carcharodontosaurus* of Morocco and *Giganotosaurus* of Argentina.

Of course, not all theropods were large predators. In fact, they were extremely diverse in size, form, and function during their 165-million-year history. For example, the majority of theropods, unlike most dinosaurs, were bipedal. This plesiomorphic trait of dinosaurs was probably related to natural selection for longer hind limbs, shorter fore limbs, modification of the pes, and lightening of the skeletal structure. Increased mobility and speed were subsequent benefits of this evolution (Chapter 6). Indeed, some theropods have skeletal features that reflect most of the adaptive characteristics seen in modern mammals and large flightless birds capable of rapid movement. The most dynamic records of theropod bipedalism and other forms of movement are their tracks, and interpretations made from such features corroborate their activities. Theropod tracks, first described by Edward Hitchcock in the mid-nineteenth century (Chapter 3), are by far the most abundant of dinosaur tracks and constitute a considerably greater fossil record than theropod skeletal remains (Chapter 14).

The past 20 years of paleontological research have also witnessed better documentation of other theropod trace fossils. Examples include nests, toothmarks, coprolites, and rare examples of gastroliths (Chapter 14), which have been accompanied by body fossils such as juvenile remains, eggs, and embryos (Chapter 8). Examination of all these data has led to a more complete picture of theropod lifestyles than can be interpreted on the basis of sometimes fragmentary skeletal data alone. Through such integrated analyses, theropods have turned out to be more complex animals than originally thought, far beyond the "eat-and-run" killing machines so entrenched as fictional stereotypes.

Resurgent interest in theropods followed spectacular finds of an Early Cretaceous theropod (*Scipionyx*) in Italy (with partial preservation of its internal organs), as well as beautifully preserved feathered theropods from Cretaceous deposits in China (including one with feathers on all four limbs). Feathered theropods, in particular, provide even more evidence to support their evolutionary connection to birds (Chapter 15). Interpretations regarding the physiological implications of feathers or feather-like structures in theropods have also prompted much scientific discussion, especially relating to thermoregulation (Chapter 8). Moreover, any conventional dividing line between theropods and birds is obscure. For example, both a dinosaur capable of flight (*Microraptor*) and a bird not quite able to fly (*Archaeopteryx*) have been hypothesized. These hypotheses sometimes create semantic difficulties whenever paleontologists try to explain differences between descendants and ancestors in theropod lineages. Indeed, by cladistic reasoning, theropods have the longest history of all dinosaur clades, beginning with the first dinosaurs about 230 million years ago through to the birds of today. As a result, this long history justifies the inordinate amount of attention theropods receive.

Definition and Unique Characteristics of Theropoda

Theropoda (*thero* = "beast" and *poda* = "foot") is a stem-based clade within Saurischia. Stem-based clades are those that have a shared common ancestor that is more recent than that of another group (Chapter 5). Theropoda is also a **sister clade** to Sauropodomorpha from the parent clade of Saurischia. A sister clade is a taxon that shares and splits from the same ancestral group as another taxon. In the case of theropods and sauropodomorphs, they had a common saurischian ancestor but then probably diverged early in their respective evolutionary histories. This clade is named after the Theropoda of O. C. Marsh (Chapters 2 and 3), who devised the term in 1881 to describe dinosaurs interpreted as meat-eaters. Marsh evidently made a mistake in creating such a name, because in the following year he authored a classification of dinosaurs wherein he named the Ornithopoda (= "bird foot": Chapter 11) on the basis of dinosaur feet that no longer resembled those of birds. Marsh's scientific reputation alone ensured that these names were cited, and they subsequently became entrenched in the paleontological literature.

More than 20 synapomorphies define Theropoda.

Within Theropoda are two (maybe three) major stem-based clades, **Ceratosauria** and **Tetanurae** (Fig. 9.1). Recent re-evaluations of Herrerasauridae (Chapter 6), which was considered a clade within Theropoda, now places some doubt on that relationship. Regardless, it will be covered in this chapter because of its historical association with theropods. Each of these three clades includes 20 or more characters that distinguish them. Some of the more important traits of theropods include (Fig. 9.2):

- A flexible jaw, indicated by an intramandibular joint.
- An extra fenestra in the maxilla.
- Lachrymal bone well exposed on dorsal surface of skull.
- Minimum of five vertebrae in sacrum.
- Manus with claws (unguals) and reduction or loss of digits IV and V.
- Slightly curved femur, which is also more than twice as long as the humerus.
- Pes with digits II through to IV; digit I reduced and separate, digit V reduced or absent.
- Well-defined (long) processes on cervical and caudal vertebrae.

Thin, low-density (mostly cancellous) limb bones and vertebrae are also a common characteristic associated with theropods. This trait probably reflects an

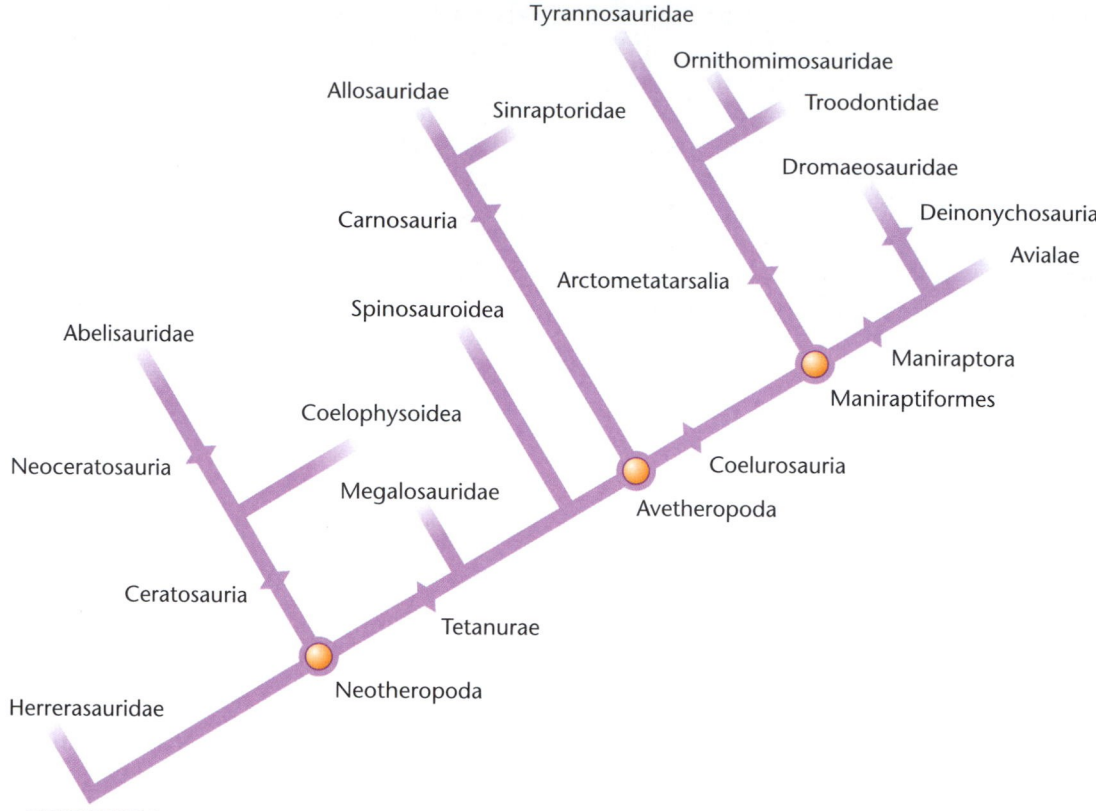

THEROPODA

FIGURE 9.1 Cladogram for Theropoda and Herrerasauridae, showing relationships of major clades and outgroups. Note that an alternative cladogram would have herrerasaurids as an outgroup sharing a common archosaur ancestor with dinosaurs, but outside of the Dinosauria.

adaptation to increased mobility but later aided in the development of flight in some theropod lineages. Related to this lightening of limb bones is another common theropod characteristic, **pneumatic** (air-filled) bones, which are found in the skull, vertebrae, and costae. Air-filled spaces in the theropod skeleton allowed room for soft-tissue air sacs, which were filled from a theropod's lungs. Air sacs in the skull functioned as lightweight support without adding extra bony tissue. This adaptation put less strain on the cervical vertebrae, which allowed some theropods to develop extremely large skulls in comparison to the rest of their bodies. Examples of such theropods include allosaurids and tyrannosaurids. (Contrast this head/body ratio to that of sauropods: Chapter 12.) Although pneumatization is not restricted to theropods, they expressed it more than any other dinosaur clade.

Large, recurved, and serrated teeth, designed for cutting through and consuming flesh, constitute a common anatomical attribute of theropods and probably represent an ancestral condition. However, they are not found in all theropods, and some species, such as *Oviraptor* and *Struthiomimus* of the Late Cretaceous, actually lack teeth. These theropods may have had beaks or other anatomical traits covering the bones that did not fossilize. They also may have been omnivorous, herbivorous, or insectivorous, but compelling evidence from stomach contents or coprolites favoring either mode is still lacking. Gastroliths, which normally are found with herbivores (Chapter 14), occur in a few theropods. These trace fossils and oddly-shaped teeth in a few theropods have led to rethinking about the previously definitive statement that all theropods were meat-eaters. Nevertheless, all dinosaur toothmarks

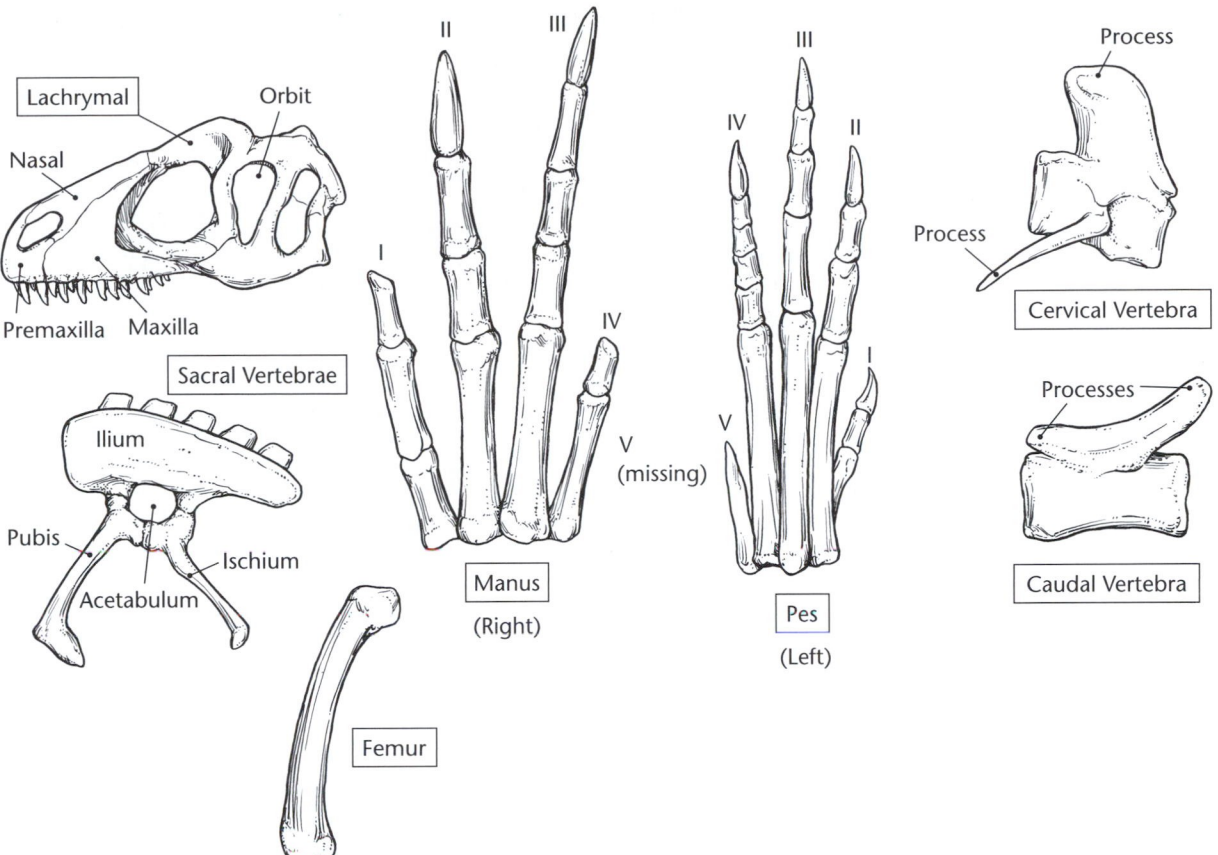

FIGURE 9.2 Important characters for Clade Theropoda: lachrymal bone, five sacral vertebrae, manus with unguals and reduction or loss of digits IV and V, curved and long femur, long and bilaterally symmetrical pes with digits II to IV and digit I separate from pes, and cervical and caudal vertebrae processes.

reported so far in bones of other dinosaurs and Mesozoic vertebrates are correlated with theropod tracemakers (Chapter 14). This evidence at least supports the assertion that, if any given dinosaur was carnivorous, it was a theropod. As consumers in most terrestrial environments from the Late Triassic through to the Late Cretaceous, many of them fulfilled important ecological roles near the top of their food chains.

In the abbreviated list of characters that distinguish theropods from other dinosaurs, most items relate to their development and refinement of bipedalism. As mentioned previously, bipedalism is probably a plesiomorphic trait in theropods, and only a few theropod species show evidence of having been quadrupedal. This obligate bipedal posture for the vast majority of theropods is interpreted on the basis of skeletal features, such as leg lengths considerably exceeding arm lengths. Theropod bipedalism is also verified by trackway evidence, which shows diagonal walking patterns that involved only two alternating pes impressions (Chapter 14). Moreover, the narrow straddle of interpreted theropod trackways correlates with anatomical characteristics that indicate theropod legs were positioned proximal to the midline of their bodies, an adaptation that aided in their efficient two-legged movement. Finally, manus imprints attributed to theropods are exceedingly rare, suggesting that they spent most of their time in an upright position, supported by their hind limbs. However, exceptions are shown by so-called "sitting traces," where theropods sat back on their haunches (metatarsals and ischiadic

FIGURE 9.3 Right manus of Late Jurassic *Allosaurus fragilis* with digits I to III (from top to bottom) and right human hand for scale. Dinosaur National Monument, Vernal, Utah.

region) and left visible impressions of their hands, too. In Alberta, Canada, one trackway of a large theropod also shows probable manus impressions where the theropod scraped the ground in front of it as it moved. This evidence also indicates a low, horizontal posture for that theropod, a condition also inferred for other theropods.

A bipedal habit means that hands were free to grasp, and a typical theropod manus with digits I through to III indeed exhibits such a capability (Fig. 9.3). For example, grasping in a human hand can still be accomplished using only digits I through to III because the first phalanx, carpal, and metacarpal of the thumb are positioned posterior to the other two digits. This arrangement allows the thumb to meet with the ventral surfaces of the other digits, and theropods were apparently capable of the same motion. Their grasping ability was more than sufficient for holding on to food items or mates. However, digit III of some large theropods of the Cretaceous, such as abelisaurids and tyrannosaurids, was so reduced that they only had a two-fingered manus. This circumstance corresponded with a relatively small humerus, radius, and ulna in each arm. Such seemingly odd traits provoke speculation about the functionality of such minimal appendages in these large theropods. Were they vestigial organs or did they serve some other, as yet, unknown purpose?

Theropods maintained their bipedalism with the aid of caudal vertebrae that were stiffened distally by long processes. These structures probably acted as a counterbalance to large skulls and consequently caused a more or less horizontal alignment of the vertebral column. Recognition of this alignment, in conjunction with other skeletal features, resulted in revisions of how theropods were depicted in both museums and textbooks. At the beginning of the twentieth century, mounts of theropod skeletons and artists' restorations put theropods in near-vertical, kangaroo-like poses. On the basis of the large amount of scientific data gained since then, from both skeletal and track data, museums now show them closer to and parallel to the ground. Of course, such a posture relates to predation, an important life habit for most theropods. A horizontally-aligned theropod conveyed greater ease of movement, and placed their arms and often large and sharp teeth at the same level as their potential prey animals.

Another unique characteristic of theropods versus other dinosaurs was the tendency of some of their members toward increased "braininess," as measured roughly by the brain-mass/body-mass ratio, and more precisely by the encephalization quotient (EQ), first discussed in Chapter 8. EQ is the cerebral-cortex–mass/total-brain–mass ratio, but the more easily calculated measurement is the brain-mass/body-mass ratio. This can be approximated for any given fossil vertebrate by measuring the volume of the braincase versus the volume of the entire body:

$$B_r = V_e/V_b \qquad (9.1)$$

where B_r is brain/body ratio, V_e is endocranial volume (normally measured in cubic centimeters, or cm³), and V_b is body volume. Another way to view this ratio from the standpoint of mass is to compare the brain mass to the total body mass as a percentage:

$$B_p = M_e/M_b \qquad (9.2)$$

where B_p is brain/body percentage, M_e is encephalic mass, and M_b is body mass. For example, the brain of a modern African elephant (*Loxodonta africanus*) is about 7500 g and its body mass is 5.0 metric tons (5000 kg). In contrast, a typical adult human brain is 1500 g with a body mass of about 0.07 metric tons (70 kg). The elephant has a brain five times more massive than the human, but a striking difference is apparent when the two species are compared using Equation 9.2:

B_p (elephant) $= 7.5 \times 10^3$ g/5.0×10^6 g $\times 100 = 0.15\%$
B_p (human) $= 1.5 \times 10^3$ g/7.0×10^4 g $\times 100 = 2.1\%$

Through this comparison, a typical human has 14 times the brain mass in proportion to its body size when matched against an elephant.

For a dinosaur example, recall that the volume of the *Tyrannosaurus* model measured in Chapter 1 (Eqns 1.3 and 1.4) was 235 cm³, but was a scale model at 1 : 30. This means that its "life-size" volume was 235 cm³ $\times 30 \times 30 \times 30 = 6.35 \times 10^6$ cm³. Using an estimated endocranial volume for *Tyrannosaurus* of 500 cm³, the brain/body ratio would have been about

$B_r = 5.0 \times 10^2$ cm³/6.35×10^6 cm³ $= 0.000079$

which, if 1.0 g/cm³ is assumed for the density of both the brain and body for *Tyrannosaurus*, is about 19 times less than the percentage calculated for an elephant. However, the tyrannosaurid is still larger than that for its possible contemporaneous prey animal, *Triceratops*. Using 300 cm³ as an endocranial volume and an estimated body volume of 8.5×10^6 cm³, *Triceratops* would have had the following brain/body ratio:

$B_r = 3.0 \times 10^2$ cm³/8.5×10^6 cm³ $= 0.000035$

which means that a typical adult *Tyrannosaurus* still had twice the brain size relative to its body size in comparison to *Triceratops*. Thus, the rhetorical question of "How large did theropod brains need to be?" is apparently answered by "Larger than those of their intended meals." Along those lines, other researchers have noticed an increase in brain/body ratios for carnivorous mammals with respect to their probable prey animals throughout the Cenozoic Era. This is a possible example of a co-evolutionary process in action (Chapter 6).

The EQs for many modern species of vertebrates have been calculated, and a plot called an **allometric line** can be drawn to best describe the average EQ of closely-related or otherwise similar modern groups. This graph can then be used to compare EQs of modern and extinct animals. For example, the average EQ of modern crocodilians can be used as a standard of 1.0 for a line to compare to dinosaur EQs. If the plot for the EQ of a particular dinosaur falls below the line, that dinosaur had a smaller brain than should be expected for an animal of similar size, and vice versa for any dinosaurs with EQs that plot above the line. When compared to the EQs of crocodilians, theropods plot above the line. Some small Late Cretaceous

FIGURE 9.4 Feathers associated with four limbs of *Microraptor*, an Early Cretaceous feathered theropod species from China. Reprinted by permission from Macmillan Publishers Ltd: Nature, Xu et al. "Four-winged dinosaurs from China", Vol. X, 2001.

theropods, such as *Troodon*, even fall in the same range as modern ostriches. However, less than 5% of all dinosaur species have had their EQs calculated. This means that considerable data are needed to test the currently accepted generalization that theropods, gram for gram, had the largest brains of dinosaurs relative to their bodies. Of course, EQ also holds implications for IQ (intelligence quotient), a subject of considerable debate among not just paleontologists but also psychologists and neurophysiologists.

Finally, theropods were different from other dinosaurs as they are the only ones known so far whose skeletal remains have shown definite evidence of feathers similar to those seen in modern birds. Long conjectured by some paleontologists and artists, the concept of feathered dinosaurs was, until recently, a poorly supported hypothesis that had little to no body fossil evidence. Discoveries of the 1990s and first decade of the twenty-first century have changed all that. As of the writing of this book, feathers have been confirmed from at least nine species of theropods (Fig. 9.4). In traditional zoology, feathers constituted one of the unique distinguishing features of birds, but now at least a few non-avian theropods have joined this formerly exclusive club. The addition of feathers to other characteristics uniting probable theropod ancestors with avian descendants also further strengthens hypotheses of their evolutionary relationships (Chapter 15). However, the functions of the feathers in theropods are still a source of dispute in dinosaur research, especially in how they relate to dinosaur physiology (Chapter 8).

Overall, theropods were different from other dinosaurs in many ways, whether through their light bones, speedy movement on two legs, carnivory, larger brains, or feathers. Such special characteristics supply reasons why theropods remain enduring symbols of dinosaurs in the public eye, but they also provide incentives for further scientific study. Such research is especially focused on how theropods evolved, where they lived (and when), and how they behaved in their everyday lives.

Clades and Species of Theropoda

Herrerasauridae

Herrerasauridae (literally "Herrera's lizard") is named after Victorino Herrera, who in 1958 discovered its eponymous genus, *Herrerasaurus*, in the Late Triassic Ischigualasto Formation of Argentina (Chapter 6). Like many dinosaur clades, it is a provisional classification subject to revision with new evidence. For example, it was considered as a clade under Theropoda, but now some paleontologists place it outside of that clade. In fact, it is a sometimes-controversial assignment within Dinosauria. Paleontologists who consider it as a dinosaur clade regard it as a basal one for Saurischia and possibly Theropoda because of:

1 cladistic analyses of its characters;
2 the stratigraphic position of its oldest members relative to other theropods; and
3 its retention of primitive traits similar to probable archosaur ancestors.

An alternative hypothesis is that herrerasaurids comprise a clade related to (yet separate from) both the Saurischia and Ornithischia, which would mean that its members are not dinosaurs. The first hypothesis has gained more support in recent years with the discovery of more postcranial elements that, when added to previous evidence, indicate a few (but not all) synapomorphies shared with the Theropoda. In light of this evidence, herrerasaurids will be treated here as a sister clade to theropods, because they are considered first and foremost as basal saurischians. In other words, "true" theropods may have more recently shared a common ancestor with sauropodomorphs (Chapter 10) than herrerasaurids.

Three Late Triassic dinosaur species placed within Herrerasauridae are *Herrerasaurus ischigualastensis* of Argentina, *Staurikosaurus pricei* of Brazil, and *Chindesaurus bryansmalli* of the western USA (Fig. 6.12). *Eoraptor lunensis* of Argentina is currently considered an ancestral dinosaur and a theropod by some paleontologists, although others omit *Eoraptor* from Herrerasauridae and Theropoda. In fact, a few do not accept that *Eoraptor* meets the minimum criteria for a dinosaur, although it certainly shares some of the traits (Chapter 6). Interestingly, recent cladistic analyses of *Eoraptor* reveal that it may be less primitive (basal) than *Herrerasaurus*, the opposite of previous assumptions.

The namesake of Herrarasauridae, *Herrerasaurus*, contains the following characters:

■ Long pubis with relation to its femur, associated with three sacral vertebrae.
■ Semiperforate to open acetabulum, with a well-developed medial wall.
■ Femur length nearly twice that of the humerus.
■ Elongate skull, nearly equal in length to its femur.
■ Serrated and recurved conical teeth.
■ Long and equally-sized metatarsals I and V on the pes.
■ Manus with five digits but digits IV and V reduced and without unguals.

A comparison of this list with that given earlier for the synapomorphies of Theropoda reveals differences in:

1 the number of sacral vertebrae;
2 digits I and V on the pes; and
3 the relative lengths of the femur and humerus.

However, it also has an intramandibular joint, serrated teeth, and long digits in the manus associated with theropods. This mixture of features is muddied further by the herrerasaurid *Staurikosaurus*, which is sometimes considered as more primitive than *Herrerasaurus* and most other dinosaurs because it only has two sacral vertebrae. However, its preponderance of other characters shared with dinosaurs does not yet necessarily exclude it on the basis of a few disparate traits.

Herrerasaurids ranged from 2 to 4.5 meters long and thus represented some of the largest predators of Late Triassic terrestrial environments. Their carnivorous habit is interpreted primarily on the basis of functional morphology. For example, they had:

1 relatively long hind limbs adapted for bipedal locomotion;
2 serrated teeth; and
3 a jaw apparatus that seemed well-suited for grasping small prey.

Additional pieces of evidence are the bones of a juvenile archosauromorph (rhynchosaur) found within the ribcage of one specimen of *Herrerasaurus*. However, no toothmarks or coprolites attributable to herrerasaurids have been identified, so little is known about whether any of them were primarily predators or scavengers. Likewise, numerous theropod and theropod-like tracks are in Upper Triassic strata from the same time as herrerasaurid skeletal remains, but none have been identified as belonging to herrerasaurid tracemakers. No nests, eggs, or embryos of herrerasaurids have been found, so little is known about their reproduction other than that they were probably egg-layers like other archosaurs.

Ceratosauria: Coelophysoidea and Neoceratosauria

Although relatively less known than Tetanurae, the stem-based clade Ceratosauria (= "horned lizard") includes abundant specimens of some interestingly varied theropods, whose remains have been found on all continents except Antarctica. Two stem-based clades are within Ceratosauria, **Coelophysoidea** and **Neoceratosauria**. Coelophysoideans were the dominant theropods soon after the beginning of the geologic range for dinosaurs. Herrerasaurids in the Late Triassic were supplanted by coelophysoideans, which are represented abundantly by *Coelophysis*. Coelophysoideans in turn were succeeded by Early Jurassic species, such as the *Syntarsus* and *Dilophosaurus*, as well as *Ceratosaurus* of the Late Jurassic.

A clade within Neoceratosauria from the Early and Late Cretaceous, **Abelisauridae**, comprises the bulk of neoceratosaurs for the latter part of the Mesozoic. Abelisaurids include *Abelisaurus*, *Carnotaurus*, *Majungasaurus*, *Masiakasaurus*, and several other relatively large theropods that mostly lived in the Cretaceous southern continent of Gondwana (Chapter 4), as indicated by specimens from South America, India, and Africa. In contrast, coelophysoideans were apparently rare or extinct by the end of the Jurassic.

Important synapomorphies of ceratosaurs include:

■ Fusion of astragalus and calcaneum in the ankle.
■ Sacrum fused to pelvis.
■ At least seven sacral vertebrae, which were fused to form a **synsacrum**.

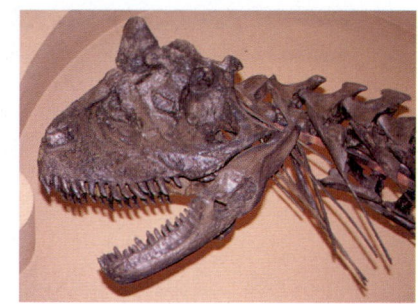

FIGURE 9.5 Early Cretaceous neoceratosaur and abelisaurid *Carnotaurus* of Argentina, showing cranial ornamentation typical of ceratosaurs. Museo de Ciencias Naturales, Madrid, Spain.

- Two fenestrae on the pubis.
- Four digits, but with digit IV reduced so that digits I through to III were the most functional.
- Two pairs of cavities (**pleurocoels**) in the cervical vertebrae.

Many ceratosaurs are also well known for the development of prominent "headgear" in some species, evident as large bony horns or crests on the dorsal surfaces of their skulls. In particular, *Dilophosaurus* of the Early Jurassic had a pair of pronounced crests that stuck out of each side of its skull. Another example was *Carnotaurus* of the Early Cretaceous, a ceratosaur that displayed orbital horns (Fig. 9.5). One proposed purpose for these seemingly non-functional features is that they were used for sexual display, but whether they are indicative of males or females is unknown.

Other than skeletal evidence, soft-tissue impressions have been found for one ceratosaur, the aforementioned Early Cretaceous abelisaurid *Carnotaurus*. These impressions display rows of low-profile conical scales interpreted as external molds of the actual skin. Feathers are currently unknown from ceratosaur remains. Nevertheless, some artists have depicted such frills on *Coelophysis* in reconstructions, despite the current lack of scientific evidence for them. One example of an Early Jurassic "sitting theropod" trace fossil, attributed to a ceratosaur, purportedly had feather impressions. Nevertheless, more detailed scrutiny revealed that the "feathers" were wrinkle marks in the sediment caused by the animal's movement.

Aside from skeletal evidence, ceratosaurs are not well understood. Theropod tracks of the Late Triassic and Early Jurassic have been tentatively linked to ceratosaur tracemakers on the basis of correlation with known body fossils of the same age. Even so, no one has yet identified tracks made by specific genera, such as *Coelophysis* or *Dilophosaurus*. No toothmarks or coprolites of ceratosaurs have been interpreted, thus little is known about whether any of them were primarily predators or scavengers. One specimen of *Coelophysis*, which had bones of another, smaller *Coelophysis* in its rib cage, was regarded for decades as evidence of cannibalism in this species. However, a more careful examination of the specimen later revealed that the bones were underneath and not inside the rib cage, meaning that (as far as we know) *Coelophysis* did not eat its own species, let alone its young. Similar to herrerasaurids, no eggs, embryos, or nests of ceratosaurs have been identified, despite the numerous body fossils found of some adult ceratosaurs (that is, *Coelophysis* and *Syntarsus*). However, this circumstance may be partially a result of the preservation bias against dinosaur eggs in general, which are not abundantly represented in the geologic record until the Early Cretaceous (Chapter 8). Abundant juvenile remains of *Coelophysis* afford a rare view of the growth and development of one species of ceratosaur, but embryos are unknown, even for this well-studied dinosaur.

Despite *Coelophysis* being the most abundant dinosaur represented by skeletal remains and the common preservation of *Syntarsus*, which is represented by more than 30 specimens, ceratosaurs are not as well-studied as tetanurans. About 25 species have been placed in Ceratosauria (Table 9.1), but some of these species are named

9

TABLE 9.1 Representative genera of Ceratosauria with approximate geologic age and location.

Genus	Age	Geographic Location
Abelisaurus	Late Cretaceous	Argentina
Aucasaurus	Late Cretaceous	Argentina
Carnotaurus	Early Cretaceous	Argentina
Ceratosaurus	Late Jurassic	Western US; Tanzania
Coelophysis	Late Triassic	Western US
Dilophosaurus	Early Jurassic	Western US; China
Elaphrosaurus	Early Cretaceous	Tanzania
Genusaurus	Early Cretaceous	France
Indosaurus	Late Cretaceous	India
Indosuchus	Late Cretaceous	India
Liliensternus	Late Triassic	Germany; France
Majungasaurus	Late Cretaceous	Madagascar
Majungatholus	Late Cretaceous	Madagascar
Masiakasaurus	Late Cretaceous	Madagascar
Noasaurus	Late Cretaceous	Argentina
Podokesaurus	Early Jurassic	Eastern US
Procompsognathus	Late Triassic	Germany
Syntarsus	Early Jurassic	Zimbabwe; South Africa; western US
Xenotarsosaurus	Late Cretaceous	Argentina

on the basis of only one or two specimens. Additionally, some of these specimens are incomplete or otherwise have scrappy remains. For example, *Dilophosaurus* is the most famous of ceratosaurs because of its starring role in the 1993 film *Jurassic Park*. Unlike most other dinosaurs in this movie, it also actually lived during the Jurassic Period. Nevertheless, *Dilophosaurus* is only known on the basis of seven specimens, all from Arizona.

Part of this neglect for all things ceratosaurian may be related to the preponderance of dinosaur paleontologists who have chosen to work exclusively on tetanurans. However, it also may be a factor of geology and geography. Most specimens of ceratosaurs are in Triassic and Jurassic strata, meaning that taphonomic processes had more time to erase their remains than later theropod lineages (Chapter 7). In terms of possible geographic bias, only a few ceratosaur species have been found so far in the USA and Europe, whereas most occur in former areas of Gondwana and in Cretaceous strata. Geologic uncertainty of preservation combined with geographic distance can create discouraging conditions for paleontologists interested in studying ceratosaurs. Regardless of the reasons for their relative neglect, more work is needed in studying ceratosaurs, especially in terms of determining their relationships to basal theropods and Tetanurae.

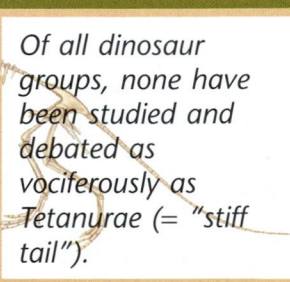

Of all dinosaur groups, none have been studied and debated as vociferously as Tetanurae (= "stiff tail").

Tetanurae: Avetheropoda and Its Numerous Clades

Tetanurae contains the theropods best known, such as *Allosaurus*, *Tyrannosaurus*, *Velociraptor*, *Deinonychus*, and *Utahraptor*. It also includes lesser-known genera, such as *Oviraptor*, *Troodon*, and *Struthiomimus*. The immediate ancestors of birds and their descendants are also placed within this clade, which has led most dinosaur paleontologists to state that birds are theropods (Chapter 15).

The overlapping of theropod traits with those of birds is the main reason why Tetanurae and its members are still the focus of much research and subject to revision with new scientific discoveries. As a result, their cladogram (see Fig. 9.1) has been revised since the first edition of this book and is only a consensus view as of this writing. It will likely displease some dinosaur paleontologists because it omits details of their phylogenetic classification or alternative cladograms. Moreover, it may be out of date by this book's publication. Indeed, the volatility of what paleontologists define as a clade for these theropods is emblematic of how dinosaur paleontology is a science that is constantly subjected to testing and modification (Chapter 2).

Some of the synapomorphies of tetanurans that unite them as a clade are:

- Dentition in the maxilla only anterior to the orbital.
- Antorbital and maxillary fenestrae, accompanied by increased pneumaticity of the skull.
- Manus with digits I through to III, but digit III is reduced.
- An expanded tibia that overlapped a reduced fibula.
- Development of a large notch (**obturator notch**) on the ischium.
- Well-developed stiffening of the caudal vertebrae by processes (**zygopophyses**) that extended anterior and posterior from the neural arches.
- Pleurocoels in the dorsal vertebrae.

Although feathers are not a characteristic required for inclusion of a theropod as a tetanuran, one clade within Tetanurae (**Coelurosauria**) does have feathered representatives. Feathered coelurosaurs found so far include *Beipiaosaurus*, *Caudipteryx*, *Microraptor*, *Protarchaeopteryx*, *Sinornithosaurus*, and *Sinosauropteryx*, all of which have been found in Lower Cretaceous strata of northeastern China. Of these feathered coelurosaurs, *Sinosauropteryx* is probably the most primitive of the feathered dinosaur specimens. The others are probably more closely related to other theropod clades within Maniraptoriformes. *Sinornithosaurus* was at first considered the nearest possible non-avian theropod relative to birds, but now several other candidates may qualify instead (Chapter 15).

Prior to the advent of phylogenetic classifications, tetanurans and a few ceratosaurs were grouped together on the basis of size and divided into two simple categories: large meat-eaters and small meat-eaters. This sort of taxonomic lumping, first proposed by Friedrich von Huene in the 1920s (Chapter 3), eventually resulted in the assignment of genera as diverse as *Albertosaurus*, *Allosaurus*, *Ceratosaurus*, *Carnotaurus*, *Dilophosaurus*, and *Tyrannosaurus* to a group called **Carnosauria** for the large theropods. The smaller theropods were relegated to Coelurosauria. Recent detailed analyses of anatomical similarities and differences have revealed that Carnosauria is questionable as a taxon. Now it is regarded as a stem-based clade and only allosaurids and sinraptorids are placed in it.

As mentioned in the previous section, the large ceratosaurs were separated into their own clade, and tyrannosaurids were recognized as coelurosaurs (albeit large ones), a difference that was intimated earlier by von Huene. **Convergent evolution**, which is an appearance of a similar change in genotype in different lineages, that results in a similarly expressed phenotypic trait (Chapter 6), is a likely explanation of how large size was selected for at different times and places for prodigious carnivores, such as those within divergent lineages of Theropoda. Natural selection for gigantism in some theropods was exemplified by the allosaurids *Carcharodontosaurus* and *Giganotosaurus* (Fig. 9.6A), which were probably closely related to one another but lived on different continents (Africa and South America, respectively) during the Early

9

(A)

(B)

FIGURE 9.6 Tetanurans as represented by allosaurids. (A) *Giganotosaurus*, a carcharodontosaurine of the Early Cretaceous of Argentina. (B) *Yangchuanosaurus*, a sinraptorid of the Late Jurassic of China. The former is on permanent display at the Fernbank Museum of Natural History, Atlanta, Georgia; the latter is currently on display in the atrium of Hartsfield-Jackson International Airport, Atlanta.

Cretaceous. These theropods may have weighed as much as 8 metric tons, which probably exceeded the mass of the most famous large theropod, *Tyrannosaurus*.

One of the mixed blessings of intensive scientific study of tetanurans is that what used to be simple becomes more complex. With dinosaurs, no group is more complicated in its classification than Tetanurae, and accordingly it has the highest number of taxa assigned to any clade within the Dinosauria (compare Table 9.2 with Table 9.1). Tetanurae in its simplest form is defined as all birds and theropods more evolutionarily related to birds than ceratosaurs, which makes it a stem-based clade. The most basal tetanuran known is *Torvosaurus* of the Late Jurassic, followed by the few members of the Megalosauridae, which includes the Middle Jurassic *Megalosaurus*, originally described by William Buckland in 1824 (Chapter 3). Megalosauridae is defined as an **outgroup**, in that it is outside of the other groups under study within Tetanurae, such as all those under the node-based clade, **Avetheropoda**. As mentioned in Chapter 5, node-based clades are those that have all of the descendants of the most recent common ancestor for two groups, where the common ancestor forms the node. Megalosauridae is also the most tenuous assignment for any clade within Tetanurae, because of its small number of members and their fragmentary fossil record.

The vast majority of tetanurans are avetheropods, and are classified on the basis of some of the following synapomorphies:

- Increased anterior extension of the pubis into a **pubic foot**.
- Pronounced cnemial process on the side of the tibia.
- Loss of digits III and IV on the manus.
- Loss of a foramen on the obturator, associated with the notch on the ischium.
- Premaxillary teeth are asymmetrical.

TABLE 9.2 Representative genera of Clade Tetanurae with approximate geologic age and location.

Genus	Age	Geographic Location
Acrocanthosaurus	Early Cretaceous	Western and central USA
Afrovenator	Early Cretaceous	Niger
Albertosaurus	Late Cretaceous	Alberta, Canada
Alectrosaurus	Late Cretaceous	China; Mongolia
Allosaurus	Late Jurassic	Western and central USA
Alxasaurus	Early Cretaceous	Mongolia
Avimumus	Late Cretaceous	Mongolia, China
Bambiraptor	Late Cretaceous	Western USA
Baryonyx	Early Cretaceous	England
Beipiaosaurus	Early Cretaceous	China
Caracharodontosaurus	Early Cretaceous	Northern Africa
Caudipteryx	Early Cretaceous	China
Chirostenotes	Late Cretaceous	Alberta, Canada
Compsognathus	Late Jurassic	France; Germany
Crylophosaurus	Early Jurassic	Antarctica
Cryptovolans	Early Cretaceous	China
Daspletosaurus	Late Cretaceous	Western USA; Alberta, Canada
Deinocheirus	Late Cretaceous	Mongolia
Deinonychus	Early Cretaceous	Western and central USA
Deltadromeus	Late Cretaceous	Niger
Dromaeosaurus	Late Cretaceous	Western USA
Dryptosaurus	Late Cretaceous	Eastern USA
Erlikosaurus	Late Cretaceous	Mongolia
Gallimimus	Late Cretaceous	Mongolia
Gigagantosaurus	Late Cretaceous	Argentina
Gorgosaurus	Late Cretaceous	Western USA; Alberta, Canada
Ingenia	Late Cretaceous	Mongolia
Khaan	Late Cretaceous	Mongolia
Megalosaurus	Middle Jurassic	Western Europe
Microraptor	Early Cretaceous	China
Monolophosaurus	Late Jurassic	China
Neovenator	Early Cretaceous	Madagascar
Ornitholestes	Late Jurassic	Western USA
Ornithomimus	Late Cretaceous	Western USA
Oviraptor	Late Cretaceous	Mongolia
Pelecanimimus	Early Cretaceous	Spain
Piatnitzkysaurus	Late Jurassic	Argentina
Protarchaeopteryx	Early Cretaceous	China
Saurornithoides	Late Cretaceous	Mongolia
Saurornitholestes	Late Cretaceous	Alberta, Canada
Scipionxy	Early Cretaceous	Italy
Segnosaurus	Late Cretaceous	Mongolia
Shenzhousaurus	Early Cretaceous	China
Sinornithomimus	Early Cretaceous	China
Sinornithosaurus	Early Cretaceous	China
Sinosauropteryx	Early Cretaceous	China
Sinovenator	Early Cretaceous	China
Spinosaurus	Early Cretaceous	Northern Africa

9

TABLE 9.2 Continued

Genus	Age	Geographic Location
Struthiomimus	Late Cretaceous	Alberta, Canada
Suchomimus	Early Cretaceous	Niger
Tarbosaurus	Late Cretaceous	Mongolia; China
Therizinosaurus	Late Cretaceous	Mongolia, Kazakhstan
Torvosaurus	Late Jurassic	Western USA
Troodon	Late Cretaceous	Western USA; Alberta, Canada
Tyrannosaurus	Late Cretaceous	Western USA; Alberta, Canada
Utahraptor	Early Cretaceous	Western USA
Velociraptor	Late Cretaceous	Mongolia; China
Yangchuanosaurus	Late Jurassic	China

Recognition of such increasingly exclusive differences between taxa may represent minutiae to a non-paleontologist, and they are in some cases minor distinctions. Nonetheless, paying attention to such small details is important in defining these clades. By now, a student of cladistics could compile all of the characters of a dinosaur, saurischian, theropod, tetanuran, and avetheropod. This listing would give a progressively more detailed hypothesis of an avetheropod's evolutionary history, beginning with archosaur ancestors in the Middle Triassic and leading up to the present. With Linnaean classification, a hierarchy of categories existed without so much of an evolutionary context. Cladistics has thus provided a hypothetical framework for unraveling relationships between dinosaurs, particularly for theropods.

Avetheropoda includes the stem-based clade Carnosauria, which has within it the allosaurids (such as *Allosaurus*, *Carcharodontosaurus*, and *Giganotosaurus*) and sinraptorids (such as *Sinraptor* and *Yangchuanosaurus*, Fig. 9.6B), as well as the stem-based Coelurosauria. Considerable diversification of theropods is represented within Coelurosauria. Coelurosaurs have several characters in their clade, but one of the most distinctive is a **semilunate carpal**, a carpal with a half-moon shape. Two coelurosaur outgroups consisting of one species each, *Compsognathus* and *Dryptosaurus*, have been proposed as more primitive coelurosaurs than **Maniraptoriformes**, a node-based clade within Coelurosauria. Maniraptoriformes includes the stem-based clade **Arctometatarsalia**, which is named for a specific novelty in metatarsal development, the middle metatarsal "pinched" between the metatarsals on either side of it (Fig. 9.7). Within Arctometatarsalia are ornithomimosaurs and tyrannosaurids, and possibly troodontids. Troodontids are somewhat problematic in their placement, and some paleontologists place them outside of Arctometatarsalia. Only about 20 years ago, such a grouping would have been considered an unlikely phylogenetic association on the basis of overall appearance. Ornithomimosaurs are the so-called "ostrich dinosaurs", because of their close resemblance to modern ostriches in overall morphology and size. Tyrannosaurids are mostly huge carnivores, and troodontids are small- to medium-sized theropods with relatively large braincases (mentioned earlier). Representatives of all three groups lived at the same time during the Late Cretaceous (including *Troodon*, *Ornithomimus*, and *Tyrannosaurus*). Nevertheless, an objective observer during that time may have had little reason to see a recent common ancestry for such apparently divergent forms, based on their superficial appearances (Fig. 9.8).

Other groups within Maniraptoriformes include what most paleontologists agree are the most unusual theropods, therizinosaurs (such as *Therizinosaurus* and *Segnosaurus*), which evidently shared ancestry with oviraptorisaurians (*Oviraptor* and *Ingenia*, for example). Oviraptorisaurians were also unusual, as they had behavior

270

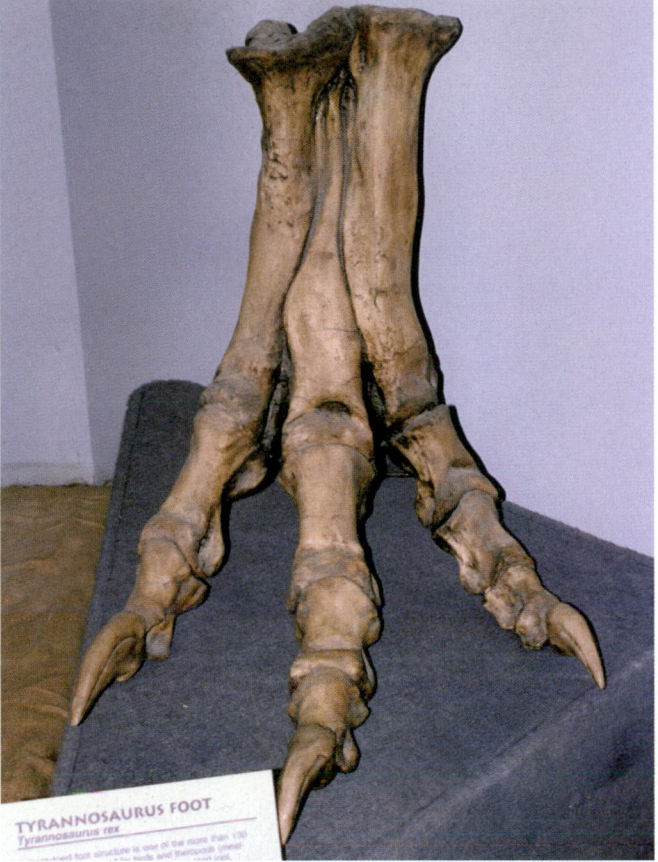

FIGURE 9.7 "Pinched" metatarsal in pes of *Tyrannosaurus*, a character of the clade Arctometatarsalia.

FIGURE 9.8 *Troodon*, a relatively small tetanuran with a relatively large brain for a dinosaur. Temporary display on loan from the Museum of the Rockies, at the Fernbank Museum of Natural History, Atlanta, Georgia.

FIGURE 9.9 *Deinonychus*, a Late Cretaceous dromaeosaur of the western USA. (A) Skeletal reconstruction of *Deinonychus*. (B) Close-up view of the upraised digit I of left pes. North Carolina Museum of Natural History.

that apparently departed considerably from expectations for theropods. Included within Maniraptoriformes is the stem-based **Maniraptora**, the lineage that led to birds. This clade diverged into two other stem-based clades, **Deinonychosauria**, which include the **dromaeosaurids**, such as *Deinonychus*, *Velociraptor*, and *Utahraptor*, and **Aviale** (Fig. 9.9). The latter is composed of *Archaeopteryx*, all fossil ancestors of modern birds, and modern birds. For the sake of simplicity, Aviale will be discussed in detail in Chapter 15.

The body fossil record for tetanurans is not only diverse, especially in comparison to herrerasaurids and ceratosaurs, but is becoming better understood through

(A)

(B)

FIGURE 9.10 *Acrocanthosaurus*, an Early Cretaceous allosaurid from Texas. (A) Skeletal reconstruction: North Carolina Museum of Natural History. (B) Large theropod track affiliated with *Acrocanthosaurus*, preserved in limestone but slightly submerged in Paluxy River, eastern Texas. Human footprint in river mud (right) for scale.

an ever-increasing database of trace fossil evidence. In some cases, these trace fossils can be directly linked to tracemakers within a specific clade or species within Tetanurae. For example, large theropod tracks in the Lower Cretaceous of east Texas, discovered by Roland Bird in the 1940s (Chapters 3 and 14), have been tentatively correlated with the similarly-sized and shaped feet of *Acrocanthosaurus*, an allosaurid within same-age strata of the same region (Fig. 9.10). Thin-toed footprints found in Lower Cretaceous strata of Colorado, which are the right size and shape for these ostrich-like dinosaurs, represent probable ornithomimid tracks. Perhaps the best fit between a track and a tetanuran tracemaker of the same age is a little-doubted tyrannosaurid track from the Late Cretaceous of New Mexico. This track is so large (85 cm long) that it could be assigned only to *Tyrannosaurus* (Fig. 9.11). Besides tracks, other tetanuran trace fossils include the following:

(A)

(B)

FIGURE 9.11 "Before" and "after" depictions of *Tyrannosaurus* foot anatomy. (A) "Before" of a large, fleshy foot represented by a probable *Tyrannosaurus* track (left foot) from the Upper Cretaceous Raton Formation, New Mexico. Cast at University of Colorado, Denver. (B) "After" ventral (sole) view of *Tyrannosaurus* foot (left again) without the flesh from the viewpoint of being stepped on. Specimen part of skeletal mount in Denver Museum of Science and Nature. Note the corresponding position of the hallux in both track and foot.

1 *Troodon* nests from the Late Cretaceous of Montana;
2 probable tyrannosaurid coprolites from the Late Cretaceous of Alberta, Canada;
3 gastroliths within the body cavity of the Early Cretaceous coelurosaur *Caudipteryx* of China; and
4 toothmarks attributed to *Allosaurus*, *Troodon*, and tyrannosaurids, among others (Chapter 14).

A wealth of other theropod trace fossils from the Jurassic and Cretaceous ensure that more links between tetanurans and the artifacts of their behavior will be made in the future.

Paleobiogeography and Evolutionary History of Theropoda

As discussed in Chapter 10, theropods were among the first dinosaurs and their appearance as body fossils slightly preceded, or was contemporaneous with, the evolution of primitive ornithischians during the earliest part of the Late Triassic, slightly less than 230 million years ago. Tracks attributed to theropods, with some dispute over the actual identity of their tracemakers, are in Middle Triassic strata. Undoubted theropod tracks are relatively abundant in Upper Triassic rocks, where in some places they co-occur with skeletal material. Throughout the remainder of the Mesozoic, theropods subsequently diversified into a myriad of sizes and forms. These range from the crow-sized *Microraptor* to the multi-ton *Giganotosaurus*, both of the Early Cretaceous, to the bizarrely headed, toothless *Oviraptor* of the Late Cretaceous.

The small (1 meter long) possible theropod *Eoraptor*, of the earliest part of the Late Triassic, has been proposed as approximating the characteristics of a theropod ancestor because it shares some traits with theropods but also lacks others that define this clade. As mentioned earlier, herrerasaurids may also approximate the earliest theropods. These are in Upper Triassic strata of both North and South America, although the South American examples are geologically older. Small theropod-like tracks, showing a three-toed and almost bilaterally symmetrical compression shape associated with most theropod feet throughout the Mesozoic Era, are also recorded from Middle Triassic strata. Interestingly, these tracks precede the body fossil record for theropods. However, many paleontologists do not accept such tracks as necessarily belonging to theropods, although some acknowledge that theropods may have originated in the latest part of the Middle Triassic. The overlapping use of skeletal and track data is certainly helpful for narrowing down when theropods first evolved during the Triassic, and such resolution should improve considerably with more discoveries of both forms of evidence.

Theropod body fossils are in Upper Triassic to Upper Cretaceous deposits on all seven continents, and so far their trace fossils are only missing from Antarctica.

Theropod track horizons are documented from Alaska, the mid-continental USA, Canada, and Mexico in North America; Argentina, Bolivia, Brazil, and Peru in South America; England, France, Italy, Poland, Portugal, Spain, and Switzerland, in Europe; India, China, Mongolia, Japan, and Korea in Asia; northern Africa, such as Morocco; and Australia, where theropod tracks are much more common than their skeletal remains. Both body and trace fossils of theropods are in facies representing a wide variety of environments: deserts, swamps, river floodplains, upper delta plains, lake shorelines, and seashores. Of course, tracks are the only reliable *in-situ* records of which environments were actually frequented by theropods, but some other trace fossils (nests and coprolites) and body fossils (eggs, and complete, apparently unmoved skeletons) certainly point paleontologists in the right direction as well (Chapters 7 and 14). The totality of evidence for theropods from such a diversity of terrestrial environments and broad latitudinal range, spanning the entire geologic range of dinosaurs, indicates their biological diversity and adaptations to numerous niches through time. Based on current scientific evidence, theropods represent evolutionary diversification more so than any other major dinosaur clade.

Theropod skeletal remains are uncommon, normally comprising 15–20% of all dinosaur remains in any Upper Triassic–Upper Cretaceous rocks formed in terrestrial environments. Furthermore, they are even more rare (although a few are known) in shallow marine deposits (Chapter 7). A study conducted in 1976 on 171 valid theropod species found that 85% of them were named on the basis of five or fewer specimens, and a single specimen represented about 40% of these species. These percentages have changed little since then, and the recent spate of new theropod species discovered and described during the 1990s and early part of the twenty-first century are primarily based on one or two specimens. The relative scarcity of theropod remains means that numerous "ghost lineages" (Chapter 6) have been proposed for species that have few known close relatives, which constitutes a challenge for phylogenetic classification schemes applied to theropods. Fortunately, a combination of taphonomic factors and perhaps social behavior resulted in some monospecific theropod bone beds, such as those of the Late Triassic *Coelophysis*, Late Jurassic *Allosaurus*, and more recently Late Cretaceous tyrannosaurids. With regard to the latter, *Tyrannosaurus* was originally considered a rare dinosaur, but now its remains are among the most commonly encountered in strata from some areas of the western USA.

As mentioned earlier, theropod species found so far support the hypothesis that they comprised the most diverse of all dinosaur clades. This diversity may reflect more rapid evolutionary changes in theropods than their other dinosaurian contemporaries. Furthermore, this rapidity may have been coupled with:

1 higher extinction rates of theropod species;
2 higher extinction rates of organisms interacting with theropods; or
3 faster changes in gene frequencies in association with environmentally-related selection pressures.

Rapid diversification in lineages is often a result of high reproductive success and rapid growth rates combined with environmental changes that encouraged the selection of specific and overlapping inheritable traits (Chapter 6). Extinctions of major archosaur groups, as recorded in Upper Triassic rocks, may have opened niches for theropods in the ecosystems of those times and thus contributed to their evolution.

Greater numbers and species of dinosaurs corresponded with the demise of other archosaurs by the beginning of the Jurassic Period, with theropods a prime example of this ascendancy (Chapter 6). Herrerasaurids of the Late Triassic were joined by abundant ceratosaurs, such as *Coelophysis*, and these were succeeded by larger ceratosaurs (*Dilophosaurus*, *Syntarsus*, and *Ceratosaurus*) by the Early and Late Jurassic. The Middle Jurassic saw the arrival of even larger and presumably more specialized carnivores, among them the first tetanurans (*Megalosaurus*). Although they never reached the enormous sizes of sauropods (Chapter 10), theropods trended toward increased body size by the Late Jurassic, an apparent example of Cope's Rule (Chapter 6). One example was *Allosaurus,* which probably weighed as much as 3 metric tons. However, one of its contemporaries, *Compsognathus*, was among the smallest of all known dinosaurs, weighing only about 3 kg.

By the Late Jurassic, the first avians evolved from theropod ancestors, represented by what is widely acknowledged by paleontologists as the earliest known bird in the geologic record, *Archaeopteryx* (Chapter 15). Following this, the Cretaceous saw numerous small birds co-existing in terrestrial environments with feathered non-avian theropods. However, some theropod lineages evolved in one important way, that is, their descendants became much larger. The largest land carnivores that ever lived are from the Early Cretaceous (*Giganotosaurus* and *Carcharodontosaurus*) and Late Cretaceous (*Albertosaurus* and *Tyrannosaurus*). Of course, not all Late Cretaceous theropods were enormous. Some of the most behaviorally interesting theropods were relatively small, such as *Velociraptor*, *Troodon*, and *Oviraptor*. Late Cretaceous theropods, large or small, feathered or scaled, toothed or non-toothed, comprised the greatest diversity of dinosaurs known.

Theropods as Living Animals

Reproduction

Some theropods are sufficiently represented by skeletal material that sexual dimorphism is hypothesized on the basis of slight but detectable size differences and a few other criteria. For example, consistent size variations in *Coelophysis* and *Syntarsus* are proposed as a result of male–female differences; the same situation has been postulated for *Tyrannosaurus*. Interestingly, the larger individuals of *Syntarsus* and *Tyrannosaurus* may be representative of females, partially because some modern female reptiles are larger than their mates as a requirement for egg-laying capacity (Chapter 8). However, currently no definitive evidence can assist in distinguishing a male or female dinosaur, other than the presence of eggs in its internal body cavity, which so far has been documented in only one theropod. Even the discovery of adult theropod

FIGURE 9.12
Crylophosaurus ellioti, an Early Jurassic carnosaur from Antarctica. Auckland Museum, Auckland, New Zealand.

remains on egg clutches (e.g., *Oviraptor* and *Troodon*) does not necessarily mean that these dinosaurs were female. Although the majority of birds that exhibit brooding behavior are female, it has also been documented in some male birds. For example, male emus (*Dromaius novaehollandiaea*), a large flightless bird native to Australia, sit continuously on egg clutches for about 55 days and thus serve as the main provider of warmth and protection. Interestingly, climatic conditions also should be taken into account when assessing the thermoregulatory significance of dinosaurs sitting on egg clutches. For example, tropical seabirds must sit on egg clutches to prevent them from overheating in the hot daytime sun. This means that their endothermy is actually helping to keep the eggs at lower temperatures.

The prominent cranial and vertebral processes in theropods, as well as feathers, were potential sexual displays or species identifiers. Among ceratosaurs, the Early Jurassic *Dilophosaurus* has twinned parietal blades, the Early Jurassic *Ceratosaurus* has a single nasal horn, and the Early Cretaceous *Carnotaurus* has horns dorsal to the orbits. Tetanurans with head ornamentation include the Early Jurassic *Crylophosaurus* of Antarctica and the Late Jurassic *Allosaurus* (Fig. 9.12). *Crylophosaurus* had a pompadour-like projection of bone, which resulted in some paleontologists informally dubbing it "*Elvisaurus*". Similarly, *Allosaurus* possessed prominent lachrymals. Spinosaurs, such as *Spinosaurus*, *Baryonyx*, and *Suchomimus*, developed long, vertically-oriented processes that emanated from their dorsal vertebrae and formed sail-like features on their backs, probably to attract potential for mates. Alternatively, these "sails" are also interpreted as thermoregulatory structures that absorbed sunlight, vented heat, or both. Finally, Cretaceous theropods with feathery integuments also argue for a display function, which is likely because most of these accouterments were not used for flight. One tetanuran in particular, *Caudipteryx*, had a caudal feather fan with differently colored, alternating bands, a common feature in modern birds who use it to attract potential mates of the same species. Another

9

Early Cretaceous theropod, *Microraptor*, has feathers on all four limbs, which may have aided in gliding, but also would have added a considerable profile to an otherwise very small dinosaur.

Eggs, embryos, and nests for a few species of theropods are documented from Late Cretaceous strata. Otherwise, little is known about the reproduction of most theropods other than speculation derived from the reproductive behavior of modern crocodilians and birds. Some dinosaur eggs have long been allied with theropods without any other corroborating evidence other than their co-occurrence in same-age strata as theropod skeletal remains. In other cases, some eggs were mistakenly assigned to non-theropod dinosaurs when actually they were of theropod origin. More theropod eggs will almost certainly be identified in the future, and they can only be reliably attributed to theropods on the basis of:

1 examination of egg interiors for embryonic remains;
2 their direct association with an adult of the same species in a nest structure;
3 better classification of eggshell types into a cladistic framework that helps to establish relationships to extant theropods (birds); and
4 eggs located within the body cavity of an adult theropod.

The last of these criteria is documented for a specimen of *Sinosauropteryx*, which had two egg-like bodies in its pelvic region. Yet another clue about theropod eggs may come from biochemical analyses of theropod eggs that closely resemble those of modern avians (Chapter 8).

The best example of a misidentification of a theropod egglayer and its eggs was corrected with new evidence about *Oviraptor*, a common theropod in Late Cretaceous deposits of Mongolia. At least two specimens have been discovered directly above nests containing egg clutches (see Fig. 8.7A). This combination of body and trace fossil evidence is nearly indisputable for its support of brooding behavior in non-avian theropods. The clutches, which normally consist of about 15 eggs, had egg forms previously assigned to the ceratopsian *Protoceratops*, which is in the same deposits. This mistaken identity lasted for nearly 70 years until an *Oviraptor* embryo was found in one of the presumed *Protoceratops* eggs. Unfortunately, numerous illustrations during that time graphically depicted *Oviraptor* crushing eggs, yolk dripping villainously from its toothless jaws. This presumption was an example of how evidence was fitted to a hypothesis, as this theropod's odd jaw apparatus was conjectured as an evolutionary adaptation for breaking eggs. In actuality, *Oviraptor* was a "good mother" theropod, although the functional morphology of its jaws is still subject to debate.

Similarly, in Upper Cretaceous rocks of Montana, a skeleton of *Troodon* was discovered on top of eight eggs attributed previously to the ornithopod *Orodromeus* (Chapter 11), and a predator–prey relationship was hypothesized for these two also. However, re-examination of the previously identified *Orodromeus* eggs revealed that they contained *Troodon* embryos. Accordingly, the proximity of the adult skeleton and eggs actually supports a parental relationship. Other theropod embryos include probable *Velociraptor* remains that were, interestingly, associated with *Oviraptor* eggs (discussed briefly in Chapter 8) and an unidentified species of therizinosaur in the Cretaceous of Mongolia.

Nest structures are only known for two theropods, *Oviraptor* and *Troodon*. A complete mound nest attributed to *Troodon*, with a distinctive upraised sedimentary rim, was also found in the Late Cretaceous Two Medicine Formation of Montana, which indicates that theropod nests might be recognizable as trace fossils without necessarily having accompanying body fossil evidence. The egg clutch size for *Troodon* was as much as 24. Eggs were laid in a spiral pattern and oriented with their long axes nearly vertical, suggesting that the mother manipulated them after they were laid in the nest structure (see Fig. 14.11). Some paleontologists have also suggested that these nests originally were covered with vegetation to aid incubation of the eggs (similar to modern crocodilian nests). However, no direct evidence of this association, such as fossil plant material, has been described so far.

The statistically significant close proximity of paired eggs within the *Troodon* clutch is strong evidence that favors dual oviducts in this theropod, which would have enabled laying the eggs two at a time. This is an excellent example of how indirect, non-skeletal evidence can be interrelated with theropod soft-part anatomy. Likewise, *Oviraptor* clutches also seem to show egg pairing, although statistical analyses on these clutches are lacking. Furthermore, the two eggs found in the pelvic region of a specimen of *Sinosauropteryx* also suggest the former presence of two oviducts. Consequently, dinosaur paleontologists are now considering the possibility of dual oviducts in at least some theropods.

Other than *Oviraptor*, *Troodon*, and *Sinosauropteryx*, no other information about egg laying or brooding behavior is currently available for theropods. Of particular interest to some paleontologists are the reproductive habits of the larger ceratosaurs and tetanurans. Questions that remain to be answered are:

1 did they build nests?
2 what did their eggs look like?;
3 what were the clutch sizes, and did they lay a few large eggs or many smaller ones, and
4 did either parent stay close to the nest after eggs were laid, or otherwise care for their young?

Hopefully, future investigations and discoveries will lend further insight into these questions and others about theropod reproduction.

Growth

Hypothesized growth sequences in theropods, from hatchlings to adults, are well supported in a few species abundantly preserved in the geologic record. For example, the best represented dinosaur species is the ceratosaur *Coelophysis bauri*, of which hundreds of individual specimens are known from the western USA. Because abundant remains of *Coelophysis* are in the same stratum and presumed to be more or less contemporaneous, a reasonable picture of their population structure can be reconstructed through biometry. These data are based specifically on the size distribution of limb and skull lengths. Another ceratosaur, *Syntarsus* of southern Africa, has provided sufficient specimens to use as another example of a growth series in a theropod species. Abundantly represented species such as these can show gradations in sizes that presumably reflect intermediate and end members of their growth.

The tetanuran *Allosaurus fragilis*, although not as abundant as *Coelophysis* or *Syntarsus*, nevertheless also has a good body fossil record in comparison to most dinosaurs and is particularly abundant in the Late Jurassic Morrison Formation from the Cleveland-Lloyd Quarry of Utah (Chapter 3). Growth series based on femur length are readily discernible from Morrison specimens of *Allosaurus* in this region. Nevertheless, whether such assemblages are representative of a contemporaneous population of *Allosaurus* is questionable and is based on a paleoenvironmental interpretation that was not originally examined in great detail.

The long-standing paleoenvironmental interpretation of the Cleveland-Lloyd deposit is that it was a "predator trap." In this scenario, a muddy area, such as a watering hole, mired a few hapless prey dinosaurs, such as the sauropod *Apatosaurus*. These hapless animals then attracted numerous predators or scavengers that likewise became stuck and died at nearly the same time. A recent taphonomic analysis in the Cleveland-Lloyd Quarry tested this hypothesis by plotting bone orientations to determine the energy level in the environment. For example, a low-energy environment, such as a watering hole, may have caused a random orientation of bones. In contrast, a higher-energy environment, such as a river, should show preferred orientations of the long bones, such as limb bones (Chapter 7). The result of the analysis was that the bones show a weak orientation along a preferred direction, which is evidence supporting some current orientation. These data thus may mean that at least some allosaurid body parts in the Cleveland-Lloyd deposit are allochthonous and represent an assemblage that was averaged over time and crossed multiple generations of allosaurids.

Bone histology has aided in the study of theropod growth, providing an approximate measure of how long theropods lived and how quickly some of them grew (Chapter 8). Based on growth lines in any given individual, a minimum age can be estimated. However, some growth lines can be absorbed during the lifetime of a vertebrate, especially early in its life. Nevertheless, a study of growth lines in *Troodon* and *Syntarsus* bones shows that they may have lived a minimum of 5 and 7 years, respectively. Recent recognition of abundant *Tyrannosaurus* bones has allowed calculations of its life history; examined specimens indicate ages of 2 to 28 years old. Having an adult age to work with, paleontologists can then calculate relative rates of growth by using the size distribution of a species in conjunction with an adult's life expectancy. For *Tyrannosaurus*, rapid rates of growth were needed for it to reach its full, mature size (at least 5 tonnes) in about 20 years.

Growth-rate calculations reveal that some theropods grew rapidly in comparison to other dinosaurs and modern crocodilians, and at rates comparable to those of extant large, flightless birds. Interpreted theropod growth rates pertain to whether they were **precocial** when young, leaving their nests and parents at an early age, or **altricial**, staying at home and depending on their parents for support. If juvenile theropods were able to "hit the ground running," then they may not have needed parental care as much as dinosaurs that were not developed enough to leave the nest for a few years. Future research should test these ideas, derived from the currently limited database, to see whether theropods in general grew up sooner than dinosaurs of other clades, or whether this trait was limited to only a few theropods.

Locomotion

Theropods collectively represent the fastest of all dinosaurs, a concept allied with hypotheses of their mostly predatory habits and reflecting their evolutionary heritage from fleet-footed bipedal archosaurian ancestors (Chapter 6).

Many suppositions about theropod locomotion were originally inferred on the basis of leg lengths and other adaptations evident in their appendicular skeletons. Additionally, a vast amount of evidence for theropod movement comes from their most abundant fossil record – tracks. Indeed, applications of formulas used for calculating theropod speeds, based on footprint length, stride length, and some predetermined parameters (Eqns 14.3 to 14.7), support the hypothesis that they were the swiftest of dinosaurs. In a few cases, they apparently approached velocities of 40 km/hour. Of course, running trackways are rare because, like most animals both ancient and modern, theropods probably conserved energy and spent much of their active time simply walking.

How fast could large theropods move? This was the subject of a study that examined possible consequences of their clumsiness during high speeds. Two researchers, one a paleontologist and the other a physicist, asked the simple question, "What would have happened to a *Tyrannosaurus* if it had been running, then tripped and fell?" Part of the answer to that question can be demonstrated through use of the equations for either momentum or force (Eqns 4.8 and 7.4). Calculations resulting from these equations show that the resulting impact of a falling tyrannosaurid would have varied in relation to its velocity or acceleration. For example, if it had been moving at 40 km/hour and weighed 6 tonnes, then its forward momentum would have been:

Step 1. $M = (6 \times 10^3 \text{ kg})(40 \text{ km/hour})$

(Converting kilometers/hour and seconds/hour to meters/second):

Step 2. $M = (6 \times 10^3 \text{ kg})(40 \text{ km/hour})(1000 \text{ m/km})/(3600 \text{ s/hour})$

Step 3. $= (6 \times 10^3 \text{ kg})(11.1 \text{ m/s})$

Step 4. $= 6.7 \times 10^4 \text{ kg} \times \text{m/s}$

This is the approximate forward momentum (M) of an adult tyrannosaurid running at such a speed. For the sake of comparison, a human who weighs 75 kg and is running at 10.2 m/s (which is about 36 km/hour) would have a momentum of

Step 1. $M = (75 \text{ kg})(10.2 \text{ m/s})$

Step 2. $= 7.6 \times 10^2 \text{ kg} \times \text{m/s}$

The injuries typically sustained by humans running this fast who then trip, as is common in races involving hurdles, are directly related to the speed and mass of the runner. Our hypothetical tyrannosaurid, because of its greater mass, had about two orders of magnitude more momentum and, hence, a much larger risk of injury from any sort of fall. As a result, one conclusion that can be reached is that *Tyrannosaurus* was not adapted to running fast, and that the dire consequences of missteps prevented a cursorial mode of hunting. Similarly, the authors of another study calculated the leg-muscle mass needed for a fast-running *Tyrannosaurus*, and they found that its legs would have comprised an absurdly high percentage (86%) of the total body mass. Consequently, these multiple lines of evidence favor the hypothesis that some large theropods, such as *Tyrannosaurus*, were better adapted for walking, although this likelihood does not completely preclude their having been active hunters.

So far, only one trackway of a large theropod, which probably weighed 1–2 metric tons, shows a relatively fast speed of about 30 km/hour, using speed estimations calculated via Equations 14.3 to 14.7. Another example of large theropod locomotion is taken from a lone probable *Tyrannosaurus* track (see Fig. 9.7) in which the bedding plane did not show any other track in a 5-meter distance. This indicates that the stride length was a minimum of 10 meters, which yields a calculated minimum speed of about 12 km/hour. If such large theropods ever ran at high speeds, the evidence was either not recorded or has not yet been found. So, in the meantime, paleontologists must be content with the numerous beautiful trackways that show large theropods walking (Fig. 9.13).

Other noteworthy examples of theropod locomotion, recorded by their trackways, give more detailed insights on theropods and the variability of their behavior:

- Stalking of prey, interpreted from a trackway in east Texas made by a large Early Cretaceous theropod (probably *Acrocanthosaurus*), possibly following the tracks of a large sauropod.
- An Early Cretaceous stampede in Queensland, Australia of a group of small theropods and ornithopods, most running in the opposite direction from a much larger therapod.

9

FIGURE 9.13 Large theropod trackway from the Upper Jurassic Morrison Formation of Colorado. (A) Overview of trackway showing pace and stride lengths; dinosaur ichnologist for scale. (B) Close-up of one track in sequence with pressure-release structure evident in left upper (outer) edge of track, indicating a pushing of the sediment by the theropod as it shifted from the left to the right foot.

(A)

(B)

- Apparent pack hunting, where tracks left by a herd of Late Jurassic sauropods were followed by tracks of a group of large theropods. A similar pack-like configuration of large theropods is inferred from multiple and equally-spaced trackways on the same bedding plane from the Early Cretaceous of Mongolia.
- Early Jurassic theropods in the western and eastern USA (Utah and Massachusetts, respectively) that stopped to sit down, leaving metatarsal and posterior body impressions.
- Limping, presumably from a limb injury, which might be expected for an animal subjecting itself to risky behavior, such as running too fast and tripping.

Probably the single most important insight gained from theropod trackway information is that theropods were apparently the most active of all dinosaurs. Despite their relative scarcity as body fossils in comparison to other dinosaurs, theropod tracks are the most abundant of all dinosaur tracks. In most places where dinosaur tracks are found, they outnumber the tracks of all other dinosaur clades combined. For example, Middle Jurassic shoreline deposits in northwestern Wyoming show thousands of theropod tracks, but not one track attributable to an ornithopod, sauropod or thyreophoran. The high activity level and mobility indicated by this wealth of data, along with their paleobiogeographic distribution, has been used to infer that theropods were physiologically different enough from other dinosaurs in that they were endothermic (Chapter 8). This is a requirement in modern terrestrial animals that stay active for long periods of time. In contrast, few modern ectothermic terrestrial animals are active on a regular basis and must spend large amounts of time soaking up sunlight. The physiological considerations of theropods, including whether they were endothermic, ectothermic, or perhaps a combination of the two in various stages of their lives, is also a subject pertinent to theropod–bird interconnections (Chapter 15).

Interestingly, recent discoveries of small, feathered theropods from the Early Cretaceous of China suggest that not all theropod locomotion was on the ground. At least two species show adaptations for tree climbing and two others were capable of either gliding or powered flight. *Epidendrosaurus ninchengensis* and *Scansoriopteryx heilmanni* are actually similar enough that they may represent one species, and both show the following features indicative of an arboreal lifestyle:

1 small body sizes;
2 very long digit III on the manus; and
3 very long fore limbs.

These features are interpreted as adaptations for climbing and living in trees, an old idea for theropods that is now gaining credence in the light of this evidence. *Microraptor gui* and *Cryptovolans pauli* are interpreted as dromaeosaurids capable of either gliding or flight on the basis of:

1 feather impressions associated with their limbs; and in *Microraptor* these are on all four limbs (the only vertebrate known to have this trait);
2 long fore limbs; and
3 a well-developed keel (sternum) in *Cryptovolans*, which would have supported flight muscles.

However, the latter is missing in *Microraptor*, which means that it is currently interpreted as a tree-climber and glider, rather than an active flyer. On the other hand, *Cryptovolans* may have been better adapted than *Archaeopteryx* for active flight, an

9

interesting twist of previous assumptions about the lifestyles of non-avian and avian theropods (Chapter 15).

Feeding

Theropod trackways not only contribute to hypotheses about their locomotion, but also directly relate to interpretations of their feeding behavior. As mentioned in the previous section, stalking, pack hunting, and smaller theropods running away from a larger carnivore have all been suggested by their tracks. Nevertheless, the starting point of discussion of theropod feeding behavior typically involves describing their teeth.

Most often, theropod teeth are categorized as ziphodont, a plesiomorphic trait in which they curved posteriorly, are serrated, conical (pointed), and have carinae, which were well adapted for slicing through soft animal tissue. Thicker and more robust examples, such as the teeth of tyrannosaurids, such as *Albertosaurus*, were also capable of punching through bone (Fig. 9.14). The meat-eating interpretation of theropod teeth is also well supported by numerous examples of toothmarked bones where the marks match known theropod teeth (Chapter 14). In addition, skeletal remains in the body cavities of a few theropods may well represent consumed animals. Probable theropod coprolites containing bone fragments, and in one case fossilized muscle tissue, were documented. These numerous lines of evidence thus corroborate hypotheses, derived originally from just teeth, that most theropods were undoubtedly carnivores. Interestingly, few theropods, such as some ornithomimids (*Ornithomimus* and *Struthiomimus*) and oviraptorids (*Oviraptor* and *Ingenia*), had no teeth or any sign of tooth sockets. Consequently, they must have lost them as an inheritable trait in their preceding evolutionary history. Although Ornithomimosauria is typically thought of as a group of toothless dinosaurs, one ornithomimid species (*Pelecanimimus*) had about 200 teeth, the most of any theropod. As a result, exceptions arise for each generalization about theropods.

One such exception is the hypothesis that some theropods may not have been carnivores. For example, therizinosaurs, such as the Early Cretaceous *Alxasaurus* in China and *Therizinosaurus*, *Segnosaurus*, and *Erlikosaurus* of the Late Cretaceous in Mongolia, represent mysteries in how they fed and what they ate. An inventory of their meal-gathering tools shows some strange traits for theropods. For example:

1 *Alxasaurus*, *Erlikosaurus*, and *Segnosaurus* had poorly-developed, leaf-shaped teeth;
2 teeth were absent from the anterior part of the premaxilla of *Erlikosaurus*; and

FIGURE 9.14 The Late Cretaceous tetanuran and tyrannosaurid *Albertosaurus*, showing its jaws filled with prominent, recurved, and serrated teeth ideally suited for slicing and dicing flesh and bone. Royal Ontario Museum, Toronto, Ontario, Canada.

3 *Therizinosaurus* had unguals on its manus that measured up to 70 cm long, greater than its forearm length.

Such prodigious unguals, analogous to the fictional character *Edward Scissorhands*, arguably could have been used for defense. However, they were certainly too unwieldy to be used in any effective way against other large, predatory theropods. Instead, they are interpreted as adaptations for feeding.

These and other therizinosaur traits are the reason why they are often considered the oddest of all theropods. With their relatively long prosauropod-like necks and stout torsos, they clearly were not well-suited for hunting and meat-eating. As a result, some paleontologists have proposed that they were best adapted for browsing on vegetation. In this hypothesis, the claws are interpreted as implements for raking tree branches. Giant ground sloths, herbivores that lived only about 12,000 years ago in North America, had similar adaptations, so this might be an example of convergent evolution. An alternative hypothesis is that such claws were used for eating insects, much like modern anteaters that use their comparable armature to tear apart termite mounds. Trace fossil evidence shows that termites had developed mounds by the Late Jurassic that resembled modern ones in Africa and Australia. Therizinosaurs show up in the geologic record by the end of the Early Cretaceous, so from an evolutionary standpoint it is appropriate that they could be termite-eaters. However, most therizinosaurs were huge in comparison to a typical modern anteater, which means that they would have had to tear apart many termite nests to gain sufficient nutrition for sustenance. Like many problems in evolution, no single solution provides a complete answer to these enigmatic theropods.

Another group of theropods recently suspected of herbivory is the ornithomimosaurs. This hypothesis is based on the following criteria:

1 ornithomimosaurs have less robust skulls than similarly-sized theropods;
2 these skulls often lack teeth and instead are assumed to have been beaked;
3 they are more common than undoubted carnivorous theropods in some Late Cretaceous assemblages; and
4 some specimens of the ornithomimosaurs, *Sinornithomimus* and *Shenzhousaurus*, have gastroliths in the areas of their abdominal cavities.

A light skull lacking teeth argues for adaptations to foodstuff that was either not fighting back or was softer than most prey animals. Although modern predatory birds do not need teeth to kill or tear into their prey animals, they also have strong skulls, high speeds aided by powered flight that can cause killing impacts, and sharp talons. Ornithomimosaurs had none of these compensating traits for effective predation. Relative abundance can also be a clue to their supposed herbivory as ecological communities tend to have many more herbivores than carnivores. Finally, the documentation of gastroliths in *Sinornithomimus* and *Shenzhousaurus* lends more credence to their having been herbivorous, because gastroliths are present largely in other undoubted herbivorous dinosaurs, presumably as an aid to grinding plant material (Chapters 10 and 11).

As far as most flesh-eating theropods were concerned, anatomical data provide clues about their sensory abilities used for discerning and acquiring either prey or corpses. Many theropod lineages, such as allosaurids, troodontids, and tyrannosaurids, have orbits positioned toward the front of the skull rather than laterally. This suggests that they had developed **stereoscopic vision**, also known as **binocular vision**. Prey also could have been detected through sound, and *Troodon* shows sufficient cavities in its skull for organs that would have allowed for sound location. Olfactory sensations were yet another way for theropods, which could have used a sense of smell to find food, particularly already-dead animals. Recall that

9

putrefaction of a body occurs within a few hours to days after death (Chapter 7). Soon afterwards, a scavenger's sensitive nose, under the right wind conditions, can detect a decaying body over great distances, often more effectively than through listening for sounds of other scavengers or attempting to spot the body. Brain endocasts and CT scans of *Tyrannosaurus* support this type of adaptation in the form of an enlarged olfactory bulb in the anterior portion of its brain. This and other evidence has led some paleontologists to suggest that this supposedly fierce predator may have been more like a six-tonne vulture. Other aspects of theropod senses used in predation can be speculated on the basis of modern predatory animals. For example, they may have been able to detect vibrations from the ground caused by herd movement. This presumably would have been easy with large herds of sauropods, ceratopsians, or hadrosaurs. They also could have "tasted" the air with their tongues, which is actually a form of smell used by some snakes and lizards. However, without any other corroborating evidence, both of these ideas remain speculation for now.

If a given theropod identified its potential prey, then a number of methods could have been used to kill it. The most often depicted way was the use of teeth and jaws to inflict fatal wounds, but in modern terrestrial carnivores this method is rarely used by itself. Modern land predators, such as large cats and bears, use fore limbs that are typically armed with sharp claws, which are combined with throwing their body weights against a prey animal while running at high speed. These techniques stop the prey long enough to deliver killing actions. Additionally, large cats will not necessarily slash with teeth or claws to kill an animal but will clamp their jaws around its neck to suffocate it. In fact, induced drowning is even possible. For example, a pair of cheetahs were once observed chasing an antelope into a water body and holding its head under water until it drowned. In this case and others, pack hunting is a very effective strategy, whereby multiple predators wear down their prey until it is too exhausted to offer any resistance to fatal wounding.

Some theropods certainly had teeth, jaws, and claws as available weapons. Unfortunately, the low preservation potential for soft-tissue wounds prevents independent confirmation of most hypotheses on theropod predation, but several features of a few theropods invoke only visions of death-dealing implements. Large, serrated teeth are one persuasive attribute, but huge, powerful arms that end in sharp unguals are another persuasive attribute of hunting. An intriguing find in the Upper Cretaceous of Mongolia was of 2.4-meter long forelimbs that ended in recurved unguals and phalanges. The rest of the body was never found, but the unusual arms were assigned to a new species, *Deinochirus mirificus*, which is currently interpreted as an ornithomimid. Like those of therizinosaurs, these arms with their prominent claws have been interpreted as possible tools for demolishing termite mounds, but they also would have served well for larger game.

Ambiguous as some skeletal evidence may be, little doubt is expressed about the most well-known feature of some dromaeosaurids, including the Late Cretaceous *Velociraptor* and Early Cretaceous *Deinonychus* and *Utahraptor*: a sharp, retracted ungual on digit II of the foot (Figs 9.9 and 9.15). This claw remained permanently above the ground surface with its point forward, strongly suggesting that it had an offensive purpose, such as disemboweling prey animals. Once again, because the preservation potential for soft-tissue damage is so low, independent verification of this claw as a killing feature is virtually absent. Nevertheless, one of the few examples in the geologic record of two dinosaur species directly interacting with one another as they died and, consequently, one of the most remarkable dinosaur discoveries ever found is a *Velociraptor* conjoined with its apparent intended prey, a *Protoceratops* (see Fig. 7.9). In this case, the digit II unguals of *Velociraptor* are clearly within the ventral (abdominal and throat) region of *Protoceratops*. The protoceratopsian also had the right forearm of the *Velociraptor* caught in its jaws, showing that it was

FIGURE 9.15 The formidable *Utahraptor* and its raptorial digit II ungual, skeletal mount with claw raised on left pes. Note its similarity to *Deinonychus* in both form and inferred function. College of Eastern Utah Prehistoric Museum, Price, Utah.

probably responding violently to its impending death. Amazingly, both animals had been killed at the same time by the depositional event that buried them. This deposition froze their positions until they were uncovered nearly 70 million years later in the Upper Cretaceous of Mongolia. This predator–prey tableau is appropriately nicknamed "The Fighting Dinosaurs."

Similar to this discovery, although less obvious in its interpretative value, is another interesting case of a possible predator–prey relationship interpreted from associated dinosaur skeletal remains. In this example, a Lower Cretaceous deposit in Montana contains the bones and teeth of at least four individuals of *Deinonychus* and one individual of the ornithopod *Tenontosaurus*. The unusual juxtaposition of a larger ornithopod with four small theropods of the same species, which normally are rare finds even as individuals, has been interpreted as the end result of pack hunting. In this scenario, the dromaeosaurids may have attacked the ornithopod and killed the much larger prey animal through a concerted and cooperative effort. The partial remains of the four predators are explained by a hypothesis that the prey animal used its superior bulk of nearly one tonne and accompanying strength to defend itself. As a result, several of the smaller predators, which weighed only about 50 to 100 kg each, may have suffered fatal injuries before the prey itself succumbed to their onslaught. Although the circumstances surrounding the remains of this presumed battle still raise more questions than answers, it is one of the few persuasive cases for pack hunting in some theropods. This idea has been long conjectured but not supported by much more than theropod trackways (mentioned earlier) and observations of pack-hunting behaviors in modern terrestrial carnivores. Further evidence supporting this predator–prey relationship of *Deinonychus* and *Tenontosaurus* are 15 localities discovered so far in the Lower Cretaceous of Montana, where *Deinonychus* teeth were found in the same vicinity as *Tenontosaurus* bones.

Of all theropods, tyrannosaurids are probably the dinosaurs best known for their eating habits, which are delineated by:

1 toothmarks and teeth in bones of their former meals;
2 contents of the coprolites attributed to them; and
3 probable former gut contents.

Of these lines of evidence, teeth and toothmarks are the most common, whereas gut contents are the most rare. Nevertheless, combined use of these data creates a remarkably complex picture of tyrannosaurid feeding preferences and their relationships to other dinosaurs. For example, *Tyrannosaurus* teeth were found in the fibula of *Hypacrosaurus* and a rib of *Edmontosaurus*, both ornithopods (Chapter 11). Probable tyrannosaurid toothmarks are also documented in bones of a theropod, the dromaeosaurid *Saurornitholestes*, as well as in bones of the ceratopsian *Triceratops* (Chapter 13) and *Edmontosaurus* (Chapter 11). The toothmarks in *Triceratops* are on its ilium, so they were most likely not death-dealing marks but rather signs of feeding after the animal was already dead. Furthermore, the toothmarks show both puncturing and scraping of the bone. This suggests that the tyrannosaurid bit deeply into the hip region of the ceratopsian, with a calculated bite force of 13,400 N, the greatest known for any animal. The tyrannosaur then pulled meat from the bone, once again indicating a dead and non-struggling food item. However, tyrannosaurid toothmarks on the caudal vertebrae of a specimen of *Edmontosaurus* show signs of healing after they were inflicted, indicating that a tyrannosaurid attempted to munch on a live prey. Other specimens of *Edmontosaurus* have toothmarks attributed to both *Tyrannosaurus* and *Albertosaurus*, showing that this herbivore was on the menu for more than one species of tyrannosaurid. Toothmarks from the dromaeosauridid *Saurornitholestes* also occur in *Edmontosaurus* bones. This combination of body and trace fossil evidence permits the beginning sketch of a Late Cretaceous food chain: *Edmontosaurus* ate land plants, and was in turn eaten by both *Saurornitholestes* and *Tyrannosaurus*, but *Tyrannosaurus* also ate *Saurornitholestes*. Consequently, *Tyrannosaurus* was at the top of this food chain.

Coprolites and probable stomach contents augment tooth and toothmark data by showing what entered and exited at least a few of their gastrointestinal tracts. Two probable tyrannosaurid coprolites are interpreted on the basis of:

1 their stratigraphic occurrence in beds containing tyrannosaurid remains, the Upper Cretaceous Dinosaur Park Formation of Alberta, Canada;
2 their large size (44 and 64 cm long); and
3 their contents, as both consist of cylindrical masses of phosphatic rock that contain many small bone fragments.

The 44-cm long specimen contains bone from a probable juvenile hadrosaur, and the larger specimen contains bone from an unidentified juvenile dinosaur. Finely-ground bone in a coprolite could be interpreted as the result of thorough chewing, but tyrannosaurid jaws were not amenable to their teeth repeatedly coming together in this way. An alternative hypothesis is that the bone consists of fragments accidentally scraped off with the flesh as the tyrannosaurid pulled with its anterior teeth. Indeed, the contents of the larger coprolite support gorging behavior rather than chewing, because it includes three-dimensional preservation of muscle tissue. Such preservation indicates a brief time for this food in the gut.

Probable former abdominal contents in another tyrannosaurid species, *Daspletosaurus*, provide even more information to augment the intriguing implications of the coprolite data. The skeletal specimen of *Daspletosaurus*, in the Upper Cretaceous Two Medicine Formation of Montana, had the vertebrae and a dentary from a juvenile hadrosaur in what was probably its gut. These hadrosaur bones show evidence of corrosion caused by partial digestion. This suggests that *Daspletosaurus* had an enzyme-producing proventriculus, the anterior portion of a two-part stomach followed posteriorly by a muscular gizzard. Such an anatomical arrangement is also seen in crocodiles and some birds, which leads to the hypothesis that some theropods shared this trait as a synapomorphy. However, much testing of this hypothesis is required because it is based on such scanty and preliminary results.

With reference to other feeding preferences, *Baryonyx* and other spinosaurs, such as *Suchomimus*, have been interpreted as piscivorous (fish-eating) theropods. This conclusion is made on the basis of:

1 fish scales found in association with a *Baryonyx* skeleton;
2 the crocodile-like skull and numerous small teeth of *Suchomimus*, well adapted for grabbing and holding fish; and
3 long arms with well-developed hook-like claws on both *Baryonyx* and *Suchomimus*.

The combination of many teeth, which are not as robust or lethal-looking as those of some other theropods, along with their unusual arms, led to the hypothesis that spinosaurs preferred a diet of fish derived from shallow aquatic environments. In this sense, these dinosaurs may have been like modern grizzly bears (*Ursus arctos*), which scoop fish out of streams.

No definitive single piece of evidence shows that a theropod actually killed another dinosaur, although the "fighting dinosaurs" of *Velociraptor* and *Protoceratops* persuasively shows the intent to kill in one instance. Nonetheless, the combination of data from functional morphology in theropod body plans and their toothmarks, trackways, and coprolites provide reasonable evidence that many, if not most, theropods were active predators. Lively debates have centered on whether some theropods were primarily predators or scavengers, and no dinosaur has received as much attention in this respect as *Tyrannosaurus rex*.

Tyrannosaurus is a prime example of a theropod that in some ways resembled a killer, but in other ways did not seem well-adapted to full-time hunting. One of the major objections to viewing *Tyrannosaurus* as a predator focuses on its ridiculously small arms, which were so short that they could not even reach its mouth. Moreover, these arms ended with two small digits, rather than the robust three-digit hands seen in other theropods. Because of this, some paleontologists suggest that *Tyrannosaurus* was primarily a scavenger. After all, such arms alone could not have possibly held on to a multi-ton struggling prey animal, and predation without the use of arms would have required a "land shark" approach. In this scenario, the large theropod would have waited for a prey animal to walk by, and then ambushed it by biting it. Of course, how a 13-meter-long, 6-tonne predator would remain unnoticed by a prey animal presents a problem for this scenario.

Modern terrestrial carnivores provide a guide to better understanding this problem. For these animals, part-time or full-time scavenging is a common means of obtaining meat. Indeed, some popularly assumed full-time predators, such as present-day large cats, have been observed chasing smaller predators away from kill sites to consume the corpse. One suggestion for *Tyrannosaurus* is that it was a part-time hunter that killed smaller animals, such as juveniles, a hypothesis supported by coprolite and toothmark evidence. However, it also may have used its large size to chase away smaller theropods from the scene of a successful hunt, an adaptation that would have been augmented considerably if more than one tyrannosaur showed up at a kill site together. Whether such Mesozoic bullying was common or not is unknown, but the feeding behavior of modern terrestrial carnivores is complex and often opportunistic. Accordingly, theropods were probably no different, getting what they could by whatever means that worked, regardless of whether their food was alive or recently dead.

9

Social Life

Although theropods were normally depicted in older illustrations and fictional stories as lone hunters that avoided other individuals of their same species except for occasional mating, scientific evidence has started to partially refute this stereotype. For example, the close proximity of multiple individuals of the same species of theropod is now considered as evidence of their being together at the time of death and burial, rather than a taphonomic coincidence. The earliest examples of this were represented in a spectacular way by remains of the Late Triassic ceratosaur *Coelophysis* at the Ghost Ranch Quarry in New Mexico, where more than 100 individuals were recovered. Another ceratosaur, the Early Jurassic *Dilophosaurus*, is rare in its worldwide distribution, but three specimens were found in the same deposit from a small area in Arizona. In comparison to other dinosaur species, the Late Jurassic tetanuran *Allosaurus* occurs in disproportionately large numbers at the Cleveland-Lloyd Quarry, suggesting that they were proximal to one another at the time of death. More recently, numerous tyrannosaurids, *Albertosaurus*, were found in an Upper Cretaceous deposit of Alberta, Canada, in what was probably a contemporaneous assemblage. Paleontologists studying the assemblage even proposed that it might represent a family structure.

As mentioned earlier, some trackways show closely spaced groups of theropods, some of them large, moving together in a similar direction. Similarly, one Middle Jurassic tracksite in northwestern Wyoming contains thousands of theropod tracks in a relatively small area, suggesting gregarious behavior. Of course, one of the potential pitfalls of track evidence is that the tracks on any given bedding plane may have been formed by repeated visits of the same animals, or at vastly different times (Chapter 14). Fortunately, a close examination of the qualitative characteristics of tracks can reveal whether they were made contemporaneously. Once these features are taken into account, a better answer as to whether some theropods moved in groups, and perhaps as family units, may emerge. Such trackways and skeletal evidence should shed more light on social dynamics of theropods.

Health

A great deal of evidence indicates that most theropods did have active lifestyles that included hunting, seeking mates, or moving together as family units. As a result, one could expect that they encountered more problems than dinosaurs whose food and mates did not move so fast. Theropods show the most evidence of health-related problems of all major dinosaur clades, although most were healthy animals. Apparently, the most common were limb injuries. For example, several theropod trackways show evidence of leg injuries in the trackmakers, where limping caused one pace length to be considerably longer than the other. An injury to the shoulder girdle, perhaps a tendon or muscle pull, is evident in one specimen of *Allosaurus*, where an overgrowth of bone on its left scapula marks the injury site. Similarly, a humerus of a *Tyrannosaurus* has a concavity symptomatic of a tendon tear. Other documented injuries among theropods include bone fractures that later healed, such as in *Albertosaurus*, *Allosaurus*, *Deinonychus*, *Syntarsus*, and *Tyrannosaurus*, and broken teeth, a common dental difficulty for many theropods.

In terms of diseases, bone infections were interpreted for one specimen each of *Allosaurus*, *Dilophosaurus*, and *Troodon*, as well as metastatic cancer for one specimen of *Allosaurus* and gout for one specimen of *Tyrannosaurus* (Chapter 6). Although healed bite marks from other theropods have been suggested to explain some bone abnormalities, no definitive examples are currently documented in the peer-reviewed literature. This is especially the case for bites inflicted by compatriots of the same species. Such an interpretation was applied to odd holes in the mandible of one specimen of *Tyrannosaurus* ("Sue": Chapter 2), but a more careful examination revealed that these were more likely from a fungal infection. Evidence of bite marks from the same species would of course verify intraspecific competition, which accordingly has important implications about theropod social interactions.

SUMMARY

Theropoda is arguably the best known and most studied of all dinosaur clades. A combination of body and trace fossil evidence for theropods speaks of their long and rich history, which started at the beginning of the dinosaurian reign in the Late Triassic and continues today through birds, their probable descendants. *Eoraptor* of the Late Triassic of Argentina may have been both the earliest known dinosaur and theropod, although not all paleontologists accept this designation. Among the first possible theropods were the herrerasaurids (e.g., *Herrerasaurus* and *Staurikosaurus*), which were affiliated with Theropoda but may represent a separate saurischian clade. By the end of the Triassic, ceratosaurs (e.g., *Coelophysis* and *Syntarsus*) showed that non-avian theropods had arrived to stay in terrestrial ecosystems for the next 150 million years or so. This was ably demonstrated by other ceratosaurs of the Jurassic (*Ceratosaurus*) and Cretaceous (abelisaurids, such as *Carnotaurus*), but most prominently by tetanurans beginning in the Middle Jurassic. Tetanurae is the most diverse of theropod clades and included some of the most famous dinosaurs, such as *Allosaurus*, *Deinonychus*, *Giganotosaurus*, *Tyrannosaurus*, and *Velociraptor*. Tetanurans also had one of their lineages contribute to the evolution of birds by the Late Jurassic, and numerous examples of feathered tetanurans have been recently found in Cretaceous strata of China. Theropods occupied the entire geologic range for dinosaurs, and their fossils occur on every continent.

Despite their widespread distribution through time and space, theropods are relatively uncommon as body fossils in comparison to most dinosaurs. Fortunately, their tracks are exceedingly abundant in places, more so than those of other dinosaur clades, and other theropod trace fossils are becoming more recognized. Theropod evolution is reflected by characteristics that show:

1 pneumatization and consequent lightening of the skeleton, especially the skull;
2 increased endocranial volume with proportion to total body volume;
3 changes in hind limb and pelvic girdle anatomy that further enhanced bipedalism and mobility; and
4 fore limb alterations that allowed for better grasping, especially with digits I through to III.

These and other characteristics show apparent tendencies toward specialization in their respective environments, and theropods were among the most diverse dinosaurs known.

Theropods are normally generalized as the meat-eating dinosaurs and are often synonymized with dinosaur predation. They were mostly carnivorous, but a few, such as some ornithomimosaurs and oviraptorsaurs,

9

SUMMARY Continued

and therizinosaurs, may have been herbivorous or insectivorous. Hypotheses about theropod feeding habits, based on functional morphology, tracks, toothmarks, gastroliths, and coprolites, are still unclear as to whether most meat-eaters were primarily predators or scavengers. Reproductive behavior is well documented for two tetanuran species, *Oviraptor* and *Troodon*, which show evidence of both nest-building and brooding behavior, but is poorly known for other theropods. Some theropod juveniles were apparently precocial, and rapid growth rates are inferred for a few species of theropods, which may relate to thermoregulatory modes similar to endothermy.

Some evidence from both monospecific assemblages of theropod bones and closely spaced (and possibly contemporaneous) trackways suggests social behavior in some theropods, such as pack hunting. In contrast, individual trackways suggest that they sometimes roamed alone. Theropods were typically healthy animals but occasionally were injured or suffered from infections. Their general healthiness would have contributed to the evolution of some long-lived theropod lineages.

DISCUSSION QUESTIONS

1. What would have been the evolutionary advantage of a reduction of digit III in a theropod manus, such as that shown by the Early Cretaceous ceratosaur *Carnotaurus* and the Late Cretaceous tetanurans (and coelurosaurs) *Albertosaurus* and *Tyrannosaurus*? If you cannot think of any advantage for it, then how could it have become a vestigial organ, like a human appendix?

2. While working as a summer volunteer on a dinosaur dig, you and several other lucky participants uncover skulls from three new theropod species. Three years later, when you read the scientific papers reporting on the skulls, you notice the following endocranial volumes estimated for each species. The reconstructed original body volumes are given in parentheses following the endocranial volumes. Calculate the brain/body volume ratios for the three. Use your three calculations to determine which one is the "brainiest" in comparison to body size. How do they compare with the brain/body ratio calculated for *Tyrannosaurus* in the chapter?
   a. Theropod Species A: 210 cm^3 (1.8 × 10^6 cm^3)
   b. Theropod Species B: 112 cm^3 (7.1 × 10^5 cm^3)
   c. Theropod Species C: 55 cm^3 (3.2 × 10^5 cm^3)

3. One hypothesis proposed for why some theropods had feathers is that they served as a sexual display for potential mates. What is an alternative hypothesis for a function of feathers in theropods that

seemingly were flightless? Moreover, what were possible additional functions of feathers for those theropods that were arboreal gliders or flyers?

4. Using the list of characters for Herrerasauridae listed in the chapter, compare it with those for the hypothetical ancestral dinosaur listed in Chapter 6. In what characters do they overlap and in which ones do they differ? Give an explanation for why some should differ and how they might reflect evolutionary processes.

5. Using the lists of characters in the chapter, compare Herrerasauridae versus Ceratosauria and Ceratosauria versus Tetanurae. Once again, give an explanation for why some characters should differ and how they might reflect evolutionary processes.

6. Explain the difference between a stem-based clade and a node-based clade, using theropods as examples. What new fossil finds for each clade could cause these classifications to change?

7. Compare and contrast the skeletal foot of *Tyrannosaurus rex* with its probable footprint in Figure 9.11. What aspects of soft-part anatomy for the foot are most obvious in the footprint that otherwise would not be evident in the skeletal foot? How could this information relate to the possible walking or running speeds of *Tyrannosaurus*?

8. Given three theropod trackways with the following estimated speeds, calculate the momentum for each respective theropod using their hypothetical masses:
   a. Theropod A: 14.5 km/hour (750 kg)
   b. Theropod B: 7.6 km/hour (1.2 metric tons)
   c. Theropod C: 23.8 km/hour (80 kg)

9. Using the characteristics of eggs, nests, and skeletal data from *Troodon* and *Oviraptor*, how could you develop a model for identifying egg clutches for other theropods without having identifiable embryonic remains? Develop a list of characteristics that would help in such identification.

10. What evidence would you need to support the following alternative hypotheses used to explain the close proximity of four specimens of *Deinonychus* with one specimen of *Tenontosaurus*? Think about both body and trace fossil evidence that would need to be present to make it more convincing.
   a. The *Deinonychus* specimens were traveling together as a group, and drowned in a stream. Their bodies were washed into the same location as a dead *Tenontosaurus*, and then scavengers ate from all five carcasses.
   b. One pack of *Deinonychus* attacked and killed the *Tenontosaurus*, but a rival pack entered the area and fought the first pack, causing the deaths of members from each pack.
   c. Scenario B occurred, but the two packs consisted of two families traveling as groups.
   d. The *Tenontosaurus* specimen was traveling with a herd, but it was separated from the herd by a pack of *Deinonychus*, which then hunted it down and killed it.
   e. The *Tenontosaurus* specimen died from an infection, then a *Deinonychus* pack scavenged the corpse and several died from eating the toxic meat.

9

Bibliography

Bailey, J. B. 1997. Neural spine elongation in dinosaurs: Sailbacks or buffalo-backs? *Journal of Paleontology* **71**: 1124–1146.

Brochu, C. A. 1999. High-resolution CT analysis of a *Tyrannosaurus rex* skull. *Journal of Vertebrate Paleontology* **19**, 3: Suppl. 34A.

Charig, A. J. and Milner, A. C. 1986. *Baryonyx*, a remarkable new theropod dinosaur. *Nature* **324**, **6095**: 359–361.

Coria, R. A. and Salgado, L. 1995. A new giant carnivorous dinosaur from the Cretaceous of Patagonia. Nature **377**, **6546**: 224–226.

Currie, P. J. 1998. "Theropods". *In* Farlow, J. O. and Brett-Surman, M. K. (Eds), *The Complete Dinosaur*, Bloomington, Indiana: Indiana University Press. pp. 216–233.

Czerkas, S. A. and Yuan, C. 2002. An arboreal maniraptoran from northeast China. *The Dinosaur Museum Journal*, Vol. 1.

Erickson, G. M., Makovicky, P. J., Currie, P. J., Norell, M. A., Yerby, S. A. and Brochu, C. A. 2004. Gigantism and comparative life-history parameters of tyrannosaurid dinosaurs. *Nature* **430**: 772–775.

Farlow, J. O., Brinkman, D. L., Abler, W. L. and Currie, P. J. 1991. Size, shape and serration density of theropod dinosaur lateral teeth. *Modern Geology* **16**: 161–198.

Hammer, W. R. and Hickerson, W. J. 1994. A crested theropod dinosaur from Antarctica. *Science* **264**, **5160**: 828–830.

Holtz, T. R., Jr. and Osmólska, H. 2004. "Saurischia". *In* Weishampel, D. B., Dodson, P. and Osmólska, H. (Eds), *The Dinosauria* (2nd Edition). Berkeley, California: University of California Press. pp. 21–24.

Holtz, T. R., Jr., Molnar, R. E. and Currie, P. J. 2004. "Basal tetanurae". *In* Weishampel, D. B., Dodson, P. and Osmólska, H. (Eds), *The Dinosauria* (2nd Edition). Berkeley, California: University of California Press. pp. 71–110.

Horner, J. R. and Lessem, D. 1993. *The Complete* T. rex. New York: Simon and Schuster.

Jerison, H. J. 1979. "The evolution of diversity in brain size". *In* Hahn, M. E., Jensen, C. and Dudek, B. C. (Eds). *Development and Evolution of Brain Size: Behavioral Implications*. New York: Academic Press. pp. 29–57.

Lockley, M. G. and Hunt, A. P. 1994. A track of the giant theropod dinosaur *Tyrannosaurus* from close to the Cretaceous/Tertiary boundary, northern New Mexico. *Ichnos* **3**: 213–218.

Lockley, M. G. and Matsukawa, M. 1999. Some observations on trackway evidence for gregarious behavior among small bipedal dinosaurs. *Palaeogeography, Palaeoclimatology, Palaeoecology* **150**: 25–31.

Norell, M. A. and Makovicky, P. J. 2004. "Dromaeosauridae". *In* Weishampel, D. B., Dodson, P., and Osmólska, H. (Eds), *The Dinosauria* (2nd Edition). Berkeley, California: University of California Press. pp. 196–210.

Norell, M., Ji, Q., Gao, K., Yuan, C., Zhao, Y. and Wang, L. 2002. Modern feathers on a non-avian dinosaur. *Nature* **416**: 36–37.

Ostrom, J. H. 1969. Osteology of *Deinonychus antirrhopus*, an unusual theropod from the Lower Cretaceous of Montana. *Bulletin of the Peabody Museum of Natural History* **30**: p 165.

Padian, K., Hutchinson, J. R. and Holtz, T. R., Jr. 1999. Phylogenetic definitions and nomenclature of the major taxonomic categories of the carnivorous Dinosauria (Theropoda). *Journal of Vertebrate Paleontology* **19**: 69–80.

Sampson, S. D., Witmer, L. M., Forster, C. A. and Krause, D. W. 1998. The evolution and biogeography of Gondwanan theropod dinosaurs; new information from the late Cretaceous of Madagascar. *Journal of African Earth Sciences* **27**: 167–168.

Sereno, P. C. and Novas, F. E. 1993. The skull and neck of the basal theropod *Herrerasaurus ischigualastensis*. *Journal of Vertebrate Paleontology* **13**: 451–476.

Tykoski, R. B. and Rowe, T. 2004. "Ceratosauria". *In* Weishampel, D. B., Dodson, P. and Osmólska, H. (Eds), *The Dinosauria* (2nd Edition). Berkeley, California: University of California Press. pp. 47–70.

Varricchio, D. J. 1993. Bone microstructure of the Upper Cretaceous theropod dinosaur *Troodon formosus*. *Journal of Vertebrate Paleontology* **13**: 99–104.

Varricchio, D. J., Jackson, F, and Trueman and Clive, N. A. 1999. Nesting trace with eggs for the Cretaceous theropod dinosaur *Troodon formosus*. *Journal of Vertebrate Paleontology* **19**: 91–100.

Xing, X., Zhou, Z. and Wang, X.-L. 2000. The smallest known non-avian theropod dinosaur. *Nature* **408**: 205–208.

Xu, X., Tang, Z.-I. and Wang, X.-L. 1999. A therizinosauroid dinosaur with integumentary structures from China. *Nature* **399**: 350–354.

Xu, X., Wang, X.-L. and Wu, X.-C. 1999. A dromaeosaurid dinosaur with a filamentous integument from the Yixian Formation of China. *Nature* **401**: 262–266.

Zhang, F., Zhou, Z., Xu, X. and Wang, X.-L. 2002. A juvenile coelurosaurian theropod from China indicates arboreal habits. *Naturwissenschaften* **89**: 394–398.

9

Chapter 10

While browsing through dinosaur books in a local used-book store, you notice a recurring theme in many illustrations from older books. The largest dinosaurs are shown submerged in water, with only their long necks and perhaps shoulders above the water surface. You recall reading in more recent books that these same dinosaurs were land dwellers. You also remember that their long necks were supposedly more adapted for eating leaves from the tops of tall trees, than as "periscopes" for an aquatic lifestyle. You look through more books and see illustrations of these same dinosaurs sitting back on their hind legs, reaching for the higher parts of trees.

What evidence supports land being the habitat for these dinosaurs? And why were their necks so long in comparison to other dinosaurs? What evidence supports that their neck length was related to tree heights? Does any evidence support that they ate vegetation from the tops of trees or that they sat on their hind legs to reach it?

Sauropodomorpha

10

Why Study Sauropodomorphs?

In 1883, O. C. Marsh provided the first, classic appraisal of *Apatosaurus*, which at that time was known more widely as *Brontosaurus*:

> *A careful estimate of the size of Brontosaurus . . . shows that when living the animals must have weighed more than twenty tons. The very small head and brain, and slender neural cord, indicate a stupid, slow-moving reptile. The beast was wholly without offensive or defensive weapons, or dermal armature. In habits, Brontosaurus was more or less amphibious, and its food was probably aquatic plants or other succulent vegetation.*

Beginning with this description, the perceptual legacy for sauropods was of large, stupid, slow, defenseless animals that frequented water bodies and ate soft foods. However, knowledge gained since 1883 about sauropods and their sauropodomorph relatives, the prosauropods, has resulted in radical revisions to most of the concepts in Marsh's original assessment of *Brontosaurus*. Nonetheless, sauropodomorphs are still not regarded as the most intelligent of vertebrates, especially when their brain size is compared to their overall body size. Regardless, they were certainly intelligent enough for evolutionary success in their respective environments for about 140 million years.

Marsh is still correct as far as size is concerned: some sauropodomorphs evolved into the largest animals that ever dwelled on land, and some may have weighed more than 50 tonnes. Because all sauropodomorphs seemingly were vegetarians, they would have included the largest terrestrial herbivores that ever lived. Their unparalleled gigantism presents an interesting puzzle for evolutionary scientists. What sorts of genetic and environmental factors caused selection for this body size in some lineages, beginning in the Late Triassic and continuing through the Late Cretaceous? Additionally, the impact of these huge herbivores on Mesozoic plants and their ecosystems must have been considerable. What types of plants could they have eaten that would have grown back quickly enough to sustain subsequent generations? Lastly, their mere movement over the land would have caused noticeable changes to the habitat. The closest modern model to the ecological and physical impact of sauropods is a herd of elephants, but this analogy seems to pale in comparison to the vision of herds of 50-tonne herbivores walking together across a Mesozoic landscape and dining on its flora.

Movement is another revised and refined concept in the study of sauropodomorphs. This applies not just to locomotion but also to other aspects of their anatomy, specifically the neck and tail. The relatively tiny heads of some large sauropods make complete sense as adaptations to their extremely long necks (one species had 19 cervical vertebrae!), but why their necks were so long is a valid evolutionary mystery to ponder. Likewise, the tails of some were as long as their necks, which collectively made some of them the lengthiest animals. So why did they also have such an extreme number of caudal vertebrae? Recent and not-so-recent research on sauropodomorph necks and tails resulted in some hypotheses that partially answer these questions. These hypotheses then encourage continuing debate on the role sauropodomorphs played in Mesozoic ecosystems and how they interacted with plants, their predators, and one another.

Sauropodomorphs thus represent evolutionary experiments on a grand scale through both time and space, because of their unparalleled sizes coupled with the longevity of their lineages. Adaptations in their skeletal architecture were astonishing, resulting in a balance of lightening and strengthening of a support system for massive muscles and organs. The fact that they not only moved easily with these adaptive features but also mated, made nests, laid eggs, browsed for food, and possibly migrated long distances is borne out of their widespread record of both bones and trace fossils. Paleontologists today examine these fossils to better understand their subsequent, lasting effect on the Earth.

Definition and Unique Characteristics of Sauropodomorpha

Sauropodomorpha (= "lizard-foot form") is a stem-based clade within Saurischia, like its sister clade Theropoda. Within Sauropodomorpha are two other stem-based sister clades, **Prosauropoda** and **Sauropoda**, which also had a common ancestor (Fig. 10.1). Prosauropods were originally interpreted as ancestral to sauropods, but recent analyses have supported their separate lineages from a common ancestor.

Sauropodomorpha is distinguished as a clade on the basis of some of the following synapomorphies (Fig. 10.2):

- Large nares.
- Distal part of the tibia covered by an ascending process of the astragalus.
- Short hind limbs in comparison to the torso length.
- Three or more sacral vertebrae.

10

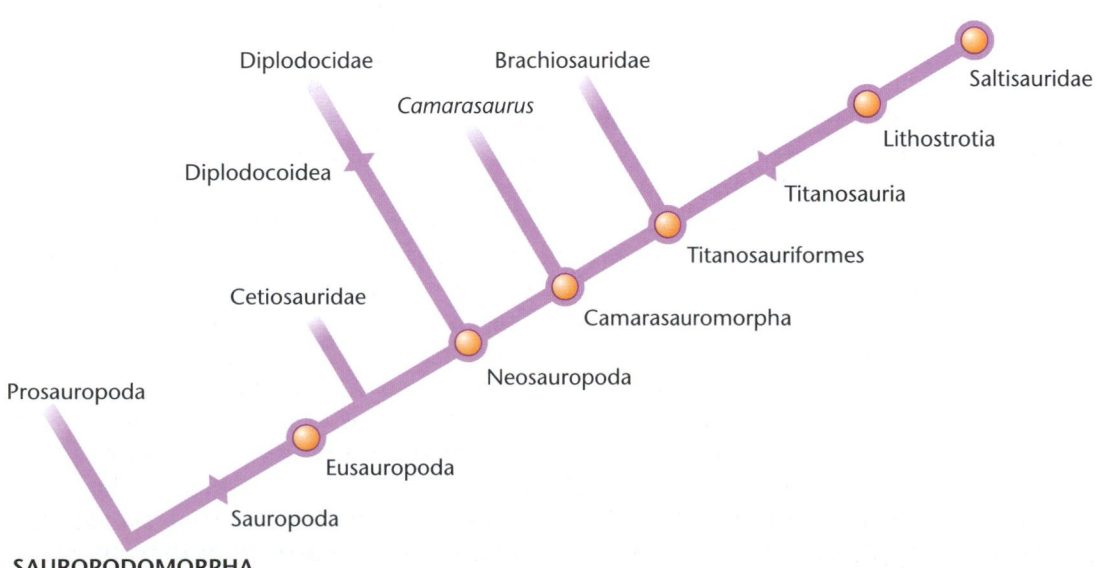

FIGURE 10.1 Cladogram for Sauropodomorpha, with major clades (Prosauropoda, Sauropoda) and clades within Prosauropoda and Sauropoda.

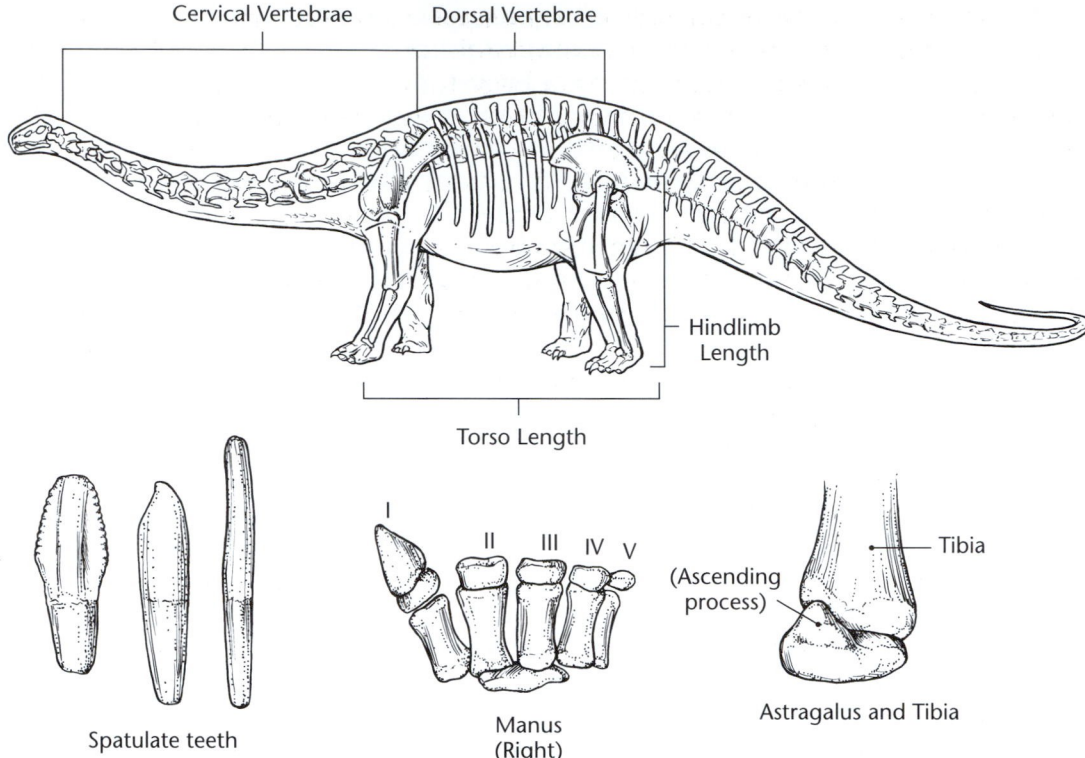

FIGURE 10.2 Important characters for Clade Sauropodomorpha: distal part of the tibia covered by an ascending process of the astralagus, short hind limbs in comparison to the torso length, spatula-like teeth with bladed and serrated crowns, 10 elongated cervical vertebrae along with 15 dorsal vertebrae (25 presacrals), large digit I on manus.

- Thin and flat (spatula-like) teeth with bladed and serrated crowns.
- Minimum of 10 cervical vertebrae that are typically elongated and 25 pre-sacral vertebrae.
- Large digit I on the manus.

Sauropodomorphs had small skulls in relation to their postcranial skeletons, with the skull being less than half the femur length. This disparity is especially apparent in larger members of the clade.

Modifications to the limbs and torsos of most sauropodomorphs reflect adaptations to load-bearing, whether it was as bipedal or quadrupedal animals. Most sauropodomorph species tended toward sizes of at least several metric tons, so such adaptations were necessary for the more massive species. None approached the smaller sizes exhibited by some theropods, such as the troodontids, compsognathids, and feathered coelurosaurs (Chapter 9). Accordingly, femurs, tibia, humeri, radii, ulnas, metatarsals, metacarpals, and phalanges were normally robust, although with some exceptions. In some cases the proximal bones reflect sites for the attachment of huge muscles (see Fig. 5.11). Those sauropodomorphs that were obligate quadrupeds should have had stout metacarpals and phalanges on the manus adapted for bearing weight. In contrast, if any of the phalanges on the manus seem more delicate and adapted for grasping, then a sauropodomorph was more likely bipedal or at least facultatively bipedal.

The relatively puny teeth and jaws of sauropodomorphs were definitely not adapted for much grinding or other oral processing of plant material. The small skulls, with their lack of evidence for attachments of large masticatory musculature, corroborate their inability to chew much food. Instead, the teeth and jaws seemed more suited for raking and shearing foodstuff before sending it down often-long necks to the rest of the alimentary canal. The long torso of most sauropodomorphs relative to their hind-limb lengths also provided more room for a gut that digested large amounts of plant material. This adaptation was reflected by prodigious body sizes beginning in the Middle Jurassic and lasting until the end of the Cretaceous.

Because sauropodomorph skulls were disproportionately the smallest in comparison to body size among all dinosaurs, the observation that they had the lowest EQs should also come as no surprise. Sauropodomorphs also had an anatomical predisposition to a taphonomic bias. Because their skulls were only held in place by a small cervical vertebra, the atlas, they were prone to detaching and becoming separated from the rest of their voluminous bodies. This resulted in numerous headless sauropodomorph skeletons. One hypothesis proposed previously for the rarity of sauropodomorph skulls was that predators preferentially ate them for their brain matter. However, this assertion is belied by the extremely small amount of brain matter that would have been gained from such discriminatory eating. A more likely explanation is that both physical and biological processes disaggregated the small, relatively delicate bones of a typical sauropodomorph skull once it was separated from the rest of the body. This left sturdy vertebrae and limb bones as the most likely candidates for burial, permineralization, and subsequent preservation (Chapter 7). Indeed, skull parts are represented in only 24 of the more than 100 genera of sauropods.

Another common characteristic of sauropodomorphs was the variety of innovative adaptations reflected by their vertebrae, including pleurocoels.

Pleurocoels were lateral spaces on the vertebrae that lessened the density of these already weighty bones, thus lightening their skeletal structure. These served a function similar to pneumatization in theropods (Chapter 9) and they similarly may have been filled with air sacs. Regardless, these structures aided in decreasing the mass per unit volume of the vertebrae. Transverse processes, neural spines, and chevrons, which were elongate processes on the ventral surfaces of the caudal vertebrae, added to the ornate appearance of many sauropodomorph vertebrae. These features provided attachment sites for musculature but in the case of chevrons also prevented damage to veins and arteries in the tail. With such complex and variable forms, sauropodomorph vertebrae are useful for identifying species, especially with regard to their centra. For example, a sauropodomorph vertebra that has the "positive" (ball) part of the centrum anterior and the "negative" (socket) part posterior is termed **opisthocoelus**. The opposite situation is called **procoelous**. If both parts are sockets, then it is called **amphicoelous**. Simple identification of such conditions in vertebrae, with no other body parts present, can immediately help paleontologists narrow down the possible choices of sauropodomorphs to which the bones belong.

An interesting logistical problem associated with the study of sauropodomorphs is related to how they are normally the largest of dinosaurs, either as solitary skeletons or in an assemblage. Consequently, they are the least likely dinosaurs to be recovered from their discovery site and studied later in more detail. When given a choice between transporting a 1-meter long theropod skeleton and a 20-meter long sauropod from a remote field area, a paleontological crew will have no difficulty coming to their decision. So, although sauropod bones are relatively common in a few areas of the world, sauropodomorphs have not received as much detailed

10

description as theropods. Fortunately, some spectacular sauropodomorph skeletons were recovered from sites in remote areas of Argentina, Tanzania, Egypt, Mongolia, and China, and these finds were brought back to laboratories and museums. As a result, paleontologists have a better understanding of sauropodomorph evolutionary history and paleobiology than would have been possible from only field descriptions.

A combination of body and trace fossil evidence unique to sauropodomorphs can aid in further exploration for their remains and can better illuminate their lifestyles. Sauropodomorph trace fossils, particularly tracks, are well documented, even from areas where their bones are uncommon (Chapter 14). In fact, both prosauropod and sauropod tracks have been reliably identified. Some of these tracks are the largest ones ever left by an animal in any environment in the history of the Earth. Wide, deep, and semi-circular pothole-like features in Mesozoic rock are difficult-to-miss clues to the former presence of a sauropodomorph (see Fig. 14.4B). In places where large numbers of them walked, the disturbed sediment was permanently deformed. Other trace fossils of sauropodomorphs include gastroliths, which are potentially very abundant when recognizable. Gastroliths have been found in at least one sauropod skeleton and are also associated with some prosauropod skeletons. Where bones or tracks were not preserved, the concentration of gastroliths in some deposits may be the only sign that sauropodomorphs were in an area. Sauropodomorph toothmarks are unknown, and some coprolites, which are appropriately rather large, have been only tenuously attributed to this group. Hence, neither of these trace fossils are currently useful for studying sauropodomorphs. In contrast, spectacular examples of sauropodomorph nests, allied with body fossils such as eggs, bones of hatchlings, and embryos, have been described recently. Such trace fossils can thus provide evidence of a former sauropodomorph presence in areas where adult bones may be absent.

Details of their fossil record show that size is not all that matters in the definition of sauropodomorphs. We gain a better picture of sauropodomorphs as real animals through the fossil remains of their:

1 often-large bodies coupled with small heads;
2 tooth and jaw arrangements;
3 limb structures;
4 complex vertebral anatomy; and
5 tracks, gastroliths, nests, eggs, and remains of young offspring.

In many instances, whether dealing with prosauropods or sauropods, paleontologists have enough material to study. Nonetheless, the discovery of new sauropodomorph species, especially with attached skulls, is always welcomed and provides yet more insight into their uniqueness in the history of animals.

Clades and Species of Sauropodomorpha

Prosauropoda

The oldest sauropodomorph found so far in the geologic record is the prosauropod *Saturnalia tupiniquim*, which was recovered from Upper Triassic rocks in Brazil. Prosauropoda includes some of the first dinosaurs ever named: *Thecodontosaurus* in 1836 and *Plateosaurus* in 1837 (Chapter 3), both of which were also from the Late Triassic. Some prosauropod bones were discovered as early as 1818 in Connecticut,

TABLE 10.1 Representative genera of Prosauropoda with approximate geologic age and where they occur.

Genus	Age	Geographic Location
Ammosaurus	Early Jurassic	Western and eastern USA
Anchisaurus	Early Jurassic	Eastern USA
Azendohsaurus	Late Triassic	Morocco
Camelotia	Late Triassic	England
Coloradisaurus	Late Triassic	Argentina
Euskelosaurus	Late Triassic	South Africa, Zimbabwe
Jingshanosaurus	Early Jurassic	China
Lessemsaurus	Late Triassic	Argentina
Lufengosaurus	Early Jurassic	China
Massospondylus	Late Triassic	South Africa
Melanorosaurus	Late Triassic	South Africa
Mussasaurus	Late Triassic	Argentina
Plateosaurus	Late Triassic	Germany
Riojasaurus	Late Triassic	Argentina
Ruehleia	Late Triassic	Germany
Saturnalia	Late Triassic	Brazil
Sellosaurus	Late Triassic	Germany
Thecodontosaurus	Late Triassic	England
Yimenosaurus	Early Jurassic	China
Yunnanosaurus	Early Jurassic	China

but they were initially misidentified as human bones. A later diagnosis showed them as rightfully belonging to *Anchisaurus* of the Early Jurassic. Regardless of any such missteps, paleontologists immediately recognized the kinship of prosauropods and sauropods. Because the former preceded sauropods in the geologic record, they were considered as ancestral to sauropods. For this reason, they were given the name "Prosauropoda," first dubbed in 1920 by Friedrich von Huene (Chapter 3). Although current evidence indicates that prosauropods were not ancestors of sauropods, like many other mislabeled dinosaur taxa the name has stuck and will likely continue to be used indefinitely. Compared to other major clades of dinosaurs, prosauropods were not only early on the scene but relatively short-lived; their range was from the Late Triassic through to the Early Jurassic. Different prosauropod species went extinct by the end of the Triassic or the end of the Early Jurassic, but the clade as a whole was relatively diverse throughout its 50-million-year-old history (Table 10.1).

Plateosaurus is perhaps the archetypical prosauropod because it is both the most abundant in the fossil record and best studied, thanks to its numerous remains in Late Triassic rocks of Europe. Its skull and postcranial elements also have some important distinguishing features (Fig. 10.3A), which include:

- Small but long skull that is relatively compressed laterally.
- Postorbital part of skull turned downward.
- Large nares but small, laterally placed orbitals.
- Relatively large olfactory bulb, as determined from endocranial casts.

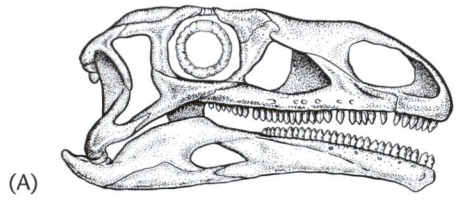

(A)   (B)

FIGURE 10.3 (A) Skull of Late Triassic prosauropod *Plateosaurus* of Germany. Anatomical features of the skull. (B) Complete specimen; Naturhistoriches Museum Basel, Basel, Switzerland. See Figure 5.3 for overall anatomy.

- Premaxilla and maxilla together containing as many as 36 teeth, whereas dentary contains as many as 28.
- Roughly serrated and leaf-like teeth.

The postcranial features (Fig. 5.3) of *Plateosaurus* include:

- Long neck in comparison to most theropods, yet shorter than those of most sauropods, with an atlas and a total of 10 cervical vertebrae.
- 15 dorsal, 3 sacral (attached to the pelvic bones), and nearly 50 caudal vertebrae.
- Weight seemingly associated more with its posterior.
- Phalangeal formula on pes of 2-3-4-5-1, with small unguals, and phalanx on digit V seemingly vestigial.
- Phalangeal formula on manus of 2-3-4-3-2, with unguals present on digits I through to III.
- Enlarged ungual on digit I that is deviated from the rest of the manus (discussed later).

Using functional morphology as a guide for examining both the skull and postcranial skeleton, the anatomical data suggest the following interpretations:

1 *Plateosaurus* was a mostly quadrupedal but facultatively bipedal animal;
2 it could raise up on its hind limbs to grasp and tear off plant materials, such as in trees;
3 it could bring plant material closer to its mouth using its clawed hands; and
4 it could shear this plant material with its teeth and jaws, with minimal chewing before swallowing.

The long neck also apparently aided in this high-browsing mode of life, as it could have reached vegetation out of the reach of other large herbivorous contemporaries in the Late Triassic. In this sense, during the Late Triassic and Early Jurassic, *Plateosaurus* and most other prosauropods might have been the ecological equivalent of modern giraffes.

In most respects, other prosauropods only varied slightly from this basic model, and the few important differences probably reflect adaptations to different environmental conditions and niches. For example, the Late Triassic *Riojasaurus* of Argentina was considerably larger (11-m long) than *Plateosaurus* (8-m long). This size difference suggests that *Riojasaurus* adapted to different plant foods or otherwise gained an advantage from being larger, perhaps as a defensive measure against increasingly larger ceratosaurian predators. The Early Jurassic *Yunnanosaurus* of China had pencil-like teeth more adapted for raking vegetation, which differed from the probable shearing function of the serrated teeth in *Plateosaurus*. *Anchisaurus*, *Sellosaurus*, and *Thecodontosaurus* had relatively larger orbits and smaller nares than *Plateosaurus*. Perhaps most importantly, they were smaller than most other prosauropods and accordingly could probably move faster. This interpretation is augmented by the ratio of metatarsal III (the middle one) to the tibia of each prosauropod species, where higher ratios should reflect abilities for faster movement. For example, *Riojasaurus* (= slower) has a ratio of about 0.4, whereas *Anchisaurus* is close to 0.7 (= faster).

Prosauropods as a clade represent a good start for the sauropodomorph lineage and their longevity in the geologic record was certainly an indicator of their success, even though sauropods surpassed them by the end of the Early Jurassic. Whether the Early Jurassic arrival of other large herbivores, such as sauropods and thyreophorans (Chapter 12), was part of an ecological displacement of prosauropods is unknown. Nevertheless, their extinction was the first one for any major dinosaur clade. This circumstance provoked comparisons to the paleoenvironmental conditions associated with later dinosaur extinctions and aided in working out any similarities in patterns or trends.

Sauropoda: Diplodocoids, Camarasauromorphs, Titanosaurs, and Others

The first discovered sauropod, *Cetiosaurus*, came from the Middle Jurassic of England and was described and named by Richard Owen in 1841. Unfortunately for dinosaur studies, he thought that it was a large, whale-like marine reptile. Not until later in the nineteenth century did investigators link it to other sauropods, such as those found in the western USA (Chapter 3). The oldest sauropods are from the Late Triassic (see Table 10.2), indicated by the fragmentary remains of *Blikanasaurus*, "*Euskelosaurus*", and *Antetonitrus* of South Africa and *Isanosaurus* of Thailand. *Blikanasaurus* and one example of "*Euskelosaurus*" were originally

TABLE 10.2 Representative genera of Sauropoda with approximate geologic age and where they occur.

Genus	Age	Geographic Location
Andesaurus	Early Cretaceous	Argentina
Alamosaurus	Late Cretaceous	Western USA
Amargasaurus	Early Cretaceous	Argentina
Ampelosaurus	Late Cretaceous	France
Antarctosaurus	Late Cretaceous	Argentina, Chile, Brazil
Apatosaurus	Late Jurassic	Western USA
Aragasaurus	Late Jurassic	Argentina
Argentinosaurus	Late Cretaceous	Argentina
Atlasaurus	Middle Jurassic	Morocco
Austrosaurus	Early Cretaceous	Australia
Barosaurus	Late Jurassic	Western USA, Tanzania
Barapasaurus	Middle Jurassic	India
Bellusaurus	Late Jurassic	China
Blikanasaurus	Late Triassic	South Africa
Brachiosaurus	Late Jurassic	Tanzania, western USA, Portugal
Camarasaurus	Late Jurassic	Western USA
Cedarsaurus	Early Cretaceous	Western USA
Cetiosaurus	Middle Jurassic	England
Datousaurus	Middle Jurassic	China
Dicraeosaurus	Late Jurassic	Tanzania
Diplodocus	Late Jurassic	Western USA
Euhelopus	Late Jurassic	China
Haplocanthosaurus	Late Jurassic	Western USA
Hypselosaurus	Late Cretaceous	France, Spain
Isanosaurus	Late Triassic	Thailand
Jobaria	Early Cretaceous	Niger
Kotasaurus	Early Jurassic	India
Lapparentosaurus	Middle Jurassic	Madagascar
Lourinhasaurus	Late Jurassic	Portugal
Malawisaurus	Early Cretaceous	Malawi
Mamenchisaurus	Middle Jurassic	China
Nemegtosaurus	Late Cretaceous	Mongolia
Neuquensaurus	Late Cretaceous	Argentina, Uruguay
Nigersaurus	Early Cretaceous	Niger
Omeisaurus	Middle Jurassic	China
Opisthocoelicaudia	Late Cretaceous	Mongolia
Paralititan	Late Cretaceous	Eygpt
Patagosaurus	Middle Jurassic	Argentina
Pleurocoelus	Early Cretaceous	Western USA
Rapetosaurus	Late Cretaceous	Madagascar
Rayosaurus	Early Cretaceous	Argentina
Rhoetosaurus	Middle Jurassic	Australia
Saltasaurus	Late Cretaceous	Argentina
Sauroposeidon	Early Cretaceous	Western USA
Seismosaurus	Late Jurassic	Western USA
Shunosaurus	Middle Jurassic	China
Supersaurus	Late Jurassic	Western USA
Titanosaurus	Late Cretaceous	India
Tornieria	Late Jurassic	Tanzania
Vulcanodon	Early Jurassic	Zimbabwe

interpreted as prosauropods, but a more careful examination of their traits indicates that they are more likely basal sauropods. Late Triassic sauropods are also identified by their probable tracks in Upper Triassic strata of the southwestern USA. Until the preceding revelations, which have only taken place in the past five years, the earliest known sauropod was the Early Jurassic *Vulcanodon* of Zimbabwe.

> *Some sauropod genera actually overlapped in size or were smaller than the largest hadrosaurs of the Late Cretaceous (Chapter 11). Therefore the generalization that "sauropods were the largest dinosaurs" does not take into account those exceptions.*

Sauropods would last from the Late Triassic through to the Late Cretaceous, and have earned numerous names that incorporate synonyms of "big": leviathans, behemoths, enormous, gigantic, colossal, stupendous, massive, and so on. These superlative expressions of awe made their way into the Linnaean names of sauropod genera, such as *Supersaurus*, *Seismosaurus*, and *Paralititan*. Although the adults of some sauropod species did reach exceptional sizes, larger than any other terrestrial animals that have ever lived, size alone is too general a characteristic to distinguish sauropods as a dinosaur group.

Some of the important traits of Sauropoda that help to distinguish it from its sister clade Prosauropoda include attributes of both the axial and appendicular skeletons. For the axial skeleton, the following are notable:

- A change of some of the anterior dorsal vertebrae into cervical vertebrae, a process called **cervicalization** by some paleontologists, resulting in 12 or more cervical vertebrae.
- Neural arches and other processes forming better support for musculature as struts.
- Sacrum with four or more vertebrae.
- Large and separate pelvic bones, including an ilium with an expanded anterior end.
- Minimum of 45 to more than 80 caudal vertebrae that were sometimes elongate, which led to some extremely long tails.

The appendicular skeleton of sauropods was characterized by the following:

- A large coracoid attached to the scapula, with a widened anterior end.
- Radius and ulna shorter than the humerus, and a femur 10–20% longer than the length of the humerus (with the notable exception of brachiosaurids).
- Denser limb bones in general.
- Large astragalus coupled with a smaller calcaneum (the latter absent in diplodocids), where the astragalus fits snugly with the tibia.
- Pes with five digits and unguals on digits I to III.
- Manus with five digits and smaller phalanges on digits II to V.
- Ungual on digit I of manus (with the exception of a few Cretaceous species in which it is missing).

A gradual change in the cranial anatomy of sauropods that became more apparent through their evolutionary history was a migration of their nares from the anterior part of the skull to farther away from the mouth. This placed the nostrils of some sauropods on top of their heads (Fig. 10.4). This feature was once interpreted as an adaptation to an aquatic lifestyle, whereby nostrils on the highest point of the head aided in breathing while the majority of the body was submerged.

FIGURE 10.4 Skulls of the Late Jurassic diplodocids *Apatosaurus* (left) and *Diplodocus* (right) showing dorso-ventral positioning of nares in relation to the anterior portion of their skulls. Dinosaur National Monument, Vernal, Utah. Contrast with skull of *Plateosaurus* in Figure 10.3.

Independent evidence has since refuted this hypothesis (as discussed later), but the nares and their association with possible soft-tissue structures is still an area of speculation and controversy in sauropod research.

The cranial capacity of sauropods was rather limited: make a fist and it will approximate the size of the largest sauropod brain, no matter what its tonnage may have been. In fact, the cranial anatomy seemingly was a continuation of the evolutionary theme explored by prosauropods, in that it was simply a conduit for passing food to the gut and air to and from the lungs. Overall profiles of sauropod skulls can be broadly divided into:

1 rounded skulls with leaf-like (but thickened) teeth, as seen in brachiosaurids and camarasaurids; or
2 elongate skulls with pencil-like teeth, exemplified by diplodocids.

Teeth in any of the sauropods were typically non-serrated, as opposed to those of most prosauropods. Evidence from sauropod skulls relating to sensory abilities are not well defined, although some of them had well-developed orbits and nares.

Sauropods used to be classified into four major clades and a polyphyletic group: Diplodocidae, Brachiosauridae, Camarasauridae, Titanosauridae, and "Cetiosauridae." Recent cladistic analyses have resulted in a reclassification of these and other sauropod groupings. Node-based clades within this new classification include Eusauropoda, Neosauropoda, Camarasauromorpha, Titanosauriformes, Lithostrotia, and Saltasauridae (Fig. 10.1). With this reorganization, Brachiosauridae and Titanosauridae are now within Titanosauriformes, Camarasauridae is within Camarasauromorpha, Diplodocidae is within Neosauropoda, and the formerly polyphyletic "Cetiosauridae" more likely constitutes a monophyletic one. In fact, *Cetiosaurus*, *Barapasaurus* (Lower Jurassic of India), and *Patagosaurus* (Middle Jurassic of Argentina) are regarded as members of this clade.

Neosauropoda includes the stem-based clade Diplodocoidea. Diplodocoidea is partially represented by diplodocids, probably the best known of sauropods. Diplodocids include familiar North American dinosaurs such as *Apatosaurus*, *Barosaurus*, *Supersaurus*, and *Seismosaurus*, and the eponymous *Diplodocus* (Fig. 10.5), all from the Late Jurassic. Cretaceous examples of diplodocoids include the Early Cretaceous *Amargasaurus* of Argentina (Fig. 10.6), *Rebbachisaurus* of Morocco, and *Nigersaurus* of Niger, as well as the Late Cretaceous *Nemegtosaurus* of Mongolia and *Rayososaurus* of Argentina.

(A)

(B)

FIGURE 10.5 *Diplodocus*, the Late Jurassic sauropod that inspired a pub song.
(A) Skeletal reconstruction; Denver Museum of Science and Nature. (B) Left pes and ankle of *Diplodocus*, showing the large ungual on digit I.

FIGURE 10.6 *Amargosaurus*, an Early Cretaceous sauropod from Argentina with unusually long vertebral processes. Trelew Museo de Paleontológica, Trelew, Argentina.

Diplodocoids can be distinguished from other sauropod clades by:

1 relatively more cervical and caudal vertebrae than other sauropods, adding up to some considerable body lengths;
2 teeth restricted to the anterior portion of the skull; and
3 nares dorsal to the orbits.

The caudal vertebrae were so numerous (70–80) and small distally that the tails tapered into a whip-like structure. The proximal caudal vertebrae had prominent, horizontally-oriented chevrons developed on the ventral parts of the vertebrae, presumably for added support and protection of blood vessels. With their cervical, dorsal, and caudal vertebrae added together and projected into a possible total length for each genus, *Seismosaurus* and *Supersaurus* may have been the longest dinosaurs. By some conservative estimates, these sauropods were about 27 meters long, but lengths of more than 30 meters have been also hypothesized.

Because other sauropods had similarly long necks and tails, such as the Middle Jurassic *Omeisaurus* and Late Jurassic *Mamenchisaurus* of China, they were originally interpreted as sharing ancestry with diplodocids. *Omeisaurus* and *Mamenchisaurus* had 17 and 19 cervical vertebrae, respectively, making them animals with necks that were about half of their total body lengths. However, cladistic analyses show that *Omeisaurus* and *Mamenchisaurus* are in clades outside of Neosauropoda, although they are in Eusauropoda.

Of the clade Camarasauromorpha, probably the most common sauropod species, and accordingly one of the best studied, is the Late Jurassic *Camarasaurus* of the western USA. *Camarasaurus* (= "chambered lizard"), so named because of its relatively spacious skull, is known from numerous well-preserved and articulated specimens,

FIGURE 10.7 Cast of nearly complete juvenile *Camarasaurus* from the Late Jurassic of Utah, USA. Dinosaur Adventure Museum, Fruita, Colorado.

from juveniles to adults. These were all discovered in the Morrison Formation of Utah, Colorado, and Wyoming (Fig. 10.7). *Camarasaurus* has a more rounded skull than diplodocids and most other sauropods, as well as spoon-like teeth. The cervical, dorsal, and caudal vertebrae were also more conservative in number, resulting in body proportions somewhat in between the extremes presented by diplodocids in length and brachiosaurids in height. Two sauropod genera that were once considered as close relatives of *Camarasaurus* were *Euhelopus* from the Late Jurassic of China and *Opisthocoelicaudia* from the Late Cretaceous of Mongolia. However, *Euhelopus* is now considered too primitive and *Opisthocoelicaudia* too derived; the former is outside of Neosauropoda, and the latter is in Saltasauridae. Regardless, Camarasauromorpha has a geologic range from the Late Jurassic through to the Late Cretaceous on the basis of *Camarasaurus* and *Opisthocoelicaudia*, among others.

Titanosauriformes includes another well-known sauropod genus, *Brachiosaurus*, the representative genus of brachiosaurids. *Brachiosaurus* (= "arm lizard"), named with respect to its long forelimbs, occurs in Upper Jurassic strata of the western USA and Tanzania. *Cedarsaurus* and *Sauroposeidon* are also brachiosaurids, both coming from the Lower Cretaceous of the western USA. Like *Camarasaurus*, brachiosaurids had more rounded skulls than diplodocids and their nares were positioned more anteriorly and below the orbits. Brachiosaurid necks also had long cervical and dorsal vertebrae with well-developed pleurocoels (Fig. 10.7). Furthermore, they had a reduced digit I in the manus, which may have been an adaptation to more weight bearing on the fore limbs. *Brachiosaurus* was perhaps the most massive and tallest of all dinosaurs, weighing as much as 50 tonnes. Less conservative estimates have placed it at 80 metric tons, a difference of nearly 60%. It also had a neck that, when extended fully vertical, was close to 13 meters high. In part, the height of *Brachiosaurus* was related to its lengthened fore limbs, where its humerus exceeded the femur length. A sort of trade-off between height and length in diplodocids and

10

brachiosaurids is evident. For example, *Brachiosaurus* had shortened caudal vertebrae and was consequently shorter overall as compared to the diplodocoid *Apatosaurus*, which it outweighed by at least 2 : 1.

Titanosauriformes contains members of the last of the great sauropods, represented within the stem-based clade Titanosauria. Titanosaurs had several distinguishing features, but one of the most surprising was in the skin of some genera: dermal armor and dorsal spines. Before the discovery of titanosaurs, sauropods were regarded as defenseless animals with no natural protection against predation other than size or staying in herds. Because titanosaurs are the only sauropods to have developed osteoderms and most lived in the Cretaceous, some paleontologists have proposed that natural selection from predation pressures led to a "Red Queen" type of coevolution (Chapter 6). In other words, their large size would have helped to counter the increasingly larger and bigger-toothed theropods, such as *Giganotosaurus* and *Carcharodontosaurus*, that lived in the same areas as these sauropods (Chapter 9). Procoelous caudal vertebrae also help to identify titanosaurs. Fortunately, this is a unique trait because no complete skeleton of a titanosaur, let alone a complete skull, has ever been found in association with their vertebrae or limb bones. Some examples of titanosaurs are: *Titanosaurus*, *Antarctosaurus*, *Argyrosaurus*, *Neuquensaurus*, and *Saltasaurus* of Argentina; *Janenschia*, *Malawisaurus*, and *Paralititan* of eastern Africa; *Phuwiangosaurus* of Thailand; and *Austrosaurus* of Australia. These sauropods collectively hail from the former southern continent of Gondwana, indicating that titanosaurs mostly proliferated and diversified there. However, one titanosaur did make it from part of Gondwana (South America) into North America – the Late Cretaceous *Alamosaurus* of the western USA. Lastly, no description of titanosaurs is complete without mentioning what likely was one of the largest sauropods that ever lived, the Early Cretaceous titanosaur *Argentinosaurus*. This gigantic sauropod was known initially through just one prodigious dorsal vertebra and a few other fragments, but the discovery of more bones has helped to reconstruct it (Fig. 10.8).

In summary, sauropods are represented by a large number of clades that were only recently well-defined by detailed cladistic analyses. The results of these analyses caused considerable changes to how sauropods were classified, which means that many genera were reassigned to clades different from their traditional taxonomic groupings. The intention of such revisions is to better understand

FIGURE 10.8 Cast and reconstruction of the Late Cretaceous titanosaurid *Argentinosaurus*, which may have been the largest dinosaur (but it had a lot of competition). Fernbank Museum of Natural History, Atlanta, Georgia.

phylogenetic relationships within Sauropoda, which in turn is augmented by the geologic ages and geographic distribution of sauropod genera.

Paleobiogeography and Evolutionary History of Sauropodomorpha

Sauropodomorphs lived on every continent, including the former subcontinent of India. Antarctica was recently added to their paleobiogeographic range, with the discovery of remains from a still-unnamed Early Jurassic prosauropod.

With evidence gained from both prosauropod and sauropod body and trace fossils, paleontologists can confidently state that sauropodomorphs comprised a long-lived and widespread clade. Extending from the earliest of saurischian history in the Late Triassic (slightly less than 230 Ma) to the end of the Cretaceous, sauropodomorphs spanned the entire geologic range of the dinosaurs, although their numbers and diversity differed considerably throughout this 165-million-year-long interval.

Sauropodomorphs were similar to other dinosaurs in that they had broad habitat ranges, indicated by their skeletal material and trackways found in fluvial, lacustrine, swamp, deltaic, and coastal marine (tidal flat and shoreline) deposits. Sauropod trackways in particular show a preference for semi-arid regions and low latitudes. However, this circumstance may just reflect the optimal preservation conditions for tracks in these environments and is contradicted in part by their body fossil record.

Sauropodomorphs have an initial fossil record that is comparatively sparser than that of their herrerasaurid contemporaries (Chapter 9). The oldest known prosauropods are *Saturnalia* of Brazil, mentioned earlier, and *Azendohsaurus* from the Carnian of Morocco. These were followed later in the Carnian by the primitive sauropod *Blikanasaurus*, which is represented only by hind-limb parts, and three fragmentary specimens of the prosauropod *Melanorosaurus*. Both of these sauropodomorphs are from South Africa. A slightly younger but still Late Triassic (Norian) sauropod is *Antetonitrus*, also of South Africa. This preliminary concentration of early sauropodomorphs argues for their probable origin in, and dispersal from, Gondwana.

Prosauropods later became important components of Late Triassic terrestrial ecosystems, best represented by the relatively abundant *Plateosaurus* and the substantial *Riojasaurus*. As discussed before, the latter prosauropod was as long as 11 meters, making it the largest known herbivore on the Late Triassic scene. Prosauropods adapted admirably to consuming plants and otherwise filling their ecological niches from the Late Triassic to their demise at the end of the Early Jurassic. Sauropods, which had a more modest beginning in the Late Triassic, later supplanted prosauropods in what were presumably similar niches. Of course, the main food plants available for sauropodomorphs changed considerably throughout the Mesozoic, especially with the advent of flowering plants by the beginning of the Cretaceous.

Most dinosaur paleontologists now accept the common ancestry of prosauropods and sauropods. Nonetheless, the previous hypothesis that prosauropods were the ancestral group for sauropods was a reasonable one in light of the perceived stratigraphic separation of the first prosauropod and first sauropod body fossils. In this case, a little bone makes a big difference: the reduced digit V on the typical prosauropod pes most likely means that it is a vestigial digit. That is, it was reduced through natural selection from an originally much larger digit. However, a typical sauropod has a fully-developed digit V on its pes, so if prosauropods were

FIGURE 10.9 The Middle Jurassic cetiosaurid *Bellusaurus*, a smaller sauropod than most. Temporary display at Fernbank Museum of Natural History, Atlanta, Georgia.

ancestral, this vestigial digit made a heroic comeback and became large again. The more likely explanation is that the common ancestor to both prosauropods and sauropods had this large digit V as a plesiomorphic trait. It then became reduced in the prosauropod lineage after it diverged from the sauropod lineage, which retained the digit during its different evolutionary journey.

The main problem with resolving evolutionary relationships between prosauropods and sauropods was the previous lack of sauropod fossils from the Late Triassic. The gap between the appearance of the first prosauropod (229 Ma) and the first sauropod (208 Ma) was a significant one, and meant that sauropods would have had a ghost lineage of nearly 20 million years. However, the identification of *Blikanosaurus*, *Antetonitrus*, and some remains of "*Euskelosaurus*" as Late Triassic sauropods and sauropod tracks, recently identified in Upper Triassic rocks from New Mexico, has helped to close this gap.

Predator–prey interactions probably contributed to the coincidence of the largest land herbivores overlapping in time with the largest land carnivores that ever lived (Chapter 9). In modern terrestrial animals today, large size and herding behavior are still deterrents to predation. Trends toward gigantism that ecologically separated the sauropods from other potential large herbivore competitors, such as ornithopods (Chapter 11), stegosaurs and ankylosaurs (Chapter 12), and ceratopsians (Chapter 13), also cannot be discounted. Interestingly, prosauropods had almost no overlap with these latter herbivorous clades of dinosaurs, which along with sauropods also may have filled the various former niches of prosauropods. Only a few sauropods were relatively small compared to other herbivorous dinosaurs (Fig. 10.9).

Another hypothesis for sauropod gigantism is that sauropods became large so that their bodies could accommodate an alimentary canal long enough to digest massive amounts of food with low nutritional quality. Moreover, their increased dimensions, as a result of this adaptation, could have made more food sources easily accessible. Longer necks meant more reach, for example, whether in a horizontal or vertical plane. Yet another hypothesis is that the extremely long necks, tails, and legs of some sauropods served as "heat vents," meaning that these body parts conveyed the excess body heat generated by movement and digestion away from the torso. This is a controversial interpretation of sauropod physiology discussed earlier (Chapter 8). Of course, a synthesis viewpoint is that an interaction of numerous factors, such as changing climates, habitat alterations, and biological factors, would have provided for confluent situations that favored the selection of larger, longer, and heavier sauropods. Clearly this extreme sort of selection has not happened to terrestrial vertebrates since the Mesozoic, thus paleontologists can confidently say that the circumstances leading to this type of evolution must have been unique.

Sauropodomorphs as Living Animals

Reproduction

The logistics of sauropodomorph mating is difficult to imagine but obviously it did happen many times. The traces of this activity in itself are either undiscovered or perhaps cover such a large scale that paleontologists do not recognize them. However, the results of successful mating and conception, eggs and nests, provide evidence that sauropodomorphs were oviparous animals, despite some conjecture (with no accompanying data) that they gave birth to live young. In fact, the two oldest reported dinosaur nests are interpreted as belonging to prosauropods in Late Triassic rocks, which indicates that oviparity was a primitive trait in sauropodomorphs. One nest, attributed to *Massospondylus* because of juvenile bones and nearby remains of the adult species, contains a clutch of six eggs that were 5.5 cm wide and 6.5 cm long. The second nest contained eggshell fragments and five hatchlings of *Mussasaurus*, although no adult remains were associated with the eggs. No trace fossil evidence of the nest structure was described in either case. Consequently, whether prosauropods buried their clutches, laid eggs in open nests and brooded them, or otherwise showed any parental care is still unknown. A possible candidate for a parent prosauropod for *Mussasaurus* was *Coloradisaurus*, which occurs in strata of the same age and region, but the certainty of this parentage is less than that for the nest associated with *Massospondylus*.

Apparent sauropodomorph eggs are rare in any rocks older than the Cretaceous Period, and Cretaceous clutches are presumed to have belonged to sauropodomorph nest sites on the basis of the sheer size of the eggs. Sauropod eggs were tentatively identified on the basis of co-association of large and abundant dinosaur eggs with the remains of the titanosaurid *Hypselosaurus* in nineteenth-century France (Chapter 3). Taking this assumption further, some paleontologists proposed that dinosaur eggs with certain shell structures, when associated with large size (more than 5000 cm^3 volume) and spherical or subspherical overall shape, are attributable to sauropods (Chapter 8). These egg types occur in Argentina, France, India, and Spain, which indicates a similarity across a broad paleogeographic range. The problem with the assumption that these eggs are from sauropods is that the hypothesis was originally based on no associated embryos or hatchlings. Such suppositions have been repeatedly falsified with other dinosaur eggs and their presumed mothers (Chapters 9 and 11).

Assuming that some eggs are from sauropods, a few clutches show interesting patterns that were initially attributed to egg laying and nest building, but are more likely the result of taphonomic factors. The typical clutch size for these large eggs is 3 to 12, and clutches are apparently randomly distributed within conical structures. These structures were originally interpreted as possible excavations made by the dinosaur prior to laying the eggs. Another pattern from the Lower Cretaceous of France apparently indicated the pathways of five or six mother dinosaurs. In this interpretation, the dinosaurs walked along, squatted, and laid eggs in semicircles with 1.3- to 1.7-meter radii. The length, width, and radii of the egg assemblages suggest sauropod tracemakers, as does the size of eggs, although the apparent "clutch size" of 15 to 20 (assumed because they described each arc) differs from those described for other suspected sauropod nests. However, these detailed interpretations are probably wrong. The latest examination of the original site showed

FIGURE 10.10 Egg clutch of titanosaurids from the Late Cretaceous of Patagonia, Argentina. Temporary display at Fernbank Museum of Natural History, Atlanta, Georgia.

that the eggs actually were in the bottoms of ancient channel depressions. This means that the eggs were likely transported from their original nests and might also have more than one species of dinosaur represented. Whether their original makers were sauropods or other dinosaurs is still unknown.

No sauropodomorph embryos were known from the geologic record until the recent find of extraordinary sauropod nesting horizons in Late Cretaceous rocks of Patagonia, Argentina, a discovery that has provided a rare look into the individual and social behaviors of sauropod reproduction.

The nests recently discovered in Patagonia, some of which are definable as mound structures with upraised rims, are in multiple horizons of the same relatively small area (about 2.5 km^2), indicating **site fidelity**. Site fidelity means that the sauropods, represented by an unidentified species of titanosaur, came back to nest in the same place over multiple breeding seasons. The eggs in these horizons number in the thousands. The embryos are exquisitely preserved and some show skin impressions, the only ones known for any dinosaur embryos. Not only do these eggs provide a rare glimpse into the life history of a titanosaurid, but they are the same size and morphology as some of the previously unidentified Cretaceous dinosaur eggs found in such widely separated places as China, India, Europe, and Africa. (Fig. 10.10). This similarity may serve as a model for comparison and thus aid in the identification of other sauropod nests and embryos in the future. Furthermore, the nest structures, which were probably constructed by the sauropods, can aid in identifying other sauropod nests in the absence of eggs or embryos. This same search image serves as a tool for identifying some theropod nests as well (Chapter 9).

Growth

Only a few species of sauropodomorphs have juvenile specimens represented in the fossil record, with the exception of the aforementioned *Mussasaurus* hatchling and titanosaur embryos. Some smaller sauropods do show juvenile affinities, but some paleontologists have noted that the bias of the geologic record favors adult sauropodomorph body fossils. One explanation for this discrepancy is that juveniles were subject to worse preservation conditions; greater predation and scavenging may have destroyed their smaller bodies. A few researchers have also examined tracks for evidence of growth series among sauropodomorphs. Tracks are a promising source of future data regarding sauropod juveniles, as more juvenile sauropodomorph tracks are being recognized on the basis of their much smaller sizes. Although they can only provide relative age series, rather than absolute ages, tracks are nevertheless a start toward interpreting life histories.

Some growth series have been outlined for prosauropods based on body fossils from monospecific bone beds of *Plateosaurus*. For sauropods, a specimen of the Late Jurassic *Camarasaurus*, discovered by Earl Douglass in the Morrison Formation early in the twentieth century (Chapter 3), turned out to be one of the first undoubted, as well as exquisitely preserved, juvenile skeletons of a sauropod ever found (see Fig. 10.5). Its completeness indicates that it must have been immediately buried before its corpse was scavenged; only a few caudal vertebrae are missing from the specimen. A growth series has been approximated for *Camarasaurus* on the basis of some juvenile remains in comparison to the more abundant adults, but this series has nowhere near the completeness of those for some theropods (Chapter 9). The Late Jurassic diplodocoid *Apatosaurus* likewise has a poorly-defined growth series, so little is known about the life history of this sauropod, too.

Some bone histology work on sauropodomorphs suggests that rapid growth occurred early in life and possibly slowed once they reached adulthood. Fast initial growth was a probable mode because otherwise long ages would have been required for sauropods, such as *Brachiosaurus* and *Seismosaurus*, to reach their weights of 50 tonnes and lengths of 30 meters. If sauropods had slow growth throughout their entire lives, some would have had to live more than 100 years. Such long lifespans seem unlikely, considering the rarity of modern animals with a similar longevity, and indeed these estimates are based on reptilian growth rates. The most recent data on sauropod bones indicate that they grew rapidly, and thus many of the largest specimens were much less than 100 years old. Offspring must live long enough to reach sexual maturity in order to reproduce, so the longer this takes, the more likely the offspring will die before reproducing. Hence natural selection would have favored sauropods that reached sexual maturity earlier rather than later. This again implies that sauropod adulthood was probably reached in considerably less than 100 years. Indeed, recent research on *Apatosaurus* suggests that they achieved about 90% of their adult size after only 10 years.

Growth rates have implications for sauropodomorph physiology, which could have been endothermic, ectothermic, or homeothermic (Chapter 8). Supporters of the homeothermic hypothesis propose that the larger sauropods were inertial homeotherms. In other words, once they were heated by either their external environment or by internal, fermenting bacteria in their huge guts (Chapter 8), they maintained this temperature for a long time because of their very low (compared to other dinosaurs) surface area/volume ratio. As mentioned earlier, heat venting might have been needed to prevent overheating, and some paleontologists have suggested that the long necks, tails, and legs served as conduits for expelling excess heat. Regardless of the mechanism, sauropods would have required large amounts of energy to grow quickly during their younger years. Heat generation, as a by-product of such growth, is a reasonable hypothesis.

Locomotion

Sauropodomorph ancestors were bipedal saurischians, and the earliest-known prosauropods may have been at least facultatively bipedal. Limb lengths indirectly indicate bipedalism in fossil vertebrates, in that the fore limbs are considerably shorter than the hind limbs. However, when the limbs are equally long, the animal could still have been bipedal, so another clue lies in the manus anatomy. When the manus is composed of lighter and more delicate bones, it is not likely to have been used for load bearing. A third anatomical clue is indicated by the insertion point for muscles at an attachment site called the fourth trochanter, which is on the femur. Bipedal vertebrates have the insertion point on the proximal half of the femur, which is the case for nearly every theropod (Chapter 9). In contrast, a prosauropod such as *Plateosaurus* has the point on the middle to distal part of the femur, suggesting that this prosauropod was in between the end members of bipedal and quadrupedal. Sauropods have the insertion point even lower than that of *Plateosaurus*, so this feature along with other anatomical evidence indicates that sauropods were obligate quadrupeds.

Probable prosauropod tracks are interpreted on the basis of their close anatomical similarity to the skeletal structure of a typical prosauropod manus and pes. Tracks identified as having been made by prosauropods are:

1 found in strata of the same ages as known prosauropod skeletons;
2 show a reduced digit V in the pes impressions; and
3 have an enlarged ungual on a deviated digit I in the manus impression.

The majority of these trackways indicate quadrupedal locomotion, although a few bipedal trackways from the Late Triassic and Early Jurassic have been attributed to prosauropods on the basis of how they reflect the pes anatomy. Their trackways show that, like all other dinosaurs, these probable prosauropod trackmakers were diagonal walkers with their limbs held in planes under their bodies. Some dinosaur ichnologists have assigned names to "prosauropod" tracks under the assumption of those trackmakers. However, as with most trace fossils, taphonomic variants often impede positive and precise identifications of tracks with specific trackmakers (Chapter 14).

The realization that the largest of the sauropods did not live in bodies of water, but rather were completely adapted for an active terrestrial lifestyle, marked a major shift in the perception of dinosaurs in general. Prominent and influential paleontologists, such as O. C. Marsh, were the sources of those enduring images of diplodocoids and brachiosaurids immersed in swamps and munching on soft vegetation, thus demonstrating the power of argument by authority over evidence-based reasoning (Chapter 2). The hypothesis of amphibious behavior for the largest of sauropods was justified mostly on the basis of their size, as animals so massive were thought to be biomechanically incapable of supporting themselves on land.

> *Under normal atmospheric conditions at sea level, air presses down on the surface of every organism at the equivalent of 1.034 kg/cm², which is also known as a standard 1.0 atmosphere (atm).*

Atmospheric pressure (see box) is caused by a column of air that goes up into the stratosphere to a height of about 8.0 km. The pressure of a fluid (either air or water) is expressed by the following equation:

$$p = hD \qquad (10.1)$$

where p is pressure, h is the height of the fluid above the pressed-upon surface, and D is the density of the fluid. For example, if the atmosphere averages a density of 1.25 g/L and is in a column of air measuring about 7.8 km above an animal's skin, the air pressure would be:

(First converting liters to cubic centimeters)

Step 1. $p = (7.8 \text{ km})(1.25 \text{ g/L})(0.00 \text{ L/cm}^3)$

Step 2. $= (7.8 \text{ km})(0.00125 \text{ g/cm}^3)$

Step 3. $\quad = (7.8 \times 10^5 \text{ cm})(1.25^{-3} \text{ g/cm}^3)$

Step 4. $\quad = 975 \text{ g/cm}^2$

which is close to the measured value of 1034 g/cm² for 1.0 atm.

In contrast, if an organism is immersed in water, the considerably greater mass of water with respect to air (Chapter 7) means that pressure will increase in a much shorter distance of overlying water in comparison to air. Fresh water is about 800 times denser than air at 1.0 g/cm³, so the height of the fluid needed to achieve the same 1.0 atm would be:

Step 1. $\quad h = p/D$

Step 2. $\quad = 1.034 \times 10^3 \text{ g/cm}^2/1.0 \text{ g/cm}^3$

Step 3. $\quad = 1034 \text{ cm}$

$\quad\quad\quad$ (Converting to meters)

Step 4. $\quad = 10.34 \text{ m}$

This value implies that at a depth of only 10.3 meters in fresh water pressure will double to 2.0 atm (remember that the pressure effect of the overlying air is added to that of the water), and triple to 3.0 atm at about 20.6 meters. Marine salinity causes seawater to be slightly denser than fresh water, at 1.02 g/cm³, which causes greater pressure with less depth.

Brachiosaurus can be used as an example of how an aquatic lifestyle would have affected sauropods. This animal was about 13 meters tall with its neck fully extended vertically. Full immersion of its torso, so that only its head was above the water, would have caused it to experience an additional 1.0 to 1.5 atm of pressure. This situation would have made movement of its lungs for breathing extremely difficult or even caused their collapse. Consequently, the physics of breathing is reason enough to reject the notion that sauropods were habitually aquatic animals that used their necks as snorkels. This is the case even if other data did not support the likelihood of sauropods having had terrestrial lifestyles.

Indeed, as more sauropod trackways were found, they also showed little or no evidence of having been formed in submerged environments. Nevertheless, the "aquatic sauropod" model was kept alive even after the first sauropod trackways were discovered in the 1930s by Roland Bird (Chapter 3). For example, sauropod trackways that had manus impressions but lacked pes impressions were regarded as evidence that the sauropod tracemakers were swimming, because their hind limbs would have been suspended in the water while their fore limbs occasionally touched the bottom. Later descriptions and interpretations of these trackways revealed that pes impressions were present but were only preserved as **undertracks**, tracks preserved below the surface the dinosaur walked on. This realization negated the need to use flotation to explain the absence of tracks (Chapter 14).

Trackway evidence also can tell a great deal about the movement of individuals and groups of sauropodomorphs. The sheer size of their tracks can make the trackways of sauropods rather easy to distinguish from those of other dinosaurs. For example, some sauropod pes impressions from Western Australia, discovered in 2001, are as wide as 2 meters. The smaller impressions may be comparable in size to the tracks of thyreophorans (Chapter 12) or ceratopsians (Chapter 13), but typical sauropod tracks have distinctive compression shapes. The manus impression is crescentic, and the pes impression is oblong and commonly shows the claw impression from digit I. Like prosauropods and all other known dinosaurs, sauropods were diagonal walkers (Chapter 14). Yet their trackway straddle can be wider than that of any other dinosaur and have a lower pace angulation than the near-180° ones expressed by theropods. Dinosaur ichnologists often call such trackways "wide gauge," in an allusion to railroad tracks, but other sauropod trackways are narrower and appropriately are called "narrow gauge". Whether wide or narrow, all sauropod trackways show that they moved with their limbs under their bodies, and did not have the sprawling, lizard-like gait hypothesized for them by scholars earlier in the twentieth century.

Sauropod trackways definitely constitute the most impressive of all dinosaur trace fossils, and where numerous individuals walked over the same area they caused mixing of the sediments on the scale of heavy machinery. Sauropod trackways are also among the longest preserved for dinosaurs, such as two in the Middle Jurassic of Portugal that are nearly 150 meters long. Most such trackways show linear movement, but at least one in North America (Utah) shows that a Late Jurassic sauropod took an abrupt right turn for unknown reasons (Chapter 14).

Feeding

According to all known evidence, sauropodomorphs were obligate herbivores. Some paleontologists have proposed that the serrated teeth of prosauropods, such as *Plateosaurus*, reflect carnivorous behavior, and others have noticed some anatomical similarities between prosauropods and therizinosaurs (Chapter 9). Prosauropod teeth that are serrated and leaf-like show little sign of wear. Because grinding would involve direct contact between tooth rows with occlusion, shearing is instead favored as the probable expressed jaw function. As mentioned before, the pencil-like teeth of most sauropods were not adapted for chewing, so they probably served for shearing and, even more likely, for raking vegetation.

Scratch marks on sauropodomorph teeth constitute evidence of abrasion by grit, composed of silicate-mineral silt and sand, which was on plants consumed by these animals. Detailed examination of these marks could provide clues about which sauropodomorphs may have grazed on low-lying plants, as plants closer to the ground would have contained more grit. Conversely, teeth with fewer scratches might indicate high-level browsing. Relatively little work has been done on this avenue of research, so strict interpretations of diet preference cannot be formed based on dental condition alone. Sauropodomorphs also may have been accidental insectivores, because the large amount of plants that they ingested probably contained many insects. Of course, insects may have also chosen habitats on plants that were not frequently consumed by sauropods, but this speculation lacks evidence.

The development of extremely long sauropodomorph necks, of which *Mamenchisaurus* had the longest at 14 meters, is often cited as an example of directional evolution related to feeding specialization. The evolutionary mechanism behind such lengthening, which was the greatest of any known vertebrate animals, has been compared to the evolution of the giraffe (*Giraffa camelopardalis*) of equatorial and southern Africa. Giraffes comprise the only modern terrestrial animal that also has a long neck in comparison to its closest relatives. Interestingly, giraffes only have seven cervical vertebrae, but like sauropods these are elongated. Because modern giraffes are well known as high-level browsers of treetops and are connected to fossil ancestors with shorter necks, development of the same adaptation was postulated for sauropodomorphs. In this hypothesis, originally short-necked prosauropods evolved into long-necked lineages of sauropods, epitomized by some diplodocoids, which used their long necks for gaining access to increasingly taller treetops of conifers.

One of the problems with this hypothesis has already been mentioned: prosauropods and sauropods had separate lineages. Consequently, short-necked prosauropods did not evolve into long-necked sauropods. However, considering that the saurischian ancestor to all sauropods probably had a shorter neck than the earliest known sauropod, then the elongation of sauropod necks did indeed happen as a result of evolutionary change. According to predictions from cladistic analyses, the ancestral sauropod would have had 10 cervical vertebrae. The lengthening of sauropod necks in various lineages could have happened in three ways:

1 elongation of cervical vertebrate that were already there;
2 development of new cervical vertebrae; or
3 conversion of some dorsal vertebrae into cervical vertebrae (cervicalization).

Evidence from the axial skeletons of numerous sauropod species indicates that all three processes could have happened, depending on the lineage.

In a continuation of this paradigm, long necks were seemingly not enough and no leafy branch was too high. Sauropods began to be depicted in the 1980s and 1990s as rearing on to their hind legs so that they could nearly double their browsing height. The evolution of longer necks was thought to have co-evolved with increased heights of the canopy in Mesozoic forests. This would be an example of a "Red Queen" process (Chapter 6) happening between conifers and sauropods.

Several criticisms of this seemingly reasonable hypothesis are now evident, which has led to some doubt to its universal applicability to the development of long-necked sauropods:

- Although conifers were the most common trees available for high-level browsing during the Late Triassic through to the Late Jurassic, they had (and still have) low nutritional yield. Large amounts of material would have been consumed to provide any significant food value. Most conifers also have low recovery rates to such massive damage, so sauropods would have quickly overgrazed their food supply and eventually starved, unless they were constantly migrating to plunder and denude new forests.
- Anatomical reconstructions and biomechanical considerations of large sauropods suggest that some could have reared up on their hind legs. Nonetheless, no trackways or any other aspect of the fossil record has been found to indicate that they actually did. Today, elephants in circuses often rear up on their hind legs, or can be trained to stand on one leg, but these behaviors are not commonly observed in the wild. In other words, just because an animal can move in a certain way does not mean that it did.
- Blood pressures needed to pump blood from the heart of a sauropod to the height gained by the head would have been more than twice that measured in modern giraffes and five to six times that of humans. Some unusual adaptations would have been needed for such an extreme function. Some adherents to the "vertical sauropod" hypothesis have even advocated extra "hearts", additional organs along the pathway of the neck to pump blood.
- Computer modeling of sauropod necks reveals their ranges of motion on the basis of articulations of the cervical vertebrae. Interestingly, the ranges show a limited vertical mobility of only 10 to 20° above a horizontal plane, but a much wider arc of about 90° for turns in a horizontal plane. In other words, these sauropods could move their heads back and forth more easily than they could move them up and down.

Detailed measurements of the sauropods' cervical vertebrae contributed the data for a computer program that animated the necks. The most significant result of this analysis was that the modeled necks could not move vertically much more than half the length of the neck because the vertebrae would lock. For example, the 6-meter long neck of *Diplodocus* could only be raised about 4 meters. As a result of this and other information, the feeding behaviors for sauropods, such as *Apatosaurus* and *Diplodocus*, are now being reconsidered. (See box on page 322.)

A cooperative venture between a paleontologist and a computer programmer modeled the necks of Apatosaurus and Diplodocus for their ability to flex into the S-shaped profile, which is often depicted in artistic renditions of sauropod genera.

They are now more often thought to be the ecological equivalent of cows and sheep, or low-level grazers, than giraffes, which are high-level browsers. Despite some sauropods having had necks longer than the entire body of the largest prosauropod (e.g., *Mamenchisaurus* neck = 14 meters and *Riojasaurus* body = 11 meters), prosauropods were probably more adapted for high-level browsing than some sauropods. The food plants that would have easily re-grown and otherwise recovered from ravenously grazing sauropods were ferns. These plants are abundant today, but according to their fossil record were also common plants during the geologic time span of the largest sauropods.

Regardless of from whatever height sauropodomorphs obtained their food, they certainly did not chew it much. Digestion in most sauropods was probably aided by having a huge alimentary canal, which would have utilized large amounts of fermentative bacteria for digestion. At least a few sauropodomorphs also required gastroliths to aid their digestion, which further ground the foodstuff that was not milled into finer material in the oral cavity. Of all dinosaur clades, gastroliths are most often identified with sauropodomorphs. In fact, some paleontologists have attempted to document these trace fossils as independent indicators of a former sauropodomorph presence in Mesozoic strata otherwise devoid of their bones or traces. Three genera of prosauropods, *Ammosaurus*, *Sellosaurus* and *Massospondylus*, had presumed gastroliths associated with their abdominal cavities. Two sauropod species, *Cedarsaurus* and *Seismosaurus*, likewise had subrounded polished stones largely within their skeleton. In the case of *Seismosaurus*, the spatial distribution of gastroliths was meticulously mapped in relation to the skeleton. Such mapping revealed a clumping of the stones in two regions of the skeleton, approximating the locations of a former crop and gizzard. This type of evidence demonstrated a potential use of gastroliths as indicators of soft-part anatomy, leading one to wonder how many sauropodomorph skeletons may have been excavated without recognition of gastroliths.

Appropriately enough, the last piece of fossil evidence considered for feeding preferences in sauropodomorphs is coprolites. Unfortunately, so far no coprolites have been positively identified or correlated with sauropodomorphs. However, presumed large masses of finely-ground fossil plant material found in same-age strata as sauropodomorphs are speculated as possible candidates. The problem with this "guilt by association" type of identification is illustrated by smaller coprolites. These may have originated from smaller, contemporary herbivorous dinosaurs, such as ornithopods, or they actually may have been deposited in large but separate numbers by a large sauropod. Although a large coprolite establishes a minimum size for a tracemaker, the same is not necessarily true for smaller coprolites. After all, some large-bodied modern herbivores produce small-sized pellets (Chapter 14). Consequently, the recognition of sauropodomorph coprolites will require examination of both small and large coprolites in context with other fossil evidence of a sauropod presence.

Social Life

The view that sauropodomorphs were gregarious animals, and that they moved about together in at least family units, if not herds, has gained popularity recently, exemplified by artistic depictions of great migrating herds of sauropods. In this case, a popular idea about dinosaurs also may be correct. Most of the data supporting the idea of herding behavior in these animals, which as individuals could have outweighed a family unit of elephants, is based on trackway evidence. However, taphonomic analyses of sauropodomorph assemblages and the previously-mentioned nesting horizons of titanosaurs are independent lines of evidence that also support the probability of sociality for some sauropodomorphs. Taken collectively, the data indicate that sauropodomorphs were more likely to have lived in groups than not. This assertion can be tested in individual cases but is now generally accepted by many paleontologists who study sauropods.

The aforementioned nesting horizons of titanosaurs argue strongly for these sauropods having participated in social groups for their brooding of young, similar to what has been hypothesized for some ornithopods (Chapter 11). The close proximity of the nests, the sheer number of eggs, and their abundance on different horizons in the same small area reflect its revisiting in different nesting seasons by large numbers of sauropods of the same species. Although only one locality so far has body fossil evidence to corroborate a sauropodomorph nesting ground, this example provides a model for testing in other places, whether their reproductive and brooding behavior was:

1 more typically solitary;
2 based on family units; or
3 involved more than one family unit living in the same area at the same time.

Taphonomy also gives an insight into sauropod sociality through the examination of monospecific bone beds. Just as in some theropod species, such as *Coelophysis baurii* (Chapter 9), a monospecific assemblage of sauropods provokes the question, "How did only bones from many individuals of the same species end up in the same place at the same time?" For example, a bone deposit of the Middle Jurassic cetiosaurid *Patagosaurus fariasi* of Argentina is composed of bones representing five individuals, including both juveniles and adults. The current answer to this question is that the assemblage may represent a family that died and was buried together during a sudden event such as a river flood. Although adherence to only one hypothesis is risky, all other answers would be far more complicated than this scenario. As a result, it is the one that has been conditionally accepted for the cetiosaurid deposit and similar monospecific dinosaur bone beds. Two adult *Camarasaurus* skeletons associated with a juvenile of the same species, in the Upper Jurassic of Wyoming, provide another case of a possible family structure, and assemblages composed of only *Alamosaurus* bones from the Upper Cretaceous are also cited as evidence of gregariousness in sauropods.

Of course, the large number of sauropodomorph trackways documented worldwide constitutes excellent additional evidence for their social behavior. Multiple trackways on the same horizon are good clues as to how some species of sauropods traveled together or otherwise had preferred directions of travel. For example, one trackway horizon in Upper Jurassic rocks of southeastern Colorado shows five equally-spaced and parallel sauropod trackways, which is strong evidence supporting the hypothesis that they were traveling together as a herd (Fig. 10.11). This hypothesis is held up by the coincident direction of the trackways, which varied in harmony with one another, and the regular spacing between the trackways. Placing a minimum distance between one another when traveling in groups is a behavior observed in large modern animals, such as elephants. The preservational condition of the tracks also is similar, indicating that they were all formed within a short span of time before the character of the substrate changed. Another trackway from the Lower Cretaceous of Texas, discovered by Roland Bird (Chapter 3), shows that as many as 30 sauropods, seemingly of the same type (species), were moving in the same direction over a relatively narrow area. Some overlap of the

10

FIGURE 10.11
Sauropod trackways in the Late Jurassic Morrison Formation near La Junta, Colorado, showing parallelism and indicating probable herding behavior.

tracks indicated that later individuals stepped on some of the immediately preceding tracks.

Other sauropod trackways from the Upper Jurassic of Portugal show that a group of seven juveniles (based on the small sizes of their footprints) were accompanied by three larger sauropods, strongly suggesting a family structure. Moreover, the similar track sizes of the seven smaller individuals advocate that they represented the same age range and may have been hatched and raised from the same clutch. Such information helps with estimations of juvenile mortality by comparing the number of eggs within sauropod clutches to trackways of probable juveniles. A presumed monospecific track assemblage could also give an independent measure of population structure. However, such interpretations have at least one caveat: the adult tracks preserved on the same horizon may represent stratigraphically younger tracks that penetrated down to the same, older layer as the juvenile tracks (see Fig. 14.8). Nevertheless, trackways left by juvenile sauropods are becoming more recognized in the geologic record, which then help in better definition of sauropod populations. Some of the best examples of juvenile sauropod tracks are in Cretaceous rocks of South Korea, where more than 100 such trackways have been discovered.

Of course, just because animals are moving in the same direction does not necessarily qualify as evidence of herding. For example, a linear geographic barrier, such as a shoreline or a cliff, could have restricted the animals' lateral movement over a long period of time. An argument against this critique is that the variation in footprint sizes suggests that the population was composed of both younger and older individuals of the same species. If a geographic barrier equally affected the movement of all animals through an area, tracks of other species might be expected as a source of track variation. Other sauropod trackways that were apparently made by the same type of sauropod at approximately the same time should provide for more testing of herding hypotheses.

Consequently, bone-bed data, nesting grounds, and trackway evidence provide a compelling argument in favor of sauropodomorph sociality, although varying degrees of sociality for each species were likely. Sociality certainly conferred a major advantage in the form of added protection against large, theropod predators, despite already-massive sizes for some individual sauropodomorphs. It also allowed groups to explore for and exploit food resources together. These animals most likely aided each other's survival, thus ensuring reproductive success and thereby passing on their traits to the next generation of sauropodomorphs.

Health

Sauropodomorphs were apparently healthy animals and the vast majority of specimens show little or no evidence of paleopathological conditions. One reported possible ailment found in specimens of *Diplodocus*, *Apatosaurus*, and *Camarasaurus* is **diffuse idiopathic skeletal hyperostosis** (better known by its medical acronym, **DISH**), a condition where ligaments running laterally alongside the vertebrae become ossified. This condition gives the vertebrae a fused appearance, although the vertebrae themselves are not actually fused. In those sauropods so affected, some of the caudal vertebrae were bonded together by this ossification, which made them stiffer structures. Hence, one explanation for the frequency of this condition is that it represents an adaptation more than a malady. Sauropod trackways show no clear evidence of tail dragging, for example, and a stiffened tail would have prevented the formation of such a trace fossil. Interestingly, despite their often-huge sizes, no sauropods studied so far have shown any unequivocal signs of stress fractures or osteoarthritis.

Sauropodomorphs were subject to consumption by carnivores, specifically by their theropod contemporaries. Evidence for their role in Mesozoic food chains is largely expressed through toothmarks on sauropod bones, such as the documented numerous toothmarks on *Apatosaurus* bones that were probably inflicted by its Late Jurassic contemporary, *Allosaurus* (Chapter 14). One assumption is that such toothmarked sauropods were the spoils of predation, but no definitive evidence supports that sauropods were the victims of predation more or less often than their corpses were scavenged. Indeed, the sheer size of some sauropod species, as well as trackway evidence showing that sauropods moved together in herds, have been proposed as protective measures that discouraged predation. In similar modern terrestrial predator–prey relationships between large herds of herbivores and pack-hunting animals, lions have been known to isolate and kill very young or weak members of an elephant herd. However, the mere presence of the herd structure, compounded with the sizes of adult elephants and their fierce protectiveness, is normally sufficient deterrent to predation. This means that a successful hunt on an elephant is rare. With this simile in mind and the little evidence that is known, sauropodomorphs probably were not subject to any more predation than other herbivorous dinosaurs. In effect, they may have been even more resistant to death through such means, despite the enormous amount of potential protein that some sauropodomorph species could have provided for predators.

SUMMARY

Sauropodomorphs are among the most well known dinosaurs and the largest land animals that ever walked the surface of the Earth. As shown by their bones, tracks, and other fossils, they had a worldwide distribution that varied regionally but lasted from the Late Triassic to the end of the Cretaceous. Sauropodomorphs originally evolved at about the same time in the Late Triassic as the theropods, their sister clade within Saurischia. They apparently diverged into two clades early in their history, the Prosauropoda and the Sauropoda.

The Prosauropoda were distinguished by their anatomical adaptations toward possible facultative bipedalism and quadrupedalism, indicated by limb proportions and foot anatomy, as well as probable herbivory, discerned from their teeth and jaws, and fore-limb anatomy. They were certainly the largest herbivores of the Late Triassic and Early Jurassic, some reaching up to 11 meters in length (*Riojasaurus*). Prosauropods were relatively common by the end of the Triassic, and are most abundantly represented by the well-studied *Plateosaurus*. Prosauropods show the earliest known evidence for nesting of any dinosaur clade in Late Triassic deposits and they left distinctive tracks that correlate well with their skeletal foot anatomy. Despite nearly 50 million years of existence, prosauropods went extinct by the end of the Early Jurassic. Sauropods and other herbivorous dinosaurs that showed up in the geologic record at about that time probably filled their ecological niches as the first high-level browsers, but some sauropods were very likely low-level grazers as well.

Sauropods have a rich fossil record that continues to amaze paleontologists, who keep finding larger body parts of these dinosaurs that push the theoretical constraints on size for land-dwelling animals. Although their fossil record probably began in the Late Triassic, the diversification of sauropods apparently did not begin until the Early Jurassic. Their evolution resulted in some impressive forms adapted to carrying more massive bodies by the end of the Jurassic and continuing through the Cretaceous. Well-represented clades of sauropods include Diplodocidae (*Diplodocus*, *Apatosaurus*), Brachiosauridae (*Brachiosaurus*), Camarasauridae (*Camarasaurus*), and Titanosauridae (*Argentinosaurus*, *Malawisaurus*, *Titanosaurus*). Cetiosauridae was regarded as a loosely-held group containing sauropod species that did not fit well into the other clades, but is now considered a clade in its own right.

Sauropods were undoubtedly land-dwelling herbivores, as shown by their tracks, teeth, and gastroliths. Some may have been high-level browsers of Mesozoic forests, but recently acquired evidence (some anatomical, some botanical) suggests that some were better adapted for low-level grazing. At least some sauropods were social animals; their monospecific bone beds, nesting grounds, and trackways imply that they congregated and traveled in herds. Sauropods seem to have been healthy animals, and although they were on the menu of some theropods there is little evidence to support that they were the objects of predation more often than scavenging.

DISCUSSION QUESTIONS

1. What evidence would be needed to show that any sauropodomorphs lived in shallow aquatic environments but came up on land occasionally, as with modern hippopotamuses? If one species of sauropodomorph was deemed to be aquatic but it was only capable of withstanding 1.5 atm of pressure, what would be the maximum depth difference between it in fresh water versus salt water?

2. What evidence would be needed to support the hypothesis that some of the long-tailed sauropods used their tails as whips? What would you look for in the skeleton itself, and what sorts of trace fossils might indicate that the tails were used in such a way?

3. Make a list of the postcranial characteristics of prosauropods and sauropods. How did they differ and how were they similar? What do the differences tell you about evolutionary changes in sauropods as opposed to prosauropods?

4. Some presumed sauropod nests, interpreted on the basis of clutches of large, spheroidal eggs, are also interpreted as having been laid in conical pits. What anatomical adaptations did sauropods possess that would have aided in their excavation of nests for laying eggs? Would members of some clades have had an anatomical advantage over others in digging nests?

5. What evidence indicates that sauropodomorphs in general were not great "chewers" of food? Cite both body and trace fossil evidence. How would undoubted sauropodomorph coprolites or former gut contents either corroborate or falsify the hypothesis?

6. If sauropods did move in herds, what kind of trackway patterns would convince you that the adults in the herd protected the juveniles from pack-hunting or individual theropods? Sketch such patterns, then compare them with those sketched by your classmates. Discuss which of the strategies they represent would have offered the best protection and why.

7. While doing volunteer work on a dinosaur excavation site, you find what you suspect are gastroliths in a Late Jurassic deposit, but there is no sauropod skeleton evident around them. How would you support the hypothesis that they are gastroliths, and not just randomly distributed stones? What evidence would convince you that your hypothesis is wrong?

8. At the same site as in question 7, with its excavation of Late Jurassic rocks, you find a suspected sauropod coprolite with identifiable fragments of ferns. How could this coprolite be correlated with potential tracemakers through the use of data on neck anatomy and tooth wear?

9. What are some hypothetical evolutionary effects of sauropods stepping on low-lying plants and grazing large amounts of these same plants? For analogies, think of modern grazing animals and the impact they have on the plants of their landscape.

10. Compare prosauropods to the therizinosaurs (Chapter 11). How are these dinosaurs similar in overall body plans? What are some key anatomical traits that you would use to distinguish them if you found either one in the fossil record?

Bibliography

Barrett, P. M. and Upchurch, P. 1995. "Sauropod feeding mechanisms: Their bearing on paleoecology. *In* Sun, A., and Wang, Y. (Eds), *Symposium on Mesozoic Terrestrial Ecosystem and Biota Short Papers,* **6**: 107–110.

Bonaparte, J. F. and Coria, R. A. 1993. Un nuevo y gigantesco sauropodo titanosaurio de la Formación Río Limay (Albiano-Cenomaniano) de la Provincia del Neuquén, Argentina. *Ameghiniana* **30**: 271–282.

Buffetaut, E., Suteethorn, V., Cuny, G., et al. 2000. The earliest known sauropod dinosaur. *Nature* **407**: 72–74.

Chiappe, L. M., Coria, R. A., Dingus, L., Jackson, F., Chinsamy, A. and Fox, M. 1998. Sauropod dinosaur embryos from the Late Cretaceous of Patagonia. *Nature* **396**: 258–261.

Christiansen, P. 1999. On the head size of sauropodomorph dinosaurs: Implications for ecology and physiology. *Historical Biology* **13**: 269–297.

Colbert, E. H. 1952. Breathing habits of the sauropod dinosaurs. *Journal of Natural History* **5, 55**: 708–710.

Colbert, E. H. 1993. Feeding strategies and metabolism in elephants and sauropod dinosaurs. *American Journal of Science* **293, A**: 1–19.

Coombs, W. P., Jr. 1975. Sauropod habits and habitats. *Palaeogeography, Palaeoclimatology, Palaeoecology* **17**: 1–33.

Cooper, M. R. 1980. The prosauropod ankle and dinosaur phylogeny. *South African Journal of Science* **76**: 176–178.

Fiorillo, A. R. 1991. Dental microwear on the teeth of *Camarasaurus* and *Diplodocus*: Implications for sauropod paleoecology. *Paleontological Contributions from the University of Oslo* **364**: 23–24.

Galton, P. M. 1985. Diet of prosauropod dinosaurs from the Late Triassic and Early Jurassic. *Lethaia* **18**: 105–123.

Galton, P. M., and Upchurch, P. 2004. "Prosauropoda". *In* Weishampel, D. B., Dodson, P. and Osmólska, H. (Eds), *The Dinosauria* (2nd Edition). Berkeley, California: University of California Press. pp. 232–258.

Gillette, D. D. 1994. *Seismosaurus, the Earth Shaker*. New York: Columbia University Press.

Lockley, M. G. 1990. Did "Brontosaurus" ever swim out to sea?: Evidence from brontosaur and other dinosaur footprints. *Ichnos* **1**: 81–90.

Marsh, O. C. 1883. Principal characters of American Jurassic Dinosaurs. Part VI: Restoration of Brontosaurus. *American Journal of Science* **Series 3, 26**: 81–85.

McIntosh, J. S. 1989. "The sauropod dinosaurs: A brief survey". *In* Padian, K. and Chure, D. J. (Eds), *The Age of Dinosaurs*. Short Courses in Paleontology, *Paleontological Society* **2**: 85–99.

Reid, R. E. H. 1981. Lamellar-zonal bone with zones and annuli in the pelvis of a sauropod dinosaur. *Nature* **292, 5818**: 49–51.

Ricqlès, A. de, 1983. Cyclical growth in the long limb bones of a sauropod dinosaur. *Acta Palaeontologica Polonica* **28**: 225–232.

Stevens, K. A. and Parrish, J. M. 1999. Neck posture and feeding habits of two Jurassic sauropod dinosaurs. *Science* **284, 5415**: 798–800.

Upchurch, P. 1995. The evolutionary history of sauropod dinosaurs. *Philosophical Transactions – Royal Society of London, B* **349, 1330**: 365–390.

Upchurch, P., Barrett, P. M. and Dodson, P. 2004. "Sauropoda". *In* Weishampel, D. B., Dodson, P. and Osmólska, H. (Eds), *The Dinosauria* (2nd Edition), Berkeley, California: University of California Press. pp. 259–322.

Weaver, J. C. 1983. The improbable endotherm; the energetics of the sauropod dinosaur *Brachiosaurus*. *Paleobiology* **9**: 173–182.

Wilson, J. A. 2002. Sauropod dinosaur phylogeny: Critique and cladistic analysis. *Zoological Journal of the Linnaean Society, London* **136**: 217–276.

Wilson, J. A. and Carrano, M. T. 1999. Titanosaurs and the origin of "wide-gauge" trackways: A biomechanical and systematic perspective on sauropod locomotion. *Paleobiology* **25**: 252–267.

Yates, A. M. and Kitching, J. W. 2003. The earliest known sauropod dinosaur and the first steps towards sauropod locomotion. *Proceedings of the Royal Society of London: Biological Sciences* **270**: 1753–1758.

10

Chapter 11

While listening to the radio, you hear a science report that discusses efforts to protect the nests of sea turtles on the barrier islands of the eastern USA, some of which are endangered species in parts of the world. As you listen to the story, you learn that sea turtle mothers will crawl on to a sandy beach, dig a hole, lay their eggs, bury them, and then return to the sea. This means that the mothers will never see their offspring, having left them to fend for themselves. Moreover, the offspring may never hatch at all because raccoons and feral hogs prey on turtle eggs.

Did dinosaur mothers act like sea turtles, or did they care for their young, not only watching them hatch but staying with them while they grew in their nests? What evidence would be needed to prove dinosaurs took care of their offspring?

Ornithopoda

11

Why Study Ornithopods?

Ornithopoda was one of the most diverse and geographically widespread of all dinosaur clades. Their success originated in the Early Jurassic and continued until the end of the Cretaceous when, as with other non-avian dinosaurs, they became extinct. Their diversity can be best illustrated by their size range – some species were as small as 1 meter but others were about 15 meters long, placing them well within the size range of theropods and around the same mass as some sauropods. They are sometimes known as the "cows of the Mesozoic" in recognition of their specialized herbivore diets, perceived passivity, and probable gregarious nature. However, some of their members, especially the hadrosaurids, had complex intraspecific behaviors more similar to birds than bovines. The most innovative adaptation ornithopods had in common with one another was the development of teeth and jaws that were arguably the most efficient grinders of plant food ever devised in terrestrial herbivores. Indeed, some ornithopods had large arrays of teeth (dental batteries) and moveable skull parts that would have surpassed those of any modern mammalian plant eater.

Ornithopods have the most complete geologic record of any major dinosaur clade. In terms of body fossils, they are represented by:

- complete skeletons ranging from juveniles to adults in some species;
- many eggs, some with embryos identifiable to ornithopod species;
- skin impressions;
- former gut contents; and
- possible internal organ preservation.

Their trace fossil record is also rich, consisting of numerous trackways, nests, and coprolites. They became so widespread throughout the Mesozoic that they left their bodily remains or traces of their behavior on every continent by the end of the Cretaceous. Perhaps the most interesting aspect of this wide geographic range was that some of their skeletal material has been found in both Arctic and Antarctic regions. Such a latitudinal distribution has been part of the basis for hypothesizing ornithopod migrations, conjuring once odd but now-familiar recreations of great herds of these dinosaurs moving with the seasons, as with modern mammalian herbivores.

Perhaps the best reason to study ornithopods is because their fossil record provides the most complete and compelling evidence available about the social lives of dinosaurs, which is only rivaled by sauropods. Preliminary data indicate that at least one species of ornithopod lived in large communal nesting grounds, where the juveniles were restricted to their nests for their formative years while the parents took care of them. Other ornithopods show elaborate head structures that probably served as resonating chambers for calling to one another, but also may have been prominent cues for gender identification. Some ornithopods had easily visible tusks, and others had large spiked thumbs and hands that could grasp, all giving rise to hypotheses about the functions of such features.

Ornithopods certainly comprised one of the most interesting of dinosaur clades and have been the source of some revolutionary hypotheses about dinosaurs in general. Their study has been at the center of the significant changes in perceptions about intraspecific behavior of dinosaurs that have taken place in only the past 25 years.

Definition and Unique Characteristics of Ornithopoda

Ornithopoda (= "bird foot"), first named as a group by O. C. Marsh in the late nineteenth century (Chapter 3), was described in early literature as bipedal ornithischians. However, considering that bipedalism is now recognized as a probable ancestral trait for all dinosaurs, saurischians included, the definition of ornithopods has been refined considerably. Ornithopoda is now distinguished by the following characters, among others (Fig. 11.1):

- Offset tooth row, where the maxillary teeth are higher (more dorsal) than those in the premaxillary, although teeth in the latter might be missing altogether.
- Occlusal surface is higher (more dorsal) than the jaw joint.
- Crescent-shaped paraoccipital process is located in the posterior of the skull.
- Premaxilla has an elongate process that touches either (or both) the prefrontal or lachrymal.

As mentioned earlier, ornithopods are ornithischians, meaning that they shared a common ancestor with thyreophorans (Chapter 12) and marginocephalians (Chapter 13). This common ancestry places the ornithopods, thyreophorans, and marginocephalians in the node-based clade **Genasauria**. Genasauria splits into stem-based and sister clades Thyreophora (ankylosaurs and stegosaurs) and **Cerapoda** (Marginocephalia and Ornithopoda). Ornithopoda itself is a stem-based clade, having two main branches to **Heterodontosauridae** and **Euornithopoda** (Fig. 11.2). The majority of ornithopod species compose the latter, and most of

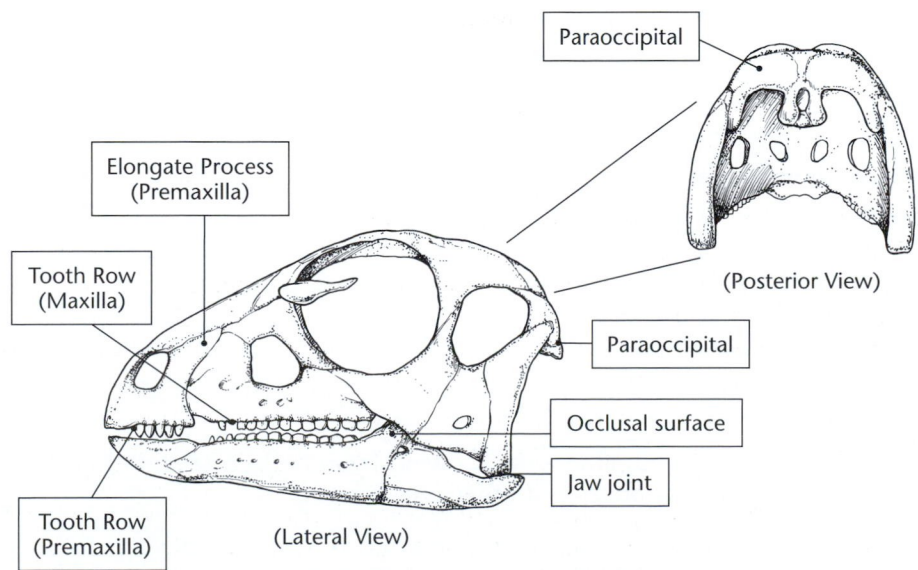

Paraoccipital

(Posterior View)

Elongate Process (Premaxilla)

Tooth Row (Maxilla)

Paraoccipital

Occlusal surface

Jaw joint

Tooth Row (Premaxilla)

(Lateral View)

FIGURE 11.1 Important characters for Clade Ornithopoda: offset tooth row; occlusal surface higher than jaw joint; crescent-shaped paraoccipital process; premaxilla with an elongate process.

11

TABLE 11.1 Representative genera of clade Heterodontosauridae and other non-iguanodontian ornithopods, with approximate geologic age and where they occur.

Genus	Age	Geographic Location
Abrictosaurus	Early Jurassic	South Africa
Agilisaurus	Middle Jurassic	China
Anabesetia	Late Cretaceous	Argentina
Bugenasaura	Late Cretaceous	Western USA
Drinker	Late Jurassic	Western USA
Echinodon	Early Cretaceous	UK
Gasparinisaura	Late Cretaceous	Argentina
Heterodontosaurus	Early Jurassic	South Africa
Hypsilophodon	Early Cretaceous	Western USA
Jeholosaurus	Early Cretaceous	China
Leaellynasaura	Early Cretaceous	Australia
Notohypsilophodon	Late Cretaceous	Argentina
Orodromeus	Late Cretaceous	Western USA
Othniela	Late Jurassic	Western USA
Parkosaurus	Late Cretaceous	Alberta, Canada
Thescelosaurus	Late Cretaceous	Western USA; Canada
Tsintaosaurus	Late Cretaceous	China
Yandusaurus	Middle Jurassic	China
Zephyrosaurus	Early Cretaceous	Western USA

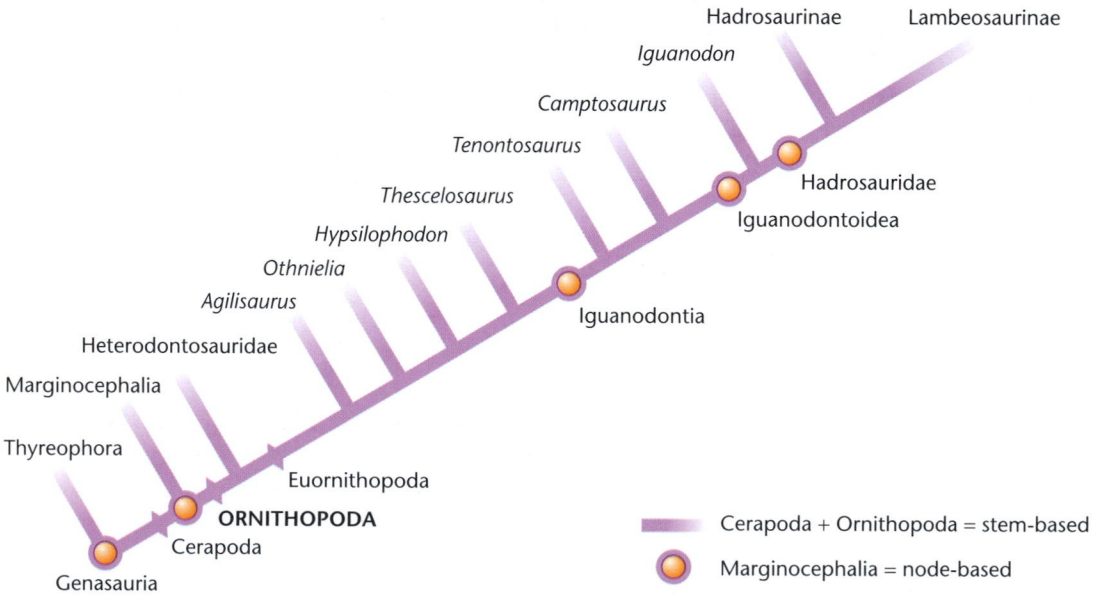

FIGURE 11.2 Cladogram for Ornithopoda, showing hypothesized relationships between Heterodontosauridae, Iguanodontia, and other ornithopods.

these are in **Iguanodontia** (Tables 11.1 and 11.2). "**Hypsilophodontidae**" is a grouping of ornithopods that was once considered as a clade within Euornithopoda, but recent analyses reveal that it is probably polyphyletic. As a result, its former members are now placed within other clades under Euornithopoda.

TABLE 11.2 Representative genera of clade Iguanodontia, with approximate geologic age and where they occur.

Genus	Age	Geographic Location
Altirhinus	Early Cretaceous	Mongolia
Anatotitan	Late Cretaceous	Western USA
Bactrosaurus	Late Cretaceous	China
Brachylophosaurus	Late Cretaceous	Alberta, Canada; western USA
Camptosaurus	Late Jurassic	Western USA; UK
Claosaurus	Late Cretaceous	Western USA
Corythosaurus	Late Cretaceous	Alberta, Canada
Draconyx	Late Jurassic	Portugal
Dryosaurus	Late Jurassic	Western USA; Tanzania
Edmontosaurus	Late Cretaceous	Western USA; Alberta, Canada
Eolambia	Late Cretaceous	Western USA
Gilmoreosaurus	Late Cretaceous	China
Gryposaurus	Late Cretaceous	Alberta, Canada
Hadrosauridus	Late Cretaceous	Eastern and western USA
Hypacrosaurus	Late Cretaceous	Western USA; Alberta, Canada
Iguanodon	Early Cretaceous	Western Europe; USA
Jinzhousaurus	Early Cretaceous	China
Kritosaurus	Late Cretaceous	Western USA
Lambeosaurus	Late Cretaceous	Western USA; Canada; Mexico
Muttaburrasaurus	Early Cretaceous	Australia
Olorotitan	Late Cretaceous	Russia
Ouranosaurus	Early Cretaceous	Niger
Parasaurolophus	Late Cretaceous	Western USA; Alberta
Probactrosaurus	Early Cretaceous	China
Prosaurolophus	Late Cretaceous	Alberta, Canada; western USA
Saurolophus	Late Cretaceous	Alberta, Canada; Mongolia
Shantungosaurus	Late Cretaceous	China
Telmatosaurus	Late Cretaceous	Romania; Spain; France
Tenontosaurus	Early Cretaceous	Western USA
Tsintanosaurus	Late Cretaceous	China
Valdosaurus	Early Cretaceous	UK, Romania, Niger
Zalmoxes	Late Cretaceous	Romania

Euornithopoda is distinguished by a synapomorphy of a prepubic process that projects anteriorly and at a high angle away from the pubis, causing an appearance similar to a saurischian pelvis. However, the Euornithopoda pubis itself is still posteroventral and together with the ischium, which is an ornithischian trait.

Within the Euornithopoda, some species have cheek teeth that lost their denticles (apparent as ridges) on tooth crowns, as well as a distinctively rod-shaped prepubis. Probably the best-known clade within Euornithopoda is Iguanodontia. Iguanodontia has some of the most abundantly represented ornithopods of the fossil record and well-known denizens of the Cretaceous, such as the hadrosaurids (clade **Hadrosauridae**). Iguanodontians were best known for their lack of premaxillary teeth and enlarged nares. These features and others, such as an elongation and flattening of the snout, contributed a general "duckbill" appearance to the anterior portion of the skull in hadrosaurids.

Accordingly, hadrosaurids, such as *Edmontosaurus*, *Maiasaura*, and *Anatotitan*, were nicknamed the "duckbilled dinosaurs."

Ornithopods were either bipeds or facultative quadrupeds in their locomotion. Some of the bipeds, such as the Early Cretaceous *Hypsilophodon*, show fore-limb/hind-limb proportions comparable to those of theropods, with the hind limbs considerably longer than the fore limbs. Such limb proportions are invoked as evidence favoring cursorial behavior, meaning that these ornithopods may have been capable of running quickly. Their small sizes and lightly-built frames relative to other ornithopods also argue for the ability of quick getaways from theropods that preyed upon them. Trackways attributable to ornithopods show either bipedal or quadrupedal modes but, like the vast majority of dinosaur trackways, only a few of them can be interpreted as representative of speeds faster than walking. However, trackways of some iguanodontians of the Early Cretaceous show quadrupedalism, where a hoof-like manus preceded a three-toed pes. In contrast, many trackways attributed to Late Cretaceous hadrosaurids reflect bipedalism, although they may have been bipedal only when needed. Also, in some cases, a lighter manus pressure may not have been preserved, thus not recording evidence of quadrupedal behavior. Ornithopod pes impressions are distinctive, with three-toed patterns that have a compression shape equally wide and long, rather than the elongate ones left by theropods. Their tracks also have thicker toe impressions and lack claw marks (Chapter 14). Some hadrosaurid tracks are also among the largest dinosaur tracks found so far in the geologic record.

The aforementioned analogy of ornithopods as "cows" of the Mesozoic is justified by conclusions that all ornithopods were herbivores and apparently had few overt defenses against predators. They lacked the body armor or otherwise thickened bone seen in thyreophorans (Chapter 12), ceratopsians (Chapter 13), and a few sauropods (Chapter 11). Thus, restorations of ornithopods have typically shown them in idyllic settings, contentedly grazing on vegetation or caring for their young, until a voracious predator appears on the scene. Although a few ornithopods do indeed show signs of either having been bitten or eaten by theropods and crocodiles (Chapters 7 and 9), only a few individuals show evidence of active predation. The apparent contradiction between the perception of ornithopods as easy fodder for carnivores and the paucity of evidence that they were so victimized leads to questions about what natural defenses they did successfully employ.

The more subtle aspects of how ornithopods related to other dinosaurs add to defining their uniqueness, not just in a cladistic sense but behaviorally. A closer look at the various ornithopod clades should reveal their diversity and the numerous interrelated facets of their geologic record should provoke numerous hypotheses.

Clades and Species of Ornithopoda

Heterodontosauridae

Heterodontosaurids (= "different toothed lizard"), which lived only during the Early Jurassic and mostly in southern Africa, derive their name from their differentiated teeth, an unusual condition for any dinosaurian clade. In fact, some of their teeth are morphologically distinctive enough that they are known only in heterodonto-saurids. These teeth, occurring in the areas of the former cheeks, have been described as "chisel-like" because they come to an edge at their crowns, which are also adorned with denticles. Other teeth that are radically different from the cheek teeth are canine-like tusks on the predentary and dentary (Fig. 11.3). Although these teeth might look menacing, and were probably visible externally when the mouth was closed,

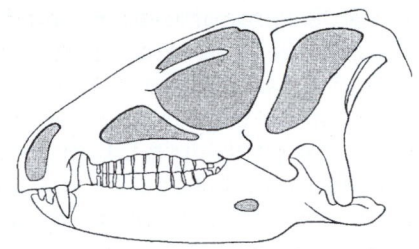

FIGURE 11.3 Skull of the Early Jurassic *Heterodontosaurus*, a heterodontosaurid of South Africa. From Cowen (2000), *History of Life*, 3e, Blackwell Science, Inc., Malden, MA, p. 220, fig. 12.10. (After Charig and Crompton.)

they were not used for predation. A more likely explanation is that they served for display within their species. A few paleontologists have suggested that they may even be indicators of sexual dimorphism and present in only one gender. Heterodontosaurid males are assumed to have been the possessors of these attributes, although this assignment is arbitrary.

Parts of the appendicular skeleton of heterodontosaurids indicate that they had well-developed arms and hands, which are interpreted as adaptations to feeding, such as pulling or holding on to plants, or even for digging. No other evidence, such as trace fossils or former gut contents, has verified these interpretations, but they are reasonably inferred on the basis of other adaptations that clearly favor an herbivorous habit. The best-known heterodontosaurid species, the eponymous *Heterodontosaurus*, had a large number of caudal vertebrae, which resulted in a tail that made up more than half the total length of the body. It also has hind limbs with fused distal elements, specifically in the tibia, fibula, and tarsals. These traits are interpreted as aids to fast movement. Such an adaptation certainly would have been advantageous for a small herbivore in an Early Jurassic world already inhabited by large ceratosaurs (Chapter 9).

Heterodontosaurids also included *Abrictosaurus*, *Lycorhinus*, and *Echinodon*, although the latter two are only identified from sparse skeletal material. This scarcity is probably related to their apparent biogeographic restriction, relatively short geologic range, and taphonomic factors in their preferred environments. However, heterodontosaurid remains have been found in arid-climate facies, which normally are conducive to preserving skeletal remains in the event of quick burial (Chapter 7). Faced with such a disappointingly small number of bones, paleontologists could then turn to trace fossils. Although ornithopod tracks from the Early Jurassic are common in some places, none have been attributed specifically to heterodontosaurids. Likewise, no other trace fossils of heterodontosaurids, such as nests, tooth marks, gastroliths, or coprolites, have been recognized. Embryonic remains of heterodontosaurids are also unknown. Thus, paleontologists who study heterodontosaurids do not have much material for interpreting the lifestyles of these seminal ornithopods.

"Hypsilophodontidae"

Because current cladistic analyses now suggest that Hypsilophodontidae is a paraphyletic grouping of basal ornithopods, the use of "hypsilophodontid" is used here to describe all euornithopods outside of the clade Iguanodontia and is not meant to connote a clade.

In contrast to heterodontosaurids, hypsilophodontids are better known through complete or otherwise well-preserved skeletons of one species (*Hypsilophodon*) and fragments of about 10 other species. Hypsilophodontids occur in strata formed during the Middle Jurassic through to the Late Cretaceous, representing a wider geologic range than heterodontosaurids. Fossil specimens are found in regions as widespread as North America, South America, Europe, China, and Australia. Morphologically, hypsilophodontids were larger than their heterodontosaurid predecessors (about 1.5 to 4 meters long) and their robust hind limbs were slightly

11

longer in proportion to their fore limbs, indicating better adaptations to bipedalism. Ossified tendons reinforced their caudal vertebrae in places, an adaptation that would have rectified the tail and formed a counterbalance during fast movement. This trait is similar to the functional morphology of dromaeosaurids (Chapter 9).

Erroneous interpretations of the foot structure of *Hypsilophodon*, particularly the position of the hallux, led some paleontologists, earlier in the twentieth century, to propose that these not-so-small dinosaurs had an arboreal habitat. This conclusion was based on an initial assessment of a hallux (digit I) that pointed posteriorly, in direct opposition to the anteriorly pointing digits II through to IV. This arrangement would have given it an **anisodactyl foot**, which is characteristic of modern birds with perching habits (Chapter 15). In perching birds, the hallux grasps a branch from behind while the other toes grasp from the front. Nevertheless, re-examinations of the foot anatomy by the mid-1970s showed that the foot was not anisodactyl, but instead had the hallux pointing anteriorly, as with other ornithopods. These insights tie in much better with other skeletal features that argue for a cursorial animal, rather than a tree climber.

Hypsilophodontid skulls underwent some slight outward modifications with respect to those of heterodontosaurids, especially in the loss of their distinctive tusks in the anterior portion. However, their skulls were among the first of herbivorous dinosaurs to show evidence of **pleurokinesis**. This condition was a jointing between the premaxilla and the rest of the skull that caused the maxilla to shift outward when the mouth closed. In humans, the maxilla stays in the same place when chewing and the lower jaw can move laterally. In hypsilophodontids (and all other euornithopods), the maxilla moved out so that a shearing motion occurred when the lower jaw occluded with the upper jaw. Pleurokinesis was accompanied by:

1 the retention of the premaxillary teeth;
2 better-developed dental batteries in the cheek teeth, which also lost their denticles; and
3 a non-ossified (horny) beak.

The latter served as an extension to the anterior part of the face and was used to crop vegetation. Overall, hypsilophodontid skulls point toward advanced adaptations to feeding that exemplified later modifications to herbivore behavior in other euornithopods.

Besides *Hypsilophodon*, an important hypsilophodontid is *Thescelosaurus* of the Late Cretaceous in North America, which retains some primitive characteristics of basal ornithopods (Fig. 11.4). This retention shows that it was a conservative form because it lived toward the end of the geologic range for basal euornithopods. In 2000, *Thescelosaurus* became famous from one well-preserved specimen that had a mineralized mass in the probable site of its heart. This concretion led to the hypothesis that it represented a four-chambered heart with a single aorta, which subsequently encouraged some critical scientific debate (Chapter 8).

Jurassic hypsilophodontids include the Middle Jurassic *Agilisaurus* and *Yandusaurus*, which come from the same formation in China. The similar and small (less than 1 meter long) *Drinker* and *Othniela* from the Late Jurassic Morrison Formation are representative hypsilophodontids from the western USA (Fig. 11.5). Ironically, these dinosaurs were named after rival paleontologists: *Drinker* was named in honor of Edward Drinker Cope, whereas *Othniela* was named after Othniel Charles Marsh (Chapter 3). Early Cretaceous genera from Australia are *Atlascopcosaurus*, *Leaellynasaura*, and *Qantassaurus*, but each genus is known from only a few fragments. Although the Early Cretaceous *Hypsilophodon* considerably predates *Parkosaurus* from the latest part of the Cretaceous, the two species closely resemble one another, indicating probable relatedness. Another Late Cretaceous species,

FIGURE 11.4 *Thescelosaurus*, a non-iguanodontian ornithopod from the Late Cretaceous. Auckland Museum, Auckland, New Zealand.

FIGURE 11.5 The diminutive ornithopod *Othniela* of Late Jurassic in western North America. Specimens in the Denver Museum of Science and Nature, Denver, Colorado.

Orodromeus, has the best-known life history of any hypsilophodontid because of some well-preserved juvenile skeletons found in deposits of the western USA.

Iguanodontia

Its namesake, *Iguanodon*, was one of the first dinosaurs to be named from the geologic record in England. Later in the mid-nineteenth century, tracks were also linked to iguanodontians reported in England. Soon after, Louis Dollo described numerous beautifully preserved *Iguanodon bernissartensis* skeletons from the Early Cretaceous of Belgium (Chapter 3). These studies led to iguanodontians as subjects of some of the first detailed scientific studies of dinosaurs.

Iguanodontia is recognized as a stem-based clade today, but its name has a long history in the study of dinosaurs.

Interpretations of *Iguanodon* and other iguanodontians have undergone much revision since then, as a result of new discoveries and re-analysis of previously discovered skeletal material. However, probably the most important hypothesis that emerged from the study of *Iguanodon* was of bipedalism in some dinosaurs, which Dollo discerned through the study of more than 30 skeletons of this species. More recently, iguanodontians have been recognized as dinosaurs that may have had complex sociality, which is evident from an extensive body and trace fossil record of hadrosaurids of the Late Cretaceous. Moreover, iguanodontians were important contributors to terrestrial ecosystems on most continents, from the Late Jurassic through to the Late Cretaceous, partially or completely replacing sauropods in ecological niches as the largest herbivores.

Probably the most primitive of iguanodontians is the Early Cretaceous *Tenontosaurus* of western North America. Some paleontologists have interpreted this iguanodontian as a hypsilophodontid or otherwise outside of Iguanodontia, which illustrates its seemingly basal status. Other basal iguanodontians include an Early Cretaceous iguanodontian from Australia, the colorfully named *Muttaburrasaurus*, as well as the Late Cretaceous *Zalmoxes* of Romania and *Rhabdodon* of France.

Some iguandontians show how cladistic interpretations of "primitive" (more basal) versus "advanced" (more derived) forms can be at odds with actual geologic ages. For example, the Late Jurassic *Dryosaurus* and *Camptosaurus* of western North America show relatively more advanced traits than *Tenontosaurus* and the other aforementioned basal iguanodontians. This contradiction is explained by paleontologists not yet knowing which iguanodontian was the most ancestral. This hypothetical common ancestor of Iguanodontia probably lived during the Middle or Late Jurassic, although later descendants, represented by *Tenontosaurus* and other species, retained its traits. Furthermore, *Dryosaurus* and *Camptosaurus* are still more primitive than the Early Cretaceous *Iguanodon*, suggesting that some iguanodontian lineages evolved more quickly than others. Interestingly, *Camptosaurus* shows some gross morphological convergence with prosauropods (Chapter 10), in that it had a small head and long neck in comparison to other ornithopods (Fig. 11.6). Some of the other Jurassic and Early Cretaceous iguanodontians looked similar to their non-iguanodontian cousins in many ways, but they did differ in one important respect – size. Some specimens of *Iguanodon* are as long as 10 meters, and by the Late Cretaceous a few hadrosaurids, such as *Shantungosaurus* of China, reached lengths of 15 meters.

Along the evolutionary journey of ornithopod lineages to Hadrosauridae is the stem-based clade Iguanodontoidea, which has *Iguanodon* as its most basal genus. *Ouranosaurus* was an Early Cretaceous iguanodontoidean of Niger that was closely related to *Iguanodon*, but it was strikingly different because of its elongate and robust processes on its dorsal vertebrae. These processes are interpreted as supports for either sail-like flaps of skin or humps like those interpreted for spinosaurs, which were theropods with similar features (Chapter 9). Despite such a novel trait, it is still considered a predecessor to the hadrosaurids.

The Late Cretaceous *Telmatosaurus* has what are considered the most ancestral traits for the node-based clade Hadrosauridae, so it is used for comparison to all other probable hadrosaurid descendants. The following characters distinguish Hadrosauridae:

FIGURE 11.6 *Camptosaurus*, a common iguanodontian of the Late Jurassic in western North America. Specimen in the College of Eastern Utah Prehistoric Museum, Price, Utah.

- Long and wide anterior portion of skull ("duckbill").
- Well-developed dental batteries.
- Increase of vertebrae to at least 8 sacral and 12 cervical vertebrae.
- Loss of digit I on the manus.
- Development of prominent unguals (similar to hooves) on the pes.
- Long fore limbs relative to hind limbs.

Eolambia, an Early Cretaceous iguanodontoidean of the western USA, was briefly considered the oldest member of the Hadrosauridae, but more detailed study revealed that it lies outside of that clade. For example, it has only seven sacral vertebrae, whereas every member of Hadrosauridae has eight.

The body and trace fossil record for hadrosaurids begins in the Early Cretaceous, setting the stage for their excellent Late Cretaceous fossil record. Hadrosaurids are among the best studied and well known of dinosaurs, partially because of their extensive fossil record but also because they have been studied for nearly 150 years. For example, the first discovery of a near-complete skeleton of a dinosaur was *Hadrosaurus foulkii* in New Jersey in 1857. Joseph Leidy described this specimen soon afterwards and inspired the then-new idea of bipedalism in dinosaurs, later confirmed by Dollo (Chapter 3). Furthermore, the excellent fossil record for hadrosaurids resulted in the naming of numerous species and a probable (or at least apparent) high diversity in comparison to other dinosaur clades. Hadrosaurids provide an approximate mirror image of earlier iguanodontian diversity in that they

FIGURE 11.7 Representative genera of a hadrosaurine and lambeosaurine from western North America and China. (A) *Edmontosaurus*, a Late Cretaceous hadrosaurine of western North America; Denver Museum of Natural History, Denver, Colorado. (B) *Tsintaosaurus*, a Late Cretaceous lambeosaurine of China; Fernbank Museum of Natural History, Atlanta, Georgia.

(A)

(B)

flourished during the Late Cretaceous, whereas more primitive iguanodontians showed a decline in both numbers and species.

Because of this abundance and diversity, the classification of hadrosaurids is nearly as complicated, detailed, and contentious as that devised for theropods (Chapter 9). To make sense of the large amount of fossil data for hadrosaurids, they will be discussed through their two most well-represented clades, Hadrosaurinae and Lambeosaurinae.

Hadrosaurines are nicknamed the "flat-headed" hadrosaurids because they have skulls that are wider than tall. Representative species include the Late Cretaceous *Anatotitan*, *Brachylophosaurus*, *Edmontosaurus* (Fig. 11.7A), *Gryposaurus*, *Maiasaura*, *Prosaurolophus*, *Saurolophus*, *Shantungosaurus*, and of course *Hadrosaurus*. In contrast, lambeosaurines are called the "hollow-crested" hadrosaurids owing to their enlarged and separated nasal chambers enclosed by bones that formed dorsal crests on their skulls. Lambeosaurines, which were also common in the Late Cretaceous, include *Parasaurolophus*, *Corythosaurus*, *Hypacrosaurus*, and *Lambeosaurus* of North America, as well as *Tsintaosaurus* of China (Fig. 11.7B). Interestingly, these two groupings of hadrosaurids were recognized long before the advent of cladistics, thus indicating how previous generations of paleontologists correctly interpreted their common ancestry despite the differences of a few anatomical features.

Paleobiogeography and Evolutionary History of Ornithopoda

Ornithopods lagged behind the saurischian clades of Theropoda and Sauropodomorpha in the geologic record, showing up in the Early Jurassic. So far, no Late

Triassic forms are yet defined by paleontologists. However, some enigmatic dinosaurs have ornithischian features and a few anatomical similarities to ornithopods. These ornithischians made their earliest known appearances in Late Triassic strata of North and South America, as well as in South Africa. Richard Owen first described some of their remains in the nineteenth century, and he wrote of their laterally compressed teeth, which come to a peak but with denticles on their tops and thin enamel on both sides. Unfortunately, Owen interpreted these dinosaurs as lepidosaurs (lizards), and little subsequent research was done on them until the 1970s.

Other primitive ornithischians are in Lower Jurassic strata of South Africa, including *Lesothosaurus diagnosticus*, the prime example used for comparison to similar skeletal material. *Lesothosaurus* and other basal ornithischians were originally grouped into a category called "**Fabrosauridae**," but this classification is now regarded as polyphyletic and has been abandoned. "Fabrosaurids" were categorized partially on the following traits:

1 light, small frames (1- to 2-meters long);
2 fore limb/hind limb ratios of about 40 : 60;
3 tibias longer than their femurs;
4 small skulls; and
5 long tails.

All of these traits indicate bipedalism and the ability to perform rapid cursorial movements. These are evolutionary adaptations that were also echoed by heterodontosaurids and "hypsilophodontids" before iguanodontians evolved toward greater sizes. As a result, it is understandable that paleontologists saw these dinosaurs as evolutionary precursors to later clades of ornithopods. Interestingly, one of these former "fabrosaurids," *Agilisaurus louderbacki* from the Middle Jurassic of China, did turn out to be the most ancestral euornithopod known, so it was recognized as a key species for understanding ornithopod evolution.

The paleobiogeography of ornithopods, as noted earlier, was apparently limited to southern Africa during the Early Jurassic. However, adaptive radiation and dispersal of ornithopod lineages had taken place by the Middle Jurassic, and by the end of the Cretaceous they occupied every continent.

Primitive ornithischians described from the Late Triassic are from North America, South America (Argentina), and South Africa. This distribution suggests that land connections between these three continents, represented by the southern continent of Gondwana (Chapters 4 and 6), permitted the dispersal of these dinosaurs prior to ornithopods proper showing up in the Early Jurassic. The heterodontosaurids, the earliest known undoubted ornithopods, also occur in the Early Jurassic of South Africa, which geographically and temporally links them to possible ornithischian relatives. The later ornithopod clades then spread throughout all other continents after the Middle Jurassic, particularly with the ascension of euornithopods from the Late Jurassic through to the Late Cretaceous.

Ornithopods occupied various habitats, their body and trace fossils indicating a former presence in fluvial, lacustrine, swamp, deltaic, and coastal marine (tidal flat and shoreline) environments. Late Cretaceous strata, associated with coal beds of western North America in particular, show abundant hadrosaurid footprints associated with many plant fossils. Some of these tracks are preserved in coal beds and are still frequently encountered by miners. This evidence indicates that at least some hadrosaurid species lived in forested areas adjacent to coal-forming swamps (Fig. 11.8).

11

FIGURE 11.8 Large track, preserved as positive relief as a natural cast on the bottom of a bed, attributed to a hadrosaurid in the Laramie Formation (Late Cretaceous) near Golden, Colorado. Note that the "heel" portion of the track was obscured by a log that was under the hadrosaurid's foot, providing evidence that the hadrosaurid was walking through a formerly forested area.

One evolutionarily based explanation for changes in ornithopod abundance and diversity in space and time is that the continued splitting of the continents combined with (or contributed to) the creation of new terrestrial ecosystems and the corresponding evolution of new plant species. In accordance with large-scale changes in vegetation, changes in the fauna feeding on the plants would have occurred, particularly during the Early Cretaceous with the onset of flowering plants. The two broadest trends in ornithopod evolution noticeable from the Early Jurassic through to the Late Cretaceous are:

1 their specialized adaptations to feeding, evident through their teeth and jaws; and
2 their increased size, which also may have been partially an adaptation to feeding.

Sauropods (Chapter 10) were also occupying some of the same environments as their ornithopod contemporaries. Nevertheless, the great differences in their sizes, readily apparent by the end of the Jurassic, suggest that they were not competing for the same foodstuffs.

At the same time when sauropods declined in species and numbers, iguanodontians in particular seemed to increase in both respects. By the end of the Cretaceous, hadrosaurids had replaced sauropods as the largest terrestrial herbivores in many areas of the world, although ceratopsians were new potential competitors (Chapter 13). The relationships between ornithopods, other dinosaurs, vegetation, and their coevolution in continental ecosystems of the Mesozoic will be further explored later.

Ornithopods as Living Animals

Reproduction

The reproductive and brooding behaviors of some species of ornithopods are probably the best documented for all dinosaurs. Consequently, their intriguing family lives warrant extended coverage here. Their stories begin with sexual attraction and end with the nurturing of juveniles, with the cycle beginning anew with each successive ornithopod generation that reached reproductive age. As products of such cycles, a few species of ornithopods left numerous clues about their reproductive lives that overturned many preconceptions about dinosaurian reproduction.

Traditionally held views of dinosaurs, based not so much on actual fossil evidence but on speculation and modern analogues, portrayed them as brutish, cold-blooded reptiles, in both the physiological and archetypical senses. Most modern reptiles, to which dinosaurs were consistently compared, lay their eggs and then leave the young to fend for themselves (Chapter 8). Sea turtles, for example, lay their eggs on land and promptly go back to sea, never seeing their potential offspring again. Unguarded nests contain potential protein for egg predators, so embryo mortality in such instances can be very high. Crocodilians show slightly more concern for the welfare of their young: mothers will stay near a nest after laying eggs and in some cases will dig out newly hatched juveniles buried in hole nests. However, crocodilians still do not take an active role in brooding and parenting, that is, they do not sit on their nests or bring food to the hatched young.

Fossil data gathered in only the past 20 years or so have changed the prior conception of little to no ornithopod parental care. Now they are considered attractive to one another, and caring animals, whose reproductive behavior was more akin to some modern birds than reptiles. However, equating ornithopods with modern birds may be an exaggeration, as ornithopods lack anatomical evidence for feathers or any feather-like structures. In fact, skin impressions from desiccated remains of hadrosaurids show variably sized knobs and scaly skin, which is associated with reptiles (see Fig. 5.8). The body and trace fossils of the theropods *Oviraptor* and *Troodon* provide far better examples of brooding behaviors in dinosaurs that resemble birds. Furthermore, these theropods are more closely related to birds than ornithopods (Chapter 9). Nevertheless, nurturing behavior was first interpreted in ornithopods, not theropods, so ornithopods have provided the historical model for comparison with all other data dealing with similar reproductive behavior in dinosaurs.

Attracting Mates

The onset of the reproductive cycle, of course, begins with attracting potential mates. Ornithopods are perhaps only surpassed by ceratopsians (Chapter 13) for dinosaurian flamboyance in their sexual advertising. The Mesozoic saw extensive and elaborate head ornamentation in some species, such as the Late Cretaceous lambeosaurines *Parasaurolophus* or *Tsintaosaurus*. For skulls of the same species, some paleontologists have even speculated that gender might be assignable on the basis of males potentially having larger headgear, which is seen in modern large mammals, such as deer. **Dewlaps**, large pieces of skin hanging from the ventral surface of the lower jaw (possessed by turkeys, for example), may have supplemented or replaced headgear for attractiveness in some ornithopods. At least one specimen of the Late Cretaceous hadrosaurid, *Maiasaura*, has a preserved skin impression of a structure consistent with a dewlap.

The tusks seen in Early Jurassic heterodontosaurids have lead to speculation that these accouterments could also have been useful in intraspecific combat and territorial disputes. Unfortunately, no actual fossil evidence of toothmarks has been reported that might record such encounters between heterodontosaurids. In terms of visual obviousness, the Early Cretaceous iguanodontian *Ouranosaurus* of Niger, with its dorsally extended processes on its vertebrae, would have been easily observed from a lateral view. Such an adaptation may have supported a sailback similar to spinosaurs (Chapter 9) or the pre-dinosaurian pelycosaurs (Chapter 6). However, an alternative hypothesis is that *Ouranosaurus* had a broad hump, like modern bison or camels, rather than a thin sail. Regardless of whether it had a sail or a hump, *Ouranosaurus* must have been easy to spot by others of its species.

Although ornithopod communication between the sexes may have been wholly conducted through visual cues, like many other animals, they also probably relied on other sensory stimuli. For example, the internal configuration of nasal cavities within the dorsally located cranial crests in some hadrosaurids may also indicate whether they were used for crooning, an auditory means of wooing. Hadrosauridine cranial crests can be categorized as solid or hollow, based on the presence or absence of bone in the crest. Solid-crested forms, such as *Saurolophus*, have large nasal chambers under the solid bone; non-crested hadrosaurids (such as *Edmontosaurus* and *Kritosaurus*) have a similar anatomical situation. In contrast, the hollow-crested forms, represented by the lambeosaurines, have tube-like structures that emanate from the nasal cavities and extend dorsally and posteriorly.

Initial analyses of these structures resulted in hypotheses that fit the former view of ornithopods and most large dinosaurs, such as sauropods (Chapter 10), as aquatic animals. In these hypotheses, the tubes were thought to be snorkels. However, more complete lambeosaurine skulls, such as those of the Late Cretaceous *Parasaurolophus* and *Tsintaosaurus*, show the tubes to be U-shaped, having no discernible openings at the top. This configuration meant that the structures did not aid in breathing underwater, effectively falsifying the snorkel explanation. Thus far, the simplest hypothesis devised on the basis of the evidence is that these tubes, which connected with the nasal cavities, were **resonating chambers** used for changing the sounds produced, by moving air through the lambeosaurine skull.

An understanding of how resonating chambers worked in these skulls requires a discussion of the fundamental physics of sound.

If a tree falls in the forest and no one is there to hear it, does it make a sound? The answer is no, because the tree does not make a sound, as sound is a mental translation made from perception of a compressional wave that traveled to the animal. Sound not only depends on a host that can perceive it but, because of its wave properties is also affected by:

1 the density of the compressed medium;
2 temperature, and
3 any obstacles that might interfere with the transmission of the sound.

For example, a compressional wave will travel faster through a denser medium: at 20°C, sound waves in water move at about 1470 m/s, but in air they move at 343 m/s. In gases, with higher temperatures, the speed of sound increases noticeably, at about 30 cm/s for each 1°C. So during times of higher global temperatures, which were common during much of the Cretaceous Period, sound traveled slightly faster. For example, on a typical equatorial day in the lowland tropics of the Mesozoic the temperature may have been 40°C, which would have caused the speed of sound to have been:

$$V = (343 \text{ m/s}) + (0.3 \text{ m/s} \times 20) = 343 \text{ m/s} + 6.0 \text{ m/s} = 349 \text{ m/s} \qquad (11.1)$$

where V is the speed of sound.

Because warmer and denser air is normally close to the surface of the Earth, sounds travel faster for ground-dwelling animals living near sea level, and slower for those at higher and colder elevations. For a Mesozoic example that relates to survival, the cracking sound made by a stalking theropod stepping on a branch would have reached the ears of an ornithopod slightly faster on a warm day near a beach than during a cold day in the mountains. This simplistic scenario is complicated by:

1 obstacles that reflect the sound, exemplified by echoes off canyon walls;
2 refraction of the sound as it passes from local air masses of different temperatures; and
3 wind direction and velocity.

With regard to the latter, animals downwind of sounds will hear them more easily than animals upwind or standing in still air. This is because the sounds have been carried to the downwind animals as compressional waves in the air moving toward them. Sound also will travel farther if it has a low frequency, with frequency defined as:

$$n = V/\lambda \tag{11.2}$$

where n is frequency, V is velocity of sound (m/s), and λ is wavelength (meters). Frequency is measured as the number of vibrations that pass a point per second, as indicated by the equation that has the length units (meters) canceled out. For example, notice how the frequencies differ for sound waves that have wavelengths of 60 cm and 9 cm at the same speed of sound:

(For 60 cm)

Step 1. $n = 343$ m/s/0.6 m

Step 2. $n = 572$ vibrations per second (Hertz)

(For 90 cm)

Step 1. $n = 343$ m/s/0.9 m

Step 2. $n = 381$ vibrations per second

The longer wavelength for the second sound also results in a lower frequency. Using the human voice as an example, a baritone will sing using lower frequency sounds (longer wavelengths) than will a tenor, and a soprano will have a higher frequency sound (shorter wavelengths) than a tenor. Modern wind instruments also illustrate these principles: a bass tuba emits lower frequencies than a flute.

In the case of *Parasaurolophus* and its unusual U-shaped cranial tube, sound would have been caused by air moving through the tube as it was inhaled and exhaled through the nares. Sound would have been compressed in the tube and thus the length would have controlled the wavelength and the frequency of any sound emitted. In other words, the longer the tube, the longer the frequency. This regulation of frequency to change the quality of a sound is **resonance**. *Parasaurolophus* had a long tube (about 2 meters) bent around like a trombone, which apparently served as a resonance chamber. Models of the tubes scaled to the same size make low-frequency sounds of about 85 vibrations per second when air is blown through them. Slightly different lengths of tubes, which might have been inherent with sexually dimorphic *Parasaurolophus* (although this dimorphism is not firmly established), correspondingly would have produced slightly different sounds. These differences could have been important from a mating perspective because they would have helped the animals to distinguish genders of the same species from a distance without requiring any visual contact. Furthermore, juvenile *Parasaurolophus* had shorter tubes, so their sounds would have had a higher pitch (frequency) that would have been distinct from the adult sounds, which certainly would have aided active parenting. Other hadrosaurids, whether they were non-crested, solid-crested, or hollow-crested, may have supplemented their enlarged nasal cavities with soft tissues that also would have affected the resonance of any sounds that they made.

Building Nests and Laying Eggs

Once mates were attracted and successful mating occurred, the development and laying of eggs, as well as nest building, were the next steps in the perpetuation of ornithopod species. Egg shapes confirmed for some hadrosaurids are spheroidal to oblate spheroids; egg sizes are 12 to 20 cm wide, with calculated volumes of 1250

FIGURE 11.9 The hadrosaurid *Maiasaura* with its nest and juveniles, based on associated body and trace fossil material from the Late Cretaceous Two Medicine Formation of Montana. Fernbank Museum of Natural History, Atlanta, Georgia; restoration on loan from the Museum of the Rockies, Bozeman, Montana.

to 4000 cm³. In the absence of *in-situ* embryos, connecting egglayers with dinosaur eggs is always fraught with risk, but a few ornithopod examples provide well-supported identifications. The most complete information regarding eggs, embryos, nest structures, and associated juveniles for an ornithopod species is represented by the Late Cretaceous hadrosaurine *Maiasaura pebblesorum* of western North America (Fig. 11.9).

Clutch sizes for *Maiasaura* consisted of a minimum of 11 eggs. However, the arrangement of the clutch and whether the eggs in a nest structure actually represent a single egg-laying episode are still uncertain. These trace fossils were sedimentary structures distinct from the surrounding sediment, so they were probably mound nests. Fossil evidence for vegetation in the nests, such as carbonized plant material or leaf impressions, is currently lacking. Some paleontologists, however, have inferred that vegetation must have been present for protection and keeping eggs at near-constant incubation temperatures. No information has conclusively shown that *Maiasaura*, or any other ornithopods, actively brooded their clutches by sitting on their nests, as proposed for smaller theropods such as *Oviraptor* and *Troodon*. *Maiasaura* must have done nest building through digging, piling, and compacting sediment into the described structures, presumably using their limbs but they may also have put their shovel-shaped snouts to good use.

> *Nest structures associated with Maiasaura eggs and juvenile remains are circular bowl-shaped depressions about 2 meters wide, and 0.8 meters deep, with slopes of nearly 30° in places.*

Nest structures occur at the same horizons in some locations, suggesting that at least some of them were contemporaneous. This coincidence implies communal nesting grounds for this species. Interestingly, no root structures or other evidence of *in-situ* vegetation is noticeable between nest structures, which probably means that the mounds were made in open areas without substantial tree cover. The nest structures are about seven meters apart from one another, which is the length of an adult *Maiasaura*. This distance may point toward a respect for living space by different *Maiasaura* parents. For example, body-length spacing (at a minimum) would have helped to prevent adults from stumbling over or into nests belonging to other adults. The recurrence of nesting horizons on different stratigraphic levels but in the same area argues persuasively for site fidelity, where the maiasaurs returned to the same area year after year for nesting and brooding. Such site fidelity for dinosaur nesting is also interpreted for Late Cretaceous titanosaurids in Argentina (Chapter 10).

The most startling hypothesis that emerged from detailed examination of these nesting grounds is that the adults took care of their young for extended periods of time after they hatched. Data supporting this hypothesis were based on the following observations and inferences:

- Juvenile remains of as many as 15 individual *Maiasaura* also occur in several of the nest structures. These skeletons reveal that some individuals were as much as 3 meters long while still in the nest, so they may have been several years old.
- Eggshell fragments are in the nest structure with the bones and have the appearance of having been broken by more than simple hatching. The small fragments could be a result of daily trampling by the juveniles. Alternatively, this circumstance may also have a non-biological cause, such as compaction by overlying sediments.
- Limb joints of the smaller juveniles show little evidence of ossified bone, meaning that they were still cartilaginous and thus not well suited for supporting their weight (Chapter 5). This undeveloped state implies a limited mobility, so the juveniles were altricial, in contrast to some theropod juvenile limbs that show cursorial abilities at an early age (precocial).
- The large size of some juveniles indicates that they were eating, despite dwelling in or near nest structures located in an area that had no evidence of nearby vegetation. The lack of nearby food sources combined with their altricial condition meant that food was brought to the juveniles, probably by their parents, so that they were able to thrive. Even if the juveniles became semi-precocial and were able to move out of the nest, they would have had little incentive to leave home while they were still being fed.

Another ornithopod, *Hypacrosaurus*, probably also had nesting grounds. A horizon in the Late Cretaceous of Montana associated with *Hypacrosaurus* remains, consisting of an overwhelming number of eggshell fragments, can be followed for 3 km. This egg horizon represents the richest known concentration of dinosaur egg material in North America. Unfortunately, the sequential evidence of egg/nest structure/embryo/juvenile/adult established for *Maiasaura* is lacking for *Hypacrosaurus*, simply because not as many of the criteria are so clearly defined. Nevertheless, *Hypacrosaurus* eggs are well defined. They have a spherical geometry and about three times the volume of those reported for *Maiasaura* (1200 cm^3 versus 3900 cm^3), although the clutch sizes have not been established. A still unidentified lambeosaurine egg clutch, also discovered in Late Cretaceous rocks of Montana, has an egg size and shape similar to that of *Hypacrosaurus* and a clutch size of 22. If this clutch size does indeed reflect a single egg-laying episode by a lambeosaurine mother, then it had an internal capacity in its reproductive tract of about 0.085 m^3, which corresponds to about 85 liters of liquid volume. Such calculations indicate the probable large size of whatever species of lambeosaurine laid the eggs. Clutch size is

11

positively correlated with body size in modern crocodilians and the same condition is reasonably assumed for dinosaurs.

Other ornithopod embryos have been reported for the Late Jurassic iguanodontians *Dryosaurus* and *Camptosaurus* of North America, although nest structures and associated egg material for these species are still unknown. The disappointing lack of definite attribution of eggs, embryos, and nest structures for heterodontosaurids and other non-iguantonian ornithopods means that hypotheses concerning the evolution of nurturing behavior in these ornithopods are still untested. For example, the small ornithopod *Orodromeus* was originally considered the nestmaker and egglayer responsible for dinosaur egg assemblages in Late Cretaceous strata of Montana. This conclusion was made on the basis of the closely associated remains of near-embryonic juveniles. However, this hypothesis was falsified when the interior of one of the eggs revealed an embryonic *Troodon*, a small theropod (Chapter 9). The presumed parenthood of *Orodromeus* was dropped as a result of this information and a lesson was learned about avoiding "guilt by association" – that is, assuming that certain eggs and nests belong to dinosaurs whose remains happen to be nearby. The actual nests and eggs of *Orodromeus* are still unknown, despite the better-than-average establishment of the reproductive habits of its contemporaneous iguanodontian relatives.

Our knowledge of parental care among ornithopods specifically, and dinosaurs in general, is mostly limited as a hypothesis to iguanodontian fossil data from the Late Cretaceous of North America. Nevertheless, these fossils and their interpretations will give future researchers a better idea of what search patterns to assume when looking for similar fossil evidence from other ornithopods.

Growth

Growth series calculated for some ornithopod species, based on bone proportions, are partially to almost complete. Moreover, these data are supplemented by bone histology relating to growth lines and LAGs (Chapter 8). The Late Cretaceous hadrosaurid *Maiasaura* of North America is again well represented in this respect, with specimens ranging from possible newborns to full-sized adults. Partial growth series also are available for the Late Jurassic *Dryosaurus* and Late Cretaceous *Hypacrosaurus* of North America. For example, *Dryosaurus* skulls show that a typical juvenile trait is small skulls correlated with large orbits. The latter trait meant that the eyes were large relative to the rest of the face. With increased growth of the skull, the orbits became proportionally smaller. Similar ontogenetic trends are also observed for *Hypacrosaurus*, where both the orbits and frontals shrunk in proportion to the rest of the skull and the face became longer. Size-frequency analyses of bones in general also can be used to estimate growth rates. For example, an analysis of *Maiasaura* juveniles indicates that they may have achieved 3-m lengths within one year. This evidence implies that these ornithopods had high metabolic rates in the early stages of life, similar to some theropods (Chapter 9).

Of course, longer limb bones are assumed to belong to older individuals of any given species, which then are used as a relative scalar for other measurements, such as LAGs and amounts of fibrolamellar bone between LAGs. LAGs are possibly a result of annual periods of slow (arrested) growth, and if this is assumed they can be used to estimate how old a given dinosaur was when it died. However, environmental factors independent of seasonal changes can also cause LAGs, so they are not absolutely reliable for calculating ages. Regardless, LAGs in the Late Cretaceous hypsilophodontid *Orodromeus* suggest that this dinosaur had intermediate growth rates. Furthermore, thick deposits of fibrolamellar bone in between LAGs comprise a characteristic of fast growth, which has been recorded for *Maiasaura* juveniles.

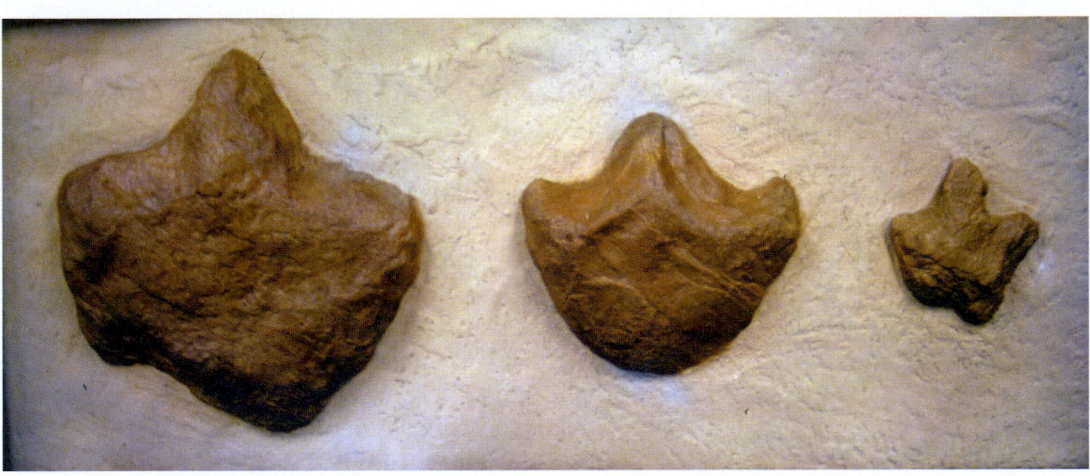

FIGURE 11.10 Morphologically similar but differently sized Late Cretaceous hadrosaur tracks, interpreted as representative of growth stages. Casts of original tracks put together for comparison, Dinosaur Track Display, University of Colorado, Denver.

When used in conjunction with size-frequency data, these deposits corroborate hypotheses about early altricial stages followed by rapid growth on the way to a precocial state. Again, these data point toward higher metabolic rates for these dinosaurs, consistent with an endothermic physiology (Chapter 8). Nonetheless, growth-rate estimates are based on limited data sets and require more samples and critical analysis before they can be considered reliable.

An independent source of estimates about growth series in ornithopods is provided by tracks made by juvenile and adult animals. As with other parts of the body, feet grow in some relative proportion with age. Therefore, a bedding plane containing tracks from a population of the same species of ornithopod should constitute a good sample of its growth series (Fig. 11.10). Caveats of using such trackway horizons for growth series estimates are:

1 they may have been made by more than one species of ornithopod in the same area with similarly shaped and sized feet; and
2 they may not have been contemporaneous, and the undertracks of adults possibly could have been transmitted to underlying (older) layers containing juvenile tracks; and
3 foot growth rates may have been different from growth rates for other body parts.

Nonetheless, if these problems are resolvable, then the sheer abundance of tracks in some instances provide censuses that typically far exceed any given ornithopod bone bed. Indeed, one study of iguanodontian tracks from a single bedding plane in an Early Cretaceous stratum of Colorado yielded hundreds of measurements showing a variety of sizes (20–48 cm long) for otherwise morphologically identical tracks. Localities with Early and Late Cretaceous strata that contain abundant ornithopod footprints, such as those in western Canada, South Korea, and other areas of the world, are also amenable to such an analysis. These data can thus provide insights into population structures of ornithopods that can supplement measurements of bones.

11

Locomotion

Ornithopods were obligate bipeds, facultative bipeds, or quadrupeds, as suggested by their appendicular skeletons and verified by their trackways. The appendicular skeletons of heterodontosaurids and other non-iguanodontians imply that these lighter and smaller ornithopods were quicker moving than iguanodontians. Of course, ornithopod trackways, particularly from the Cretaceous, are very common in some parts of the world and constitute an excellent fossil record of their locomotion, which can be used in conjunction with body fossils. In contrast, tracks and trackways attributable to heterodontosaurids and other non-iguanodontians are still not well documented, although some 10–15 cm long ornithopod tracks in Lower Cretaceous strata of Spain are interpreted as belonging to such smaller ornithopods.

One difficulty in distinguishing the tracks of these generally smaller ornithopods in deposits worldwide lies in their possible morphological and dimensional similarity to theropod tracks in same-age strata. For example, a typical heterodontosaurid pes is close in size and shape to the feet of some small theropods that lived during the Early Jurassic. Likewise, the pes of *Hypsilophodon* also has a narrow-toed three-toed outline that probably produced elongate tracks, which can be easily confused with those of some theropods or bipedal prosauropods (Chapter 14). Because many three-toed Jurassic tracks have been described from the geologic record, it is reasonable to predict that at least some of those currently correlated with theropods were actually made by smaller ornithopods. If more accurate distinctions can be made, more detailed analyses of non-iguanodontian locomotion can be used to test the conceptions derived from skeletal data.

One example of how speedy, smaller ornithopods have been interpreted from trackway evidence is the Early Cretaceous tracks associated with a so-called "dinosaur stampede" preserved in a Lower Cretaceous stratum in Queensland, Australia (Chapter 14). The tracks indicate that the smaller ornithopods and some theropods mostly ran in the opposite direction of a large theropod at top speeds of 4–5 m/s (14–18 km/hour). However, whether these ornithopods were non-iguanodontians is unknown; skeletal data from strata in the same area are lacking in this respect.

The equant, three-toed, and occasionally very large feet of most iguanodontians are distinctive and correlate easily with numerous tracks in Cretaceous deposits.

Iguanodontian tracks show that iguanodontians were either bipedal or quadrupedal, and seemingly all were diagonal walkers. Thus far, no iguanodontian running trackways have been found; their calculated top speeds, based on track and hip-height measurements, are about 2 m/s (7 km/hour). This trackway evidence agrees with hypotheses of iguanodontian locomotion based on skeletal proportions, which also indicate relatively slow-moving animals. For example, iguanodontian fore limbs typically are only about 70% the length of the hind limbs, so some paleontologists argue that these ornithopods were not capable of running quadrupedally. To visualize this biomechanical concept, think of how horses or deer have nearly equal fore limb and hind limb lengths, which helps them to run more efficiently. The simplest solution to running, or any other movement at higher speeds, was thus achieved bipedally. Consequently, bipedal trackways made by iguanodontians can be tested independently through examination of pace length, footprint length versus hip height, and pressure-release structures to see whether they indicate running speeds (Chapter 14).

One of the best-documented examples of iguanodontian trackways is in the area of Morrison, Colorado, in the same area where "The Great Dinosaur Rush" began in the 1870s (Chapter 3). About 50 years after that, road construction removed Early Cretaceous rocks, unveiling dozens of iguanodontian tracks mixed with some theropod tracks (probably from ornithomimids) on a bedding plane (Fig. 11.11).

FIGURE 11.11 Abundant iguanodontian tracks on a bedding plane of the Lower Cretaceous Dakota Formation, Morrison, Colorado. Note the parallel movement of one larger individual next to a smaller individual.

The iguanodontian tracks represent numerous individuals that walked quadrupedally along a shallow marine shoreline. Some were apparently traveling alone, but others were walking together in the same direction. At least two trackways show a smaller iguanodontian walking parallel to a larger one, which is evocative of a parent and juvenile moving together. This information, along with ornithopod trackway data from other areas, argues for sociality and herd behavior in at least some iguanodontians.

Probably the most enduring antiquated concept of ornithopod locomotion is that some of them were either fully or semi-aquatic. This hypothesis was first proposed in the early part of the twentieth century with the discovery by the Sternberg family (Chapter 3) of hadrosaurids that were "mummified." These hadrosaurids were skeletons preserved with skin impressions nearly all the way around their bodies. This mode of preservation was most certainly a result of arid conditions and desiccation soon after death (Chapter 7), which caused an interesting look: the skin impressions between the phalanges seemed stretched, giving the feet a "webbed" appearance. This apparent condition, interpreted without its taphonomic subtext, was originally considered as evidence of an aquatic adaptation. As a result, other features of hadrosaurids (and iguanodontians in general) then were fitted to this hypothesis. For example, the lack of dermal armor and other obvious defenses against predation led early researchers to propose that these large ornithopods used the supposed safety of water bodies to swim away from voracious theropods. Of course, this hypothesis did not take into account some of the 10-meter long crocodiles, such as *Deinosuchus*, that lived in those water bodies and ate hadrosaurids during the Late Cretaceous (see Fig. 7.1). The mistaken assumption of aquatic iguanodontians was similar to the one made with sauropods, which were deemed too large to have lived on dry land (Chapter 10). Thus, the artistic recreations of large ornithopods and sauropods from earlier in the twentieth century show them in or near bodies of water (Chapter 1).

The aquatic-habitat hypothesis is unsupported, not only by trackway evidence showing that iguanodontians traveled on emergent land, but also by anatomical details such as the following:

- The ossified tendons that reinforced the tails kept them stiff, not flexible. In contrast, modern aquatic species of large reptiles, such as crocodilians, are very dependent on their flexible tails for propulsion.
- The manus and pes of a typical iguanodontian are much too small in proportion to the rest of the body to have overcome the resistance of the water that would have dragged on their considerable bulks. This situation is regardless of whether they had webbed feet or not.
- The hind limbs show little evidence that they were better adapted for swimming than those of similarly sized theropods. This is not to say that iguanodontians could not swim. After all, similarly bulky elephants can swim long distances. But if a primary mode of defense was to swim away from predators, they should show adaptations that are clearly superior to those of their supposed theropod harassers.

As mentioned earlier, the hollow cranial crests of lambeosaurines also were considered to be snorkel-like aquatic adaptations, but the discovery that the "snorkel" did a U-turn (and thus would have quickly caused drowning) nullified that hypothesis. Finally, the "duckbill" of hadrosaurids, with its lack of anterior teeth and its distinctive shape, so characteristic of these ornithopods, was also cited as evidence for a semi-aquatic habit. Also, soft plants typical of coastal areas and lakes were thought to be the food most suitable for these ornithopods. Such a scenario ignored these dinosaurs' impressive dental batteries, which are now recognized as among the best adaptations to feeding on tough, fibrous foods ever devised in the history of terrestrial herbivores.

However, some iguanodontian remains are found in nearshore marine or other lowland facies, which means that a few were at least proximal to water bodies. Consequently, the aquatic-ornithopod hypothesis is not completely falsified, but it is poorly supported, which means that paleontologists will conditionally accept that most ornithopods, like most dinosaurs, were terrestrial animals.

Feeding

All ornithopods were obligate herbivores and unlike some other herbivorous dinosaurs, such as sauropodomorphs (Chapter 10), their teeth and jaw structures show that they were efficient chewers of plant material. In fact, features in the skull reflecting adaptations for processing large amounts of vegetation are diagnostic of Ornithopoda. For example, a few primitive ornithschians are discounted as ornithopods by paleontologists partially on the basis of their lack of inset teeth. This characteristic also implies that these ornithischians did not have cheeks for temporarily storing food. More advanced adaptations to cutting and grinding vegetation first show up in the Early Jurassic heterodontosaurids, which had their cheek teeth proximal enough to form rudimentary dental batteries.

Other more derived non-iguanodontians show a progression toward more grinding ability, evidenced by:

1 the loss of denticles on the teeth, which were in heterodontosaurids;
2 skulls with pleurokinetic ability; and
3 development of dental batteries.

Dental batteries were later greatly improved upon by Cretaceous iguanodontians, such as hadrosaurids, so that the many interlocking teeth became broad surfaces for the milling of tough plant material. Daily growth lines in these teeth show that they were replaced every 200 days or so, indicating major wear from constantly processing plant fibers. Indeed, this amount of periodic wear would have caused exceedingly sophisticated adaptations for consuming plants that contained much incorporated silica or external grit.

Possible former gut contents were reported from one specimen of the Late Cretaceous hadrosaurid *Edmontosaurus*, which consisted of needles, seeds, and twigs from conifers. A few large (about 30-cm diameter) irregularly shaped coprolites, attributed to hadrosaurids (*Maiasaura*, specifically) interpreted from the Late Cretaceous of Montana, also contain ground-up conifer wood tissue (Fig. 14.14). This coprolite evidence, in combination with the aforementioned gut contents from a similar species, corroborates a dietary choice of difficult-to-chew foodstuff, which certainly could have been handled by the dental batteries of both animals. However, two qualifications should be kept in mind before making any grand generalizations about ornithopod feeding preferences:

1 the "stomach contents," although within the body cavity of an *Edmontosaurus*, alternatively could have been washed into the open body after death; and
2 *Maiasaura* may not have necessarily made the coprolites, although they are closely associated with *Maiasaura* body fossils and no other likely tracemaker has been proposed.

One of the more interesting paleoecological side notes gained from analysis of the possible *Maiasaura* coprolites is that these products of ornithopod digestion also provided food for other animals in the Cretaceous ecosystem, namely dung beetles. The coprolites contain distinctive burrows, comparable to those made in modern fecal material by dung-eating beetles. This insight gained from dinosaur coprolites helps to better understand the flow of energy and matter between conifers, beetles, and dinosaurs during a thin slice of time from the Late Cretaceous.

Gastroliths are apparently either absent from or rarely associated with ornithopod remains. However, when Barnum Brown first discussed these dinosaur trace fossils in the early part of the twentieth century, he used polished stones found with a skeleton of the Late Cretaceous hadrosaurid *Claosaurus* as an example (Chapter 3). This proposed association was later rejected as too unconvincing. Indeed, no definitive examples of gastroliths are documented for ornithopods, despite their rich body fossil record with numerous near-complete skeletons. Either this lack is genuine and ornithopods did not have gastroliths, or they were present but overlooked in excavations of their remains. If the former were correct, then an absence of gastroliths in ornithopods could be explained by compensating adaptations of the teeth and jaws to grind plants in the mouth rather than in the gut. Additionally, heterodontosaurids and other non-iguanodontians might be expected to have gastroliths because their dentition was not as advanced as that of iguanodontians. Interestingly, this generalization does not hold up well for at least one species of ceratopsian that has both well-developed dental batteries and gastroliths (Chapter 13). Consequently, any future discoveries of ornithopods that contain stones consistent with gastroliths might provide more questions than answers.

Social Life

As mentioned earlier in various contexts, a few species of ornithopods, particularly some from the Late Cretaceous, exemplify the best-known social lives of all dinosaurs. Indeed, acquisition and dissemination of this knowledge contributed in a major way to changes in both scientific and public perceptions of dinosaurs in general. The aforementioned nesting grounds of the iguanodontians *Maiasaura* and *Hypacrosaurus* in the Late Cretaceous of Montana comprise the most persuasive

11

evidence of large communities, perhaps with thousands of individuals of the same species. Furthermore, a mass burial site of *Maiasaura* in Montana has an estimated 10,000 individuals; in this case, these dinosaurs possibly succumbed to the suffocating effects of volcanic ash from a nearby eruption (Chapter 7). Anatomical features, such as humpbacks, dewlaps, and cranial crests that are hollow in some instances, suggest intraspecific competition and communication. Lastly, Early and Late Cretaceous ornithopod trackways suggest family structures and show remarkable consistencies in their directionality. In some track horizons, tracks of as many as 80 ornithopods indicate movement in the same direction. All of these data collectively led to the reasonable proposal that some species of ornithopods moved as herds and possibly migrated.

The preceding evidence provides at least a preliminary hypothesis to address the question "How did ornithopods defend themselves against predators?": sheer numbers. As mentioned previously, the early explanation for ornithopod protection built upon the presumption that large herbivorous dinosaurs lived near or in water bodies, and swam away from rapacious predators. After this hypothesis was placed in doubt, alternatives were sought. In one instance, some paleontologists originally regarded the enlarged and pointed digit I (thumb) on the manus of *Iguanodon* as a possible weapon against large theropods. However, simply because an anatomical feature looks like a weapon to humans does not mean that this was its primary function. Indeed, no independent evidence has supported this hypothesis. For perspective, an *Iguanodon* using its thumb as a primary defense against a theropod of equal or larger size is comparable to a human using a small knife against a mountain lion (*Puma concolor*). Based on current evidence, "strength in numbers" and vocalizations to communicate potential threats were more likely defenses than were individual ornithopod responses.

Health

Ornithopods, like most dinosaurs, were mostly healthy animals, but a few specimens have evidence of paleopathological conditions. For example, one specimen of *Camptosaurus* shows fractures in the caudal vertebrae and ilium. Fractured caudal vertebrae also have been reported in hadrosaurids. Some paleontologists have speculated that these fractures are related to mating, caused by stress placed on the caudal region of a receptive (or not so receptive) female by an enthusiastic and vigorous male weighing several tonnes. Unfortunately, independent data to test this somewhat ribald speculation have yet to be found. Another example of a bone fracture in ornithopods is in a specimen of *Iguanodon*, which had an injured ischium. Two other specimens of *Iguanodon* show evidence of osteoarthritis in bone overgrowths called **osteophytes**, which are visible in the anklebones of these dinosaurs. Nonetheless, osteoarthritis is actually rare in dinosaurs, despite this reported occurrence in two individuals of the same ornithopod species.

Some ornithopods do show evidence of having been prey for theropods and were at least scavenged by crocodiles, as indicated by both toothmarks and coprolites. Late Cretaceous hadrosaurids in particular seem to have been on the menu for theropods both large and small. As far as predation is concerned, the famous specimen of *Edmontosaurus* (Fig. 11.7A), nicknamed "The One That Got Away," shows a large-diameter but healed bite mark in its caudal vertebrae. This, and another *Edmontosaurus* with a tyrannosaurid tooth embedded in a healed rib, comprise undoubted indicators of at least two tyrannosaurids that preyed on live hadrosaurids. A convincing but still equivocal example of a predator–prey relationship between a theropod and ornithopod is a find of several of the Early Cretaceous dromaeosaur *Deinonychus* surrounding a single iguanodontian *Tenontosaurus*, as described in detail earlier (Chapters 7 and 9).

SUMMARY

Ornithopods have arguably the best fossil record of any major dinosaur clade and are well represented by body fossils, including eggs and growth series from embryos to full-sized adults in some species; and trace fossils, such as trackways, nests, and coprolites. As a result of this combination of quality and quantity of fossil material, lifestyles of some species of ornithopods are among the best understood of all dinosaurs. This understanding has lead to revisions of some stereotypes of dinosaurs in general. Recent conceptions about dinosaurs resulted from the study of ornithopods and include intraspecific visual and vocal communication, nurturing behavior, formation of communal nesting grounds, and herding.

The fossil record for primitive ornithischians, which may have been linked to ornithopods begins during the Late Triassic, but the first undoubted ornithopods are the Early Jurassic heterodontosaurids of South Africa. These ornithopods, exemplified by *Heterodontosaurus*, were small and probably quick-moving bipedal herbivores with differentiated teeth as their most distinctive characteristic. Soon after the arrival of these primitive ornithopods on the Jurassic scene, other non-iguanodontians, such as *Hypsilophodon* and *Orodromeus*, dispersed throughout most continents from the Middle Jurassic through to the Late Cretaceous. These relatively larger bipedal ornithopods are mostly understood through a limited body fossil record, and their most striking evolutionary development was pleurokinesis, which aided considerably in processing vegetation.

By far the most successful clade of ornithopods in diversity, evolutionary specializations, and representation in the fossil record through sheer numbers was Iguanodontia. Iguanodontians eventually lived in most terrestrial environments on all continents during the Late Jurassic through to the Late Cretaceous. The Early Cretaceous *Iguanodon* is the most famous genus of the iguanodontians, but the hadrosaurids, which probably also began in the Early Cretaceous, are the best known of all ornithopods. Hadrosaurids, which include hadrosaurines (*Hadrosaurus*, *Maiasaura*) and lambeosaurines (*Lambeosaurus*, *Parasaurolophus*), are some of the most ornate and socially complex of dinosaurs. These dinosaurs were large (some as long as 15 meters), quadrupedal or bipedal animals that developed specialized adaptations to feeding, such as extensive and continually replaced dental batteries. They also have the best-understood reproductive cycles of all dinosaurs, epitomized by the nesting grounds interpreted for *Maiasaura* and *Hypacrosaurus*.

11

DISCUSSION QUESTIONS

1. What anatomical constraints cause human jaws to be incapable of pleurokinesis? If humans were capable of pleurokinesis, how would their chewing be changed and what effect might this have on food choices?

2. Examine Tables 11.1 and 11.2 and count the number of genera that occur with each of the following time divisions: Early Jurassic, Middle Jurassic, Late Jurassic, Early Cretaceous, Late Cretaceous. What general trend in ornithopod diversity can you hypothesize on the basis of these data? What are possible sources of error in your analysis? (Hint: Making a bar graph with the number of genera on the y-axis and time divisions on the x-axis may help illustrate your point.)

3. A currently accepted hypothesis about *Maiasaura* is that they lived in communal nesting grounds where many individuals of the same species occupied an area at the same time and returned to it annually. What new evidence could change the current interpretation to one that would indicate that at least some *Maiasaura* mothers may have lived a solitary existence, raising their juveniles without others of their species nearby?

4. Hum a popular song for a minute, then abruptly open your mouth while trying to continue to hum the song.
   a. How did the sound change, especially with regard to its frequency?
   b. How did its wavelength change in accordance with the frequency?
   c. Based on your preliminary experimentation, do you think that some hadrosaurids would have been more likely to have opened or shut their mouths when using low-frequency sounds?
   d. Would the hadrosaurids have been inhaling or exhaling while making their sounds? On what basis can you make such an assumption?

5. During a warm day (35°C) of the Early Cretaceous in western North America, a herd of iguanodontians are browsing through a coastal forest when a pack of dromaeosaurs begins stalking them. The lead dromaeosaur, only 15 meters away from an iguanodontian at the back of the herd, steps on a twig from an angiosperm. The twig breaks with a loud crack that is heard by the iguanodontian, alerting it to the presence of the predators. Two seconds after hearing the sound, it makes a loud warning sound to the others in the herd.
   a. Using the given data and information presented previously in this book, what are probable species representatives of the dromaeosaur and iguanodontian that lived in the same area and at the same time?
   b. What evidence from the fossil record supports that the species of dromaeosaur was a pack hunter that preyed upon the species of iguanodontian?

DISCUSSION QUESTIONS Continued

    c. How much faster did the sound of the breaking twig travel to the iguanodontian than if the temperature had been 25°C?

    d. Assuming that the iguanodontian farthest away from the one at the back of the herd is 150 meters distant, how much time would have elapsed between the breaking of the twig and this iguanodontian receiving the warning sound?

6. The chapter provides the values of 0.085 m^3 and 85.5 liters for a lambeosaurine egg clutch. How would you arrive at these figures? What are some of the assumptions made in the calculations and what are possible sources of error in the extrapolation?

7. List the anatomical traits that characterize Heterodontosauridae and Iguanodontia, as well as any other distinguishing information known for them. What general evolutionary trends can you pick out that occurred for ornithopods from the Early Jurassic through to the Late Cretaceous? What could have caused those trends?

8. What evidence would you find the most convincing that some hadrosaurids did not take care of their young? List all of the possible forms of evidence and rank them in order from most reliable to least reliable, while explaining why you ranked them this way.

9. Current hypotheses about ornithopods do not support the previously held ones about some species living semi-aquatic lifestyles. What new evidence would better support the notion that at least some species were semi-aquatic on a regular basis?

10. A dinosaur paleontologist finds a healed bite mark from a large theropod in an adult specimen of *Edmontosaurus*. How could that paleontologist figure out when the ornithopod was bitten? In other words, how could the paleontologist answer the question as to whether it was attacked as a young juvenile rather than a full-grown adult?

Bibliography

Brett-Surman, M. K. 1997. "Ornithopods". *In* Farlow, J. O. and Brett-Surman, M. K. (Eds), *The Complete Dinosaur*, Bloomington, Indiana: Indiana University Press. pp. 330–346.

Carrano, M. T., Janis, C. M. and Sepkoski, J. J., Jr. 1999. Hadrosaurs as ungulate parallels: Lost lifestyles and deficient data. *Acta Palaeontologica Polonica* **44**: 237–261.

Colbert, E. H. 1951. Environment and adaptations of certain dinosaurs. *Biological Reviews of the Cambridge Philosophical Society* **26**: 265–284.

Cotton, W. D., Cotton, J. E. and Hunt, A. P. 1998. Evidence for social behavior in ornithopod dinosaurs from the Dakota Group of northeastern New Mexico, U.S.A. *Ichnos* **6**: 141–149.

Dodson, P. 1975. Taxonomic implications of relative growth in lambeosaurine hadrosaurs. *Systematic Zoology* **24**: 37–54.

Forster, C. A. 1990. Evidence for juvenile groups in the ornithopod dinosaur *Tenontosaurus tilletti* Ostrom. *Journal of Paleontology* **64**: 164–165.

Forster, C. A. 1997. "Hadrosauridae". *In* Currie, P. J. and Padian, K. (Eds), *Encyclopedia of Dinosaurs*. New York: Academic Press. pp. 293–299.

Galton, P. M. 1971. The mode of life of *Hypsilophodon*, the supposedly arboreal ornithopod dinosaur. *Lethaia* **4**: 453–465.

Horner, J. R. 2000. The bone histology of the hadrosaurid dinosaur *Maiasaura peeblesorum*: Growth dynamics and physiology based on an ontogenetic series of skeletal elements. *Journal of Vertebrate Paleontology* **20**: 109–123.

Horner, J. R. and Makela, R. 1979. Nest of juveniles provides evidence of family structure among dinosaurs. *Nature* **282**: 296–298.

Horner, J. R. and Dobb, E. 1997. *Dinosaur Lives*. San Diego, California: Harcourt Brace and Company.

Horner, J. R., Weishampel, D. B. and Forster, C. A. 2004. "Hadrosauridae". *In* Weishampel, D. B., Dodson, P. and Osmólska, H. (Eds), *The Dinosauria* (2nd Edition). Berkeley, California: University of California Press. pp. 438–463.

Lockley, M. G., Hunt, A. P. and Matsukawa, M. 1999. Three age groups of ornithopods inferred from footprints in the Mid-Cretaceous Dakota Group, eastern Colorado, North America. *Palaeogeography, Palaeoclimatology, Palaeoecology* **147**: 39–51.

Nadon, G. C. 1993. The association of anastomosed fluvial deposits and dinosaur tracks, eggs, and nests: implications for the interpretation of floodplain environments and a possible survival strategy for ornithopods. *Palaios* **8**: 31–44.

Norman, D. D. 1989. "Ornithopod dinosaurs; relationships, structure, and habits". *In* Padian, K. and Chure, D. J. (Eds), *The Age of Dinosaurs. Paleontological Society, Short Courses in Paleontology* **2**: 58–70.

Norman, D. B. 2004. "Basal Iguanodontia". *In* Weishampel, D. B., Dodson, P. and Osmólska, H. (Eds), *The Dinosauria* (2nd Edition). Berkeley, California: University of California Press. pp. 413–437.

Norman, D. B. and Weishampel, D. B. 1985. Ornithopod feeding mechanisms: their bearing on the evolution of herbivory. *American Naturalist* **126**: 151–164.

Norman, D. B., Sues, H.-D., Witmer, L. M. and Coria, R. A. 2004. "Basal Ornithopoda". *In* Weishampel, D. B., Dodson, P. and Osmólska, H. (Eds), *The Dinosauria* (2nd Edition). Berkeley, California: University of California Press. pp. 393–412.

Stokes, W. L. 1987. Dinosaur gastroliths revisited. *Journal of Paleontology* **61**: 1242–1246.

Weishampel, D. B. 1981. Acoustical analysis of potential vocalization in lambeosaurine dinosaurs (Reptilia, Ornithischia). *Paleobiology* **7**: 252–261.

Weishampel, D. B. 1984. Evolution of jaw mechanisms in ornithopod dinosaurs. *Advances in Anatomy, Embryology, and Cell Biology* **87**: 109.

Chapter 12

You visit the gift shop of your local natural history museum and, as you reach into a bin of plastic models of dinosaurs, pterosaurs, and mammoths, your hand encounters an upraised ridge on one model. You see that it is Stegosaurus and the "ridge" is actually a row of plates on its dorsal surface. While you are waiting at the checkout counter, you notice a poster that depicts a stegosaur in life, with its plates overlapping and arranged in two rows. You look at the model in your hand and see that it only has a single row of plates that do not overlap.

Which restoration is most likely, based on the available evidence? What was the function of the plates? How could the arrangement of the plates have an impact on interpretations about their function?

Thyreophora

Why Study Thyreophorans?

Thyreophora (= "shield bearer") consists of both the ankylosaurs and stegosaurs (Tables 12.1 and 12.2). Thyreophorans are sometimes col-

TABLE 12.1 Representative genera of Ankylosauria with approximate geologic age and where they occur. An = Ankylosauridae, No = Nodosauridae. Basal thyreophorans *Emausaurus*, *Scutellosaurus*, and *Scelidosaurus* of the Early Jurassic were not included

Genus	Age	Geographic Location
Anamantrax (No)	Late Cretaceous	Western USA
Ankylosaurus (An)	Late Cretaceous	Western USA, Alberta, Canada
Cedarpelta (No)	Early Cretaceous	Western USA
Crichtonsaurus (An)	Late Cretaceous	China
Dracopelta (No)	Late Jurassic	Portugal
Edmontia (No)	Late Cretaceous	Western USA, Alberta, Canada
Euoplocephalus (An)	Late Cretaceous	Western USA, Alberta, Canada
Gargoyleosaurus (An)	Late Jurassic	Western USA
Gastonia (An)	Early Cretaceous	Western USA
Gobisaurus (An)	Early Cretaceous	Mongolia
Hylaeosaurus (No)	Early Cretaceous	UK, France, Spain
Maleevus (An)	Late Cretaceous	Mongolia
Minmi (An)	Early Cretaceous	Australia
Mymoorapelta (An)	Late Jurassic	Western USA
Niobrarasaurus (An)	Late Cretaceous	Western USA
Nodosaurus (No)	Late Cretaceous	Western USA
Panoplosaurus (No)	Late Cretaceous	Western USA
Pawpawsaurus (No)	Early Cretaceous	Western USA
Pinacosaurus (No)	Late Cretaceous	China, Mongolia
Panoplosaurus (No)	Late Cretaceous	Alberta, Canada.
Polacanthus (An)	Early Cretaceous	UK
Saichania (An)	Late Cretaceous	Mongolia
Sarcolestes (An)	Middle Jurassic	UK
Sauropelta (No)	Early Cretaceous	Western USA
Shamosaurus (An)	Early Cretaceous	Mongolia
Silvisaurus (An)	Early Cretaceous	Western USA
Struthiosaurus (No)	Late Cretaceous	Austria, France, Romania
Talarurus (An)	Late Cretaceous	Mongolia
Tarchia (An)	Late Cretaceous	Mongolia
Tianzhenosaurus (An)	Late Cretaceous	China

TABLE 12.2 Representative genera of Stegosauria with approximate geologic age and where they occur.

Genus	Age	Geographic Location
Chialingosaurus	Middle Jurassic	China
Chunkingosaurus	Late Jurassic	China
Dacentrurus	Late Jurassic	Western Europe
Dravidosaurus	Late Cretaceous	India
Hesperosaurus	Late Jurassic	Western Europe
Huayangosaurus	Middle Jurassic	China
Kentrosaurus	Late Jurassic	Tanzania
Lexovisaurus	Late Jurassic	UK, France
Monokonosaurus	Early Cretaceous	China
Stegosaurus	Late Jurassic	Western USA
Tuojiangosaurus	Late Jurassic	China
Wuerhosaurus	Early Cretaceous	China

loquially known as the "armored dinosaurs" in recognition of their abundant, well-developed external bony parts, such as knobs and spikes. The only other dinosaurs that possessed body armor were titanosaurids (Chapter 10), thus this trait was apparently unusual outside of Thyreophora. Because of their impressive osteodermal accouterments and the inference that all thyreophorans were herbivores, they are often used as an example of animals that engaged in an evolutionary "arms race" with their potential predators, particularly theropods (Chapter 9). Fossil evidence suggests that defenses by sauropods (Chapter 10) against theropods included extreme size, and those by ornithopods (Chapter 11) included large numbers. In contrast, thyreophorans stayed within relatively small size parameters (by dinosaur standards) and made themselves individually difficult to attack. They not only used body armor for protection, but some also had clubbed and spiked tails. Their body armor probably resulted in sacrificing mobility. Coupled with short limbs, their armor may have kept them close to the ground and their ventral surfaces relatively safe from assault.

Despite the defensive emphasis of such prominent features, thyreophorans probably had multiple purposes for their more visual osteoderms, such as display for sexual attraction, forms of intraspecific competition or, in the case of some stegosaurs, body heat regulation. Unfortunately, little more is known about thyreophoran lifestyles than through their skeletal remains. Some trackways from ankylosaurs and stegosaurs have been identified, but no eggs, nest structures, toothmarks, or coprolites have been tied to thyreophorans. Consequently, an approach that uses multiple lines of evidence must be limited to considering the functional morphology and biomechanics of thyreophorans. Because this information is all that paleontologists have for forming their hypotheses about thyreophorans, much of what is presented in this chapter may change considerably in the future with new discoveries. Regardless, thyreophorans provide an extreme in dermal armor never seen in any other land-dwelling vertebrates. For this reason alone, study of how they evolved to such a unique status should give insights on the evolution of plant–herbivore and predator–prey relationships.

Definition and Unique Characteristics of Thyreophora

Thyreophora is a stem-based clade within Ornithischia and Genasauria, and a sister clade to Cerapoda within Genasauria. Thyreophora contains two stem-based clades

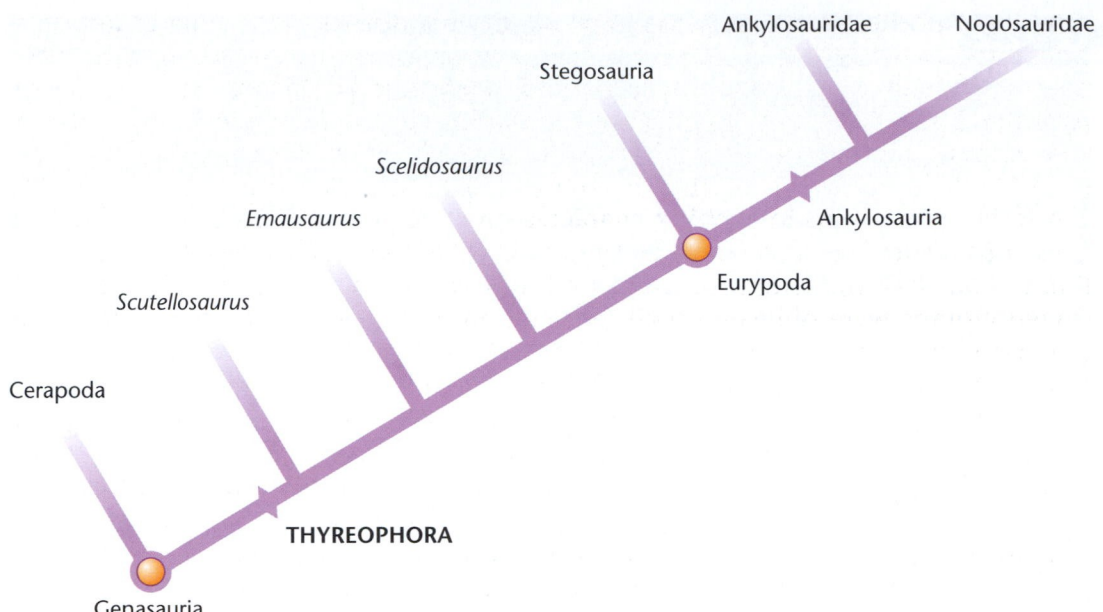

FIGURE 12.1 Cladogram showing interrelationships between basal thyreophorans (*Scelidosaurus*, *Scutellosaurus*, and *Emausaurus*) and other clades within Thyreophora, particularly Ankylosauria and Stegosauria.

of importance (Fig. 12.1): **Ankylosauria**, which includes **Nodosauridae** (nodosaurids) and **Ankylosauridae** (ankylosaurids); and **Stegosauria** (stegosaurs). Members of Thyreophora have been known since near the beginning of dinosaur studies. The nodosaurid *Hylaeosaurus* was named by Gideon Mantell in 1833, followed several decades later by T. H. Huxley's naming of the nodosaurid *Polacanthus* and ankylosaurid *Acanthopholis* (Chapter 3). Richard Owen also described the basal thyreophoran *Scelidosaurus* from a well-preserved specimen in 1861. By comparison, stegosaurs were among the later discoveries. Public knowledge of their existence commenced with the naming of *Stegosaurus* in 1877 by O. C. Marsh. A few years before the first description of *Stegosaurus*, fragments of the European stegosaurs *Lexovisaurus* and *Dacentrurus* had been studied but were not allied with a stegosaur origin until early in the twentieth century.

As with most clades, long, detailed lists of derived characters and cladograms are used to define Thyreophora. However, this clade can be described more simply by contrasting it with most other dinosaur clades.

For the purposes of this discussion, the following two criteria, especially the first, are key features for thyreophorans:

1 dermal armor, typically as scale-like **scutes**, in rows parallel to the midline of the body on both lateral and dorsal surfaces of the torso; and
2 well-developed postorbital process associated with the jugal and a **palpebral** (supraorbital bone).

Although the most recognizable trait of a thyreophoran is its dermal armor, this feature is not diagnostic by itself. In fact, an assumption that dermal armor was a foolproof identifier of a former thyreophoran presence led to an interpretation that they lived in South America because isolated osteoderms had been found in some Cretaceous deposits there. Much later, these structures were instead found in association with titanosaurids (Chapter 10). As a result, thyreophoran identifications are now applied more cautiously, and they are still poorly known in South

America. Another qualifier is that not all thyreophorans were equally armored. Ankylosaurs had hundreds of scutes over a majority of the surface areas of their bodies, whereas stegosaurs only had localized patches and plates. Thus, an initial hypothesis that dinosaur remains are thyreophoran can be made when scutes or any other osteoderms are found in continental Jurassic or Cretaceous deposits, but must be followed by more stringent identifications and analyses.

Another general distinguishing characteristic of thyreophorans is a squat stature caused by short legs in relation to the rest of the body. These legs also have fore limbs about half the length of the hind limbs (a ratio of about 35 : 65). Nearly all thyreophorans were obligate quadrupeds because of these proportions. One basal thyreophoran, *Scutellosaurus* from the Early Jurassic of the southwestern USA, was possibly bipedal because of its relatively shorter fore limbs. However, other features, such as its large number of osteoderms, a long torso and fore limbs, and big manus, argue more for quadrupedalism. This mixture of features may mean that it was descended from bipedal ancestors and represents a transition between the two types of locomotion (Chapter 6). Secondary quadrupedalism is typical for quadrupedal dinosaurs in general, but was certainly a necessary adaptation for ankylosaurs in particular, once they developed more extensive body armor.

Thyreophorans were ornithischians but their deeply inset cheek teeth also help to unite them with Genasauria. Thyreophoran teeth represented variations on a main theme: they began as laterally compressed teeth crowned by apical denticles, which was also typical of basal ornithischians (Chapter 11). They then became less developed in later thyreophorans. Ankylosaurs and stegosaurs both had cheek teeth but they were more medial (inward). The result was broad shelves on their dentaries and maxillas, which gave them space for large cheeks.

Ankylosaurs and stegosaurs also had beaks in the anterior portions of their skulls with a horny covering called a **rhampotheca**. In all but one genera of stegosaur, the Middle Jurassic *Huayangosaurus* of China, the premaxillary teeth are missing. Likewise, ankylosaurids and all but the most primitive nodosaurids, such as the Early Cretaceous *Silvisaurus* of Kansas, have no premaxillary teeth. This implies that their beaks were the main tools used for cropping plants. Cheek teeth in stegosaurs had triangular profiles and broad crowns but long roots. The crowns were broadest at their bases, forming a **cignulum** (small shelf), and were distinguished by numerous vertically oriented ridges. Ankylosaur cheek teeth had a comparable morphology but were more poorly-developed versions of stegosaur teeth. Overall, thyreophoran teeth reflect their common ancestry, although some changes through time are evident in their lineages. These differences are attributable to evolutionary trends in response to available foodstuff in Mesozoic terrestrial ecosystems.

Clades and Species of Thyreophora

Ankylosauria: Ankylosauridae and Nodosauridae

Ankylosauria derives its name from the encasement of its members in dermal armor (*ankylos* = "fused," *sauros* = "lizard"), which is why many paleontologists describe them as "tanks." This military allusion is apt in terms of the function of the armor, as certainly defense may have been one function, but its use for offense and species recognition is also possible. The following characters help to identify members of Ankylosauria (Fig. 12.2):

- A broad, laterally compressed skull with armor covering the cranial sutures and the supratemporal fenestra.
- Deeply inset cheek teeth, with cingulum on dentary and maxillary teeth.

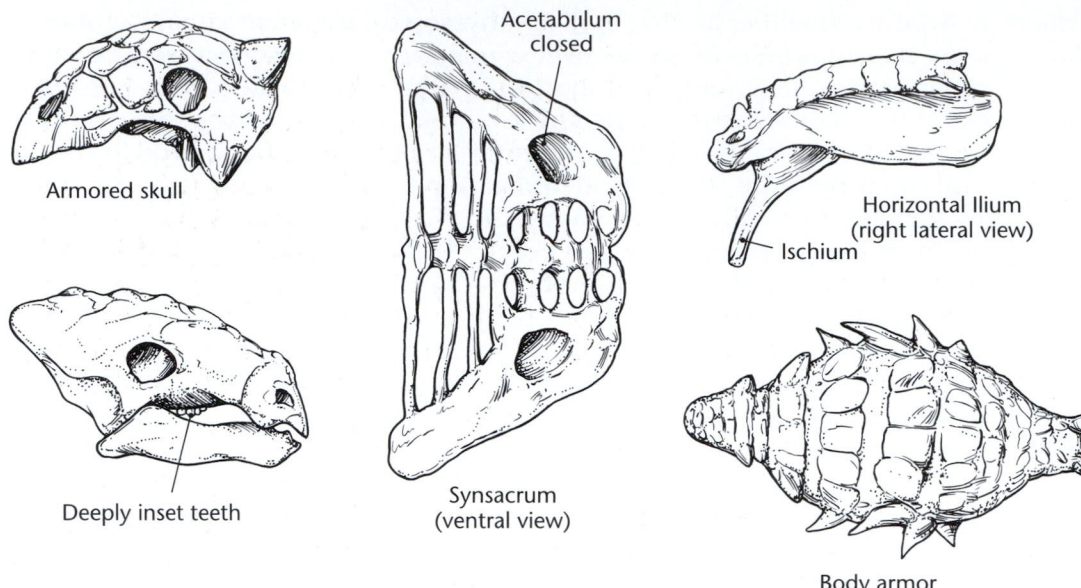

FIGURE 12.2 Defining character traits of Clade Ankylosauria: broad, armored skull with deeply inset cheek teeth; synsacrum; horizontal ilium; closed acetabulum; and body armor.

- Fusion of the precaudal third and fourth dorsal vertebrae with the sacrum into a structure called a **synsacrum**.
- A horizontally oriented ilium (evident from a lateral view) and a reduced pubis.
- Separation of the pubis from the acetabulum, with a closing of the acetabulum so that it is more like a cup for fitting the femur, rather than an open hole.
- Abundant body armor (Fig. 12.3).

Most of these features reflect adaptations that provided for protection but also allowed an ankylosaur's body to move about freely, even though armored. For example, the synsacrum provided support in the axial skeleton for body armor that covered the dorsal surface. This trait appeared in combination with a secondarily closed acetabulum. The latter is a unique feature in dinosaurs, which are partially defined by having an open acetabulum as a primitive trait (Chapters 1 and 5). The horizontally-oriented ilium also provided attachments for the muscles needed to accommodate a body laden with osteoderms. Furthermore, ankylosaur limb bones are thicker in proportion to limb lengths than most dinosaur limb bones. This means that their limbs were adapted for upright and compact ankylosaur bodies, which had great weights relative to volume. Ankylosaur fore limbs and hind limbs ended in a broad manus and pes, each with stout toes terminated by unguals (hooves). Unlike some other major dinosaur clades, digit counts for each foot differed for some ankylosaurs. For example, a five-toed manus is known for the Early Cretaceous *Shamosaurus* and *Sauropelta*, as well as the Late Cretaceous *Talarurus* and *Pinacosaurus*, but this same number has not been documented conclusively in other ankylosaurs. Numbers of digits in the pes of ankylosaurs range from five (*Sauropelta*) to four (*Nodosaurus*, *Talarurus*) to three (*Euoplocephalus*). Consequently, ankylosaur tracks are difficult to distinguish from similarly-sized stegosaur or ceratopsian tracks on the basis of manus–pes digit numbers alone (Chapter 14).

Ankylosauridae and Nodosauridae, the two clades comprising Ankylosauria, are difficult to distinguish from one another.

FIGURE 12.3 Closely spaced osteoderms of a typical Late Jurassic ankylosaur, which likely provided some excellent protection against predators from the same time. College of Eastern Utah Prehistoric Museum, Price, Utah.

Ankylosaurs can be classified into two clades by looking at either end of one. For example, most skulls of ankylosaurids have:

1 nares that face anteriorly;
2 horns at the posterior dorsal and ventral corners; and
3 a triangular profile when viewed by looking down on the dorsal surface.

In contrast, nodosaurids had:

1 laterally placed nares;
2 lacked horns; and
3 more rounded profiles to their skulls (Fig. 12.4).

Skull interiors were also markedly different. Ankylosaurid nares led into complicated and sinuous nasal chambers, whereas nodosaurids only had a single tube associated with each naris that connected with the throat.

At the other end of these dinosaurs, the tails were also different in the two clades. Ankylosaurids had long processes of the distal caudal vertebrae that reinforced the tail. This provided a handle for a bony club, composed of two pairs of large and small osteoderms (Fig. 12.5). Nodosaurids lacked such modifications to their

12

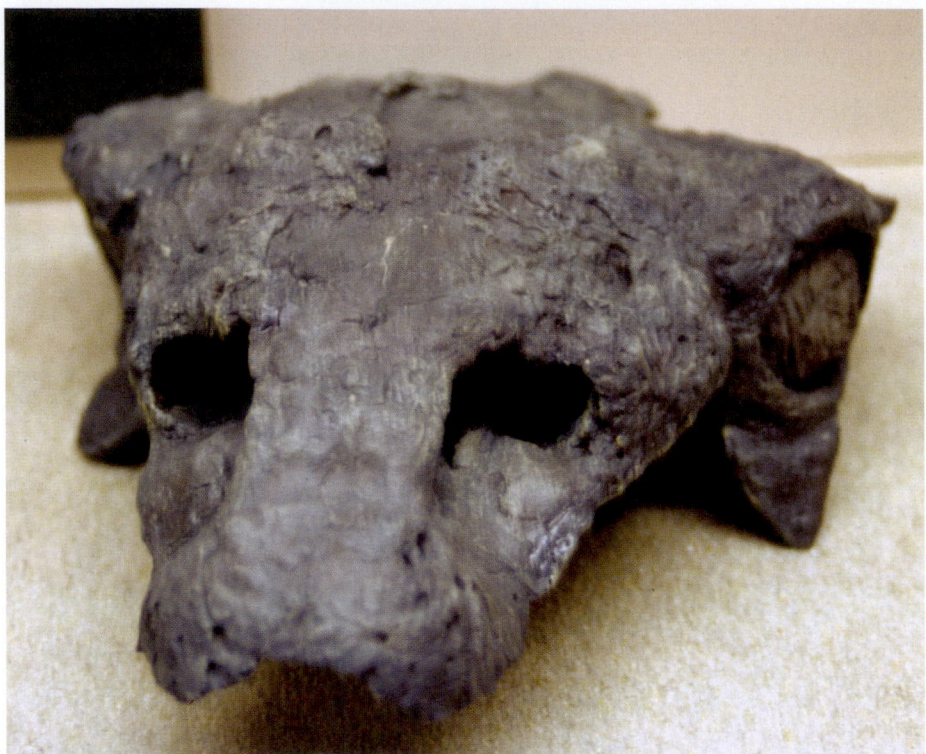

FIGURE 12.4 Late nodosaurid skull, showing typical traits for a skull of its clade: laterally placed nares, hornless, and a rounded shape. Compare with skull of *Gargoyleosaurus* in Figure 12.6. College of Eastern Utah Prehistoric Museum, Pice, Utah.

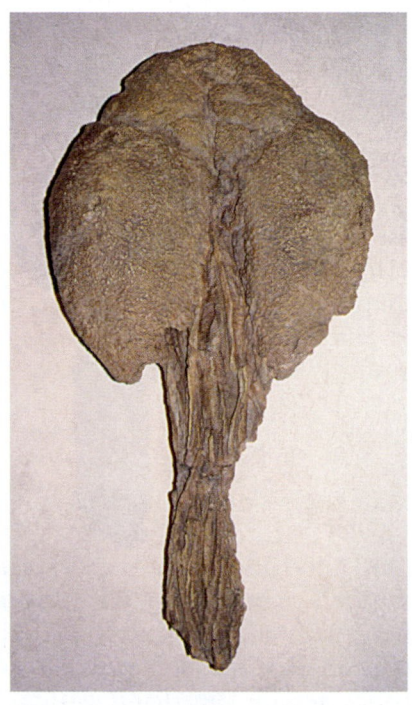

FIGURE 12.5 Tail club of the Late Cretaceous ankylosaurid *Ankylosaurus*, composed of paired osteoderms. Denver Museum Science and Nature, Denver, Colorado.

FIGURE 12.6 *Gargoyleosaurus parkpini*, a Late Jurassic ankylosaurid from the Morrison Formation of Wyoming, USA. Denver Museum of Science and Nature, Denver, Colorado.

caudal vertebrae and as a result lacked clubs. Some nodosaurids did carry a large spike that projected horizontally from their sides, which would have deterred most predators from random biting. An interesting observation made about some of these spikes is that they show muscle-attachment sites at their bases and could articulate with one another. These features support the hypothesis that the spikes could have been moved in a scissor-like motion. Such a function would have had a deleterious effect on any living flesh caught between them, such as phalanges on a theropod manus.

Ankylosaurs made their earliest known appearance in the fossil record in the Middle Jurassic, represented by the basal ankylosaur *Sarcolestes* of England, which was succeeded by the Late Jurassic *Dracopelta* of Portugal and *Mymoorapelta* and *Gargoyleosaurus* of western North America (Fig. 12.6). The earliest known and most basal nodosaurid is the Early Cretaceous *Cedarpelta* of Utah; no Jurassic nodosaurids are so far known. *Cedarpelta* differs from all other nodosaurids because of its general paucity of cranial osteoderms, meaning that later members of its lineage were more adorned. The large number of Early Cretaceous ankylosaurs bespeaks of their later diversification: *Hylaeosaurus* and *Polacanthus* of western Europe; *Gastonia*, *Pawpawsaurus*, *Sauropelta*, and *Silvisaurus* of western North America; and *Shamosaurus* of Mongolia. The Early Cretaceous *Minmi* of Australia was once considered a transitional form between ankylosaurids and nodosaurids, but is now classified as a primitive ankylosaurid. This confusion is understandable in the light of how this genus is known only from two partial skeletons, of which only one has a skull. Indeed, the fragmentary nature of the single specimens that define many ankylosaur species means that their designations as "ankylosaurid" or "nodosaurid" have been subject to revisions with each new find and more detailed cladistic analyses.

12

371

FIGURE 12.7 Ankylosaur track (probably the pes) in the Lower Cretaceous preserved as natural cast on bottom of a stratum and cross-cut by a theropod track; Cañon City, Colorado.

Late Cretaceous species of ankylosaurs represent evolutionary trends toward increased size and degree of armoring. For example, the Late Cretaceous *Ankylosaurus magniventris* of North America, the namesake of both Ankylosauria and Ankylosauridae, was arguably the largest known thyreophoran, with a total length of nearly 9 meters and an estimated weight of 4 tonnes. Other sizable ankylosaurids of the Late Cretaceous, such as *Maleevus*, *Saichania*, *Talarurus*, and *Tarchia*, mainly hail from Mongolia, as does one nodosaurid, *Pinacosaurus*. North American ankylosaurs of the Late Cretaceous are a mixture of nodosaurids, such as *Nodosaurus* and *Panoplosaurus*, and a few ankylosaurids besides *Ankylosaurus*, such as *Edmontia* and *Euoplocephalus*.

Ankylosaur trace fossils are more common than originally thought. Six trackways in the Upper Cretaceous El Molino Formation of Bolivia and gastroliths in a specimen of *Pinacosaurus* in Mongolia were the only trace fossils reported up to the year 2000. Fortunately, these have been supplemented since by discoveries of numerous trackways from western Canada, Colorado (Fig. 12.7), western Europe, and parts of Asia. These discoveries are probably owed to better search images for paleontologists of these once-elusive tracks. However, ankylosaur nests and coprolites remain unidentified from the fossil record, so hopefully these trace fossils will be found in the near future, so that they may add to a fuller understanding of ankylosaurs.

Clade Stegosauria

Clade Stegosauria is named on the basis of the shingle-like appearance of the dermal plates on its back (*stegos* = "roofed" and *sauros* = "lizard"). Two overt characteristics help to distinguish them from other thyreophorans:

1 **parascapular spines**, which are osteoderms evident as spikes on the shoulder regions; and
2 **parasagittal plates**, which are the dermal armor restricted to two rows of vertically-oriented plates, parallel to but also lateral to the axial skeleton.

Spikes also normally occur toward the posterior of stegosaurs and are often seen as additions to the caudal vertebrae. In contrast, the plates are typical of the anterior portion of the body but are still postcranial.

> *Perhaps the simplest way to be introduced to stegosaurs as a group is to look at their star member, the Late Jurassic* Stegosaurus stenops *(Fig. 12.8A).*

Stegosaurs are also identifiable by their small skulls relative to their body sizes, similar to sauropods (Chapter 10). In these small skulls were correspondingly diminutive brains, constituting some of the smallest EQs known for dinosaurs. Despite this shortcoming, stegosaurs as a clade lived minimally from the Middle Jurassic through to the Early Cretaceous, a span of about 70 million years.

Stegosaurus, by far the most popularly known and best-studied thyreophoran and stegosaur, was first discovered near Morrison,

(A)

(B)

FIGURE 12.8 *Stegosaurus stenops*, the most famous of thyreophorans and stegosaurs. (A) Juvenile specimen (a rare find), Denver Museum of Science and Nature, Denver, Colorado. (B) Cast of adult *S. stenops*, Fernbank Museum of Natural History, Atlanta, Georgia.

Colorado, and soon afterward in 1877 was given its name by O. C. Marsh. *Stegosaurus* was about 9 meters long and probably weighed 1.5 to 2.5 tonnes, so it was a relatively large thyreophoran. It is most easily recognized by:

1 horizontally positioned, paired, and pointed osteoderms (spikes) emanating from the distal caudal vertebrae; and
2 prominent osteoderms evident as vertically oriented plates on its dorsal surface and just lateral and parallel to the axial skeleton.

The *Stegosaurus* tail spikes and back plates were previously interpreted with different positions and have undergone many permutations in artistic re-creations and museum reconstructions. In older illustrations, the tail spikes are commonly shown as upright, and they may vary considerably in number, from four to eight – the correct number is four. The dorsal plates have also been reconstructed as a single row of non-overlapping plates, as a single row of offset and overlapping plates, or as two rows of opposing plates on either side of the spine. Analysis of several specimens uncovered recently, and of previous material, has revealed that the overlapping and offset arrangement is most likely the correct anatomical position for these plates. *Stegosaurus* also possessed numerous scutes that covered its throat, a form of additional protection to a potentially vulnerable area.

The much-ridiculed brain of *Stegosaurus* was indeed small; analyses of braincase endocasts reveal that it possessed a volume of about 60 cm^3. However, its olfactory bulbs were large compared to the rest of the brain, suggesting that *Stegosaurus* may have had an acute sense of smell. An oft-told "fact" about *Stegosaurus* is that its small brain was aided by a "second brain," located in the hip region, which helped control the motor functions of the tail and legs. This story began in the late nineteenth century with O. C. Marsh's observation of a large space in the sacral vertebrae. Later speculation led to the hypothesis that this space held a nerve bundle that operated the posterior half of the animal, which was communicated as a "second brain" in the sacrum. Unfortunately, this popularized label was repeated for decades in both serious and trivial publications that mentioned *Stegosaurus*. Later, a more thorough analysis revealed that the sacral spaces had more than enough volume to allow the passage of neural canals. This meant that the remaining space could have been filled with some other tissue. One proposed explanation is that it was a glycogen body used for storing carbohydrates; such an organ might have been used to supplement energy to the nervous system. Of course, this hypothesis also may be wrong, but it is certainly closer to being correct than the "second brain" idea.

The majority of known stegosaur genera are from China, consisting of the Middle Jurassic *Huayangosaurus* and *Chialingosaurus*, Late Jurassic *Chunkingosaurus* and *Tuojiangosaurus* (Fig. 12.9), and the Early Cretaceous *Wuerhosaurus* and *Monkonosaurus*. *Dacentrurus* and *Lexovisaurus* are two European stegosaurs from the Late Jurassic. Africa and Asia are represented by the Late Jurassic *Kentrosaurus* of Tanzania and stegosaur material from Thailand, respectively. Early Cretaceous stegosaurs outside of China include *Regnosaurus*, first described by Mantell in 1841 (Chapter 3), and *Craterosaurus* of the UK, as well as a few fragments from Lower Cretaceous strata of Argentina, Australia, and South Africa. *Dravidosaurus* of India is the only Late Cretaceous stegosaur positively identified so far, but even this was once re-interpreted as a plesiosaur (marine reptile: Chapter 6).

Stegosaur trace fossils, like ankylosaur trace fossils, were once considered rare but are now more commonly recognized. For example, tracks tentatively linked to early stegosaurs have been reported from the Early Jurassic of Australia, France, and Morocco. In addition, a few isolated Late Jurassic stegosaur tracks were also described from Utah, and Late Jurassic tracks located in Australia were also preliminarily identified as stegosaurian. Sadly, the latter tracks were never studied because

FIGURE 12.9 *Tuojiangosaurus*, a Late Jurassic stegosaur from China. Temporary display at Fernbank Museum of Natural History, Atlanta, Georgia.

thieves stole them before paleontologists could properly describe them (Chapter 2). In general, stegosaur tracks should be easier to tell apart from most other quadrupedal dinosaur tracks because of their unusual digit combinations – the manus has five toes and the pes has only three toes. However, this digital arrangement may have been the same in a few ankylosaurs, thus warranting some caution when attempting to identify the trackmakers. No gastroliths have been reported yet in association with a stegosaur skeleton, so how stegosaurs processed their food efficiently with such poorly-developed teeth remains a mystery. What they ate is also unknown, as no stegosaur coprolite or former gut contents have been interpreted from the geologic record. Like ankylosaurs, stegosaurs will be better understood when more of their trace fossils are recognized in tandem with their body fossils.

Paleobiogeography and Evolutionary History of Thyreophora

Thyreophoran skeletal remains have been found on every continent except Antarctica, but they are most common in the Cretaceous strata of North America and Asia.

Stegosaurs are firmly documented only in western North America, eastern Africa, western Europe, eastern Asia, and Australia. Moreover, their temporal distribution is limited to the Early Jurassic (at the earliest) through the Late Cretaceous, although Cretaceous stegosaurs are comparatively rare in relation to their Jurassic predecessors. Ankylosaurs are known from Middle Jurassic through to Late Cretaceous strata, and well-defined specimens have been recovered in North America, Europe, Asia, and Australia. Their remains also have been found in high-latitude localities, such as Alaska and Australia, meaning that a few species may qualify as "polar dinosaurs." Poorly-defined ankylosaur material is reported from South America, but well-documented ankylosaur trackways have been described from Upper Cretaceous strata in Bolivia. The latter discovery is the best evidence for ankylosaurs in South America, and it bodes well for future discoveries of more thyreophoran body and trace fossil finds in this and other southern continents.

Based on paleoenvironmental analyses of facies associated with thyreophoran body fossils, stegosaur remains occur in fluvial, near-shore marine, and lakeshore deposits. Ankylosaurs are in similar facies but specifically those indicating semi-arid conditions. Of course, taphonomic considerations dictate that the burial site of a dinosaur is not necessarily representative of where it lived (Chapter 7). Certainly in the cases of stegosaurs and some ankylosaurs, their skeletal material was buried in aquatic environments that an armored animal should have rarely frequented. However, the completeness and orientation of some ankylosaur skeletons in the Cretaceous of Mongolia argue that many of these skeletons underwent little or no transport. In taphonomic studies of these dinosaurs, the environments are hypothesized as semi-arid or arid with sparsely vegetated dunes. These dunes could have collapsed on hapless ankylosaurs; alternatively, they could have been buried by sandstorms (Chapter 7). Some of these ankylosaur skeletons show the bodies positioned horizontally to inclined, with legs underneath and heads pointed up. This evidence collectively suggests animals that were attempting to excavate themselves. An alternative hypothesis, that these 2- to 3-tonne dinosaurs were burrowers, is not taken seriously. In contrast to their Asian relatives, North American ankylosaurs seemed to have preferred moister areas, which correspondingly had more vegetation.

Interestingly, the previously low number of ankylosaur or stegosaur tracks reported from the geologic record was used as an indirect paleoenvironmental indicator. This is because if any of these animals dwelled in uplands well above coastal areas, preservation of their tracks would have been less likely. However, this argument lacks evidence, thus wherever extensive stegosaur or ankylosaur trackways are found, the hypothesis is easily falsified. Thus far, known ankylosaur tracks were made in sediments deposited on lakeshores and other coastal environments, which indicates that these ankylosaurs were not upland species.

As far as evolutionary lineages are concerned, Genasauria is the clade under which both ornithopods and thyreophorans are placed and both of these clades had a common ornithischian ancestor. Ornithischians as a clade began during the Late Triassic, but no thyreophorans are known from that time. Based on current evidence, thyreophorans probably did not develop until the Early Jurassic. The most basal thyreophorans are the Early Jurassic *Scelidosaurus* of England, *Scutellosaurus* of the southwestern USA, and *Emausaurus* of Germany, which are all considered as outgroups from the clades Ankylosauria and Stegosauria. Of these three, *Scutellosaurus*, a small (1- to 2-meters long), long-tailed ornithischian, is probably the most primitive. Moreover, it bears some resemblance to the primitive ornithischian *Lesothosaurus* (Chapter 11), which comes from Lower Jurassic strata of South Africa. Cladistic analyses, however, confirm that *Lesothosaurus* falls outside of Genasauria and is thus not as closely related to *Scutellosaurus* as the latter is to *Emausaurus* and *Scelidosaurus*. One major feature that distinguishes thyreophorans from other primitive ornithischians is abundant osteoderms (scutes) on the dorsal and lateral parts of the body.

Scelidosaurus, first described by Richard Owen in 1863 (Chapter 3), was closer to the ideal of a thyreophoran in that it was a robust (about 4 meters long) dinosaur with limb proportions more apt for a quadruped. It also had numerous osteoderms, which helps to identify it as an undoubted thyreophoran, although it cannot be placed strictly within either Ankylosauria or Stegosauria. Interestingly, nearly 130 years passed until the description of another basal thyreophoran from the Early Jurassic, *Emausaurus*.

Of the thyreophorans, ankylosaurs were apparently the most conservative in their evolutionary histories. Once they had evolved their extensive body armor and other distinctive ankylosaur features, they stayed within the parameters of this basic body plan throughout the entirety of their geologic range. Nevertheless, they certainly underwent some diversification during the Late Cretaceous and especially trended toward larger sizes. Stegosaurs, which did not have such a long geologic range, probably underwent rapid evolution from the Early to Middle Jurassic before their greatest success during the Late Jurassic. Although still present during the Cretaceous, they were seemingly sparsely distributed and especially rare during the Late Cretaceous. Because so few thyreophoran fossils are known from between the stratigraphic intervals of the stegosaur- and ankylosaur-bearing zones, little is known about their lineages or the probable factors influencing their natural selection.

Thyreophorans as Living Animals

Reproduction and Growth

Evidence relating to thyreophoran reproduction is almost unknown and represents a huge gap in knowledge about their paleobiology. No nests or eggs, let alone embryos, from either ankylosaurs or stegosaurs have been discovered. In fact, of all the dinosaur eggs that have been described, none have been allied in even a general sense to thyreophorans.

A little more is known about thyreophoran growth because of some skeletal material from the Late Jurassic for stegosaurs and the Late Cretaceous for ankylosaurs. Juvenile and other subadult thyreophorans are rare finds. For example, only one juvenile of *Stegosaurus*, the most famous thyreophoran from the geologic record, has been found so far (Fig. 12.8A). Other stegosaurs with subadults represented are *Lexovisaurus* and *Dacentrurus* from western Europe and *Kentrosaurus* from Tanzania. Unfortunately, none of these three genera are abundant enough to reconstruct growth series like those interpreted for some theropods and ornithopods (Chapters 9 and 11). The best represented growth series for thyreophorans is provided by a single species of ankylosaurid, the Late Cretaceous *Pinacosaurus* of China, for which groups of as many as a dozen juveniles have been found together. The implications of this monospecific assemblage as applied to ankylosaur sociality are discussed later.

Of course, courting preceded reproduction, and some thyreophorans may have rivaled ornithopods (Chapter 11) and ceratopsians (Chapter 13) in advertising their wares. Starting in the 1970s, dinosaur paleontologists have proposed that the dorsal plates of stegosaurs and spikes of ankylosaurs were not used exclusively for defense. These accessories may also have helped with species identification, as well as in the attraction of and competition for mates. The plates on *Stegosaurus* covered very little of its skin (Fig. 12.8A), so they were not useful for preventing attacks on its dorsal surface from large theropods. However, they certainly would have presented a visually obvious profile to other stegosaurs. A few paleontologists insisted that these plates were defensive structures and could be flexed to a horizontal position to shield more of the spine. However, the surface area covered by such a movement was minimal, and because *Stegosaurus* had no lateral body armor it would have been easily attacked from the sides. Additionally, presently known specimens of stegosaurs show no evidence of muscle attachments to the basal parts of plates, which indicates that they were fixed in the skin above the spine and not capable of movement. Ankylosaurs also had vertical to subvertical spines that potentially provided for dashing profiles. Even stegosaurs' tail spikes and ankylosaurs' tail clubs could have been raised to alert members of the opposite gender.

These impressive dermal attributes may have had a serious side in intraspecific competition. Modern horned mammals (e.g., ungulates, such as deer) commonly joust with their antler racks when competing for mates. Similarly, a male hippopotamus will open its mouth to show off huge and deadly tusks before engaging in combat with others of its species. Observations of these animals and comparisons to thyreophorans, in particular ankylosaurids and nodosaurids, have led to hypotheses that shoulder spikes, skull horns, and tail clubs may have been used for territorial displays, intimidation, or just fighting. The apparent ability of some nodosaurid spikes to move in a scissor-like manner also suggests their possible use in communication with one another. This communication could have been

FIGURE 12.10 Porous and venous texture associated with dorsal osteoderm on *Stegosaurus*, indicating high degree of vascularization. Close-up of specimen in Denver Museum of Science and Nature, Colorado.

augmented with vocalizations; ankylosaurids, with their complex nasal passages, may have been capable of a variety of sounds in a manner similar to lambeosaurines (Chapter 11). Lastly, osteoderms of some ankylosaurs were well vascularized, leading some paleontologists to propose that these dinosaurs were capable of "blushing" by bringing blood close to the surface of the skin. Such a mechanism would have made these ankylosaurs somewhat pinkish in hue. Unfortunately, as in most cases involving possible dinosaur coloration schemes, no supplementary data confirm or negate visions of pink thyreophorans (Chapter 5).

Like other anatomical attributes, stegosaur plates were not just used for show; they apparently also served as thermoregulation structures. Evidence for this function is found in the arrangement of the plates as staggered (offset) and overlapping rows, which would have been most effective for making increased surface area available for air circulating between the plates. Experimental data derived from testing the arrangements of metal plates in association with a heated tube showed that the offset rows worked best for transporting heat. As mentioned earlier, the offset arrangement was later confirmed by anatomical evidence, thus independently supporting the hypothesis. The plates themselves are extremely vascularized, indicating that a good amount of blood flowed through them (Fig. 12.10). In this sense, the plates could have been used for cooling by venting heat away from the body, or for heating by acting as passive solar collectors. How these functions related to stegosaur endothermy or ectothermy is unknown, considering that neither physiology has been firmly established for any thyreophoran (Chapter 8). Regardless, this relatively new insight into stegosaur physiology helps to show that osteoderms in general may have been structures used either all at the same time or individually for multiple purposes.

Late Cretaceous ankylosaur trackways in Bolivia provide evidence of at least one fast-moving thyreophoran. Paleontologists calculated a speed of 11 km/hour on the basis of stride length, footprint size, and known ankylosaur skeletal parameters (Chapter 14).

Locomotion

Until recently, only skeletal data provided material for hypotheses about thyreophoran locomotion. These data led to interpretations that both ankylosaurs and stegosaurus were weighed down by their dermal armor and had relatively wide bodies, similar to modern elephants or rhinoceroses. Such a condition is called **graviportal**, meaning that these animals were probably slow-moving and not capable of running any appreciable distances, in contrast to cursorial theropods. Because trackways are still sufficiently uncommon for thyreophorans, particularly for stegosaurs, such hypotheses cannot be cross-checked in most instances. Like all dinosaurs, ankylosaurs were diagonal walkers, although their stout and relatively inflexible bodies caused straddles that were wide relative to their stride lengths.

Similar to the hypotheses that have been made about sauropods, stegosaurs are thought to have had the capability of raising their front legs and sitting back on their rear legs and tails. This supposition is based on their calculated center of gravity, where the majority of their weight was in the hip region. Such an arrangement thus led to their graviportal state. With this weight distribution, a simple lifting of the front legs should have brought the rear down. In stegosaurs, the tail then would have acted as a prop for a tripodal stance. Although this scenario seems reasonable in terms of Newtonian physics and observations in modern graviportal animals (such as elephants), it was unlikely in stegosaurs. One major limiting factor is that stegosaur tails were held straight behind them through an interlocking of dorsal plates with the caudal vertebrae. This meant that the tail was less likely to have been used as part of a tripod stance, so stegosaurs were more likely adapted for staying close to the ground.

Feeding

Almost no direct information is available for determining what thyreophorans ate. Nevertheless, their teeth and jaws clearly show adaptations to herbivory. Because of their low height with relation to their sauropod (Chapter 10) and ornithopod (Chapter 11) companions during much of the Mesozoic, thyreophorans were most likely low-level browsers. An ability of stegosaurs to raise up on their hind legs to increase their browsing height was proposed as a possible adaptation to high-level browsing, but this scenario is unlikely. Similarly, there is little doubt that ankylosaurs were anything but low-level browsers, and even the most imaginative paleontologists refrain from depicting ankylosaurs as sitting up or otherwise behaving in a sprightly manner.

The advent of flowering plants by the Early Cretaceous was no doubt a factor in thyreophoran food choices throughout the Cretaceous, but Jurassic ankylosaurs and stegosaurs must have eaten ferns and other low-lying undergrowths in forest and shrubland ecosystems. The anatomical arrangement of the teeth and portions of thyreophoran skulls supports this supposition. For example, ankylosaur teeth are small, but their beaks were well developed for cropping plants. Some possible differences in dietary preferences between ankylosaurids and nodosaurids have been suggested on the basis of their beak shapes. Nodosaurids had narrow beaks, which would have nipped at plant stems and leaves more precisely. In contrast, ankylosaurids had wider beaks that would have grabbed swaths of vegetation. Ankylosaurs also had larger than normal **hyoids**, which were bones in the throat that would have supported a tongue. This tongue would have moved chewed food to the cheeks for temporary storage, followed by movement from the cheeks back into the mouth. Ankylosaurs were therefore capable of much chewing, although an examination of only their teeth would surely lead to a different conclusion.

No toothmarks or coprolites attributable to thyreophorans have been reported from the geologic record. Likewise, gastroliths have been correlated only with remains of the nodosaurid *Panoplosaurus*. Using sauropods as a model for comparison, most thyreophorans should have had gastroliths to aid their digestion of plant material because their teeth and jaws were poorly-adapted for grinding or otherwise processing their food. Future excavations of thyreophoran skeletal remains should test this hypothesis. However, if such investigations prove fruitless, then paleontologists will have to provide alternative explanations for why gastroliths were not required.

12

One hypothesis proposed for how ankylosaurs digested food is that their considerable hindquarters provided space for massive amounts of fermentation in the hindgut. This hypothesis is supported by the wide space allowed by the costae (ribs) toward the posterior of most ankylosaurs. This space gave sufficient room for a gastrointestinal tract holding symbiotic anaerobic bacteria that would have aided in the breakdown of cellulose in plants. One of the metabolic by-products of such breakdown is methane. So if ankylosaurs and other herbivorous dinosaurs employed such digestive systems, the Mesozoic atmosphere likely would have had a noticeably different scent to that of today.

Social Life

Again, very little is known about thyreophoran sociality, other than reasonable hypotheses based on functional morphology and proximity of skeletal remains. Few thyreophoran skeletons are found adjacent to one another; most are found as isolated individuals. This recurring circumstance, in stark contrast to the monospecific accumulations of theropod, sauropod, ornithopod, and ceratopsian remains, seems to be evidence that most thyreophorans roamed either as individuals or in small groups. This hypothesis is supported so far by the lack of many thyreophoran trackways presumed to have been made by the individuals of the same species traveling together at the same time.

As a result, numerous, closely-associated juvenile specimens of *Pinacosaurus*, in the same Upper Cretaceous deposit of Mongolia, constitute an unusual find. Monospecific assemblages that are also sorted by age (interpreted on the basis of their similarly small sizes) are evidence of possible age-related herding behavior for this ankylosaurid. Interestingly, no adults have been reported near these juvenile assemblages. These data encourage future investigations of and speculations about other monospecific thyreophoran assemblages, such as the ones that have come out of Tanzania, composed of fragments of the Late Jurassic stegosaur *Kentrosaurus*.

Health

Thyreophorans, like most dinosaurs, were apparently healthy animals. Thyreophoran remains have yet to show many traces of predation or scavenging by theropods or other large Mesozoic carnivores. One exception to this is the reported association of theropod teeth around the remains of the aforementioned juvenile *Pinacosaurus*, although this occurrence is interpreted as the result of scavenging. The paucity of evidence for predation or scavenging means that live and dead thyreophorans were rarely subjected to consumption, which would correlate with the heavy protection wielded by some species. One example of an apparently uneaten thyreophoran is *Stegosaurus*. Contrary to common depictions of the Late Jurassic *Stegosaurus* in mortal combat with the tetanuran *Allosaurus*, no toothmarks by *Allosaurus* have been found in *Stegosaurus* bones. Likewise, *Stegosaurus* is frequently shown defending itself by swinging the caudal spikes, but no dinosaur remains, of a predator or a *Stegosaurus*, show any signs of being attacked with such a weapon. A similar defense against theropods has been proposed for ankylosaurids, but again hypotheses about such behaviors await more evidence.

The fusion of osteoderms to the distal caudal vertebrae in ankylosaurids resulted in tail clubs, but this fusion is considered as a normal, inheritable adaptation rather than an acquired paleopathologic condition. Because nodosaurs lacked tail clubs, caudal maladies should be more obvious. Nonetheless, a fusion of some caudal vertebrae in a specimen of the nodosaurid *Edmontia*, interpreted as the result of an injury, is the only one reported thus far. Multiple specimens of the Early Cretaceous ankylosaurid *Gastonia* show depressions in their osteoderms that were probably caused by disease, although there is no evidence that these caused the animals any major problems. This paucity of evidence for ill health may be an artifact of small sample sets, but it also could be a sign that most thyreophorans avoided such trouble.

SUMMARY

Thyreophorans, which were herbivorous ornithischian dinosaurs that lived from the Early Jurassic through to the Late Cretaceous, are generally known as the "armored dinosaurs" because of their well-developed osteoderms that formed parasagittal rows along their bodies. Members of two clades, the Stegosauria and Ankylosauria, are most representative of thyreophorans. Ankylosauria is further divided into the Ankylosauridae and Nodosauridae. Stegosaurs are distinguished by their:

1 small heads;
2 parascapular spikes;
3 armor just adjacent to their spines; and
4 posterior tail spikes.

In contrast, ankylosaurs had:

1 armor all over their dorsal (and in some cases, ventral) surfaces;
2 skull sutures covered by dermal plates;
3 a synsacrum; and
4 numerous other features.

Ankylosaurids are mainly distinguished from nodosaurids on the basis of skull profiles and the presence or absence of a tail club.

Little definitive information is known with regard to thyreophoran behavior and evolution because the majority of data come from body fossils that are, in many cases, incomplete and fragmentary. Nevertheless, sufficient skeletal material for some genera and complete specimens of others, such as those of the Late Jurassic *Stegosaurus* and Cretaceous ankylosaurs, have provided windows into thyreophoran behavior and evolution. Primitive forms of thyreophorans that do not fit into either Stegosauria or Ankylosauria, such as *Scutellosaurus* and *Scelidosaurus*, demonstrate a beginning of their clade at least in the Early Jurassic in Europe and North America. Ankylosaurs, which lived from the Middle Jurassic to Late Cretaceous, were a successful thyreophoran group that diversified throughout the Cretaceous and had a near worldwide distribution. In contrast, stegosaurs were limited to possibly the Early Jurassic through to the Early Cretaceous (having reached their peak in the Late Jurassic) of mostly northern continents.

Thyreophorans were apparently healthy animals and probably had complex behaviors that challenge perceptions of them based on their relatively small brains. Osteoderms of various sorts (plates, spikes, and clubs)

12

SUMMARY Continued

probably served multiple purposes, such as protection, courting, intraspecific competition, and thermoregulation. Feeding was accomplished mostly through low-level grazing and browsing. Some specializations of skeletal features, such as deeply inset teeth, large hyoids, and expanded hindgut regions, meant that some thyreophorans were capable of processing huge amounts of food. Locomotion is little understood because of the scarcity of thyreophoran trackways, which means that hypotheses about social behavior are also based on limited data sets of body fossil information. However, the co-occurrence of remains of more than one individual in the same deposits, such as the Late Jurassic stegosaur *Kentrosaurus* and Late Cretaceous ankylosaur *Picanosaurus*, suggests that at least a few thyreophorans lived in social groups.

DISCUSSION QUESTIONS

1. What would have been a defensive disadvantage for *Stegosaurus* or other stegosaurs if their tail spikes had pointed upward (vertically) rather than outward (horizontally)?
2. The author states that ankylosaurs probably weighed more per volume than other dinosaurs because of their armor. For example, *Ankylosaurus* is estimated to have been about 10 meters long and weighed about 4 tonnes. How would you calculate an estimate of its density versus that for a 10-meter long ornithopod (which did not have any body armor)? What information would you need first before attempting to calculate such an estimate?
3. What independent information besides functional morphology would be required to convince you that ankylosaurids used their tail clubs as weapons for either interspecific or intraspecific combat?
4. Examine Tables 12.1 and 12.2 and represent in a bar graph the number of genera that occur within each of the following time divisions: Early Jurassic, Middle Jurassic, Late Jurassic, Early Cretaceous, Late Cretaceous.
   a. What general trend in changes of thyreophoran diversity through time can you hypothesize on the basis of these data and the bar graph?
   b. Compare the number of genera in each time interval versus place (region of the world). Do any biases in both time and place emerge from such an analysis?
   c. What are possible sources of error in your analysis?

DISCUSSION QUESTIONS Continued

5. A dinosaur paleontologist states, "Ankylosaurs reflect an arms race against increasingly more efficient theropod predators." You decide to test this statement by looking up information on which theropods lived in the same areas and times as some ankylosaurs.
    a. Pick out the ankylosaur genera from Table 12.1, including their time intervals and geographic locations.
    b. Look at Tables 9.1 and 9.2 and try to link the ankylosaurs to the theropods that lived in the same time intervals and geographic areas. How many linkages did you find?
    c. You decide to state that these linkages represent predator–prey relationships. What are possible sources of error in your analysis?
    d. Assuming you are right, predict what new data from the geologic record, either through body or trace fossils, would support your assertions.
6. Because thyreophoran remains are rarely found in the fossil record as representing more than one specimen, some paleontologists might suggest that most thyreophorans lived as individuals, who only got together to mate. What information presented in the chapter or possible new data would contradict such an assumption?
7. Check the assumption that humans were more "brainy" than stegosaurs through the following calculations.
    a. Given a braincase endocast from a specimen of *Stegosaurus* of 60 cm^3, how much would it have weighed if the original brain had a density of 1.2 g/cm^3?
    b. Given a braincase endocast from a typical adult *Homo sapiens* of 1200 cm^3, how much would it weigh, assuming the same density?
    c. Of a total original body weight of 1.8 tonnes, the brain of *Stegosaurus* composes what percentage of its total body weight?
    d. Assuming a weight of 65 kg for a specimen of *Homo sapiens*, the brain composes what percentage of its total body weight?
8. What independent information, besides functional morphology, would be needed to convince you that stegosaurs could raise their front legs and sit back on their rear legs? Describe the appearance of the fossil finds that would corroborate such a hypothesis.
9. Based on descriptions presented in the chapter, sketch the manus–pes pair for footprints made by a stegosaur and an ankylosaur. How do the resultant tracks differ from those made by other quadrupedal dinosaurs discussed so far? How might the Late Jurassic tracks for ankylosaurs differ from the Late Cretaceous tracks?
10. Provide a hypothesis for why thyreophoran eggs and nests have not yet been identified. How could environmental factors have affected the preservation potential of nest sites? What other taphonomic factors might have limited their preservation?

12

Bibliography

Barrett, P. M. and Upchurch, P. 1995. *Regnosaurus northamptoni*, a stegosaurian dinosaur from the Lower Cretaceous of southern England. *Geological Magazine* **132**: 213–222.

Carpenter, K. 1997. "Ankylosaurs". *In* Farlow, J. O. and Brett-Surman, M. K. (Eds), *The Complete Dinosaur*, Bloomington, Indiana: Indiana University Press. pp 307–316.

Carpenter, K. (Ed.) 2001. *The Armored Dinosaurs*. Bloomington, Indiana: Indiana University Press.

Carpenter, K., Miles, C. and Cloward, K. 1998. Skull of a Jurassic ankylosaur (Dinosauria). *Nature* **393, 6687**: 782–783.

Coombs, W. P., Jr. 1986. A juvenile ankylosaur referable to the genus *Euoplocephalus* (Reptilia, Ornithischia). *Journal of Vertebrate Paleontology* **6**: 162–173.

Coombs, W. P., Jr. 1995. Ankylosaurian tail clubs of middle Campanian to early Maastrichtian age from western North America, with description of a tiny club from Alberta and discussion of tail orientation and tail club function. *Canadian Journal of Earth Sciences* **32**: 902–912.

Dong, Z.-M. 1990. "Stegosaurs of Asia". In Carpenter, K. and Currie, P. J. (Eds). *Dinosaur Systematics: Approaches and Perspectives*. Cambridge, UK: Cambridge University Press. pp 255–268.

Galton, P. M. and Upchurch, P. 2004. "Stegosauria". *In* Weishampel, D. B., Dodson, P. and Osmólska, H. (Eds), *The Dinosauria* (2nd Edition). Berkeley, California: University of California Press. pp 343–362.

Gangloff, R. A. 1995. *Edmontonia* sp., the first record of an ankylosaur from Alaska. *Journal of Vertebrate Paleontology* **15**: 195–200.

Gasparini, Z., Pereda-Suberbiola, X. and Molnar, R. E. 1996. New data on the ankylosaurian dinosaur from the Late Cretaceous of the Antarctic Peninsula. *Memoirs – Queensland Museum* **39**: 583–594.

Kirkland, J. I., and Carpenter, K. 1994. North America's first pre-Cretaceous ankylosaur (Dinosauria) from the Upper Jurassic Morrison Formation of western Colorado. *Geology Studies* **40**: 25–42.

Le Loeuff, J., Lockley, M. G., Meyer, C. and Petit, J.-P. 1999. Discovery of a thyreophoran trackway in the Hettangian of central France. *Comptes Rendus de l'Academie de Sciences – Serie IIa: Sciences de la Terre et des Planetes* **328**: 215–219.

McCrea, R. T., Lockley M. G. and Meyer, C. A. 2001. Global distribution of purported track occurrences. *In* K. Carpenter (Ed.), *The Armored Dinosaurs*. Bloomington, Indiana: Indiana University Press. pp 413–454.

Molnar, R. E. and Frey, E. 1987. The paravertebral elements of the Australian ankylosaur *Minmi* (Reptilia; Ornithischia, Cretaceous). *Neues Jahrbuch für Geologie und Palaeontologie. Abhandlungen* **175**: 19–37.

Norman, D. B., Witmer, L. M. and Weishampel, D. B. 2004. Basal Ornithischia. *In* Weishampel, D. B., Dodson, P. and Osmólska, H. (Eds), *The Dinosauria* (2nd Edition). Berkeley, California: University of California Press. pp 325–334.

Norman, D. B., Witmer, L. M. and Weishampel, D. B. 2004. Basal Thyreophora. *In* Weishampel, D. B., Dodson, P. and Osmólska, H. (Eds), *The Dinosauria* (2nd Edition). Berkeley, California: University of California Press. pp 335–342.

Page, D. 1998. Stegosaur tracks and the persistence of facies; the Lower Cretaceous of Western Australia. *Geology Today* **14**: 75–77.

Salgado, L. and Coria, R. A. 1996. First evidence of an ankylosaur (Dinosauria, Ornithischia) in South America. *Ameghiniana* **33**: 367–371.

Sereno, P. C. and Dong Z.-M. 1992. The skull of the basal stegosaur *Huayangosaurus taibaii* and a cladistic diagnosis of Stegosauria. *Journal of Vertebrate Paleontology* **12**: 318–343.

Vickaryous, M. K., Maryanska, T. and Weishampel, D. B. 2004. Ankylosauria. *In* Weishampel, D. B., Dodson, P. and Osmólska, H. (Eds), *The Dinosauria* (2nd Edition). Berkeley, California: University of California Press. pp 363–392.

Vickers–Rich, P. and Rich, T. H. 1993. Australia's polar dinosaurs. *Scientific American* **269**: 50–55.

12

Chapter 13

Triceratops was a large and formidable animal, with its long horns (like a rhinoceros) and stout body. But according to all of the books you read, it was a peaceful plant eater. Only one thing you learned about Triceratops bothered you, and that was the pictures of it fighting Tyrannosaurus. In these pictures, this "peaceful" vegetarian was usually shown using its postorbital horns to gore the soft, ventral surface of its carnivorous adversary.

What evidence has been found in the fossil record to indicate that Triceratops, or any other ceratopsian, actually used its horns against predators? If they did not use them for defense, what was their function? What interactions (if any) did Triceratops have with Tyrannosaurus? Did Tyrannosaurus actually hunt or eat Triceratops?

Marginocephalia

13

Why Study Marginocephalians?

Clade Marginocephalia, consisting of the relatively rare pachycephalo-saurs but also the very common ceratopsians, includes one of the most famous dinosaurs known, the Late Cretaceous *Triceratops* of North America. However, perhaps less known is that Marginocephalia is a group of dinosaurs that rivals Theropoda (Chapter 9) and Ornithopoda (Chapter 11) in its diversity. Largely because of the ceratopsians, Marginocephalia is also among the best-represented of Late Cretaceous dinosaurs through their skeletal remains, first discovered in the late nineteenth century and still regularly uncovered today. Ceratopsians are exceed-ingly common in museums as a result of their abundance and completeness. Three marginocephalian genera (*Protoceratops*, *Psittacosaurus*, and *Triceratops*) rank in the top ten of all dinosaurs owned by museums. Unlike most dinosaur clades, many of its members were described on the basis of plentiful fossil material. More than 50% of ceratopsian genera and species are named from nearly complete or com-plete skeletons.

Marginocephalians were among the most endemic of all dinosaur clades in that their body and trace fossils have only been found in North America, Asia, and Europe. They also are the most limited in geologic time, nearly all comprised of Cretaceous examples, meaning that they were relative latecomers in Mesozoic ecosystems. This circumstance of limited biogeography and time was probably linked, thus affect-ing their evolutionary history until their extinction at the end of the Mesozoic.

Of all dinosaurs, marginocephalians are best known for their heads, which went to extremes in the development of thick or broad skulls. Indeed, a few ceratopsian species have the largest heads of any known land-dwelling animal. On these thick or otherwise enormous heads were some of the most elaborate and gaudy acces-sories seen in any dinosaur, such as numerous horns and spikes with various shapes, bosses, and broad frills. In terms of marginocephalian locomotion, rare ceratopsian tracks, recognized recently, fill a supposed gap in their paleontological record but contribute to ensuing controversies about their locomotion based on skeletal ana-lyses. Pachycephalosaur tracks are so far either undiscovered or unrecognized, but at least their robust skull parts were taphonomically predisposed toward being preserved.

Some of these ornate animals may have traveled in herds, as suggested by mono-specific bonebeds and paleobiogeography. Such gregariousness would have created stunning vistas on Late Cretaceous landscapes, but also would have had a consid-erable impact on plant life. On a more individual basis, intraspecific competition in dinosaurs is perhaps most vividly conjured by the image of thick-skulled pachy-cephalosaurs running into one another at high speeds. Added to this scenario are similar ones of jousting ceratopsians, deterring either predation or rivals by simply turning their heads for full visual effect. Thus, the heads of marginocephalians are

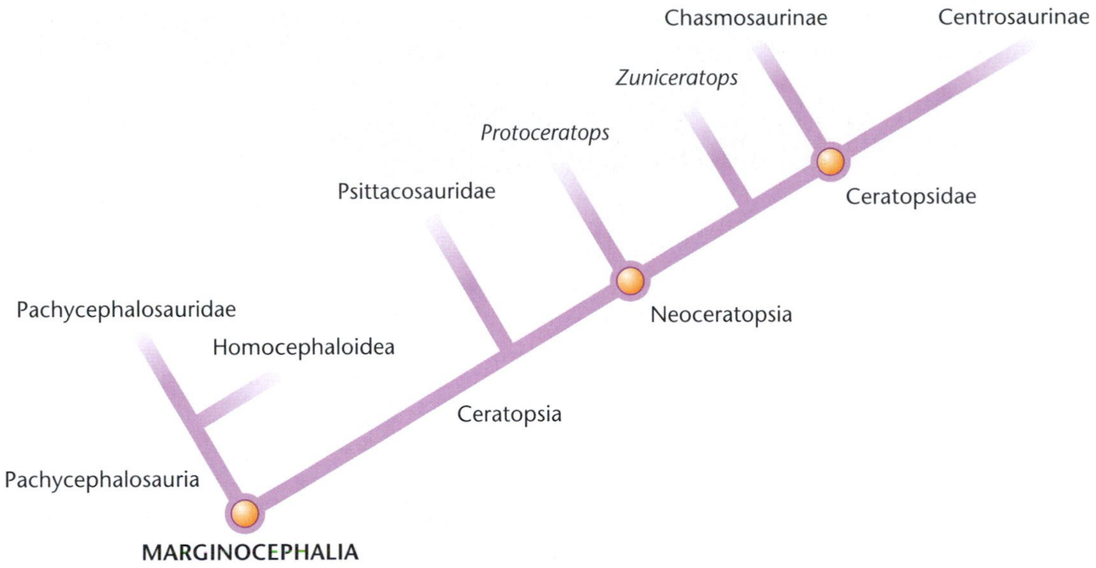

FIGURE 13.1 Cladogram showing interrelationships between clades of Marginocephalia, particularly Pachycephalosauria and Ceratopsia.

what define them and also determined how they conducted their business during the Cretaceous.

Definition and Unique Characteristics of Marginocephalia

Marginocephalia (*margino* = "margin" and *cephalia* = "head") is currently regarded as a node-based clade and a sister clade to Ornithopoda (Chapter 13); both of these share the node-based clade Cerapoda (Fig. 13.1). Two sister clades of Marginocephalia are **Pachycephalosauria** and **Ceratopsia**, which collectively are composed of many genera but with most being ceratopsians (Tables 13.1 and 13.2). *Stenopelix*, a poorly-known Early Cretaceous ornithischian from Germany, was once considered as the most basal form of marginocephalian, but is now classified as a pachycephalosaur. As mentioned earlier, nearly all marginocephalians discovered

TABLE 13.1 Representative genera of Pachycephalosauria with approximate geologic age and where they occur.

Genus	Age	Geographic Location
Goyocephale	Late Cretaceous	Mongolia
Gravitholus	Late Cretaceous	Alberta, Canada
Homalocephale	Late Cretaceous	Mongolia
Micropachycephalosaurus	Late Cretaceous	China
Ornatotholus	Late Cretaceous	Alberta, Canada; western USA
Pachycephalosaurus	Late Cretaceous	Western USA; Alberta, Canada
Prenocephale	Late Cretaceous	Mongolia
Stegoceras	Late Cretaceous	Alberta, Canada; western USA
Stenopelix	Early Cretaceous	Germany
Stygimoloch	Late Cretaceous	Western USA
Wannanosaurus	Late Cretaceous	China
Yaverlandia	Early Cretaceous	England

13

389

TABLE 13.2 Representative genera of Ceratopsia with approximate geologic age and where they occur.

Genus	Age	Geographic Location
Achelousaurus	Late Cretaceous	Western USA
Anchiceratops	Late Cretaceous	Alberta, Canada
Archaeoceratops	Early Cretaceous	China
Arrhinoceratops	Late Cretaceous	Alberta, Canada
Avaceratops	Late Cretaceous	Western USA
Bagaceratops	Late Cretaceous	Mongolia
Brachyceratops	Late Cretaceous	Western USA
Centrosaurus	Late Cretaceous	Alberta, Canada
Chasmosaurus	Late Cretaceous	Western USA; Alberta, Canada
Diceratops	Late Cretaceous	Western USA
Einiosaurus	Late Cretaceous	Western USA; Alberta, Canada
Leptoceratops	Late Cretaceous	Western USA; Alberta, Canada
Liaoceratops	Late Cretaceous	China
Monoclonius	Late Cretaceous	Alberta, Canada; western USA
Montanoceratops	Late Cretaceous	Western USA
Pachyrhinosaurus	Late Cretaceous	Alberta, Canada; Alaska, USA
Pentaceratops	Late Cretaceous	Western USA
Protoceratops	Late Cretaceous	China, Mongolia
Psittacosaurus	Early Cretaceous	Mongolia, China, Thailand
Styracosaurus	Late Cretaceous	Western USA; Alberta, Canada
Torosaurus	Late Cretaceous	Western USA; Western Canada
Triceratops	Late Cretaceous	Western USA; Canada
Udanoceratops	Late Cretaceous	Mongolia
Zuniceratops	Late Cretaceous	Western USA

so far are from the Cretaceous, the Late Jurassic ceratopsian *Chaoyangsaurus* of China being the only exception. This time constraint distinguishes this important clade from theropods, sauropodomorphs, ornithopods, and thyreophorans (Chapters 9–12).

Members of Clade Marginocephalia are defined by several important character traits (Fig. 13.2).

■ A narrow shelf of bone on the parietal and posterior part of the squamosal, which projects from the skull posterior.
■ An abbreviated posterior portion of the premaxillary as it adds to the palate.
■ A shortened pubis accompanied by widely-spaced hip sockets.

Of these synapomorphies, the most important is the shelf of bone associated with the parietal and squamosal, an easily recognizable feature. Based on this trait and others, Marginocephalia unites the superficially disparate clades of Pachycephalosauria and Ceratopsia. Pachycephalosaurs are only represented by about 14 genera, thus their proposed cladistic classification consist of only a few options. In contrast, Ceratopsia is composed of a daunting number of genera and species within genera, and some genera closely resemble one another. Consequently, their cladistics present a major challenge for the few dinosaur paleontologists who work with ceraptosians.

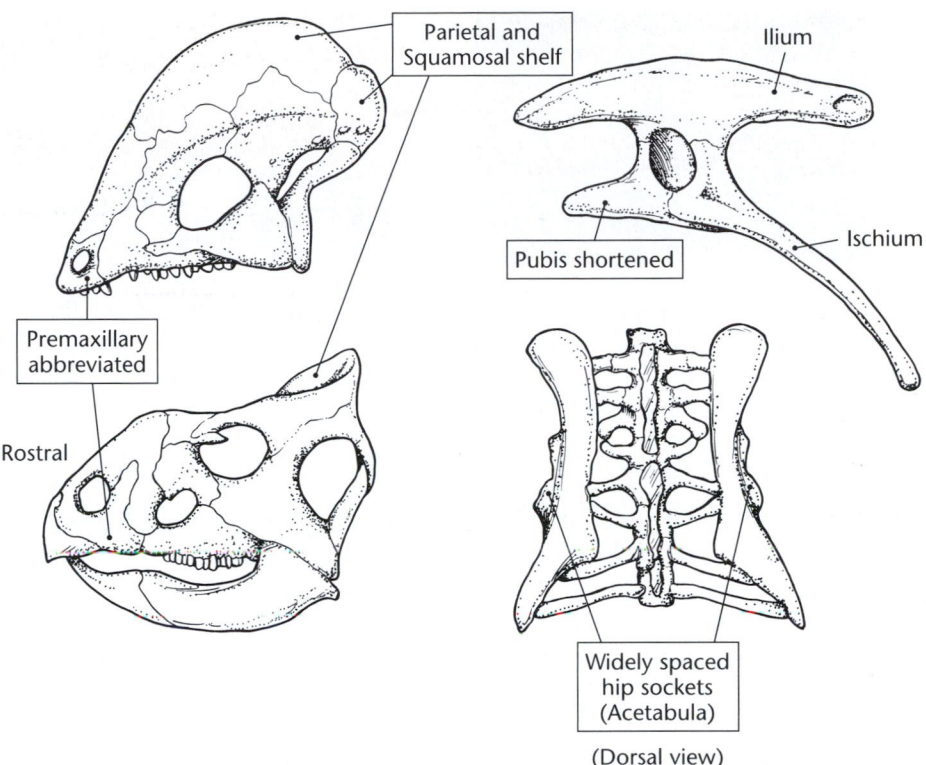

Parietal and Squamosal shelf

Ilium

Premaxillary abbreviated

Rostral

Pubis shortened

Ischium

Widely spaced hip sockets (Acetabula)

(Dorsal view)

FIGURE 13.2 Character traits of Marginocephalia: narrow shelf of bone on the parietal and posterior part of the squamosal; abbreviated posterior portion of premaxillary; and shortened pubis accompanied by widely spaced hip sockets.

Problems with ceratopsian taxonomy are attributable to two factors:

1 A very good body fossil record for ceratopsians, including numerous complete specimens for some species; and
2 overzealous naming of ceratopsian species in the past 100+ years, which resulted in many synonymies.

To illustrate how both facets have interacted, consider the following example. As many as 16 species names have been applied to the abundant remains of the genus *Triceratops*, despite a geologic range for the genus of less than 3 million years. Two explanations for this abundance of species are possible:

1 Evolutionary and genetic factors, such as interaction of environmental change, geographic isolation of small populations, genetic drift, and mutations, may have contributed to extremely rapid speciation for *Triceratops* (Chapter 6); and
2 Dinosaur paleontologists since the late nineteenth century may have used all newly-discovered features on a *Triceratops* skeleton to justify new species names.

Of these, the latter factor is the most likely, and recognition in more recent years of this human-induced diversity has decreased the number of ceratopsian taxa. Thus, among taxonomists the "lumpers" triumphed in part over the "splitters" (Chapter 5). This is not to say that *Triceratops* did not undergo speciation over 3 million years

13

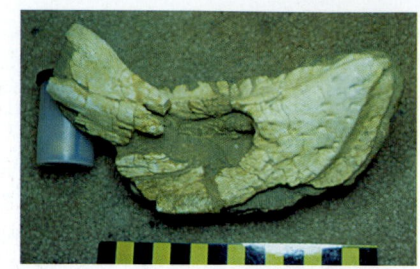 

(A)　　　　(B)

FIGURE 13.3 Typical ceratopsian lower jaws. (A) Dentary and predentary of *Protoceratops*, Late Cretaceous of Mongolia; Dinosaur Adventure Museum, Fruita, Colorado. (B) Unidentified ceratopsian dentary and predentary, Late Cretaceous, western North America; College of Eastern Utah Prehistoric Museum, Price, Utah.

or that most taxonomy of ceratopsians is scientifically unjustified. Nonetheless, the great number of genus and species names that have been given to them may be overstating their actual diversity.

All known marginocephalians were herbivores, as evidenced by the teeth and jaws of both pachycephalosaurs and ceratopsians. Pachycephalosaurs had heterodont dentition, with canines in the anterior part of the mouth and cheek teeth that were small, laterally compressed, and crowned with denticles. These teeth and their medial placement within the jaw resemble those of thyreophorans, partially reflecting their common ancestry within Genasauria (Chapter 12). In contrast, ceratopsians had cheek teeth develop into dental batteries that could have competed with those of some iguanodontians (Chapter 11). The anterior part of the mouth consisted of sharp beaks covered by horn (keratin) with a premaxillary **rostral** bone above it occluded with a pointed predentary and dentary below (Fig. 13.3). The members of each clade were so different in their eating apparatuses that contemporaneous species must have been specialized for different food sources. Most forms were probably low-level browsers because of their low heights or propensity toward quadrupedalism.

Both marginocephalian clades also differed considerably in their locomotion. Judging from the short fore limbs seen in some specimens, pachycephalosaurs were apparently bipedal, although limb proportions for some species are unknown. Likewise, although the manus for a few specimens is known, no pachycephalosaur pes has yet been found. Consequently, pachycephalosaur foot anatomy cannot be compared to tracks that might have been made by these animals. Such information would certainly help to test assumptions about obligate bipedalism. Some basal ceratopsians were also obligate bipeds, but the majority of ceratopsians were heavily built and obligate quadrupeds. Only a few might have been facultative bipeds. Advanced ceratopsian foot anatomy consisted of five toes on the manus and four on the pes; later quadrupedal forms had shortened but stout phalanges that ended in hooves. However, ceratopsian manus tracks should only show impressions from four toes (digits I through to IV), because digit V is markedly reduced in comparison to the other toes.

Undoubted marginocephalian trace fossils are so far only represented by ceratopsian trackways from Late Cretaceous strata in the western USA (but nowhere else) and gastroliths in *Psittacosaurus*, a primitive ceratopsian. Ceratopsian tracks are identified on the basis of the following data:

- They match the aforementioned foot anatomy (Fig. 13.4).
- Size comparisons of track compression shapes also match those of known ceratopsian foot sizes, especially when flesh is "added" to the foot.

FIGURE 13.4 Ceratopsian track interpreted from the Laramie Formation (Late Cretaceous) near Golden, Colorado.

- They occur in strata formed in continental environments that are age-equivalent with known ceratopsian remains.

The gastroliths that occur in *Psittacosaurus* are localized masses of about 50 similarly-sized, rounded stones found within the rib cage of more than one specimen. The fulfillment of such specific criteria for gastroliths leaves little doubt that these stones are indeed trace fossils related to ceratopsian digestion. As far as other trace fossils related to feeding are concerned, no toothmarks (or beakmarks) have been described for marginocephalians, nor have any coprolites been linked to them (Chapter 14).

Numerous dinosaur nests in Late Cretaceous strata of Mongolia, often containing bountiful egg clutches, were originally attributed to the ceratopsian *Protoceratops*. Updated analyses and new evidence have rendered this identification suspect or in some cases falsified it entirely (Chapter 9). Consequently, paleontologists are beginning with a clean slate with regard to the body fossils (eggs and embryos) and trace fossils (nest structures) associated with marginocephalian reproduction. In contrast, possible trace fossils of marginocephalian behavior related to mating or territorial disputes, such as scars or fractures inflicted on one another, as evident in healed injuries in bone, may provide more insights on how pachycephalosaurs and ceratopsians interacted with one another.

13

Clades and Species of Marginocephalia

Pachycephalosauria

The most easily recognizable and well-known trait of pachycephalosaurs is an extremely thick, flat or domed bony skull (*pachy* = "thick" and *cephalia* = "skull"). For example, *Pachycephalosaurus* had a skull that was about 22 cm thick. This remarkable feature was formed by fusion of the frontals and parietals and the addition of

(A) (B)

FIGURE 13.5 Skulls of Late Cretaceous pachycephalosaurids. (A) *Stegoceras* as represented only by its fused frontal and parietals forming a high dome, a typical trait of pachycephalisaurids; College of Eastern Utah Prehistoric Museum, Price, Utah. (B) *Pachycephalosaurus*, showing deeply inset cheek teeth, domal dorsal surface, and fringing osteoderms on dentary (ventral), nasal (anterior), and squamosals (posterior). Specimen is cast of original, Denver Museum of Science and Nature, Denver, Colorado.

thick deposits of bone to this unit. Despite the impressively large and convex heads this fusion caused in some pachycephalosaurs, they did not have a corresponding increase in cranial capacity. In fact, their EQs are only slightly higher than not-so-brainy dinosaurs such as thyreophorans.

In the early twentieth century, Lawrence Lambe (Chapter 3) was the first to examine pachycephalosaur material, and he named *Stegoceras* (Fig. 13.5A). *Pachycephalosaurus* (Fig. 13.5B), the inspiration for the naming of the clades Pachycephalosauria and Pachycephalosauridae, was not discovered until about 40 years later. Soon after, recognition of some already discovered but mislabeled pachycephalosaur remains helped to relate them. Later discoveries of new species of pachycephalosaurs led to the insight that they belonged to two different groups. So far, all but two species have been found in Late Cretaceous strata of North America and Asia; the exceptions are *Stenopelix* and *Yaverlandia*, which are both from the Lower Cretaceous of Germany and the UK (Table 13.1).

Pachycephalosauria can be divided into two clades, **Homocephaloidea** *(nicknamed the "flat-headed" dinosaurs) and* **Pachycephalosauridae** *("dome-headed" dinosaurs), with a few outgroups represented by* Wannosaurus *and* Goyocephale.

Homocephaloids (see box) are only defined by two genera, *Homocephale* and *Ornathotholus*. They are distinguished by flat ventral surfaces on their pitted skulls and supratemporal fenestrae, whereas pachycephalosaurids have smooth (non-pitted), dome-shaped skulls and lack these fenestrae. Some paleontologists hypothesize that the homocephaloids represent a more primitive status, with traits that evolved into those seen in pachycephalosaurids. However, the small amount of fossil material and closeness in geologic ages for the various pachycephalosaurs make resolution of such relationships difficult. The largely complete skulls for some species show some basic characteristics of their group, as well as accouterments such as osteoderms that formed bosses (see *Pachycephalosaurus*; Fig. 13.5B) or small horns (*Stygimoloch*) that fringed the skull in various places.

The oldest known pachycephalosaur, and probably the most basal, is *Stenopelix valdensis*, followed by *Yaverlandia bitholus*, both from the Early Cretaceous. Despite its geologic age relative to other pachycephalosaurids, *Yaverlandia* has some advanced features, such as fused frontals and a moderately domed skull. These observations led to the conclusion that the evolutionary history of pachycephalosaurs may extend back into the Late Jurassic. Unfortunately, the skeletal material for both *Stenopelix* and *Yaverlandia* is so fragmentary that they cannot be classified as either homocephaloid or pachycephalosaurid. *Wannanosaurus*, which comes from the Upper Cretaceous strata of China, is the best-defined primitive pachycephalosaur. Its large supratemporal fenestrae were hypothesized to have diminished within the Pachycephalosauridae lineage, but instead it retained these into the Late Cretaceous. The Late Cretaceous *Goyocephale* of Mongolia is another of the pachycephalosaurs that does not fit easily into either Homocephaloidea or Pachycephalosauridae. Some paleontologists proposed that it may represent an ancestral form of homocephaloid, but it is now considered as an outgroup preceding them.

Homocephale, the namesake of the homocephaloids and one of the most completely known pachycephalosaurs, also lived during the Late Cretaceous of Mongolia. Homocephalids were apparently restricted both geographically and temporally, as they only occur in Upper Cretaceous deposits of Asia. Pachycephalosaurids are similar in that all genera are Late Cretaceous, but with most from North America (e.g., *Pachycephalosaurus*, *Sphaerotholus*, *Stegoceras*, and *Stygimoloch*). All others are from Asia, with three specifically from China (*Micropachycephalosaurus*) and Mongolia (*Prenocephale* and *Tylocephale*), although the few fragmentary remains of *Micropachycephalosaurus* make its position within Pachycephalosauridae unverifiable.

Pachycephalosaurs are interpreted as small- to medium-sized obligate bipeds on the basis of the admittedly limited data derived from fore limb/hind limb proportions. (Fig. 13.6). Some forms also are known to have had long tails. The preserved tails have distal caudal vertebrae encased by a meshwork of ossified tendons that

FIGURE 13.6 *Pachycephalosaurus*, reconstructed as an entire skeleton but with little more to inspire this other than a very bony skull. North Carolina Museum of Natural History, Raleigh, North Carolina.

13

stiffened that half of the tail. This adaptation was probably related to locomotion and helped to keep the distal part of the tail pointing away from the body as a counterbalance, similar to that of some theropods (Chapter 9). *Stegoceras* is the most commonly encountered pachycephalosaur genus in the geologic record, although the majority of its material comes from pieces of its skull. Nevertheless, enough of these exist that a few paleontologists have been able to hypothesis sexual dimorphism in one species, *Stegoceras validum*, discussed in more detail later. The largest pachycephalosaur known is *Pachycephalosaurus*, which would have been about 8 meters long, including its tail. Some other species, such as the aptly named *Micropachycephalosaurus*, were less than a meter long.

Ceratopsia

The prominent horns of *Triceratops* were what first caught the attention of O. C. Marsh when he saw its skull in 1887, leading him to first identify it as a fossil bison. Later, more cautious examination revealed that the horned skull belonged to a dinosaur, and it was promptly renamed in 1889 to *Triceratops horridus* (= "horrid three-horned face"). Marsh is credited with naming this famous dinosaur and the Family Ceratopsidae (using the Linnaean system), but the first ceratopsian named from the geologic record was *Monoclonius*, described by his rival Edward Cope in 1876. Despite the presence of these heavyweights of dinosaur paleontology at the forefront of describing ceratopsians new to science, by far the most important single contributor to the study of ceratopsians was John Bell Hatcher (Chapter 3). Although he only lived for 42 years, Hatcher discovered 50 ceratopsian skeletons. He also wrote the majority of *The Ceratopsia*, the classic treatise on these dinosaurs, which was published in 1907, three years after he died. The study of these horned dinosaurs would not be nearly as advanced as it is today without Hatcher's seminal contributions, and his legacy is apparent whenever people see these dinosaurs depicted in fact and fiction.

A dinosaur belonging to Ceratopsia has several important characters:

- A rostral bone anterior to the maxilla that paired with a predentary to form a sharp beak.
- A frill formed by the parietals that hung past the rest of the skull.
- Cheeks that extend laterally and posteriorly, giving the skull a triangular shape when viewed from above its dorsal surface.
- A palate positioned high in the skull.

Notice that all of these features deal with novelties in the skull, which re-emphasizes the importance of how marginocephalians in general, and ceratopsians in particular, are largely identified by their cranial anatomy. In other words, a headless ceratopsian skeleton found in the field would present a greater challenge to classify than a disembodied head.

Ceratopsia is split into two sister clades, **Psittacosauridae**, based only on the genus *Psittacosaurus*, and **Neoceratopsia**, which includes all other ceratopsian genera. The oldest known ceratopsian is *Chaoyangsaurus* from the Middle–Upper Jurassic of China, but the most primitive ceratopsian is *Psittacosaurus*, a small (less than 2 meters long) ceratopsian that occurs abundantly as near-complete or complete specimens in Lower Cretaceous rocks of Asia. Ten species of *Psittacosaurus* have been identified so far, which either means that it diversified during the Early Cretaceous, or (more likely) was the victim of much taxonomic splitting. The latter might be attributable to the large amount of skeletal material available for study, which lends itself well to identifying small anatomical differences. Regardless, *Psittacosaurus* can be distinguished from other ceratopsians through:

FIGURE 13.7 Comparative anatomy between skulls of two small ceratopsians. (A) Early Cretaceous psittacosaurid *Psittacosaurus* of Asia, the oldest known ceratopsian and namesake of its clade. (B) Late Cretaceous neoceratopsian *Protoceratops*, also of Asia.

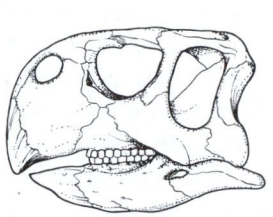
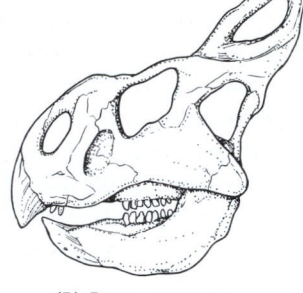

(A) Psittacosaurus   (B) Protoceratops

1 its nares, which were elevated away from the rostral;
2 a loss of digit V on its manus, which left it with only three fingers; and
3 a loss of its antorbital fenestra (Fig. 13.7A).

Psittacosaurids are thought to have retained more primitive features than the neoceratopsians. They certainly had simpler-looking skulls, consisting of a barely detectable parietal shelf overhanging the occipital and a lack of horns. Its proposed membership in Ceratopsia was doubted until specimens were found with rostrals which, along with its triangular skull and toothless beak, confirmed its common ancestry with neoceratopsians. It is also one of the few ceratopsians interpreted as a facultative biped, and it currently is the only ceratopsian found with undoubted gastroliths. The unique aspects of psittacosaurids are interpreted as being a reflection of their more primitive ornithischian ancestry.

Neoceratopsians are the geologically youngest of the major groups of dinosaurs, with their geologic range restricted mostly to the Late Cretaceous. Although they only occur in North America and Asia, they are well represented by numerous body fossils of high variety and completeness. Their young geologic age and the excessively bony skulls of some species favored their preservation over older dinosaurs with less bone (Chapter 7). For some species that occur in Upper Cretaceous strata of Mongolia, the exceptional preservation provided by these deposits also contributed much of what is known about neoceratopsians today.

Neoceratopsians are some of the most variable dinosaurs known.

Neoceratopsians are best distinguished from the psittacosaurids by the considerable size increase in their skulls in proportion to their postcranial skeletons. This tendency resulted in a few species competing for the largest skull ever owned by a land-dwelling animal. Some frills reached lengths of 1.5 meters and total skull lengths approached 3 meters! One adaptation for supporting such huge heads was a fusion of the anterior cervical vertebrae, and another was the development of stout fore limbs. By default, most neoceratopsians must have been obligate quadrupeds with only a few smaller forms having a possibility of facultative bipedalism.

The cladistic classification of neoceratopsians has been a point of contention in recent years, mostly as a result of a large amount of character data that conflict with traditional classification schemes. For example, a venerable gradistic name was applied to a grouping of some of the smaller (1–3 meters long) and more basal neoceratopsians, Family **Protoceratopsidae**, with its star member *Protoceratops* (Fig. 13.7B). This was then used to fit its members into a clade by the same name, but subsequent cladistic analyses have failed to support its monophyletic grouping. Consequently, numerous basal neoceratopsians precede the other major clade within Neoceratopsia, **Ceratopsidae**, which also is based on a former family name.

13

Non-ceratopsid neoceratopsians are an interesting and varied group with a long history in dinosaur paleontology. For example, one of the most abundantly represented dinosaurs in the geologic record is *Protoceratops andrewsi*, a small-frilled, unornamented neoceratopsian that only reached about 2.5 meters in length. This neoceratopsian was postulated as the possible inspiration for the griffin, a mythical beast with lion and bird features whose legend arose out of central Asia, where skeletons of *Protoceratops* occur abundantly (Chapter 3). *Protoceratops* has been nicknamed the "sheep of the Gobi" because its specimens are common in Late Cretaceous deposits of Mongolia. Moreover, they were speculated to be defenseless fodder ("Mesozoic mutton") for their carnivorous contemporaries. *Protoceratops* is famous as one of the dinosaurs discovered in the original Mongolian expeditions led by Roy Chapman Andrews of the American Museum of Natural History (Chapter 3). For nearly 70 years, *Protoceratops* was credited as the egglayer of the numerous clutches found in the Gobi, until the discovery that the eggs were actually laid by its supposed egg predator, *Oviraptor* (Chapter 9). One specimen of *Protoceratops* contradicts its species' reputation as a passive, docile animal in that it may have fought a *Velociraptor* to the death (see Fig. 7.9). Enough skeletons of *Protoceratops* have been studied that some paleontologists think that they can discern sexual dimorphism within the species. The form with a more vertically raised frill has been arbitrarily identified as the "male" form. Other basal neoceratopsians share characteristics with *Protoceratops*, but it is recognized as a more derived form than some of the others.

As fascinating as *Protoceratops* and other smaller neoceratopsians are, ceratopsids stand out from the Late Cretaceous dinosaurs in many ways. They were among the largest herbivores of their time and the largest ornithischians of the entire Mesozoic, some having reached 8-meter lengths and weights of 7 to 8 metric tonnes. More importantly, with regard to their biology and aesthetics, ceratopsids also carried the most beautiful "advertisements" ever seen in dinosaurs. Their visually appealing traits included single or multiple horns that were hooked or straight, positioned either on or above the nares and orbits (both seen in *Triceratops*), or on the periphery of the frill. Broad or long frills, as well as parietal fenestrae, seen as large holes in the frill in some species, were other attributes that best express ceratopsid diversity. Other details of the skeletal anatomy that distinguished them from their neoceratopsian predecessors were:

1 a femur longer than the tibia; and
2 10 sacral vertebrae that fused to form a cohesive unit in the hip, which helped to support some considerable musculature.

The many varied members of this group can be further categorized on the basis of squamosal lengths into the clades **Chasmosaurinae** (= long squamosals) and **Centrosaurinae** (= short squamosals). Chasmosaurines include *Anchiceratops*, *Arrhinosaurus*, *Pentaceratops*, the previously mentioned *Triceratops*, *Torosaurus* (Fig. 13.8), and *Chasmosaurus* (Fig. 13.9). Two primitive genera of the chasmosaurines are *Pentaceratops* and *Chasmosaurus*. Among the traits that distinguished them from more advanced genera in their clade are very large fenestrae in their parietals. *Avaceratops*, *Brachyceratops*, *Centrosaurus*, *Monoclonius*, *Pachyrhinosaurus*, and *Styracosaurus* are all centrosaurines, the most basal of which are *Avaceratops* and *Brachyceratops*. These two are not crowned by the horns on the frills seen in more advanced forms such as *Styracosaurus* (Fig. 13.10). All genera of chasmosaurines and centrosaurines found so far occur only in Upper Cretaceous strata of western North America; the centrosaurines have an even narrower geologic range and geographic distribution. As discussed later, centrosaurines are

FIGURE 13.8 Cast of skull for the Late Cretaceous neoceratopsian (and ceratopsid) *Torosaurus latus*, from the Hell Creek Formation of South Dakota, a candidate for the largest skull possessed by any land animal. Utah Field House of Natural History, Vernal, Utah.

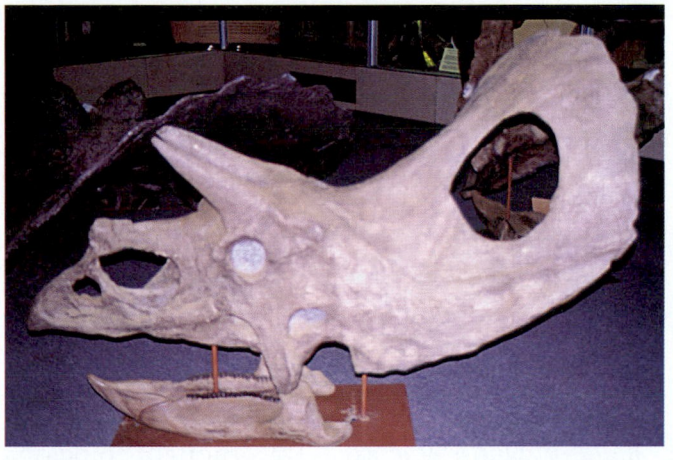

(A)   (B)

FIGURE 13.9 *Chasmosaurus*, a Late Cretaceous neoceratopsian and ceratopsid of North America. (A) Frontal view, showing legs positioned underneath the plane of the body (erect posture). Cast of skeleton, College of Eastern Utah Prehistoric Museum, Price, Utah. (B) Lateral view.

FIGURE 13.10 *Styracosaurus*, a Late Cretaceous ceratopsian from North America that apparently could never have had enough horns. Dinosaur Adventure Museum, Fruita, Colorado.

13

especially relevant to hypotheses of how ceratopsid evolution happened during the Late Cretaceous.

Paleobiogeography and Evolutionary History of Marginocephalia

Marginocephalian paleobiogeography is relatively simple, as almost all of its representatives have been found in North America, China, and Mongolia, with only three errant specimens discovered in Thailand, Germany and England (Table 13.1).

Two dinosaur specimens from South America and Australia have been proposed as ceratopsian remains, but these classifications are based on fragmentary material and have not been confirmed. Similarly, a skull fragment from the Cretaceous of Madagascar was initially thought to belong to a pachycephalosaur. It was named *Majungatholus*, and the paleobiogeographic range of pachycephalosaurs was briefly extended to Africa. However, the later discovery of more complete material from another specimen of *Majungatholus* revealed that it was actually a theropod (Chapter 9). This subsequently disproved the extension of pachycephalosaurs, and by default, marginocephalians, to Africa.

The relatively restrictive distribution for marginocephalians is used as evidence favoring endemism. In this scenario, marginocephalian ancestors developed in certain ecosystems in specific places, then their pachycephalosaur and ceratopsian descendants did likewise with later radiations. For example, the preferred environments for neoceratopsians in the western part of North America, based on occurrences of complete skeletal remains, were lowland areas associated with fluvial floodplains or upper delta plains that were forested. However, these environments differed from those of their smaller ceratopsian predecessors, such as *Psittacosaurus* and *Protoceratops*. These dinosaurs multiplied in the semi-arid conditions of the Late Cretaceous in Mongolia. The common ancestry of these neoceratopsians is evident from anatomical data, but their distinctive forms and occurrence in facies representing diverse environments reflect adaptations to their respective environments. Because the smaller Asian ceratopsians have more primitive features relative to the larger North American ceratopsids, the hypothesis is that ceratopsid ancestors migrated from Asia to North America around the mid-Cretaceous Period and later evolved into distinctive forms in their new environments. Once in North America, ceratopsids achieved considerable latitudinal distributions. For example, chasmosaurines have been discovered in Late Cretaceous deposits from Mexico to Alaska.

Neoceratopsian dispersal and evolutionary routes may seem straightforward, but pachycephalosaur evolution is fraught with difficulties. One problem with working out pachycephalosaur paleobiogeography lies in the relative rarity of pachycephalosaur material. Additionally, the materials found so far consist mostly of skull fragments, so the majority of pachycephalosaur genera are described on the basis of thick skull fragments and little else. The corresponding absence of postcranial pachycephalosaur material for many of their species thus can be attributed to selective transport and burial. After all, a thick skull was more likely to be preserved than a thin limb bone (Chapter 7). This explanation is supported by the occurrence of pachycephalosaur material in Cretaceous facies representing fluvial environments (floodplains) or arid regions (dunes of deserts). Paleontologists interpret pachycephalosaurs as upland animals that did not normally inhabit lowland coastal areas. This circumstance certainly would have placed them in different habitats from their hadrosaur and ceratopsian contemporaries, although a few Mongolian species lived in the same areas as smaller ceratopsians.

The evolutionary divergence of marginocephalian ancestors into the pachycephalosaurs and ceratopsians is difficult to work out because of the lack of specimens that clearly demonstrate when this split took place. Nearly all marginocephalian species known so far are in Early or Late Cretaceous strata, and mostly in North America and central Asia. These data suggest that the split between pachycephalosaurs and ceratopsians took place early in the Cretaceous or during the latest Jurassic. Moreover, this split most likely took place in an area connected to both North America and Asia at the time. As mentioned earlier, the migration of ceratopsians to North America is one explanation for their diversification into ceratopsids, and a similar situation may have occurred with marginocephalians in general.

Some Late Cretaceous ceratopsids, along with the associated dinosaurs sharing their deposits, exemplify one of the best-documented and persuasive arguments for evolutionary sequences seen in dinosaurs. Additionally, these sequences could be tied to broad-scale environmental change (sea-level fluctuations) interpreted in the Upper Cretaceous Judith River and Two Medicine Formations of Montana. Ceratopsids are particularly well suited for studying evolution because of:

1 their abundance;
2 the taphonomic biases that favor preservation of their skulls; and
3 the detailed morphological variations on these skulls, which may reflect genotypic change (Chapter 6).

Consequently, the fact that the people working on this sequence concentrated their efforts on ceratopsid remains is not surprising.

Paleontologists and geologists have documented how the rise or fall of sea level, for the seaway that covered the mid-continent region of North America during the Cretaceous, affected ceratopsian habitats. With the seaway to the east and the ancestral Rocky Mountains newly uplifted to the west by plate convergence during the Late Cretaceous (Chapter 4), dinosaurs inhabited a narrow area of North America that contained lowland terrestrial habitats. Whenever sea level rose, the habitats became more restricted, which should have caused local environmental pressures on dinosaur populations. In contrast, when sea level dropped, lowland terrestrial habitats were expanded and could more easily accommodate migrating populations of dinosaurs. Therefore, with repeated cycles of the sea level rising and falling, dinosaur faunas should have undergone speciation through the course of geologic time in this region. The timing of the sea-level fluctuations in the formations was independently determined through a combination of relative age-dating techniques, such as superposition, lateral continuity, and biologic succession, as well as radiometric age dates derived from volcanic ash deposits (Chapter 4). Thus the paleontologists were able to place the changes they observed in ceratopsid species and some ornithopods within the context of geologic time, which in this case was over the course of about 12 million years (84–72 Ma).

The ceratopsids that figured in the proposed evolutionary sequence were centrosaurines, which showed a stratigraphically defined sequence of (oldest to youngest) *Styracosaurus albertensis* → *"Styracosaurus"* (an unnamed species slightly different) → *Einosaurus procurvicornis* → *Achelousaurus horneri* → *Pachyrhinosaurus canadensis*. This sequence is within deposits representing a small portion of the 12-million-year framework noted previously, with each species separated from the other by time periods of less than 500,000 years. If the identified fossil species are indeed representative of biological species, then evolution of these dinosaurs must have been relatively rapid.

13

Of the marginocephalians available for study, ceratopsians, and ceratopsids in particular, offer the best opportunities for similar studies of dinosaur evolution. Using an approach modeled on the one just reviewed, paleontologists and geologists interested in future work on evolutionary sequences within dinosaur lineages might want to focus on these dinosaurs, rather than a few spectacular specimens of other clades.

Marginocephalians as Living Animals

Reproduction and Growth

Despite the excellent body fossil record for ceratopsians, little is known about marginocephalian reproduction, although some speculations about their pre-mating behaviors have gained considerable attention. Similar to some ornithopods, especially hadrosaurs (Chapter 11) and thyreophorans (Chapter 12), marginocephalians are subject to hypotheses concerning the probable sexual advertisements they carried as part of their bodies. Ceratopsids in particular are regarded as the divas of dinosaurs because of their broad and ornate head shields, bedecked with horns, bosses, and knobs. These head shields may have provided protection against the occasionally encountered predators, but more likely they served an everyday use as easy visual-recognition cues within their respective species. Included in this recognition process must have been not only potential mate identification, but also intraspecies rivalry. Intraspecies competition expressed through combat has been long hypothesized for marginocephalians on the basis of similar behavior in mammalian males that have prominent headgear and aggressively compete for females in a herd (such as bighorn sheep, moose, and elephants). Furthermore, given the thick, bony skulls of pachycephalosaurs and ceratopsians, and the prominent nasal and antorbital horns of some ceratopsians, these features certainly could have aided in such competition.

Pachycephalosaurs are perhaps the dinosaurs most often used as examples for intraspecies competition. Both popular and serious scientific publications alike show them butting heads with one another. Although this inferred behavior prompts their comparison to bighorn sheep, pachycephalosaurs weighed considerably more, as much as two tonnes, a fact that should be kept in mind during discussion of their behavior. Evidence in favor of head-butting includes:

1 thick skulls;
2 robust occipital condyles at the back of the skull, thus cushioning the skull in its articulation with the first cervical vertebra; and
3 tightly interlocking dorsal vertebrae.

The physics behind combat scenarios might help to illuminate what types of forces would have been involved in head butting.

The latter trait would have reinforced the back posterior to the neck and skull, preventing lateral movement of the vertebrae (which encased the spinal cord) during any forceful impact. However, other observations that would supplement these data, such as obvious dents in pachycephalosaur skulls or other signs of trauma, are still forthcoming or not clearly defined.

Using a hypothetical example of two male *Pachycephalosaurus*, each weighing about 1.5 tonnes (1500 kg), and the familiar formula for calculating force ($F = ma$: see Eqn 4.8

in Chapter 4), the force generated by a single pachycephalosaurid running (and accelerating) at 5 m/s^2 would be considerable:

Step 1. $F = (1500 \text{ kg})(5 \text{ m/s}^2)$
Step 2. $= 7,500 \text{ N}$

The impact of two objects of the same mass and at the same acceleration causes additive force, which would have generated 75,000 N of force in this example. Using the formula for stress ($\sigma = F/A$: see Eqn 4.9 in Chapter 4) and assuming that the skulls impacted on a point-to-point contact in a small area, such as 100 cm^2 (10 × 10 cm), the total stress generated would have been

Step 1. $\sigma = 15,000 \text{ N}/0.1 \text{ m}^2$
Step 2. $= 1.5 \times 10^5 \text{ N/m}^2$

This is a frightening amount of stress, corresponding to about 15 N/cm^2. For example, compare this figure to that derived for "foot stress" in Eqn 14.1, Chapter 14.

However, these are idealized calculations based on many assumptions. The speeds could have varied, the masses could have differed, the contacts may not have been applied at directly opposing directions and to such small areas, deceleration from the impact has been discounted, and so on. Calculating the kinetic energy of a running pachycephalosaur would derive a more realistic estimate. Kinetic energy is calculated by the following formula:

$$K_e = 0.5(m)(v)^2 \qquad (13.2)$$

where K_e is kinetic energy, m is mass, and v is velocity in meters per second. For one of the pachycephalosaurids in our example, the kinetic energy would have been:

Step 1. $K_e = 0.5(1500 \text{ kg})(5 \text{ m/s})^2$
Step 2. $= 0.5 \times 7500 \text{ kg} \times \text{m/s}^2$
Step 3. $= 18,750 \text{ N}$

For one pachycephalosaurid to completely stop the other, the force must be absorbed over a given distance that varies with the material doing the absorbing. This relation is expressed by the formula:

$$A_e = Fd \qquad (13.3)$$

where A_e is the energy absorbed and d is deceleration distance. To bring the pachycephalosaurid to a screeching halt, the calculated force needed is expressed by:

$$F_a = m(v)^2/2d \qquad (13.4)$$

where F_a is the absorbed force. The safety airbag in an automobile can illustrate this principle: it absorbs a driver or passenger, which results in the person traveling a greater distance into the airbag than if he or she hit the steering wheel or windshield. This means that less force is transmitted to the person by impacting the airbag.

In our pachycephalosaurid example, assuming deceleration distances of 0.0015 meter (15 mm) for bone (after all, bone does not compress very easily) and 0.1 meter

(10 cm) for flesh (think about the skull sinking into the side of the rival pachy-cephalosaurid instead of contacting the other skull), the absorbed forces would be:

Example 1 (bone)

Step 1. $F_a = (1500 \text{ kg})(5 \text{ m/s})^2/2(0.0015 \text{ m})$

Step 2. $= 7500/0.003$

Step 3. $= 1.25 \times 10^7 \text{ N}$

Example 2 (flesh)

Step 1. $F_a = (1500 \text{ kg})(5 \text{ m/s})^2/2(0.1 \text{ m})$

Step 2. $= 37,500/0.2$

Step 3. $= 1.87 \times 10^5 \text{ N}$

Consequently, an impact into flesh would have generated two orders of magnitude less force than a head-to-head collision. In other words, a head-to-flesh collision would have survival advantages for both animals, not just the one doing the ramming, and especially if ramming was a frequently expressed behavior. Of course, the deceleration distance would not have been only for the head but would have been passed down the entire length of the body, increasing the distance and correspondingly decreasing the force.

The point being made here is that when two massive animals run toward one another and their body parts directly impact, they could have less easily absorbed the force on bone than flesh. This sort of force applied to bone should have left marks, no matter how thick or spongy the bone. It also quite likely would have exceeded the structural limits of the dinosaurs' spinal columns for absorbing the impact behind the skull. Based on the realities represented by these calculations and the previously mentioned information, the likelihood that pachycephalosaurs actually rammed into one another head-to-head is doubtful. Furthermore, the selection pressures caused by simultaneously inflicted paralysis or death surely would have caused this behavior to quickly disappear. After all, dead or paralyzed animals cannot pass their genes on to a successive generation. Thus, rather than stating that pachycephalosaurs did not use their heads for defensive or pre-mating purposes at all, a good compromise hypothesis is that if any ramming happened it was directed to softer areas.

Similarly intriguing features in ceratopsid skulls are apparent healed wounds which offer evidence for intraspecific competition in ceratopsians. These wounds are visible as scars or defects that seem to have been applied to the skull while the animal was alive. One example in a *Triceratops* skull is a hole that passes through the jugal and has a diameter similar to that for the distal end of a *Triceratops* horn. Other ceratopsid genera reported with similar skull defects are *Diceratops*, *Pentaceratops*, and *Torosaurus*. The coincidence of such scars found in the skulls of ceratopsids that have substantial horns is currently considered as a reasonable basis for a cause-and-effect hypothesis. One statement that can be made about these possible trace fossils is that they are related to intraspecies combat, although what may have prompted the fighting in individual cases is still uncertain. In extant species, fighting may be triggered by competition for mates, establishing territory, asserting dominance, or illnesses, such as rabies, that cause paranoia or other aggressive behavior. The styles of intraspecific combat behavior also would have varied with the skull morphology of the ceratopsids. For example, large-frilled chasmosaurines, such as *Chasmosaurus*, may have only had to turn their heads, giving a rival a full frontal view for intimidation (see Fig. 13.9A). In contrast, the short-frilled centrosaurines, such as *Centrosaurus* or *Styracosaurus*, did not have such obvious tools of intimidation, so they may well have locked horns more often than their large-frilled relatives. Regardless, actual evidence supporting these generalized behaviors is still scanty and subject to further critical review.

As far as actual reproductive behavior is concerned, little is known about marginocephalians. Presumed ceratopsian eggs and nests, attributed to *Protoceratops*, were discovered in Late Cretaceous rocks of Mongolia in the 1920s by the American Museum of Natural History expeditions to that region (Chapter 4). However, later analyses revealed that at least some of the eggs belonged to the theropod *Oviraptor* (Chapter 9), thus rendering all other similarly identified "protoceratopsian" eggs and nests as suspect. Since these corrections were made, no undoubted ceratopsian nest or embryonic remains within an egg have been identified. However, 15 hatchlings of *Protoceratops*, found together in a deposit in Mongolia, is persuasive evidence favoring the proximity of a nest. Still, no ceratopsian eggs have been definitely linked with hatchlings. No eggs, embryos, or hatchlings of pachycephalosaurs have been recognized, although a few skulls of juvenile *Stegoceras* have been described. These limited data mean that the reproductive habits of marginocephalians as a clade are poorly understood.

Nevertheless, embryonic and other juvenile remains have been identified for *Bagaceratops*, *Breviceratops*, *Psittacosaurus*, and *Protoceratops*, all coming from Cretaceous strata in Mongolia and China. Furthermore, a recent and spectacular find of one adult and 34 juvenile *Psittacosaurus*, in Lower Cretaceous rocks of China, provides the most convincing evidence of parental care in ceratopsians. The 35 dinosaur skeletons were all complete and concentrated in a bowl-shaped depression with an area of 0.25 m². The juveniles were fairly grown (21 cm long) and the same size, strongly suggesting that they were from the same generation. This situation argues for a parent *Psittacosaurus* in close association with a large brood of its young at the time of their burial, the latter of which must have been nearly instantaneous to preserve them so well (Chapter 7).

Growth series have been described for a few ceratopsians, most notably *Psittacosaurus* and *Protoceratops*. Many specimens represent these Cretaceous ceratopsians from Mongolia and China (more than 100 just for *Psittacosaurus*) and most of the specimens are complete. As a result of this teeming abundance, growth series have been interpreted on size analyses of these ceratopsians, which reflect changes in ontogeny. *Psittacosaurus* was rather small as far as dinosaurs are concerned, reaching about 2 meters long as an adult; *Protoceratops* was not much larger, with some individuals reaching 2.5 meters long. Measured skull lengths of *Protoceratops* have a range of 5 to 50 cm, reflecting a minimal tenfold increase from juvenile to adult. In this growth series, the smaller forms tend to look alike, but a two-part split into large-frilled and small-frilled forms is apparent in skulls longer than about 25 cm. This divergence in forms within the growth series for this species is considered to be among the best-documented evidence for sexual dimorphism in dinosaurs. Presently, the large-frilled adult specimens are designated as "males" and the small-frilled ones as "females." Of course, the opposite may be true in these sex determinations, but more compelling evidence from the fossil record, such as a mother *Protoceratops* brooding its egg clutch, is still forthcoming. Sexual dimorphism has also been proposed for a few ceratopsids (*Centrosaurus*, *Chasmosaurus*, and *Triceratops*). The data sets for these genera are good, but are not as robust as for *Protoceratops*.

Locomotion

Marginocephalian locomotion is still a contentious issue in some respects, partially because of lack of evidence or too much contradictory evidence. For example, at

13

the time of writing, no distal parts of the hind limbs from pachycephalosaurs had been recovered from the geologic record. Consequently, dinosaur paleontologists know little about pachycephalosaur locomotion. Because of the relatively short arms recovered from a few specimens, they probably were obligate bipeds but may also have been facultative quadrupeds. There are two major problems with a lack of metatarsals, tarsals, and pes phalanges, or unguals in association with the rest of a pachycephalosaur skeleton:

1 if found by themselves without nearby material identifiable to a pachycephalosaur, such limb elements may be allied with other dinosaur species, such as ceratopsians; and
2 tracks left by pachycephalosaurs cannot be correlated with foot morphology, so their tracks may be already known but misidentified.

On the other hand, ceratopsians represent a different dilemma in interpretation of their locomotion, which centers on their posture. Older restorations of ceratopsians, such as in paintings by Charles Knight (Chapter 1) and skeletal mounts of ceratopsians in museums, show a sprawling posture that belies the modern view of dinosaurs with erect postures. More recent mounts and illustrations of ceratopsians have them fully erect, limbs directly under their bodies, as in all other dinosaurs (see Fig. 13.10). However, this newer orthodoxy has been challenged on the basis of skeletal evidence that favors a more sprawling posture for the fore limbs, or semi-sprawling. Restorations of the articulation between the humerus and the rest of the skeleton have been problematic: when attempts are made to straighten the humerus vertically, the best fit is achieved with the elbows sticking outside the plane of the body. Additional analyses were provided by paleontologists who, in cooperation with computer scientists, recently made three-dimensional scans of a *Triceratops* skeleton and attempted computer animations of the skeleton walking, although these are not as conclusive.

Trackway evidence for ceratopsians has not clearly resolved the controversy, but at least it has provided yet another perspective. The tracks, which so far have been found in only Late Cretaceous strata and in one area of the world (Colorado and Utah), have compression shapes that correlate with known neoceratopsians' foot anatomy (Fig. 13.4). Although trackways are much shorter than many other dinosaurs, they show typical manus–pes pairs, diagonal walking patterns, and a relatively wide straddle. However, the straddle is narrow enough that the paleontologists who originally reported them concluded that the ceratopsians were not sprawling, but walking erectly. The approximate speed of the trackmakers is also calculated to have been about 4 km/hour. However, an interesting detail about the manus–pes pairs is that the manus placement deviates with a noticeable pitch from the direction of travel and is slightly outside the pes impressions. Some paleontologists argue that this foot placement is consistent with a semi-sprawling posture for the fore limbs, considering that other quadrupedal dinosaurs left manus–pes impressions that were precisely aligned (reflecting fully erect postures). However, the largest drawback to this track evidence is its scantiness: very few trackways are available for analysis and they contain few tracks. Once more trackways of ceratopsids are found and analyzed, this controversy may be resolved.

Feeding

All marginocephalians were probably herbivores, ascertained from their teeth, jaws, and other skeletal adaptations. For example, because they had deeply inset cheek teeth, they could have had cheeks used for temporarily storing masticated plants while chewing. Pachycephalosaurs had heterodont dentition comparable to that of other dinosaurs interpreted as herbivorous. However, this dentition was weakly developed in comparison to the dental batteries seen in some other ornithischians, such as hadrosaurids (Chapter 11). Ceratopsians represent a departure from pachycephalosaurs in that their rostral and predentary formed beaks ideal for slicing through vegetation. Their cheek teeth also developed into dental batteries but ones more inclined toward vertically oriented shearing, rather than horizontal (back and forth) grinding. This movement must have caused much wear and so required the development of batteries that continually replaced worn-out teeth. Indeed, tooth-replacement rates estimated for *Triceratops* were 50 to 100 days, with as many as six teeth in line to replace any one already in the jaw. The prodigious head shields of some ceratopsids were originally interpreted as having evolved for the support of massive jaw musculature, which was most likely present for such active chewing. Nevertheless, muscle attachments to frills may also have been a secondary consequence of sexual selection in favor of large frills, rather than an adaptation primarily for chewing (Chapter 6).

The food eaten by ceratopsians is suggested to have been palms and cycads, which were non-flowering plants, and small shrubs or trees of angiosperms. However, paleobotanists have pointed out that fossils of these plants are not abundant in areas where ceratopsid bones are most commonly found. Instead, they suggest that low-level flowering plants were the preferred food for these dinosaurs. To add fuel to the controversy, ceratopsid tracks in Upper Cretaceous sandstones of Colorado (mentioned earlier) co-occur with impressions of palm and angiosperm leaves (Fig. 13.11).

FIGURE 13.11 Impression of palm frond in Upper Cretaceous Laramie Formation, which co-occurs with ceratopsian tracks in the same part of the formation and thus indicates a co-occurrence with a potential ceratopsian food source; Golden, Colorado.

This confluence provides good supplementary evidence of close proximity between these herbivores and their proposed foodstuffs. As mentioned earlier, no toothmarks, stomach contents, or coprolites are known from marginocephalians, although gastroliths are reported in more than one specimen of the Early Cretaceous ceratopsian *Psittacosaurus* of Mongolia. As a result, paleontologists interested in marginocephalian diets must content themselves with inferences based on fossil plants contemporaneous with marginocephalians, as well as the functional morphology of skeletal adaptations to feeding.

One idea regarding ceratopsid feeding habits that has broader implications in evolutionary theory is that ceratopsid browsing, over geologically significant time intervals, was partially responsible for the evolution of flowering plants. Their co-occurrence with herbivorous hadrosaurids and abundant fossils of flowering plants during the Late Cretaceous, coupled with the demise of other low-level browsing herbivores such as stegosaurs (Chapter 12), are used as evidence for this hypothesis. In such a scenario, ceratopsids migrating as hungry herds would have caused selection pressure in favor of faster-growing flowering plants. Co-evolution also would have occurred when deposition of ceratopsid feces dispersed undigested seeds of these flowering plants. This form of evolutionary bribery is still present today, in which indigestible seeds are covered by delicious fruits. No undoubted ceratopsid coprolites or even stomach contents have been found, so this idea, although reasonable, is still speculative.

One problem with ceratopsids and hadrosaurs driving the evolution of flowering plants is the endemism of Late Cretaceous ceratopsids: how did the evolution of flowering plants take place on continents where these large herbivores were either absent or uncommon? In the case of ceratopsids and flowering plants, much more evidence is needed, as well as increased cooperation between paleobotanists and vertebrate paleontologists, before such questions can be answered.

Social Life

Because so little material has been recovered of pachycephalosaurs (and those recovered are all individual finds) and there is a lack of trace fossils, virtually nothing is known about their sociality. The thick skulls discussed earlier as functional "battering rams" imply that other pachycephalosaurs were the targets of such ramming, which required occasional proximity. However, with no further evidence, any other conclusions reached about the social lives of pachycephalosaurs would be speculative.

On the other hand, ceratopsians show plenty of body fossil evidence for having been social creatures and were quite likely gregarious. Monospecific bone beds of neoceratopsid remains include those of the Late Cretaceous centrosaurines *Achelousaurus, Centrosaurus, Einosaurus, Pachyrhinosaurus,* and *Styracosaurus* of North America, as well as the chasmosaurines *Anchiceratops* and *Chasmosaurus.* In the case of a bone bed of *Centrosaurus* in Alberta, Canada, an estimated 300 individuals occur in what seems to be a contemporaneous deposit. This extraordinary find implies that a *Centrosaurus* herd died in a sudden catastrophe, such as a river flood. Abundant skeletons of *Psittacosaurus* and *Protoceratops* from the Late Cretaceous of Mongolia and China are also suggestive of thriving populations closely associated with one another. The aforementioned wounds in various ceratopsids, which were possibly caused by individuals of the same species, are also indicative of social interaction, albeit of a violent type. Lastly, trackways reported for ceratopsids show that several individuals were walking through the same areas through time, although trackways of multiple individuals are not nearly as well documented as for theropods (Chapter 9), sauropods (Chapter 10), and iguanodontians (Chapter

11). Nevertheless, the combination of all known fossil evidence for ceratopsians is sufficient to conclude that many species were group-oriented animals, rather than lone individuals fending for themselves against voracious theropods.

Health

Like most dinosaurs, marginocephalians led fairly healthy lives, with some exceptions, indicated through paleopathological evidence in a few ceratopsians. As mentioned earlier, *Triceratops* and other ceratopsids may have inflicted wounds on one another through intraspecific competition. There were other problems, possibly related to accidents associated with everyday life in the Mesozoic:

- Bone overgrowths in a scapula and mandible of *Triceratops* may have resulted from tearing of tendons that attached to each bone.
- Healed fractures of the costae (ribs) in the centrosaurines *Pachyrhinosaurus* and *Centrosaurus* are attributed to intraspecific flank butting.
- Stress fractures in phalanges of *Centrosaurus*, *Pachyrhinosaurus*, *Styracosaurus*, and *Triceratops* were possibly related to rapid or repetitive movement, such as extended locomotion.

All of these conditions provide good evidence that ceratopsids had active lives that occasionally placed stresses on their bodies. Probably the most interesting of these conditions are the rib fractures, which are possible trace fossils that provide supplemental data to other information and hypotheses related to the social interactions of ceratopsids (Chapter 14).

Although predators of marginocephalians are mostly inferred on the basis of their contemporaneous associations with theropods or other large carnivores, a few Late Cretaceous examples stand out, providing clear evidence of how some marginocephalians provided food for theropods. The most famous example of a ceratopsian that was on the menu of a specific theropod was a *Protoceratops* that was apparently embroiled in conflict with a *Velociraptor* when they were both buried (see Fig. 7.9 and Chapters 7 and 9). Moreover, several instances are known of *Triceratops* eaten by tyrannosaurids, most likely *Tyrannosaurus rex* (Chapter 9). In one well-defined specimen, the toothmarks are in bones from the hip region of the *Triceratops*, which in life would have been covered by thick layers of skin and muscle. Consequently, these toothmarks could have been inflicted after the animal was already dead. This in turn does not disprove an alternative hypothesis of tyrannosaurid scavenging, rather than supporting a strict interpretation of predation.

Despite the diner–dinner relationship established between *Tyrannosaurus* and *Triceratops*, no data support the hypothesis that these two animals directly fought one another in mortal combat. Numerous popular treatments showing these two behemoths spilling one another's blood, with the tyrannosaurid's near-useless arms doing it no good against its horned rival, are actually based on no more evidence than the following:

1 they lived during the same time;
2 they dwelled in the same habitats; and
3 a few tyrannosaurids ate what might have been already-dead ceratopsians.

Nevertheless, the admittedly small data set showing such consumption will at least serve as a predictor for future studies of toothmarks on ceratopsid bones, which will help to establish either predator–prey or scavenger–carrion relationships in Late Cretaceous ecosystems.

SUMMARY

Marginocephalians were herbivorous and mostly Cretaceous ornithischians, represented by two main clades, Pachycephalosauria and Ceratopsia. Marginocephalians are best denoted as a clade by a narrow shelf of bone formed by the parietal and squamosal, which gives them a distinctive frill that hangs over the posterior part of the skull. Pachycephalosaurs are distinguished by the fusion of their frontals and parietals into a thick skull, and are further subdivided into groups largely on the basis of whether the skull was flat (homocephaloids) or domed (pachycephalosaurids), among other features. These thick skulls were presumably used in intraspecific competition when these dinosaurs rammed their heads into one another, although this may not represent the complete picture of their function. Ceratopsians are identified by a rostral bone on the anterior part of the mouth, a novelty that with the predentary formed a distinctive beak used to crop vegetation. Parietal frills and horns developed to such an extent in ceratopsians that some of its members had the largest and most ornate skulls of any known terrestrial animals. Ceratopsians reached their peak of diversity throughout the Late Cretaceous, as evidenced by the numerous members of Ceratopsidae, which contained both chasmosaurines (long-frilled ceratopsids) and centrosaurines (short-frilled ceratopsids). Intraspecific competition and other forms of social interactions, such as herding, were also likely among ceratopsians.

The paleobiogeographic distribution of marginocephalians shows evidence of endemism in that pachycephalosaurs only occur in Europe, Asia, and North America, and ceratopsians are only found in Asia and North America. Basal ceratopsians, which included *Psittacosaurus* and other smaller ceratopsians, were very common in parts of central Asia and well represented in Late Cretaceous deposits of Mongolia in particular. Migration of more basal ceratopsians from Asia must have occurred during the mid-Cretaceous, enabling the evolution of the larger ceratopsids of North America during the latter part of the Cretaceous. Evolution of ceratopsids may have been partly related to biogeographic isolation of populations in the western part of North America or to the types and distributions of food plants. Ceratopsids, in particular, evolved dental batteries to supplement their beaks and jaws, which were efficient slicing tools for dealing with possible new food sources, such as flowering plants.

DISCUSSION QUESTIONS

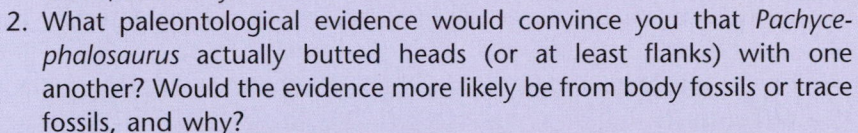

1. What paleontological evidence would convince you that *Tyrannosaurus* and *Triceratops* actually fought one another? Would the evidence more likely come from body fossils or trace fossils, and why?

2. What paleontological evidence would convince you that *Pachycephalosaurus* actually butted heads (or at least flanks) with one another? Would the evidence more likely be from body fossils or trace fossils, and why?

3. Given the following speculative masses and speeds for each genus of pachycephalosaur, calculate the kinetic energy generated by each.
   a. *Homocephale*: 145 kg, 2.5 m/s
   b. *Stegoceras*: 25 kg, 4 m/s
   c. *Stygimoloch*: 75 kg, 3.7 m/s

4. Examine the genera lists in Table 13.1 and 13.2, then omit all of the genera that only come from Mongolia to judge the impact of expeditions to Mongolia on what is known about marginocephalians.
   a. What percentage of the original genera is left after making such omissions?
   b. Which taxa are affected the most by your omission: pachycephalosaurs, non-neoceratopsians, or the neoceratopsians?
   c. Taking this a step farther, speculate how evolutionary scenarios of marginocephalian evolution would be changed by the omission of Mongolian-only species.

5. The evolutionary sequence summarized for Late Cretaceous centrosaurines in the western part of North America seemed very persuasive, but what are some alternative hypotheses to explain the observed changes through time of the centrosaurine taxa? What new evidence would either falsify or modify the previous hypothesis?

6. Review the factors given in the chapter explaining why pachycephalosaurs and ceratopsians are so differently represented in the geologic record despite their geographic proximity and overlapping geologic ranges. What are some other explanations, and what new (or previously found) evidence would support each explanation?

7. Test the speculation that centrosaurines locked horns with one another more often than chasmosaurines by classifying the genera mentioned as having skull defects into each clade. Is any trend noticeable? What evidence, either from body fossils, trace fossils, or both, would convince you further of your conditional acceptance or rejection of your hypothesis?

8. The chapter refers to healed wounds in the parietal frills and rib fractures of ceratopsids as trace fossils. Why are these features trace fossils rather than body fossils?

9. What supplementary evidence associated with *Protoceratops* would help to determine which of the two body types (small-frilled and

13

DISCUSSION QUESTIONS Continued

large-frilled) was a male or female, besides a "female" form found sitting on a clutch of eggs? Alternatively, how could the assumption that the "female" form was brooding possibly be wrong?

10. With regard to the idea that ceratopsids and hadrosaurids may have been partially responsible for the co-evolution of flowering plants, think about how the evolution of birds capable of full flight during the Cretaceous may have also contributed in a similar way. How might birds have helped, especially in those areas not frequented by ceratopsids and hadrosaurids? What other animals today have an effect on the geographic distribution of flowering plants (besides humans, of course)?

Bibliography

Brinkman, D. B., Ryan, M. J. and Eberth, D. A. 1998. The paleogeographic and stratigraphic distribution of ceratopsids (Ornithischia) in the upper Judith River Group of Western Canada. *Palaios* **13**: 160–169.

Dodson, P. 1991. Morphological and ecological trends in the evolution of ceratopsian dinosaurs. *Paleontological Contributions from the University of Oslo* **364**: 17–18.

Dodson, P. 1996. *The Horned Dinosaurs: A Natural History*. Princeton, New Jersey: Princeton University Press. p. 346.

Dodson, P. 1997. "Neoceratopsia". *In* Currie, P. J. and Padian, K. (Eds). *Encyclopedia of Dinosaurs*. New York: Academic Press. pp. 473–478.

Dodson, P., Forster, C. A. and Sampson, S. D. 2004. "Ceratopsidae". *In* Weishampel, D. B., Dodson, P. and Osmólska, H. (Eds), *The Dinosauria* (2nd Edition). Berkeley, California: University of California Press. pp. 494–513.

Farlow, J. O. and Dodson, P. 1975. The behavioral significance of frill and horn morphology in ceratopsian dinosaurs. *Evolution* **29**: 353–361.

Forster, C. A. 1996. Species resolution in *Triceratops*: Cladistic and morphometric approaches. *Journal of Vertebrate Paleontology* **16**: 259–270.

Forster, C. A. and Sereno, P. C. 1997. "Marginocephalians". *In* Farlow, J. O. and Brett-Surman, M. K. (Eds), *The Complete Dinosaur*, Bloomington, Indiana: Indiana University Press. pp. 317–329.

Giffin, E. B. 1989. Notes on pachycephalosaurs (Ornithischia). *Journal of Paleontology* **63**: 525–529.

Goodwin, M. B. and Horner, J. R. 2004. Cranial histology of pachycephalosaurs (Ornithischia, Marginocephalia) reveals transitory structures inconsistent with head-butting behavior. *Paleobiology* **30**: 253–267.

Goodwin, M. B. and Johnson, R. E. 1995. A new skull of the pachycephalosaur *Stygimoloch* casts doubt on head butting behavior. *Journal of Vertebrate Paleontology* **15**, 3: Suppl. 32.

Hailu, Y. and Dodson, P. 2004. "Basal Ceratopsia". *In* Weishampel, D. B., Dodson, P., and Osmólska, H. (Eds), *The Dinosauria* (2nd Edition). Berkeley, California: University of California Press. pp. 478–493.

Lockley, M. G. and Hunt, A. P. 1998. Ceratopsid tracks and associated ichnofauna from the Laramie Formation (Upper Cretaceous, Maastrichtian) of Colorado. *Journal of Vertebrate Paleontology* **15**: 592–614.

Lull, R. S. 1908. The cranial musculature and the origin of the frill in the ceratopsian dinosaurs. *American Journal of Science* **25**: 387–399.

Maryanska, T. and Osmólska, H. 1974. Pachycephalosauria, a new suborder of ornithischian dinosaurs. *Palaeontologica Polonica* **30**: 45–102.

Maryanska, T., Chapman, R. E. and Weishampel, D. B. 2004. "Pachycephalosauria". *In* Weishampel, D. B., Dodson, P. and Osmólska, H. (Eds), *The Dinosauria* (2nd Edition). Berkeley, California: University of California Press. pp. 464–477.

Paul, G. S. and Christiansen, P. 2000. Forelimb posture in neoceratopsian dinosaurs; implications for gait and locomotion. *Paleobiology* **26**: 450–465.

Sues, H.-D. 1997. "Pachycephalosauria". *In* Currie, P. Jand Padian, K. (Eds.) *Encyclopedia of Dinosaurs*. New York: Academic Press. pp. 511–513.

Sullivan, R. M. 2003. Revision of the dinosaur *Stegoceras lambe* (Ornithischia, Pachycephalosauridae). *Journal of Vertebrate Paleontology* **23**: 181–207.

Xing, X., Makovicky, P. J., Wang Xiaolin, W., Norell, M. A, and Hailu, Y. 2002. A ceratopsian dinosaur from China and the early evolution of Ceratopsia. *Nature* **416**: 314–317.

13

Chapter 14

While you are watching an old movie about dinosaurs, you notice that they all drag their tails and their legs are sprawled out to the sides of their bodies. Moreover, the carnivorous dinosaurs swallow their prey whole. You turn to a friend and say, "See how those dinosaurs are behaving? That's all wrong." Your friend is skeptical, because, he says, movie directors and scriptwriters probably did a lot of research into how dinosaurs behaved.

What evidence do you have to support your statement without using any information from dinosaur skeletons?

Dinosaur Ichnology

14

Dinosaur Ichnology: The Real Fossil Record for Dinosaurs?

Welcome to the "other" fossil record of dinosaurs, trace fossils. Dinosaur ichnology, which is the study of their trace fossils, particularly tracks, provides an enormous amount of information that, until about 30 years ago, was viewed as more of a curiosity than an integral part of dinosaur studies. Many dinosaur books, especially those written prior to the 1990s, only treated dinosaur trace fossils as a sideline to skeletal data. In contrast, today, several dinosaur books and hundreds of research articles focus solely on dinosaur tracks. Likewise, nests, coprolites, toothmarks, and gastroliths have also received more attention for their worth in interpreting dinosaur anatomy, behavior, and paleoecology. Nonetheless, considering that they were initially identified in the early nineteenth century (Chapter 4) and are acknowledged sources of scientific information, dinosaur tracks and other trace fossils still do not receive the coverage they deserve. As a result, they will be covered in detail here, so that their value might be better appreciated.

So do dinosaur trace fossils constitute the "real" fossil record for dinosaurs, not only outnumbering but also exceeding the scientific value of skeletal material? Such a provocative question would have been answered "no" without hesitation only 30 years ago, but now warrants a "maybe" that may eventually evolve to a "yes."

Dinosaur Tracks

Importance and Applications of Dinosaur Tracks

The most abundant and important dinosaur trace fossils are dinosaur tracks. Tracks have all of the advantages of most other trace fossils:

1 they are potentially more abundant than other dinosaur body fossils;
2 they may be preserved in rocks that do not normally preserve dinosaur body fossils; and
3 they directly reflect dinosaur behavior where it happened.

The aspects of dinosaur behavior that can be interpreted from tracks include, but are not limited to:

- Where and in what environments they preferred to roam.
- Individual or group movements in these environments.
- Interactions among dinosaurs of the same species or different species.

- Approximate speeds of movement.
- Their most likely postures.
- Nuances of individual behavior.

A few parameters of dinosaur environments that can be elucidated from their tracks include the relative moistness of the sediment they were traversing, whether the original formation of the tracks affected any other organisms, and how sediments, or even bones of other animals (including dinosaurs), were affected by dinosaur trampling.

In the absence of body fossils, whose distribution and preservation were dependent on different taphonomic factors (Chapter 7), dinosaur tracks in Mesozoic strata also could be used in biostratigraphy as indicators of a former dinosaurian presence. Such information is especially important for determining whether the ancestors of dinosaurs had evolved by the Middle Triassic (Chapter 6) or whether any dinosaurs lived past the end of the Cretaceous Period (Chapter 16). Dinosaur paleobiogeography is also better defined by adding dinosaur track data to the skeletal record. So far, dinosaur tracks have been found on six continents, only excluding Antarctica (Chapters 9 to 13). Finally, well-defined footprints reveal dinosaur soft-part anatomy that is not normally preserved in their body fossils. Thus, tracks help "flesh out" dinosaur limbs more than is possible by just looking at their limb bones (Chapter 5).

Dinosaur tracks, more often than any other dinosaur fossils, tell us what a dinosaur was doing on a given day in the Mesozoic, and in most cases also tell us exactly where they did it.

Definitions of Track Terms

> *A **track**, or footprint, is an impression made by an appendage of an animal while it was alive.*

Although most fossil tracks were formed and preserved in originally soft sediment, solid materials compressed or fractured by the weight of a moving animal also constitute tracks. Crushed eggshells in a nest on the ground, twigs broken underfoot in a forest, or trampled bones in a watering hole are all tracks. This is because they leave visible impressions made by appendages of animals. Likewise, claw marks left in solid substrates are tracks, too. When a domestic cat (*Felis domestica*) scratches furniture to get attention, a modern aardvark (*Orycteropus afer*) tears apart a termite nest in search of a meal, and a grizzly bear (*Ursus arctos*) rakes a tree with its claws to mark its territory, these animals are also leaving tracks.

Individual tracks can provide many quantitative measurements, especially if clear impressions of digits are made. Among these measurements (Fig. 14.1) are:

- Overall length and width of the track
- Number of digits
- Digit widths and lengths
- Angles between digits
- Depth of penetration of the track or individual claws (if the latter are present)
- Number of fleshy pads associated with digits
- Width and height of any visible zone of deformation around the track.

For dinosaurs, the number of digits that could have touched the ground while they walked varied from two, in some dromaeosaurs such as *Deinonychus* (Chapter 9), to five, present in some sauropodomorphs, stegosaurs, and ceratopsians (Chapters 10, 12 and 13). But the numbers also varied for the manus and pes tracks from

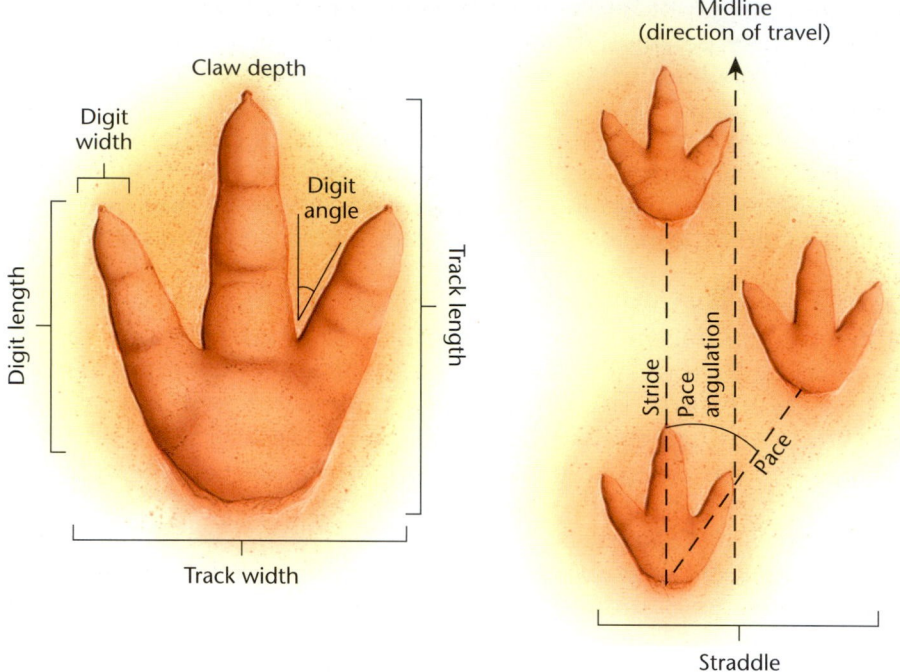

FIGURE 14.1 Measurable parameters that can be derived from a well-preserved dinosaur track and trackway, assuming bipedalism. Note diagonal pattern to the trackway, which is typical for those made by dinosaurs.

individuals of some quadrupedal dinosaurs. For example, stegosaurs show five digits on the manus but only three on the pes. The number of digits in the pes is unknown for some dinosaurs, such as pachycephalosaurs (Chapter 13), so their tracks are unrecognizable until skeletal data provide some basis of comparison.

A **trackway** is defined as a series of two or more successive tracks made by the same foot, which is ideal for making an important measurement, **stride length**. Stride length is the distance between each track made by the same foot in a trackway. Tracks made by appendages from successive opposite sides, such as right–left or left–right in a bipedal animal, can comprise a partial trackway, and the measurement gained from the distance between these impressions is the **pace length**. The **straddle** is the width between the tracks on each side of a trackway, measured directly as the distance between the outsides of tracks on the left and right sides of the trackway. Straddle is also approximated by **pace angulation**. This is an angle described by the pace of one side in comparison to the overall stride. For example, a pace angulation of 180° is a straight line, made as if the animal walked on a tightrope. This is typical of theropod trackways and is also very common in modern birds. In contrast, a pace angulation of 120° represents more of a sprawling movement and thus reflects a greater straddle. High pace angulations and correspondingly narrow straddles of most dinosaur trackways verify skeletal inferences that dinosaurs walked with their legs underneath their bodies in an upright posture (Chapter 1). A semi-sprawling posture has been interpreted for some ceratopsians, but this is still subject to debate (Chapter 13). Regardless, nearly all dinosaur trackways show a diagonal pattern if a line is drawn from one pes to the next. Because almost all of these trackways indicate walking speeds (discussed later), they can be described as diagonal walkers. Dogs, cats, cattle, sheep, deer, and many other modern mammals leave such trackway patterns.

Trackway patterns for both bipedal and quadrupedal animals can be observed in modern environments and compared to patterns left by both bipedal and quadrupedal dinosaurs. As discussed in previous chapters, animals can have two modes of locomotion, obligate or facultative. Obligate means that they can walk only in a certain way, whereas facultative means they have the ability to walk a different way from normal if required. For example, seemingly all theropods were obligate bipeds (Chapter 9). In contrast, some ornithopods were facultative quadrupeds (Chapter 11), in which they walked bipedally most of the time but switched to a quadrupedal locomotion at other times. Bears that walk on their hind legs are thus demonstrating that they are facultative bipeds. Humans are obligate quadrupeds early in their development and learn to become obligate bipeds later in life.

Trackway patterns of quadrupedal animals are potentially more complicated than those of bipedal ones. These can vary considerably, depending on:

1 an animal's adaptations for its most efficient motion; or
2 changes that occurred in its behavior while it was making the trackway.

For example, a commonly encountered trackway pattern for a quadrupedal mammal is made by it walking and leaving a visible pes track just to the back of or overlapping a manus track, resulting in right–left pairs of foot impressions (Fig. 14.2A). This is the most common pattern observed in quadrupedal dinosaur trackways, seen for tracks of some ornithopods (Fig. 14.2B; Chapter 11), sauropods (Chapter 10), thyreophorans (Chapter 12), and ceratopsians (Chapter 13). However, the trackway pattern changes a great deal when an animal picks up its pace by trotting or

(A) (B)

FIGURE 14.2 Manus–pes placement in typical walking pattern by quadrupedal animals. (A) Manus–pes pair from dog (*Canis domesticus*) trackway preserved in modern sidewalk, Emory University, Atlanta, Georgia. (B) Manus–pes pair from iguanodontian trackway in Lower Cretaceous bedding plane near Morrison, Colorado.

14

running. Trackways caused by a running quadruped typically have a considerably longer stride length in relation to the footprint size and greater spacing between the same-side manus–pes pairs. Likewise, when an animal otherwise makes movements that are more like those of animals that normally bound (hop), pace, or gallop, the resultant tracks will reflect these behaviors also.

So far, only one trackway made by a quadrupedal dinosaur has indicated running, a Late Cretaceous ankylosaur trackway from Bolivia that was described recently. On the other hand, at least eight running theropod trackways have been documented. This evidence supplements anatomical data to support the hypothesis that bipedal dinosaurs, in general, were capable of moving faster than quadrupedal dinosaurs. No dinosaur trackways found so far are interpreted as demonstrating bounding, pacing, or galloping. Based on current trackway evidence, nearly all dinosaurs just walked, which is the most common locomotion mode for modern animals, too.

One of the more mysterious-looking trackways formed by a quadrupedal animal occurs when the animal places its same-side pes into the preceding manus print, which can give the appearance that the animal was walking bipedally. This placement, called "**direct register**" by some modern trackers, is observable in some trackways made by modern feral cats and foxes. One of the ways to detect direct-registered tracks is to look for a pes impression within the manus print, which can be visible if the manus print is larger. However, if the pes is larger than the manus, it can obscure or obliterate the manus impression that immediately preceded it. Some quadrupedal dinosaurs did have differently-sized feet, and in most cases the rear feet were larger, although brachiosaurids had larger feet in the front than the rear (Chapter 12). Direct register has been documented in some sauropod trackways, although many of their trackways also show distinct manus and pes impressions (Chapter 10).

Some trackers or hunters refer to a well-worn, unvegetated path through a field or woodland area made by repeated trackways of deer and other mammals as a **trail**. However, this term is also applied to surface traces left by the movement of legless animals such as worms, snails, or snakes. Trails in modern environments are typically caused by only one or two species of animals. Thus, if clear tracks are preserved, the trailmaker usually can be correlated with its trail. In this sense, no dinosaur trail has ever been described, which may be a function of their low preservation potential or lack of distinctiveness. For this reason, our discussion of dinosaur trackways will concern their numerous well-preserved tracks and trackways.

Hunting and tracking animals has probably been a part of the human experience for the past 100,000 years. As a result, the usefulness of tracks as a source of data about animal behavior was probably tested long before the earliest known language was invented, and certainly predates modern scientific methods.

Stories Told by a Single Footprint

Evidence that ancient peoples were aware of animal tracks can be inferred from one of the oldest known art forms, macaroni (not to be confused with pasta). This art form consists of finger tracings in clay, made on the walls of caves in Spain and France during the Late Paleolithic, about 30,000 years ago. Some of these tracings include line drawings of animals, and the association of these representations with many hand and footprints are interpreted as imitations of animal tracks. Likewise, the indigenous peoples ("aborigines") of Australia made rock art in northern, central, and southern Australia that prominently featured identifiable animal tracks. The earliest form of this art style is panaramittee, which also dates from about 30,000 years ago.

The San ("bushmen"), who comprise the modern indigenous peoples of the Kalahari region of southern Africa, still use tracking for their hunting. Their methods are thoroughly scientific, as tracks are observed and questions are asked immediately about them:

- What animals made these?
- Which directions were they moving?
- How fast were they moving, and where do they vary their speeds?
- How many animals were in the group?
- How many of the animals were adults and juveniles?
- Which animals were male and which were female?
- Were any of them sick or injured?

Hypotheses are then formed to explain the data. For example, the tracks may indicate that six animals, two adult females, one adult male, and three juveniles, moved slowly while grazing on some vegetation just a few hours ago, then they simultaneously turned to look at the source of a sound, were frightened by a predator, and ran for cover in the nearest grove of trees. The hypotheses are then tested for their veracity and possibly falsified, but if they are continually falsified, the hunters and their families starve. San hunters thus have good incentive to make careful observations and modify their hypotheses in the light of new data, and their methodology has more immediacy for these hunters than it would for, say, a tenured scientist at a university. Native American tribes, such as the Apache, and the aforementioned native Australians are among the indigenous peoples who have been also renowned historically for their tracking skills.

The old Chinese saying, "The longest journey begins with a single step," attributed to Chinese philosopher Lao Tzu (about 600 BCE), also can be applied to the description and interpretation of a trackway. Any single track can be examined in detail and is a potential storehouse of information about the animal that made it. This is not just for identifying the trackmaker and associated measurements, but also for interpreting the behavioral dynamics of the animal. For example, the micro-topographic changes imparted to a substrate by an animal's foot are among the qualitative data from a single track that can be used to infer detailed interpretations of behavior. Because changes in foot movement involve applying pressure against the substrate around, inside, and underneath a track, as well as releasing that pressure, the resultant deformations can be called **pressure-release structures**. By describing such features, a track can reveal the general direction of the animal's movement and indicate whether that animal stopped and looked in a certain direction, made an abrupt change in direction, moved backward, or was carrying something in front of it or on its back (Fig. 14.3). These same criteria, related to such variations in behavior, can then be applied as models for dinosaur tracks (Figs. 9.13B and 14.4).

One of the most important principles for understanding the morphology of a single track is pressure, or stress. The force applied per unit area associated with a track depends on the mass of an animal in combination with gravitational force (at 9.8 m/s^2: Chapter 1), and the area of the foot making contact with the substrate. For one example, take a standing, vertically oriented bipedal animal, such as a human, with equal distribution of its weight on each foot. This human causes stress on the substrate below each foot almost equal to half of their weight divided by the area of the foot. Let us assume a rectangular shape for a man's foot (28 cm long, 9 cm wide, in typical running shoes) and a current weight of 66 kg. The stress applied per foot as a bipedal animal (s_f, herein called foot stress) is

14

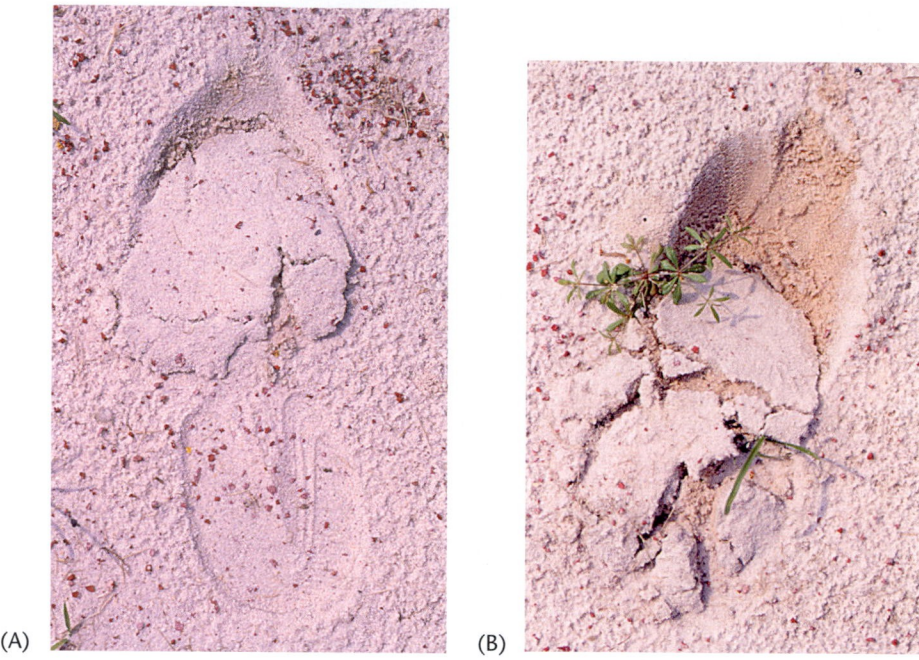

(A)   (B)

FIGURE 14.3 Different pressure-release structures caused by different behaviors, transmitted by the right foot of the same person walking in firm sand. (A) Moving straight forward. (B) Making an abrupt right turn.

(A)   (B)

FIGURE 14.4 Different pressure-release features caused by displacement of sediment from movement of a dinosaur. (A) Large theropod track in sandstone, Late Jurassic, Utah. (B) Sauropod track in sandstone, Late Jurassic, Utah. Theropod was moving straight forward, whereas sauropod had began to make an abrupt right turn.

$$s_f = (F/2)/A \qquad\qquad (14.1)$$

Step 1. = (66 kg × 9.8 m/s²/2)/252 cm²
(Converting cm² to m²) = (647 N/2)/0.0252 m²

Step 2. = (323.5 N)/0.0252 m²
= 12,837 N/m²

This number may seem high, but realize that it is the amount of stress the man would cause if his weight were distributed equally over a square meter. Perhaps a more meaningful measurement is stress as applied to a square centimeter. Using 1 m² = 1.0 × 10⁴ cm² makes the foot-stress about 1.3 N/cm². Because the area of the shoe imprint is actually more oval than rectangular, this represents less area. As a result, a correction can be accomplished by multiplying the value by 0.8, which is an approximate conversion factor for an inscribed oval within a rectangle; this changes the value to 1.6 N/cm². This corrected value increases even more once shoes are removed. This is because the surface area of the foot in contact with the substrate decreases but the weight does not (other than subtracting the weight of the shoes, of course).

Also, this value represents a mean for the given surface area. Consider that the weight of a standing person is not distributed evenly over the entire area of the foot but is mostly on the heel (metatarsal), which is directly underneath the main part of the body weight. Because most dinosaurs had their metatarsals elevated above the ground during life, the stress imparted by their feet was mainly on the phalanges. A few dinosaur tracks show metatarsal or "heel" impressions, but these are uncommon. The weight of a bipedal dinosaur was distributed over two feet, whereas a quadrupedal dinosaur had it divided by four feet that were probably unequally sized. These variations require estimation of the areas for the manus and pes. The areas for all four feet then should be added to calculate a modified general formula for cumulative stress (s_c) caused by an animal on an underlying substrate, which is identical to Equation 3.8 (Chapter 3):

$$s_c = F/A \qquad\qquad (14.2)$$

For dinosaurs, a value for this stress is calculated by measuring the area of a dinosaur's tracks. These data are then used in combination with a weight estimate for the probable tracemaker. However, this will be only a rough estimate of the foot stress and probably represents a maximum value. After all, the weight distribution of any given dinosaur was also over more of a horizontal plane than a human. Humans and penguins are the only bipedal animals known to have their head, spine, and metatarsals aligned in a more-or-less vertical plane, perpendicular to the ground surface.

Any movement of an animal also causes stress, especially where it typically pushes off the substrate, thus generating force applied to an area of that substrate. This happens whether the animal was moving forward, backward, laterally, or jumping upward. Evidence for this stress can be shown by visible zones of deformation in the substrate, which in the right substrate, such as a firm mud or sand, will preserve the effects of the movement as pressure-release structures. Normally the movement of an animal is straight forward but this cannot be assumed. Walking backwards, backing up, or moving laterally can happen for various reasons.

An individual track and all of its information can lead to a hypothesis about the track itself. Of course, the best way to test this hypothesis is to study successive tracks (if preserved) and look for corroborating or contradicting evidence. If the next track provides different information, then new questions can emerge about why the two tracks differ:

14

- Did the animal change its behavior from one track to the next?
- Did the behavior stay the same but the substrate change, so that a successive track was preserved differently from the preceding one?
- Did the animal distort or otherwise damage its own track as it pulled its foot out of the substrate?
- Did an animal of the same species follow another, causing overlapping tracks?
- Were the tracks modified by physical processes (weathering), which obscured their original forms?

All of these questions, and the hypotheses attached to them, begin with observations made on a single track. Looking for such signs is an excellent exercise in observation that can carry over into everyday life (Chapter 2), prompted by the seeking of subtle clues about what may have happened in the recent or not-so-recent past.

Finally, a single well-preserved track with skin or toe-pad impressions can provide details about the soft-part anatomy of an appendage. A dinosaur track that shows skin impressions is still a trace fossil, although it does directly reflect a body part. A detailed body impression made by a tracemaker in association with a trace fossil is a **bioglyph**. Skin impressions not associated with tracks or those made by an already-dead dinosaur have occasionally been called trace fossils. Nevertheless, only in cases where the dinosaur was still alive and indicating behavior can this designation be made. In addition to skin impressions, toe-pad impressions are also commonly preserved in dinosaur footprints, which help to determine (along with skeletal data) what dinosaur feet looked like (Chapter 5). The toe pads often correspond with phalanges, so footprints with well-preserved toe pads can be compared to known phalangeal formulas of dinosaur groups, which helps in identifying which dinosaurs made which tracks.

What about not-so-well-preserved dinosaur tracks, ones that do not show individual toes, let alone skin impressions? In fact, vague depressions in Mesozoic rocks can often provoke arguments about whether they represent tracks. Even so, measurements still can be taken of a possible track's dimensions and its overall size and shape compared to known dinosaur tracks. A geometric outline of a possible track can be categorized through a compression shape and is a useful guide for trackers who do not have "perfect" substrates for preserving detail, such as firm mud or sand. For example, with modern animals, felines leave round compression shapes, whereas canines leave oval compression shapes, all within well-documented size ranges. In dinosaurs, the compression shapes varied considerably, especially as the Mesozoic progressed, but those for major groups of dinosaurs are distinguishable from one another through length-to-width ratios and overall outlines. Through using such methods, several examples of formerly disregarded "potholes" or "bulges" in strata turned out to be dinosaur tracks.

Track Taphonomy

The preservation of a dinosaur track in a substrate, so that it is visibly identifiable as a track more than 65 million years after it was formed, can be complicated. As with body fossils, tracks are susceptible to weathering, erosion, and erasure by other organisms. These factors are especially pronounced on the surface where the animal was moving, and nearly all tracks formed on an exposed surface quickly disappear through natural processes. Consequently, the same circumstances conducive to preserving a dinosaur body fossil apply to their tracks as well. They need to have been buried quickly. In this scenario, most dinosaur tracks had a distinct advantage, as most tracks were probably "buried" (under the surface) as soon as the animal made them.

An impression made by the appendage of an animal on a substrate below the surface where the animal was moving is an **undertrack**. Heavy animals, such as adult sauropods, did not necessarily make undertracks. For example, small horseshoe crabs, not much larger than the average cockroach, also left visible undertracks in the geologic record. The only requirement for the formation of an undertrack is that the impressions made on one layer of sediment by the weight and movement of the animal deformed an underlying layer. This is the case no matter how thick or thin the upper layer may have been. An indistinct or otherwise incomplete outline of a track is one clue that it may be an undertrack, but evidence for undertrack preservation also is provided by tracks that show crosscutting by invertebrate burrows. The track had to have been already buried in unlithified sediment to be later modified by the burrowers.

Factors that affect the preservation of undertracks in sediments are:

1 the amount of water in the sediment, which affects its relative firmness;
2 grain size; and
3 cohesiveness.

The latter factor typically depends on the amount of clay minerals in the sediment, which can help sand grains to stick together. In general, fine-grained sediments, such as clay and silt, with only enough water between grains to make the sediments cohesive, are the best for preserving detailed tracks.

Coarse-grained sediments preserve little more than the compression shape of the track. Sediments with high or low amounts of water either preserve no impressions or leave impressions that tend to collapse inward. Local variations in substrate conditions, where one patch of sediment is more moist and loose than an adjacent patch, will cause different impressions by the same animal, literally from one step to the next. Similarly, a trackway where an animal walked over a substrate that was either well packed or cemented might leave no visible impression at all, which results in "missing" tracks if these areas bracket softer substrates (Fig. 14.5). Missing tracks in dinosaur trackways have been interpreted incorrectly, such as some being used as evidence for swimming dinosaurs, incredible stride lengths, or amazing leaps.

FIGURE 14.5 Differences in tracks as a function of substrate firmness, illustrated by a juvenile human female, weighing about 30 kg, making a trackway on a beach where tracks "disappear" in the middle because of firmer sand in that area.

14

The vast majority of dinosaur tracks probably formed through a wide variety of behaviors that imparted different stresses to the sediment, which were then preserved as undertracks in a variety of sediments with different grain sizes and initial water contents. The tracks therefore did not preserve many of the numerous details revealed by fresh surface tracks made by modern animals, such as all of the pressure-release structures. In some cases, the changes in movement by a trackmaker were transmitted through the layers underneath the track surface and were recorded by pressure-release structures (see Fig. 9.13B). Nonetheless, such features should not always be expected in fossil tracks.

The track preservation discussed so far only has concerned molds (negative images) of dinosaur feet, but many dinosaur tracks are also preserved as natural casts (positive images). For casts, the original footprints were filled with sediment that differed in grain size from the original substrate, such as a sand cast of a footprint made in an underlying muddy surface. Three scenarios have been proposed for how casts of dinosaur tracks were formed:

1 tracks on a muddy surface were filled with sand that later lithified;
2 tracks were impressed first as undertracks on a muddy surface, which were later exhumed and filled with sand that lithified; and
3 the casts may be undertracks, where the pressure of a dinosaur foot caused sand to squeeze into an underlying muddy layer.

In the last case, the sand formed an outline of the dinosaur foot that later lithified.

Experimental evidence for the taphonomy of tracks, including how tracks are preserved in modern environments, is currently minimal. Nevertheless, track preservation is a worthy field for further study, considering that understanding it helps to evaluate whether:

1 fossil tracks observed on a rock surface are contemporaneous; or
2 they consist of a mixture of tracks from different levels and thus different times.

Classification of Dinosaur Tracks

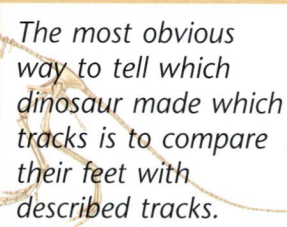

The most obvious way to tell which dinosaur made which tracks is to compare their feet with described tracks.

Although the skeletal components of a dinosaur appendage show its general overall form, the bones normally do not include the fleshy parts (Chapter 6). As a result, their skeletal feet do not indicate their total width and length, let alone the parameters for individual digits. This situation can be illustrated by comparing a human skeletal foot to a track made by the same human. A track can provide a minimum size and other parameters for the original track-making appendage. Moreover, the geometry of a reconstructed skeletal foot can also give a minimum size to expect for a track made by that animal. Conversely, if a dinosaur walked through a soft substrate, it might have sunk deeply enough to make metatarsal impressions, resulting in a much larger track than that made simply by phalanges and tarsals.

Potentially many dinosaur species are still undiscovered, which only adds to the difficulty of matching a dinosaur trackmaker with its tracks. The tracks of these unknown dinosaurs may have already been found, but relatively few members of their original populations were preserved (Chapter 7). Thus, any comparison of track data, which are more abundant in some areas than skeletal data, to only known species of dinosaurs will likely result in mismatches. Yet another problem is distinguishing the tracks of juvenile dinosaurs of one species from those of small adults from another (but morphologically similar) adult species.

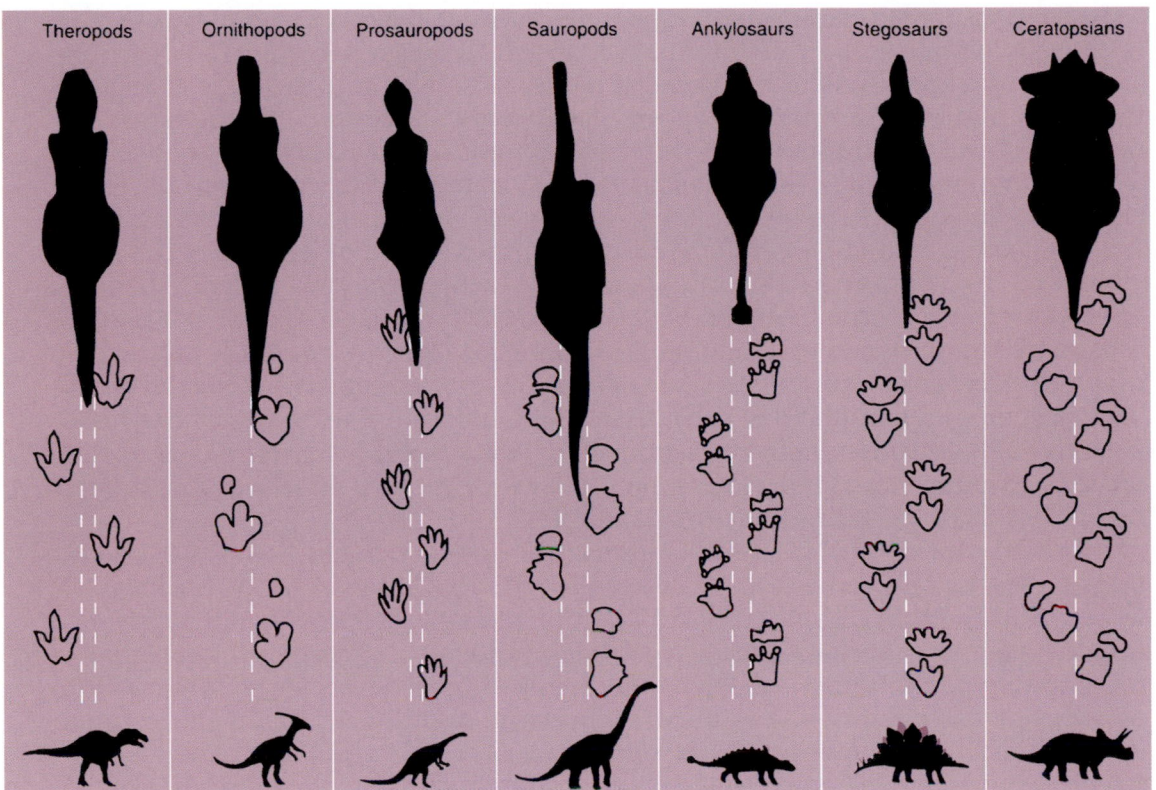

FIGURE 14.6 Typical track morphologies interpreted for theropods, ornithopods, prosauropods, sauropods, ankylosaurs, stegosaurs, and ceratopsians.

With those caveats in mind, several of the criteria applied to tracks can be used for successful correlation of dinosaurian tracemakers with fossil tracks. Criteria for individual tracks are number of toes, track size (including width-to-length ratios), and presence or absence of claws. Study of trackways will require the preceding information plus the number of feet used in locomotion. Using these characteristics as a guide, dinosaur tracks and trackways can be broadly allied to theropods, prosauropods, sauropods, ornithopods, ankylosaurs, stegosaurs, and ceratopsians (Fig. 14.6). Without careful descriptions of the tracks, they are too easily attributed to the wrong dinosaur tracemakers. For example, both theropod and ornithopod tracks have a tridactyl (three-toed) pattern, and they overlap in their size ranges. However, examination of width-to-length ratios and other specific descriptive criteria can help prevent such mistakes.

Some track types are more common than others. Theropod tracks are exceedingly common in some strata throughout the entire geologic range of dinosaurs, but ankylosaur and stegosaur tracks (Chapter 14) are rarely reported from any strata. A hypothesis for the disparity in track abundance between theropods and most other dinosaurs is that theropods, as active predators or scavengers, moved about a great deal more in their ecosystems in search of food than herbivorous dinosaurs (Chapter 9). However, occasional discoveries of horizons trampled by numerous herbivorous dinosaurs, such as sauropods and ornithopods, provide notable exceptions to such generalities (Chapters 10 and 11).

Paleontologists face a minor dilemma with dinosaur tracks compared to skeletal remains. This is because, in the vast majority of cases, trace fossils cannot be given names according to the exact species of animals that made them. This is especially true for instances where the tracemaker is otherwise unknown. One of the main advantages of the Linnaean binomial nomenclature (Chapters 1 and 5) was that it improved communication of information about particular organisms, modern or fossil, by providing standardized species names. A similar methodology is applied to dinosaur tracks, in which they are given names according to their distinctive forms, called **ichnotaxa** (= "trace names," plural for ichnotaxon). With an ichnotaxonomic nomenclature, dinosaur paleontologists can communicate more effectively about tracks by giving them ichnogenus and ichnospecies names, such as *Grallator* and *Megalosauripsus brinoensis*. Even though some paleontologists have attempted to connect certain dinosaur tracks with dinosaur genera, such as *Megalosauripsus* with the theropod *Megalosaurus*, the tracemaker should not be confused with the trace name. In ichnology, naming an ichnotaxon on the basis of its supposed tracemaker genus or species is a dangerous practice, because it mixes description with interpretation and is liable to lead to a false interpretation. For example, paleontologists in the nineteenth century originally interpreted some trace fossils made by invertebrates as body fossils, such as algae. Some of their ichnogenus names still reflect those mistaken body-fossil affinities.

Another problem with naming ichnotaxa for dinosaur tracks is that the conditions for track preservation, such as substrate type and behavior of the trackmaker, were so variable that a single trackway made by the same dinosaur could yield multiple ichnogenera. So tracks made by modern animals in a variety of substrates, reflecting myriad different behaviors and preservation modes, can provide models of comparison for testing the validity of ichnotaxa. Additionally, many of the same morphologically distinctive dinosaur tracks have been given different names by different authors. Similar synonymies have happened with biological species names based on skeletal data (Chapter 5). Finally, when track size is used as a reason for splitting a morphologically similar ichnotaxon into multiple new ichnotaxa, then the possibility that the track sizes are simply from juvenile and adult dinosaurs of the same species cannot be discounted. This diversity of ichnogenus names is one reason why "track diversity" may be a misleading indicator of actual biological diversity of dinosaurs, and such interpretations should be scrutinized carefully. The best ways to avoid the preceding problems are:

1 to make detailed descriptions of individual tracks;
2 only name a track if it is part of a well-preserved trackway;
3 review valid ichnogenera given to dinosaur tracks in previous studies; and
4 make sure that a potentially new ichnogenus of dinosaur track has a geological context to it, including a stratigraphic position and geographic location.

Unfortunately, the nomenclature of fossil vertebrate tracks, especially for dinosaurs, remains one of the most unwieldy subdisciplines in ichnology because such advice has not been followed in the past, although many of these mistakes happened in the nineteenth century.

Most dinosaur paleontologists keep in mind the vagaries and potential pitfalls of naming tracks, and few want to wade through the confusing scientific literature on dinosaur track names. Consequently, they will simply use dinosaur clade names for tracks made by members of those clades. Thus, "theropod tracks," "sauropod tracks," and so on, are the most commonly encountered descriptors for reported dinosaur tracks, despite the fact that they combine description with interpretation. Nevertheless, use of the criteria for identification is normally sufficient for correlating descriptions of dinosaur tracks with at least broad taxonomic categories. More specific designations (such as "allosaurid tracks" or "hadrosaurid tracks") can be reasonably made later in consideration of known skeletal data supporting the presence of such dinosaur taxa in the same-age strata. Nonetheless, one of the exciting facets of dinosaur track discoveries is that they may represent tracks from previously unknown dinosaurs. Such discoveries can cause much spirited discussion and debate among dinosaur paleontologists, as they attempt to resolve the apparent discrepancy between the body and trace fossil records of dinosaurs.

Individual Dinosaur Behaviors Indicated by Tracks

Trackways made by individual dinosaurs can provide much qualitative and quantitative data, which can then be used to tell a great deal about how these dinosaurs were behaving.

The best method for describing individual dinosaur trackways, also applicable to multiple trackways on the same bedding plane, is to construct a track map. These maps are similar to those used to look at distributions of skeletal components at a site (Chapter 3). The map should be a scale representation of the trackway as seen from above where the dinosaur was walking, and should include a legend, scale, and direction indicator (Chapter 4). These maps provide important information for interpreting a trackway in terms of its completeness. Moreover, they help as a tool for visualizing the spatial relations of the dinosaurian tracemaker in the context of its original environment.

Qualitative information can also be derived from a mapped trackway, relating to whether a dinosaur lived in a particular environment. For example, was the dinosaur well adapted to that habitat, and did it reflect its paleoecologic relationship to that environment (Chapter 10)? More specific idiosyncrasies can also be detected, such as where an individual dinosaur:

- Stopped to rest.
- Had trouble walking through a difficult-to-traverse substrate.
- Abruptly changed its direction of movement.
- Had a noticeable limp, probably related to an injury.
- Was following another dinosaur, perhaps as a predator.
- Was avoiding another dinosaur, perhaps as prey.

All of the preceding examples have been hypothesized for dinosaurs on the basis of trackway data.

Quantitative methods applied to dinosaur footprint data have illuminated the behaviors of individual dinosaurs. Probably the best-known application is with regard to calculating how quickly dinosaurs moved. The speeds of individual dinosaurs have been a point of curiosity for paleontologists almost for as long as dinosaurs have been studied, but a technique for estimating speed, based on a combination of skeletal and footprint data, was only developed a little over 25 years ago. Based on empirical data from living animals, this mathematical application uses stride length and footprint size in dinosaur trackways, together with the leg length (hip height) measured from dinosaur skeletons.

An easily made observation is that, compared to a slow walk, people's feet become spaced farther apart when they walk quickly or run. Also, when a shorter person runs alongside a taller person, the shorter one must take more steps in the same distance despite their equal speed. The discrepancy in the mean leg lengths of adult men versus adult women also translates to differences in mean stride length; calculations from representative samples of men and women show mean values of 1.46 m and 1.28 m, respectively.

These observations of stride lengths in association with leg lengths can be expressed for the land-dwelling animals walking on the Earth's surface, through a general relation called relative stride length (s_r):

$$s_r = s_l/l_l \tag{14.3}$$

where s_l is stride length and l_l is leg length. Relative stride length is a dimensionless number, as the units cancel out (because length is in both the numerator and denominator). Thus, it has no units of measurement associated with it.

Stride length is measured directly through a trackway, such as the one illustrated in Fig. 14.1. However, measuring leg length is a potential problem because the size of the dinosaur that made the trackway cannot be directly observed. Nevertheless, the leg length for a dinosaur can be estimated by using another dimensionless number, 5.0, as a constant in the following equation:

$$l_l = f_l (5.0) \tag{14.4}$$

where f_l is footprint length. Paleontologists have calculated ratios of leg height versus foot length for dinosaurs by measuring the leg length from the ground surface to the acetabulum in dinosaur skeletons, as well as the total skeletal foot length. The ratio is derived simply as

$$l_l f = l_l/f_l \tag{14.5}$$

Some paleontologists have calculated ratios close to 4.0, but others use 4.5 to 5.5. These variations depend on the dinosaur groups used or their sizes (e.g., small theropods = 4.5, large theropods = 5.5). Consequently, a compromise figure of 5.0 is used here for the sake of illustration.

For example, given a Jurassic theropod track 45 cm in length, the leg length could have been:

$$l_l = 45 \text{ cm} \times 5.0 = 225 \text{ cm}$$

Using different constants of 4.0 and 5.5 would have derived leg lengths with a range of 180 to 248 cm, respectively. Either way, this theropod probably had a leg length that was taller than some professional basketball players, and most people probably could have walked between its legs without stooping. Consequently, footprint length can be used as a quick method for visualizing the size of a dinosaurian tracemaker.

If the same theropod had a stride length of 307 cm, then its relative stride length was

$$s_r = 307 \text{ cm}/225 \text{ cm} = 1.4$$

If the theropod was moving faster, then the value would have correspondingly increased to greater than 1.4 as a function of stride length. So this is a relative method of working out that one theropod was moving faster than another theropod of the same size.

Although relative stride length is a good way to "equalize" dinosaur trackway data, it is only as valuable as, say, relative age dating, as compared with absolute age dating (Chapter 3). What paleontologists would definitely like to know is how fast a dinosaur was moving in some absolute measurement, such as meters per second, which can be translated to kilometers per hour. Scientists who have studied the movements of tetrapods on land have noted that increased body size results in larger animals moving faster than smaller animals, even if their relative stride lengths are identical. This means that a small theropod with a relative stride length

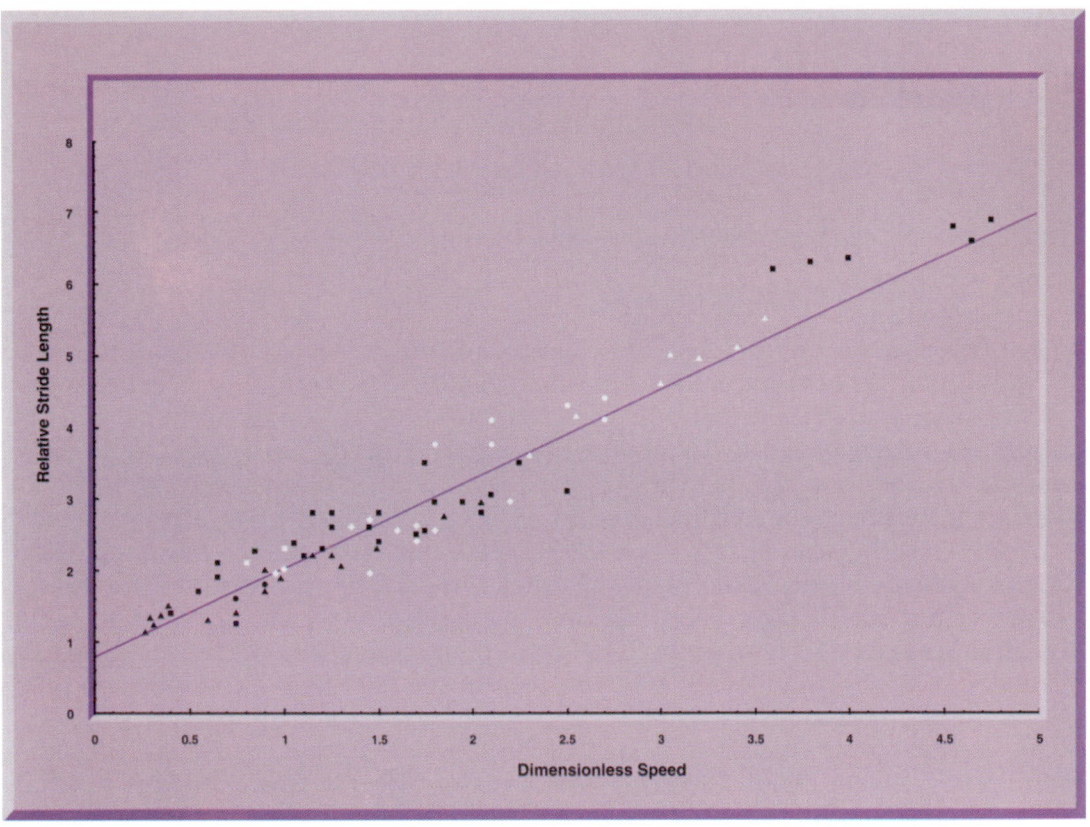

FIGURE 14.7 Plot of relative stride length versus dimensionless speed for different animals based on data derived from living cursorial vertebrates in terrestrial environments. (After Alexander, 1976, 1989.)

of 1.4 would have moved slower than our exemplified large theropod. This is related to gravitational force (Eqn 1.1, Chapter 1) providing a boost for momentum (Eqn 6.4, Chapter 6) because of the greater mass of a larger animal. As a result, a formula that takes into account the acceleration placed on to a body by gravity helps to equalize the speed of small and large animals alike through yet another dimensionless quantity, imaginatively called **dimensionless speed**:

$$v_d = \frac{v}{\sqrt{(l_1)(9.8 \text{ m/s})}} \tag{14.6}$$

where v_d is dimensionless speed. Because v is in m/s, l_l is in meters, and gravitational acceleration is in m/s², all of the units cancel out and a dimensionless number is left. When the walking or running speeds of modern animals such as dogs, cats, rhinoceroses, elephants, ostriches, kangaroos, and humans are measured, these measurements can be used with the leg lengths of the respective animals in the calculation of dimensionless speeds. When plotted against relative stride lengths, the derived line is usable as a general model for the similarity of animal movement on land (Fig. 14.7). This plot allows comparison of relative stride lengths, calculated from fossil trackways, to dimensionless speeds, which in turn correspond to the speed of the tracemaker.

For example, using our Jurassic theropod again, its relative stride length of 1.4 corresponds to a dimensionless speed of 0.45. Using algebra to rearrange Equation 14.6, the estimated speed of the theropod was

Step 1. = 0.45
Step 2. = 0.45 (4.7 m/s)
Step 3. = 2.1 m/s

Converting meters per second to kilometers per hour:

Step 4. = (2.1 m/s) × (3600 s/h) × (1 km/1000 m)
Step 5. = 7.6 km/h

To put this speed in perspective, the Olympic record for the men's 20-km race-walk (as of the writing of this book) is a mean speed of 15.5 km/h, which is more than twice as fast as our theropod and was sustained over a 20-km distance. This example demonstrates a slowly-moving theropod. Indeed, most measurements of dinosaur trackways reflect that they behaved just like most animals during the majority of their lifetimes: they walked, but not briskly. If we treat these calculated speeds as hypotheses, they can be cross-checked with the qualitative data mentioned before, such as pressure-release structures and other evidence of considerable stresses to a substrate that might have been imparted by a running versus a walking animal. Running animals also tend to become more digitigrade. This means that only the distal ends of the phalanges make impressions in the sediment, although such prints also could be undertracks.

The fastest known speeds indicated by dinosaur trackways, using the preceding methodology, were about 40 km/h. This speed seemingly was restricted to small- and medium-sized theropods (Chapter 9). However, this situation may be a result of biased sampling, because large theropod trackways are relatively rare. So far, the trackways made by quadrupedal dinosaurs, such as sauropods (Chapter 10), ornithopods (Chapter 11), and ceratopsians (Chapter 13), with the exception of one ankylosaur trackway mentioned earlier (Chapter 12), only show relatively slow walking speeds. If these animals were capable of running at speeds approaching those of some theropods, trackway evidence has not yet revealed such abilities.

Dinosaur Group Behaviors Indicated by Tracks

When dinosaur tracks are found as multiple trackways on the same horizons, hypotheses can be made about how dinosaurs related to one another, intraspecifically or interspecifically. Track maps are especially useful for formulating these types of interpretations. In combination with other qualitative and quantitative information, track maps can be used to either verify or dispute dramatic scenarios of multiple dinosaurs in close association with one another.

Intraspecific behavior related to gregariousness is best supported by track evidence. For example, did dinoaurs prefer to travel with many individuals of their own species or were they "lone dinosaurs"? If many herbivorous dinosaurs of the same species lived together as a group and traveled together, they were said to have behaved as a herd. In contrast, a grouping of carnivorous dinosaurs similar to a herd is termed a pack. These are conceptually analogous to modern mammals, such as caribou and wolves, respectively, that show the same behaviors. Birds of the same species displaying group behavior are **flocks**, regardless of their eating habits, and the same term has been applied to dinosaur groups in recognition of some of their behavioral similarities to modern birds (Chapter 15).

Some dinosaur trackways do indeed suggest either herding or pack-hunting behavior (Chapters 9 to 13). Track maps clearly show a preferred directionality by different individual dinosaurs of the same or different species, which may reflect herds migrating or predator–prey relations. However, these track data should be examined critically for the possibility that a geographic or ecological barrier, such as a cliff or water body, could have caused a preferred path for these dinosaurs. The latter interpretation is supported by parallel trackways that show 180° orientations to one another with regard to direction of movement.

Quantitative methods help to determine whether unidirectional trackways represent group behavior by:

1 calculating their speeds, keeping in mind that animals moving together will typically do so at the same speed; and
2 measuring the spacing between each trackway and their degree of parallelism.

Modern animals of the same species moving together will space themselves at nearly equal intervals to avoid "invading each other's space," and accordingly will move in formation. Indeed, a bedding plane exposure of the Late Jurassic Morrison Formation in southern Colorado shows five sauropod trackways that are parallel and equally spaced, all apparently moving at the same speed and even turning in harmony with one another (Chapter 10). Similarly, multiple parallel theropod trackways in a Cretaceous stratum in Mongolia argue for a group movement of these animals.

One of the most interesting interpretations of group behavior in dinosaurs of different species is a "stampede" recorded in an Early Cretaceous bedding plane in Queensland, Australia. At this site, a large number of small theropod and ornithopod trackways show unidirectional movement away from the trackway of a single large theropod. Some of the speeds, calculated from footprint and stride lengths, indicate that most of the smaller dinosaurs were running in the opposite direction to the large theropod. The interpretation is that a group of small theropods, perhaps of the same species (based on similarities of footprint size and morphology), was startled by the approach of a large theropod and ran away from it in a panic. Nevertheless, the exact timing of the larger dinosaur arriving on the scene relative to all of the smaller ones is uncertain; some of the smaller theropod tracks register within the larger one's tracks and thus post-date the latter.

Such examples may seem exciting and dynamic, but scientists have to critically examine the track evidence for what is actually preserved, rather than what is presumably preserved. This caution is especially warranted with reference to more dramatic interpretations of group behavior. Knowing that tracksites may represent the cumulative actions of animals traversing an area over the course of a few days or months diminishes the probability that any given set of tracks actually relates to one another through cause and effect. Yet another "reality check" is that the preservation requirements of dinosaur tracks allow for the probability of undertracks. This means that a bedding plane containing dinosaur tracks may have specimens impressed into it by animals that lived much later (perhaps by years) than the other animals that made tracks on the same surface (Fig. 14.8).

However, once these possibilities are accounted for, the usefulness of dinosaur tracks far outweighs their problems. Understanding the fundamental concepts of tracking and how they relate to dinosaur tracks provides a better basis for critically examining the case examples covered in other chapters. In many of these cases, dinosaur tracks have supplemented other dinosaur fossil data or have been the sole piece of evidence for dinosaurs, data that relate directly to the evolution, paleobiogeography, and paleobiology of major groups of dinosaurs (Chapters 10 to 15).

Dinosaur Behaviors Not Indicated by Tracks

Dinosaur tracks are valuable as sources of doubt or outright falsification of either new or long-held hypotheses of dinosaur behavior based only on skeletal data. Incorrect interpretations of dinosaur tracks also can lead to problems with

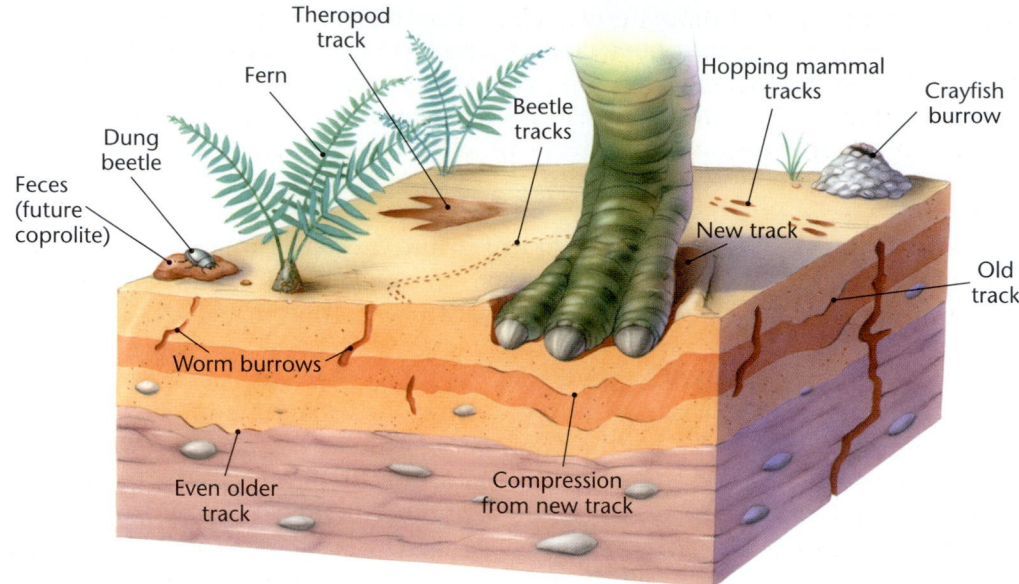

FIGURE 14.8 How varying depths of tracks and preservational modes can result in a track assemblage on a bedding plane where the tracks are not contemporaneous.

evaluating dinosaur behavior. Regardless of whether the hypotheses emanate from dinosaur body fossils or tracks, careful examinations of dinosaur tracks have helped to demonstrate that the following dinosaur behaviors were either rare or have not been clearly demonstrated.

Habitual Tail Dragging

Although often seen in older, popular depictions of dinosaurs, tail dragging seldom happened. Only a few rare instances have been found of a dinosaur's tail impressed on the same substrate as its feet, which in these cases showed that a dinosaur temporarily had a "tripod" stance or stooped to a low profile. In the past 20 years or so, the revelation that dinosaurs did not drag their tails has finally had an influence on dinosaur art. Well-informed artists now show dinosaurs of all sizes and shapes with their tails held straight behind them in nearly horizontal planes parallel to the rest of a dinosaur's axial skeleton. Skeletal data have also corroborated that dinosaur tails were horizontal and held off the ground. For example, some theropods and ornithopods had ossified tendons that stiffened the tail to assume such a posture. Before this revelation, some museum preparators in the early part of the twentieth century actually broke the caudal vertebrae of mounted dinosaur skeletons to make the vertebrae lie on the ground behind the main skeleton. Because dinosaurs were once considered as large reptiles, the prevalent view held was that dinosaurs must have been "tail-draggers" as well, but a closer look at their tracks would have revealed otherwise.

Swimming or Immersion in Water

Original interpretations of some dinosaurs, especially huge sauropods (Chapter 10) but also a few hadrosaurids (Chapter 11), held that these animals were adapted for an aquatic lifestyle. For sauropods, the reasoning was that an animal of such tonnage could not support its own weight on land, so it needed the buoyancy of water to cope. Track data were, in some cases, fitted to this hypothesis. For example, Roland Bird (Chapter 4) described a sauropod trackway where only manus impressions were preserved. Bird proposed that the rear feet were involved with swimming above a seafloor and only the front feet touched down. However, re-examination of this trackway showed that the pattern was attributable to partial preservation (under-tracks), and that the pes impressions did not sufficiently impart enough stress to be preserved equally with the manus impressions. Other sauropod tracks, as well as those of every other dinosaur, do not show conclusive evidence of swimming or an otherwise aquatic habit. This is not to say that all dinosaurs did not occasionally venture into water bodies or that they did not swim. In fact, more "swimming theropod" trackways have been interpreted in recent years with more convincing evidence of such behavior for those dinosaurs. Nevertheless, the paucity of trackway evidence for swimming in most dinosaurs encourages healthy skepticism of such interpretations.

Obligate Quadrupeds Rearing up on Their Hind Legs

Although some male quadrupedal dinosaurs probably were capable of temporarily rearing up on to their hind legs to assume a mating position behind a female (Chapter 8), no trackway evidence has shown that they actually did this. For example, anatomical reconstructions of sauropods (Chapter 12) and stegosaurs (Chapter 14) provide evidence that the center of gravity for some species may have been far enough toward the rear of their bodies that a temporarily bipedal posture was possible. This behavior was advanced as a hypothetical adaptation for herbivorous animals to reach the tops of tall-standing vegetation, or for making the animal appear larger to a threatening predator. Some illustrations and one mounted skeleton even show these dinosaurs, some of them an estimated 20 or more tonnes, in near-vertical postures. However, just because a dinosaur could have done a certain behavior does not mean that it actually did, and there is no known evidence of this behavior in sauropod or stegosaur tracks. If obligate quadrupedal dinosaurs did assume "vertical" positions, the rear feet should have doubled the foot stresses transmitted to the underlying substrate. As a result, tracks made by this behavior should be immediately obvious to a trained observer.

Intraspecific Competition or Interspecific Confrontation

So far, no dinosaur trackways provide conclusive evidence of dinosaurs of the same species competing with one another or predators pursuing and attacking a prey species. A disproportionate amount of attention has been paid to predator–prey relationships between dinosaurs, relative to the possibility of certain dinosaurs competing within their own species. The latter subject is now receiving more study (Chapters 9 to 13), but track evidence has not yet shed light on speculations about how males of a certain dinosaur species competed with one another for females. Examples include the much-publicized hypothesis that pachycephalosaurs charged one another at high speed and butted heads like modern rams (Chapter 13). Such behavior would be supported by two trackways composed of nearly equal-sized tracks with vectors heading toward one another on a 180° plane. The trackways also would show mathematically defined running speeds for both tracemakers, followed by abrupt ending of the trackways as they meet.

As noted previously, a few trackways strongly suggest that some theropods deliberately followed sauropods. However, only one trackway has been presented as recording a possible attack by a large theropod on a sauropod, which is in the Lower Cretaceous Glen Rose Formation of Texas (Chapter 9). In this interpretation, originally made by Roland Bird, the theropod's trackway ends when it converges with the sauropod trackway at one point, which Bird explained was where the theropod leaped on to the left side of the sauropod. However, the ending of the theropod trackway can also be explained as a lack of preservation of subsequent tracks. Bird's hypothesis would be supported by "push-off" marks (pressure-release structures) in the final tracks of the theropod that were formed as it leaped. The sauropod also should have shown a change in its weight distribution or another dramatic change in its behavior. After all, if it had a hungry carnivore hanging on to its flank, the sauropod should have responded to such a stimulus. Consequently, its tracks potentially should reflect such responses, but they do not.

Incredible Stride Lengths

Coal mines in Late Cretaceous strata of North America have been the source of numerous hadrosaurid trackway discoveries (Chapter 11), but an overzealous interpretation by Barnum Brown (Chapter 4) of one particular trackway caused a small paleontological controversy. Brown promoted the bipedal trackway as evidence for an enormous stride length in the tracemaker; he measured the distance between the tracks as nearly 5 meters apart. These data implied that the dinosaur was either the largest bipedal dinosaur ever discovered or that it was traveling at a very high speed. Subsequent analyses of the trackway showed that Brown had overlooked the possibility of missing tracks in the sequence, and, sure enough, one was found in between the originally measured tracks. This oversight meant that Brown measured stride length as pace length, effectively doubling the former parameter. Consequently, the trackway indicated an average-sized hadrosaurid that walked at a normal speed. One way Brown could have avoided the misinterpretation would have been to look for pressure-release structures that indicated the direction of movement for each individual footprint. This procedure would have quickly determined whether he was measuring from a left footprint to a right footprint (pace) or a left footprint to a left footprint (stride). In some case, such mistakes are understandable, as the outlines of some ornithopod and theropod tracks are nearly bilaterally symmetrical, which makes discrimination of right and left footprints more difficult.

Hopping, Galloping, or Walking Backward

Modern animals show a wide variety of behaviors in their locomotion that depend on their anatomical adaptations for movement or responses to environmental stimuli. For example, horses change from walking to trotting to cantering to galloping in what appears to be one smooth, continuous sequence. Such a continuum has not yet been recognized in a dinosaur trackway. Neither have other aberrant modes of locomotion, such as hopping, galloping, or reversing direction, been found. Typical track patterns observed in modern mammals, in addition to qualitative data (pressure-release features) associated with individual tracks, can be used for comparison with dinosaur tracks to test hypotheses of dinosaur behavior.

Paleoecological Information Gained from Dinosaur Tracks

Dinosaurs of all sizes generated stresses on substrates through their locomotion, and certain parameters of these substrates are discernible by examining their tracks. Substrate conditions that have already been mentioned are water content and sediment cohesiveness; however, sufficiently repeated stresses can change those conditions. Sediment disturbance by organisms is called **bioturbation**, an extremely common process produced by invertebrates and vertebrates in nearly every sedimentary environment on the Earth's surface. The products of bioturbation, if preserved in the geologic record, are **bioturbate textures**. The term **ichnofabric** is synonymous with bioturbate texture if it occurs in sediment, but also refers to the products of bioerosion, which occurs in solid substrates (e.g., rock, wood, and bone). The movement of animals in and on a substrate produces a large variety of ichnofabrics. For example, earthworms and other organisms churn sediments every day, a necessary process for nutrient cycling in ecosystems and one that probably played a major role in Mesozoic environments as well.

Sediment was disturbed by dinosaur locomotion in accordance with:

1 the activity level and size of any given dinosaur;
2 the number of dinosaurs; and
3 the amount of time a horizon was available for locomotion.

For example, a single juvenile hadrosaurid that stayed in its nest for the first three years of its life, followed by burial of the nest site by a river flood, had minimal time and opportunity to significantly impact sediments. In contrast, a herd of mostly adult sauropods following a frequently used route that was exposed for a long time, such as along the shoreline of a lake, could have caused sediment impacts unrivaled by any other known land animals in the history of the Earth. Indeed, some horizons in strata from the Late Triassic to the Late Cretaceous provide evidence of how dinosaurs affected significant areas of sedimentary surfaces. The advent of large body size and probable group behavior in certain dinosaurs, such as sauropods and ornithopods, also was a factor in the increased effect of dinosaur locomotion on sediments by the end of the Jurassic through the Cretaceous.

Some dinosaur paleontologists use the term "dinoturbation" to describe the process of sedimentary disruption by dinosaurs. As is evident from its etymology (from the Greek, *deinos* = "terrible" and *turbos* = "mix"), the term is actually misapplied. Most dinosaurs were not the size of large sauropods; a few, such as *Microraptor*, were crow-sized (Chapter 9). As a result, "terrible mixing" is a misnomer, even in Richard Owen's original application of the prefix "dino" (Chapter 4). If "dinoturbation" was applied only to repeated deformation of sediment caused by large animals, such as modern elephant herds, then it would be appropriate. Unfortunately, the originators of this term were specifically applying it to sedimentary disturbance by dinosaurs only, restricting it to disturbances of Mesozoic sediment. In other words, dinosaurs, just like earthworms, simply bioturbated.

In some instance, however, dinosaur bioturbation was indeed awe-inspiring. The effect of the repeated passage of large animals on a substrate is often called **trampling**. Trampling by large dinosaurs not only changed the character of the host sediment, but also had a short-term ill effect on the plants growing underfoot. It had a permanent, potentially fatal effect on any living animals that suffered the stress imparted by such a heavy animal. For example, crushed marine clams and snails have been found within some sauropod tracks (Fig. 14.9). Similarly, some dinosaur bones also show fractures that are likely related to trampling by their contemporaries. The mollusks probably lived on a shallow sea bottom and thus died in place when the unknowing dinosaur stepped on them. The crushed dinosaur bones most likely belonged to already dead individuals that were either on or near the surface traversed by the stomping dinosaurs.

Considering the effect of modern elephant herds on ecosystems and landscapes, dinosaur trampling probably resulted in similar changes in some areas. One of the most likely changes would have been the response of land plants. For example, fast-growing or otherwise flexible species of plants would have been favored in their reproduction versus those that recovered slowly, or not at all, from trampling. Tracks thus provide a perspective on how a journey of a thousand steps (or more) by a dinosaur could have been partially responsible for changing the dinosaur's world in the short-term. These short-term changes in turn contributed to the development of plant communities of the Mesozoic that continued into the Cenozoic, meaning that the steps of dinosaurs left a more subtle but broader "footprint" on modern ecosystems.

14

FIGURE 14.9 Sauropod track on top of a gastropod (snail) shell, preserved in shallow-marine limestone from the Late Jurassic of Switzerland. View is from underneath, which means that the gastropod was stepped on by the sauropod.

Dinosaur Nests

Nest Definitions

Dinosaur nests are extremely important trace fossils for understanding dinosaur reproduction, brooding, and sociality. A **nest** is a biogenic structure that may or may not contain a clutch of eggs, but it is best represented by an arrangement of eggs or eggshells in definite patterns. A nest can also be any semicircular depression with a raised rim that was originally used to hold eggs or young. Dinosaur nests are trace fossils in that they can be defined exclusive of body parts, including eggs or juvenile remains. In contrast, eggshells are body fossils because they were originally part of the developing embryo (Chapter 8). Dinosaur nests sometimes contain hatchlings or other remains of juveniles, which assists in the identification of the parental species, such as some theropods, sauropods, and ornithopods (Chapters 9 to 11). All dinosaur nests described so far were made on a ground surface. This type of construction is similar to those of some modern reptiles and a few flightless birds and shorebirds, rather than the arboreal nests of many modern birds (Chapter 15). Only a few small non-avian theropods may have been arboreal (Chapter 9), but whether they built nests in trees is unknown.

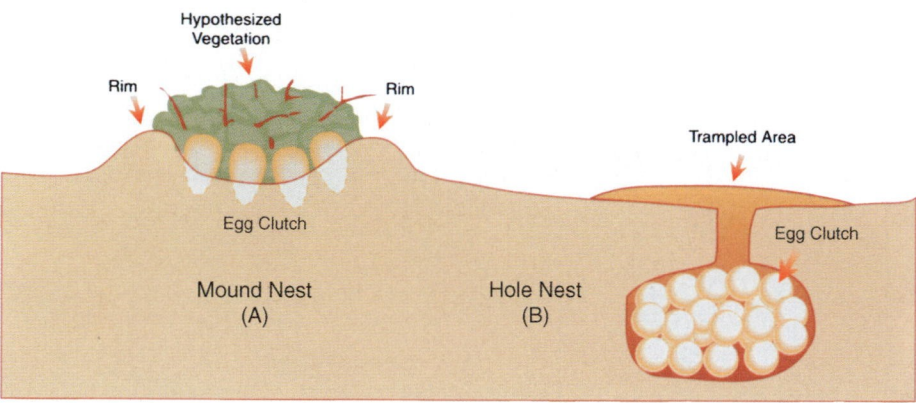

FIGURE 14.10 Nest types associated with modern reptiles, shown in cross-sectional view. (A) Mound nest. (B) Hole nest.

Once nests are found in the field, a thorough investigation requires, at a minimum, description and interpretation of the following parameters:

- Types of eggs in the nest (if present).
- Distribution, orientation, and spatial relationships of any eggs in the nest.
- Geologic setting for the eggs and nests, such as the stratigraphic unit in which they occur and their exact geographic location.
- Facies and paleoenvironments associated with the eggs and nests.
- Type, shape, and size (width and depth) of the nests.
- Spatial relationships of nests to one another in both area (horizontal distribution) and stratigraphic range (vertical distribution).

Modern analogues, in the tradition of uniformitarianism (Chapter 3), provide the basis of comparison for description and interpretation of dinosaur nests. For example, nests observed among modern crocodilians are of two major types: mound nests and hole nests (Fig. 14.10). Mound nests are composed of vegetation piled by a mother crocodilian into which a tunnel is hollowed out for passage. The mother lays her eggs inside the structure, and the nest then provides both protection and incubation. For the latter, methanogenic bacteria generate enough warmth through the decomposition of the surrounding vegetation that the eggs stay heated without direct intervention by the mother. Modern alligators, a few crocodiles, and numerous birds produce such nests. Currently no firm fossil evidence supports the hypothesis that dinosaurs used vegetation in their nests, although some paleontologists have inferred this nest-building adaptation on the basis of modern analogues. Most birds will sit on their eggs and do not rely solely on vegetation for insulating their eggs.

In contrast, a hole nest is an excavation dug by a mother crocodilian into sand, typically on a river bank or shoreline of a water body; the mother deposits and buries the eggs in the nest. Sea turtles employ a similar strategy, laying their clutches in egg chambers that are typically 30–50 cm deep and then burying them. Combinations of mound and hole nests are another possibility for dinosaur nests. Dinosaurs may have partially buried their eggs in sediment and covered the eggs with vegetation. Evidence that eggs were moved and buried by the mothers after laying is provided by egg clutches with elongate eggs oriented vertically and in definite patterns (Fig. 14.11). Moving the eggs into vertical orientations probably required the use of the fore limbs and considerable manual dexterity. It is difficult to fathom

14

FIGURE 14.11 Dinosaur egg clutch of the Late Cretaceous theropod *Troodon* associated with a nest structure (not shown). Orientation implies that eggs were moved by the mother after egg-laying. Cast of original clutch from Museum of the Rockies, Bozeman, Montana; the clutch is upside-down for display purposes.

how such precise positions could have been achieved by the use of snouts or other anatomical parts.

As viewed from above a nest site, dinosaur egg clutches typically assume four main patterns:

1 concentric circles;
2 spirals;
3 parallel rows that were doubled in some cases; and
4 poorly defined clusters.

Circular patterns are probably the result of post-laying movement, whereas the linear arrangements may have been from a mother dinosaur periodically laying eggs as it moved forward. Clusters represent less organization and probably are attributable to clutches laid in one depression without appreciable movement by the mother. The latter mode is observed in crocodilian and turtle egg-laying, where masses of eggs are in close but chaotic association with one another.

Nest Taphonomy

Common environments for ground nests can be nearly anywhere on land for reptiles and birds, but for the nests to be preserved in the geologic record they must be close to where sediments can bury them. Although this typically happens near water bodies, eolian deposition has been invoked for the burial of some dinosaur nests, such as those in Cretaceous strata of Mongolia (Chapter 7). Indeed, the majority of dinosaur nests are associated with fluvial, lacustrine, or marine shoreline facies, which indicates that at least a few species of dinosaurs nested in these areas. However, they also may have nested in upland regions that had lower preservation potential. Floodplains of rivers were probably the best environments during the Mesozoic for dinosaur nest preservation.

An additional factor favoring nest preservation is the sheer number of nests at a nesting site. For example, some modern crocodilians and ground-dwelling birds, such as penguins, can have hundreds or thousands of nests within a relatively small area. Birds in particular exhibit colonial nesting behavior, creating their nests in close proximity to one another (Chapter 15). A similar behavior has been interpreted for a titanosaurid (Chapter 10) and the hadrosaurid *Maiasaura* (Chapter 11). Lastly, some reptiles and birds that build ground nests show site fidelity, returning to nest in the same areas over generations. Even a small percentage of nests in a colonial nesting site for dinosaurs could have been preserved under the right conditions. As a result, site fidelity could have resulted in an increased number of preservable nests within the same area over the course of many years.

Probably the most difficult situation for finding dinosaur nests is in those cases where apparently no eggshell or juvenile skeletal material was preserved. Several scenarios, which use some of the taphonomic principles covered earlier for eggs, explain such an absence:

- Eggs were non-mineralized and primarily composed of organic material, proposed for most pre-Cretaceous dinosaur eggs.
- Burial rates for mineralized egg material were not rapid enough to prevent dissolution and other degradation processes affecting the material.
- Mineralized eggs were consumed or transported away from the original nest site through biological processes, such as egg predation.
- Juvenile dinosaurs that required parental care over several years pulverized the eggshells by stepping on them.
- Juvenile bones, being smaller and more easily transported, were weathered and eroded and thus not preserved.
- Juveniles were ready to leave the nest immediately after hatching (precocial), so any of their remains that might be preserved may be far away from the original nest site.
- Dinosaur egg material is preserved, but has not been recognized by the paleontologists who find it.

At least one experienced dinosaur paleontologist, Charles Gilmore of the Smithsonian Institution, was a victim of the last scenario. In 1916, Gilmore found a thick, extensive horizon of dinosaur eggshell material in the Late Cretaceous Two Medicine Formation of Montana, but referred to them in his field notes as "bivalve fragments." Although Gilmore eventually corrected himself, no one noticed the error in the original notes until 70 years later, when Jack Horner (Chapter 4) went to the same spot Gilmore described. Horner soon realized that the "bivalves" comprised one of the richest dinosaur eggshell zones in North America. This anecdote is an example of how good descriptions in paleontology can last forever, but interpretations can always change. Horner also had the benefit of having seen many dinosaur eggshell fragments before he rediscovered Gilmore's find, so he more easily applied the appropriate search pattern for finding them (Chapter 2).

All is not lost in the absence of recognizable egg material, however. Because at least some former dinosaur nests were biogenic sedimentary structures, they are potentially distinctive from their surrounding sediment as either large-scale disruptions of bedding or differences in sediment. Such structures are either composed of a different grain size or different colored sediment, or they stand out in relief as upraised, bowl-like structures that are more resistant to weathering. Such mound nests have been described for the small theropod *Troodon* (Chapter 9), titanosaurids (Chapter 10), and the hadrosaurid *Maiasaura* (Chapter 11). Other possible criteria for recognizing nests in the absence of eggs are tracks of either the egglayers or egg

14

predators, although these have not yet been reported in association with dinosaur nests.

In many cases, the mere concentration of dinosaur eggs or eggshell fragments is sufficient for interpreting the former presence of a dinosaur nest. Experimental evidence from modern bird nests shows that eggshell fragments normally remain in the general locality of the nest. Similar results have been yielded by tests of this taphonomic model on places where dinosaur eggshell fragments are abundant. Other experiments have been done on avian eggs. When they are placed in aquatic environments (continental or marine), eggs in close proximity to their original nest site have shown preservation that depends on the closely-associated depositional environment. Further testing of these models should show whether post-laying transport of eggs or eggshell fragments is a major problem for the interpretation of former nest sites. Nonetheless, current evidence points toward the presence of concentrated egg remains as a reliable indicator of nearby dinosaur nests.

Dinosaur Toothmarks

Toothmark Definitions

A toothmark is an impression left by the bite of an animal with teeth, regardless of what was being bitten (Chapter 8). Toothmarks are trace fossils, whereas the medium they bit into are typically body fossils. Dinosaur toothmarks, first described and interpreted in 1908 (Chapter 3), have been so far only reported from bones; no dinosaur toothmarks in fossil plant material are currently known. For those dinosaurs that fed on other animals, some left distinctive toothmarks on their bones. Whether these pieces of evidence are representative of feeding preferences for some dinosaurs is inconclusive. A firmer understanding of feeding preferences, especially for different species of dinosaurs, is contingent on the discovery of more toothmarks or supplementary clues from stomach contents or coprolites.

Toothmarks are described by their:

1 dimensions, such as length, width, and depth;
2 shape;
3 spacing from one another; and
4 relation to their host substrate.

Toothmarks considerably longer than deep are generally interpreted as scrape marks, whereas those that are deeper than long are considered to be puncture marks. Puncture marks provide the best opportunity for identifying the tracemaker. In some cases, these toothmarks perfectly match the "dental records" of teeth from the jaws of an identified species of dinosaur. Evenly spaced and parallel toothmarks provide strong evidence for the tooth row and occlusal surface of the tracemaker, and thus can be compared to the tooth spacing and tooth types in dinosaur jaws from same-age strata. More than one row of such multiple toothmarks can point toward a dinosaur that fed on an already dead animal. This is especially the case if the toothmarks originate at the end of a limb bone or wherever else muscles could have been torn away from a bone. In this scenario, the feeding dinosaur stripped meat from the bones at its easiest points of separation, a strategy also employed by many modern carnivores. Additionally, the position of the toothmarks in tandem with the anatomical orientation of the consumed animal can re-create how a dinosaur fed, such as when a lateral, proximal, or distal portion of a bone contains toothmarks (Fig. 14.12).

FIGURE 14.12 Four sets of toothmark rows on left ilium of a Late Jurassic sauropod, *Apatosaurus*. Toothmark spacing corresponds with tooth row of *Allosaurus*, a large theropod found in same-age strata of *Apatosaurus*. All sets begin at distal part of bone as scrape marks that trend caudally; feeding was performed through pulling of muscles from insertion point on ilium on a carcass lying on its right side. The allosaur was taller than the width of the recumbent apatosaur so that it could move its head horizontally for this "nipping" motion. Rightward trend of scrape marks suggests that allosaur was on the dorsal side of the apatosaur. Specimen in the Dinosaur Adventure Museum of Fruita, Colorado.

Studies of toothmark orientations demonstrate no apparent preferred directions. This suggests that most toothmarks were accidental by-products of meat eating and that feeding did not preferentially include bone material. Furthermore, no bias for feeding on specific bones, such as tibias, fibulas, or femurs, is indicated so far. One explanation proposed for the rarity of sauropod skulls relative to their other skeletal parts is that theropods discriminately consumed these bones (Chapter 10). However, the relative fragility of sauropod skulls in relation to their other skeletal parts and other taphonomic factors provide more likely explanations.

Toothmarks have been attributed to specific dinosaurs on the basis of their close resemblance to known tooth anatomy, especially denticles on serrations, and spacing. Some reported examples from the Late Cretaceous include *Troodon* toothmarks in ceratopsian bones, *Saurornitholestes* toothmarks in bones of an ornithomimid and *Edmontosaurus*, and *Tyrannosaurus* toothmarks in neoceratopsian, hadrosaurid, and *Saurornitholestes* bones. Of these examples, a direct correlation between "dinner" and "diner" species through toothmark evidence is of *Tyrannosaurus* toothmarks in *Edmontosaurus* and *Triceratops* bones (Chapters 9, 11, 13). Similar toothmarks attributed to a closely-related species of tyrannosaurid, *Albertosaurus*, have also been interpreted from *Edmontosaurus* bones, meaning that it was possibly consumed by at least three species of theropods.

Of course, theropod toothmarks do not necessarily mean that these same theropods preyed upon and killed any of the eaten dinosaurs. After all, the specimens may have been already dead from other means when they were munched. In contrast, toothmarks from theropods that show post-wound healing have been reported for at least two specimens of the possibly oft-victimized *Edmontosaurus* (Chapter 13). These indicate successful escapes for a preyed-upon dinosaur, or a failed hunt for the predator, depending on one's perspective. Additionally, healed toothmarks in theropod bones that match those of the same species may point toward intraspecific competition. This was proposed for tyrannosaurids at one time, but this hypothesis was later falsified (Chapter 9).

One of the more important pieces of information derived from toothmarks is an estimate of minimum biting force of a carnivorous dinosaur. The stress applied by a tooth is a result of force generated by the biting animal to the surface area in contact with the tooth. Penetration depth and the competence of the penetrated substrate can be a general indicator of such stress, for example. Experiments with modern bone material can be used as a model for biting force, which was done with *Tyrannosaurus* in mind. In these experiments, bovine bones were placed in a vise that measured stress. Stresses were then applied to duplicate the depth of penetration mirrored by *Tyrannosaurus* toothmarks in fossil bones. The calculated force was 13,400 N, which is greater than that for any known animal, living or dead. Such jaw strength in *Tyrannosaurus rex* provides good reason why it still deserves its original title of "King of the Tyrant Lizards," regardless of whether it was primarily a predator or scavenger (Chapter 9).

Toothmark Taphonomy

The preservation of toothmarks may be analogous to the same principles of under-track preservation: the deeper the toothmark, the more likely it was preserved. Bites that only superficially penetrated flesh have a lower preservation potential than deeper bites that penetrated bone, because there is a general lack of soft-tissue preservation. Consequently, wounds that could have easily killed a prey species, such as disembowelments, severed arteries, or other soft-tissue traumas, have extremely low preservation potential in the fossil record. Indeed, all reported dinosaur toothmarks are in bones and some clearly show that the chewed animal was consumed after it was already dead. Rare instances of herbivorous mammal toothmarks in fossil vegetative material from the Cenozoic Period have been documented, but no similar traces left by herbivorous dinosaurs are known. Such fossils probably would have to be deep bites or gnawings within woody tissue that were buried rapidly to facilitate preservation. Additionally, the toothmarks of larger prey animals and dinosaurs with longer teeth and greater biting force were probably preferentially preserved. In other words, toothmarks from *Allosaurus* in bones of sauropod bones should be more commonly represented than those from *Microraptor* in small mammal bones. Among feeding traces, the least likely for preservation are those left by toothless dinosaurs such as *Oviraptor* or *Ornithomimus*. However, modern toothless predatory birds can leave distinctive marks on bones of their prey, so toothless dinosaurs may have left similar marks.

Despite the large number of theropods, most interpreted as carnivores, their toothmarks are rarely reported in comparison to toothmarks caused by modern carnivorous mammals. Several hypotheses are proposed to explain this discrepancy:

1 most theropods may have fed differently from modern mammals by swallowing their prey whole or in large chunks, which would have resulted in no preserved toothmark;
2 they could have thoroughly chewed their food, which would have left no single, identifiable toothmarks;

3 bones broken by theropods, even if left unconsumed, had more exposed surface areas that weathered more quickly than unbroken ones; as a result, their preservation potential was lower; and

4 numerous theropod toothmarks are present but under-reported.

With regard to the latter, dinosaur researchers may have focused their attentions on describing the anatomical information from dinosaur bones. Meanwhile, they either overlooked or decided not to describe toothmark evidence. These same researchers also may have only recovered "museum quality" skeletal material from the field, rather than broken pieces of bone that could contain abundant toothmarks.

Evidence indicating that dinosaur toothmarks may be more common than previously supposed comes from the Late Cretaceous Dinosaur Park Formation of Alberta, Canada. One study of more than 1000 dinosaur bones revealed that as much as 14% of all hadrosaurid bones contained toothmarks. Such results are encouraging for future research on dinosaur toothmarks. Toothmarks, which in combination with identifiable dinosaur teeth from same-age strata and associated trace fossils, such as gastroliths and coprolites, can thus contribute to a better understanding of dinosaur feeding.

Dinosaur Gastroliths

Description, Taphonomy, and Relation to Soft-Part Anatomy

Gastroliths (from the Greek *gastro* = "stomach" and *lithos* = "stone"), first mentioned in Chapter 3, are stones used primarily to help in the mechanical breakdown of food within a digestive tract. Another simple, colloquial term for gastroliths is "gizzard stones." This name comes from observations of some modern birds, such as chickens, which will swallow mineral grains several millimeters in diameter. These grains then reside in their gizzards and aid in digestion of food by helping to grind tough food material. Because birds do not have teeth, they need their gizzards to break down their food mechanically. The muscular action of the gizzard and the grinding caused by the mineral material helps to increase the surface area of the food for easier digestibility. Other modern animals besides birds that swallow stones include crocodiles, which use them for ballast in regulating their buoyancy in water, rather than employing them as a digestive aid. Because these stones are not body parts and they reflect behavior of the animal, they are trace fossils. In the case of dinosaurs, gastroliths are trace fossils associated specifically with feeding behavior.

Gastroliths were first described and interpreted from a Late Cretaceous hadrosaurid (*Claosaurus*) by Barnum Brown early in the twentieth century (Chapter 3). Later, in 1931, Friedrich von Huene found them in association with bones of the Late Triassic prosauropod *Sellosaurus*. In 1942, William Lee Stokes described stones found with Late Jurassic sauropod remains. Surprisingly, they have been studied little since then, perhaps because so many paleontologists have regarded them with a high level of skepticism. The reluctance to identify gastroliths is understandable, because they are difficult to distinguish from other stones that are smooth, polished, and rounded because of physical and chemical (abiotic) wear. The latter types of stones can be found as bedload in stream bottoms, rather than in the former gut of a dinosaur. Furthermore, these trace fossils, more so than most other dinosaur trace fossils, are susceptible to secondary reworking by sedimentary processes. For example, possible pathways for gastrolith taphonomy could have included:

14

FIGURE 14.13 Typical gastroliths with rounded and polished appearance, from the Lower Cretaceous Cedar Mountain Formation of Utah. Denver Museum of Science and Nature, Denver, Colorado.

- Rounding in a stream.
- Consumption by sauropod.
- Rounding and polishing in the gizzard.
- Loss during life, through regurgitation or defecation.
- Loss after death.
- Accidental consumption by a theropod eating a sauropod.
- Burial near the place where it was lost by the sauropod.
- Transport away from place of loss, by either running water or the theropod carrying the gastrolith in its gut.
- Reconsumption by another sauropod.
- Exposure and reworking of gastrolith, followed by reburial.

Evidence for gastroliths in dinosaurs consists of the numerous polished stones associated with dinosaur body fossils, the most oft-cited examples being found within the thoracic cavity region, ventral to the cervical and dorsal vertebrae and anterior to the sacral vertebrae. These stones are typically smooth, polished, rounded, and flat to semispherical (Fig. 14.13). The sizes of the stones can vary in proportion to the size of the host animal. Gastroliths in smaller dinosaurs may be less than 1 cm in diameter, but some found in sauropod skeletons may be as large as 10 cm in their longest dimension.

Gastroliths are typically silica rich, such as those composed of chert (SiO_2) or quartzite, a durable metamorphic equivalent of sandstone composed of quartz (also SiO_2). Limestone and other calcareous rocks, such as dolomite, are unlikely candidates for gastroliths. Stomach acids have low pH values and easily dissolve such rocks, effectively defeating the purpose of swallowing them in the first place. Additionally, chert, quartzite, and other silica-rich rocks are typically harder than calcareous rocks. Consequently, they would be more likely to survive extended periods of grinding in a digestive tract.

Gastroliths also should be concentrated in relatively small areas within the confines of the dinosaur skeleton. Paleontologists who encounter gastroliths should carefully map their distributions and orientations in relation to the skeleton, which can aid in telling the former position of either the crop or gizzard of the dinosaur. Unfortunately, most excavators of dinosaur skeletons have not recorded such information.

The current hypothesis explaining the association of polished stones with dinosaur remains is that dinosaurs swallowed stones for the same reason as modern birds, to help with digestion. Because no dinosaur species has been convincingly demonstrated as habitually aquatic, the digestion hypothesis is favored over the stones' use as ballast, as in crocodiles. The rounding, smoothness, and high degree of polish of gastroliths are all attributed to their repeated grinding within the gizzard, as well as the possible effects of gastric acids as a chemical weathering agent. A commonly invoked alternative hypothesis for the juxtaposition of gastroliths and dinosaur skeletons is that these polished stones were placed in the skeleton through physical processes after death. In this hypothesis, the body cavity of a dead dinosaur was opened sufficiently for sediment to accumulate in it. This sediment may have included rounded, polished stones from a stream bed near the body. This hypothesis must be seriously considered and requires much evidence to falsify because of the very common association of dinosaur skeletal remains with fluvial deposits, which typically contain such rounded stones.

Fortunately, at least one method has been used to distinguish these different types of stones: laser light-scattering, which measures the amount of light coming from a laser and scattered by a polished surface on each stone. Probable gastroliths, which were taken from the thoracic cavity of a dinosaur, and rocks initially identified as gastroliths on the basis of their physical resemblance to the former show a high degree of scattering. Stones from fluvial environments show less scatter, whereas those from rocky shorelines have a scatter nearly intermediate to that of actual gastroliths and fluvial stones. Of course, the possibility that gastroliths were already rounded and polished by a stream bed when a passing dinosaur swallowed them cannot be ignored. Another check for gastrolith authentication is to examine the sediment that composes the rock surrounding a dinosaur skeleton. If the host rock does not contain larger, rounded, polished stones, but the interior of the dinosaur skeleton does, then gastroliths cannot be eliminated as a possibility.

Dinosaurs with well-supported evidence for gastroliths include some sauropodomorphs, both prosauropods and sauropods (Chapter 10); a nodosaurid (Chapter 12); psittacosaurids (Chapter 13); a few theropods (Chapter 9); and possibly a few ornithopods (Chapter 11). Because gastroliths are normally associated with herbivores and the theropod specimens are seemingly carnivores, the presence of gastroliths in their gut regions is somewhat of a mystery (Chapter 9). Gastroliths in psittacosaurids are also slightly odd because these dinosaurs had well-developed dental batteries that should have easily ground up their roughage. Thus, gastroliths provide additional data that can question certain assumptions about the lifestyles of dinosaurs that ordinarily might be made on the basis of body fossil evidence alone.

In that same vein, gastroliths are an important consideration when exploring for dinosaurs in areas that normally have not yielded dinosaur fossils. In the absence of other evidence for dinosaurs in continental Late Triassic, Jurassic, or Cretaceous facies, possible gastroliths, concentrated deposits of well-polished, rounded stones in otherwise fine-grained (muddy or sandy) deposits, can be the first clue to a former dinosaurian presence. Bones may not be preserved and tracks may not have been impressed deeply enough to avoid erosion, but chert cobbles are durable enough to have withstood repeated travel before burial. This is regardless of whether they traveled on the bottom of a stream or passed through the gut of a dinosaur. After burial, silica-rich gastroliths also have an advantage in preservation because they are not as susceptible as bones to dissolution by acidic groundwater (Chapter 7).

14

Dinosaur Coprolites

Coprolite Definitions

Coprolites comprise some of the most important evidence related to the feeding habits of extinct organisms. Coprolites (from the Greek *copros* = "feces" and *lithos* = "stone") are fossilized remains of the solid or semi-solid fecal material produced by an animal. Coprolites are trace fossils because they are not composed of body parts from the tracemaker and they reflect the tracemaker's behavior. Coprolites, like gastroliths, provide information about ancient diets, habitats, and the presence of dinosaurs in areas otherwise lacking dinosaur body fossils or more common trace fossils such as tracks.

Coprolites have a long history of study. In 1823, William Buckland (Chapter 3) identified probable vertebrate coprolites from marine Cretaceous deposits of England, and in one 1835 publication he coined the name still used for fossil feces. In 1844, Edward Hitchcock also reported coprolites from Late Triassic and Early Jurassic rocks that contained other trace fossil evidence of dinosaurs. However, he still thought at the time that he was studying the trace fossils of ancient birds and did not make the connection between these coprolites and dinosaurs. C. E. Bertrand of Belgium first interpreted a dinosaur coprolite in 1903. This coprolite was from Cretaceous rocks that contained numerous iguanodontian skeletons and some theropod remains. However, later workers suggested that these coprolites also could have been made by Cretaceous crocodilians. Bertrand's study and the detractors of his work illustrate the difficulty of linking a coprolite with its tracemaker. Nevertheless, he demonstrated the potential for associated coprolites with known occurrences of dinosaur body fossils and other trace fossils in the same strata.

A coprolite may contain body fossils, such as bacteria, plant fragments, or bones, or in very rare cases may preserve soft tissues. These body fossils included in coprolites make them extremely valuable for the interpretation of feeding behavior. Another fringe benefit of coprolites is that they may also contain other trace fossils. Burrows formed during secondary feeding on the digested material occur in some coprolites. Modern examples of these burrows are readily observable when insects, such as flies and dung beetles, have done their work on fresh feces. This coprophagy (see box), regardless of how unappetizing it might seem, is a necessary part of life on Earth, because it is responsible for much nutrient cycling in both terrestrial and marine environments. Traces of coprophagous behavior with regard to dinosaur coprolites can give valuable insights about nutrient cycling in Mesozoic ecosystems. This bestows an added importance to coprolites as sources of paleoecological information. For example, Late Cretaceous coprolites that were probably made by hadrosaurids have been interpreted to also contain dung-beetle traces (Fig. 14.14).

> *Coprophagy* literally means, "eating feces" (from the Greek, *phagos* = eat).

Coprolites are recognized largely through their morphological resemblance to modern feces. However, "pseudocoprolites," inorganically formed deposits that superficially resemble coprolites, also have been interpreted from the geologic record. Accordingly, a careful description of a suspected coprolite must be made before interpreting it as one. Coprolites are described through a near-standard terminology that recognizes their shape, size, surface features, and content. Shapes can generally be divided into two categories, pelletal and cylindrical, with pelletal shapes approaching flattened spheroids and cylindrical shapes being more elongate. Relative flattening of a coprolite can indicate:

FIGURE 14.14 Coprolite containing abundant ground-up conifer remains attributed to the hadrosaurid *Maiasaura*, Upper Cretaceous Two Medicine Formation, Montana.

1 firmness of the original fecal material;
2 height of the defecating animal above a surface;
3 post-depositional compaction; or
4 any combination of these factors.

Fecal firmness is influenced by composition. For instance, fecal material with low water contents will tend toward greater firmness, which can also indirectly indicate dehydration in the animal or drought conditions. The height of the animal above a surface was certainly a factor in determining the shape of fecal material for some dinosaurs. Some sauropods may have been as much as 10 meters above the ground when they defecated. Fecal flattening may also occur among animals that follow one another in a herd and step on the deposits of their predecessors. Post-depositional compaction by sediments overlying a coprolite is also a possibility if the coprolites were still soft when they were buried.

Size is an important consideration for dinosaur coprolites. One simple assumption for animal feces is: the larger the pile, the larger the animal. A large-diameter cylindrical coprolite certainly implies a correspondingly large tracemaker. Nevertheless, caution should be applied to individual, small coprolites, for these do not necessarily correlate with a smaller tracemaker. One consideration is that a coprolite may only represent one piece detached from a formerly larger and integrated fecal mass. Individual but entire dinosaur coprolites also can be quite small (<10 cm length) compared to the body size of their purported tracemakers. An example of this seemingly anomalous correlation between animal size and fecal size can be observed in modern mule deer (*Odocoileus hemionus*) and elk (*Cervus canadensis*) of North America. These are animals that can weigh nearly 100 kg but their individual pellets are often less than 1 cm in diameter. Masses of small, individual pellets can be merged into large deposits by compaction, through an initially high water content that enables them to ooze together, or by both processes. So a large mass of small pellets could be used as an indicator of a large tracemaker. Indeed, some masses of fossil vegetative material have been interpreted as dinosaur coprolites that formed from the collection of many originally small-diameter pellets. Coprolites can thus be used reliably to indicate a minimum size of the tracemaker, whereas interpretations of maximum size should be treated with more skepticism.

Surface features of coprolites can potentially indicate soft-part anatomy and functions associated with the lower part of the gastrointestinal tract that otherwise would not be preserved in most body fossils. For example, longitudinal striations on a coprolite indicate the minute folding on the intestinal wall of the tracemaker, and distinctive pinched ends of coprolites are signs of a healthy anal sphincter. For coprolites that can be correlated with dinosaurian tracemakers, such data can help to better define the form and function of a dinosaur's posterior internal organs.

The contents of a coprolite are the final piece of evidence that paleontologists need to assure themselves that they are studying a formerly digested meal and not

14

a mass of rock that simply looks like feces. Both fossil plants, such as conifers, and bones have been discovered as ground-up material in localized masses recovered from the geologic record. When such material is correspondingly contained within masses that morphologically match those of known coprolites, their identification as a coprolite is reasonable. Conversely, a coprolite-like body that lacks plant or bone material should not be assumed to be a coprolite. Likewise, a mass of finely fragmented plant or bone material, without the requisite size and shape associated with a coprolite, also should not be identified as a coprolite. A dinosaur coprolite must have both the form and composition of feces, not one or the other. Moreover, it should occur in strata of the proper ages and environments for dinosaurs.

As an example of detailed interpretations that can be made for dinosaurian diets from coprolites, evidence for consumption of a juvenile hadrosaurid by a tyrannosaurid was interpreted from a large (44-cm long) cylindrical Late Cretaceous coprolite from Alberta, Canada. The coprolite is inferred as originating from a tyrannosaurid tracemaker because of:

1 its unusually large size;
2 the inclusion of numerous small bone fragments of a juvenile hadrosaurid; and
3 its occurrence in strata known to contain body fossils of tyrannosaurids, such as *Albertosaurus* and *Tyrannosaurus*.

Another large (64-cm long) coprolite, also discovered in an Upper Cretaceous stratum of Alberta, contained three-dimensional impressions of muscle tissue and finely-ground bone. Also attributed to a tyrannosaurid, this coprolite indicates a brief digestive period for the tracemaker, rapid diagenesis, and burial of the fecal mass. Otherwise, the muscle tissue would not have been preserved.

Coprolite Taphonomy

Coprolites are trace fossils with potentially complex histories. Like gastroliths, they are susceptible to secondary reworking, so where a coprolite is found may not necessarily represent where it was deposited by the original tracemaker. Preservation of coprolites depends on their original composition, water content, the location where they were deposited, and the method of burial. For example, coprolites made by carnivorous dinosaurs (most theropods) were more likely to be preserved than those made by herbivores (some theropod and all non-theropod dinosaurs). This is because the bone material of the consumed prey animals gave them a high mineral content, dahllite (Chapters 5 and 8). The dahllite of the consumed animal served as a nucleation site for phosphatization, which can be a relatively rapid diagenetic process. Phosphatization is also aided by some bacteria, which were already present in large amounts in the gut and feces of the animal. After defecation, a good preservation environment for coprolites then would have been a floodplain associated with a river, where feces deposited on a dry part of the floodplain dehydrated slightly before being rapidly buried by a river flood. Other environments where coprolites were likely to have been preserved include "watering holes" (ponds), swamps, streams, and muddy areas associated with estuaries or lakes.

One scenario, proposed as ideal for coprolite preservation, is explained as follows:

1 Production by a carnivore tracemaker
2 Deposition either on land or in water
3 Slight dehydration (firming)
4 Shallow burial (probably rapid following deposition)
5 Limited coprophagy of fecal material (either while buried or exhumed)

6 Pre-fossilization through rapid phosphatization
7 Exhumation (through winnowing) and reworking into concentrated deposits
8 Final burial, further phosphatization or other diagenetic alteration.

As mentioned, theropod tracemakers are more likely than herbivorous dinosaurs to have had their feces preserved because of the included bone material. Characteristics of carnivore coprolites include visible bone fragments and a phosphatic composition, the latter of which can be detected by looking at the coprolite under a microscope or geochemical tests. However, a theropod tracemaker should never be blindly assumed for a suspected dinosaur coprolite, especially in light of numerous discoveries in recent years of coprolites attributed to herbivorous dinosaurs. Plant remains within an herbivorous dinosaur's coprolite should be preserved as carbonized or silicified material that may be visible macroscopically.

Dinosaur feces that have been preserved were deposited on emergent or shallow-water environments, such as fluvial floodplains. The feces were most likely dropped on a dry part of the floodplain and dehydrated slightly before burial by a river flood. Feces deposited on land should have firm, well-defined shapes but may show slight flattening from impact, depending on their initial water content. They also may show coprophagy traces from insects. In contrast, feces deposited in water should be more amorphous and have few or no coprophagy traces, as dung-consuming insects are mostly terrestrial.

Like most fossils, rapid burial is essential for preservation of coprolites but the depth of burial could have been relatively shallow. Rapid burial is indicated by a coprolite that retains much of its original form and surface ornamentation, indicating lack of extended exposure to the weather or insects. Once buried, rapid phosphatization would be suggested by the aforementioned retention of form and nucleation of secondary phosphates around original bone fragments. After an indeterminate time, coprolites may have been exhumed by running water or wind, which could have formed a concentrated deposit of coprolites. Evidence for exhumation and perhaps some movement of coprolites is seen through fracture surfaces on coprolites and co-occurrence with bones that also show breakage. Final burial of coprolites may have been followed by further phosphatization or other diagenetic alteration. As many as 225 million years after the dinosaur digested its meal, humans may then discover its remains.

Coprolites can therefore provide information about dinosaurs that goes beyond strict definitions of paleodietary preferences. They also can illuminate insect–dinosaur interactions, nutrient cycling in dinosaurian ecosystems, preferred environments for dinosaurs, sedimentary processes in those environments, and diagenesis. Not only are dinosaur coprolites the end products of dinosaur feeding habits, they are commodities that can give important and integrative information about how dinosaurs related to their environments.

Miscellaneous Dinosaur Trace Fossils

Any indirect evidence left by a dinosaur in the geologic record, exclusive of body parts, is a trace fossil, so the potential for discovering and describing more than tracks, nests, toothmarks, gastroliths, and coprolites is likely. A few other types that have been reported, but not yet discussed further, include:

- Mended ribs or holes in parietal shields in ceratopsians, which are interpreted as the result of intraspecific competition (fighting) between rival ceratopsians (Chapter 13). This behavior is common in terrestrial mammals, especially those with prominent headgear, so it is not unreasonable to hypothesize that horned dinosaurs may have also used their heads on each other.

■ Oddly mixed and broken small dinosaur bones concentrated in a deposit, which are interpreted as undigested food vomited by a theropod. As more evidence links non-avian theropods to birds (Chapter 15), one convergent behavior they may have shared was the formation of cough pellets (Chapter 8).

■ Large scour marks that cut across bedding, present in the same horizons as abundant sauropod tracks that are interpreted as urination traces. The volume and height of a urinating sauropod certainly would have caused considerable erosion of sediment and covered a large area. This means that the structure would have good preservation potential in the same strata as dinosaur tracks. However, it is unknown whether sauropods actually urinated or if they excreted semi-liquid wastes more similar to that of birds.

No doubt, other dinosaur trace fossils await their discovery by paleontologists, but they will need to be armed with good imaginations that are well grounded in scientific evidence provided by many observations of modern traces. Modern traces serve as analogues for comparison, of course, but more importantly establish search patterns that, for example, help with spotting poorly-preserved tracks, vague nest structures, tooth scrapings in bone, or deposits of gastroliths and coprolites. One expectation in dinosaur ichnology is that behavioral experiments with modern animals and their resultant traces will be more linked with those made by dinosaurs. Moreover, computer modeling is sometimes used to recreate anatomical traits and ranges of motion for dinosaurs, which then correspond to known tracks and trackways. Nevertheless, the future of dinosaur ichnology also might lie in the past. Much can still be learned from modern trackers, such as the San of the Kalahari, who use an integrated approach in their tracking more akin to ecology. Can dinosaur paleontologists some day look at a dinosaur trackway and state the following hypothesis: "It was a male adult, left-side dominant, ate only a few hours before making the tracks, was having trouble with its breathing because of a broken rib, and was stopping frequently to sniff the air for predators"? Maybe not, but the expansion of dinosaur ichnology beyond merely noting the presence of tracks or other traces has already happened, and will continue to grow with other aspects of the study of dinosaurs to provoke new hypotheses about them.

SUMMARY

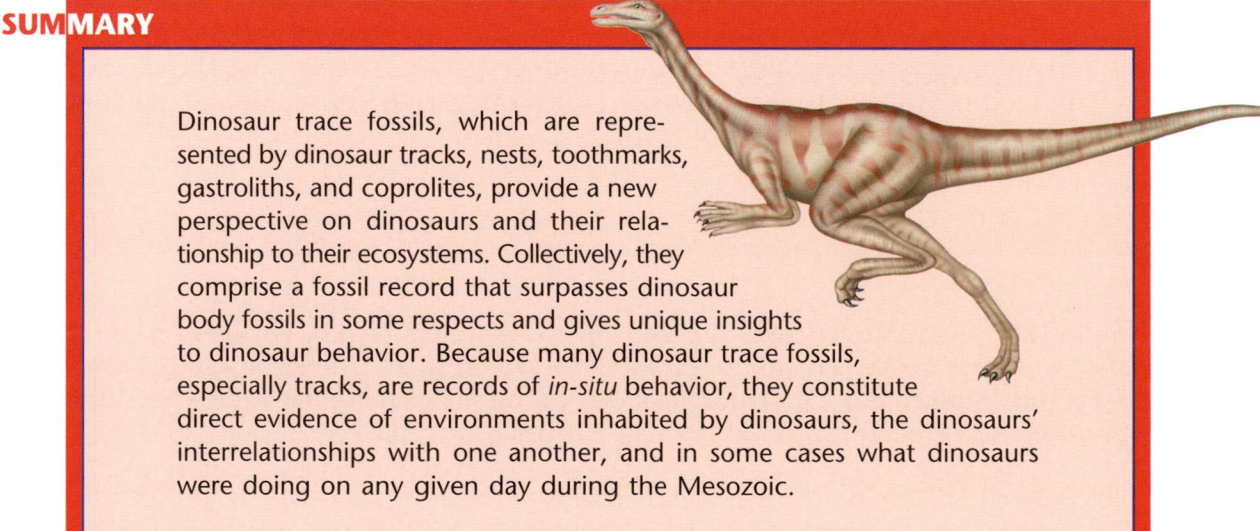

Dinosaur trace fossils, which are represented by dinosaur tracks, nests, toothmarks, gastroliths, and coprolites, provide a new perspective on dinosaurs and their relationship to their ecosystems. Collectively, they comprise a fossil record that surpasses dinosaur body fossils in some respects and gives unique insights to dinosaur behavior. Because many dinosaur trace fossils, especially tracks, are records of *in-situ* behavior, they constitute direct evidence of environments inhabited by dinosaurs, the dinosaurs' interrelationships with one another, and in some cases what dinosaurs were doing on any given day during the Mesozoic.

SUMMARY Continued

Track descriptions involve making many detailed observations that become the basis for interpretation, a skill that humans have applied in everyday life for thousands of years through tracking. Observations that can be made about tracks consist of qualitative and quantitative data. Qualitative data include firmness of the original substrate, pressure-release structures, and type of preservation. Quantitative data include track length, width, number of toes, depth, pace, stride, and straddle. Studies of how modern animals move are also useful for interpreting how dinosaurs moved (such as walking, galloping, hopping), especially when applied to the mathematical estimations of dinosaur speeds deduced from dinosaur trackways. The lessons learned from the taphonomy of tracks also help to evaluate whether dinosaur trackways actually reflect their interpreted behaviors. Paleontologists communicate with one another about tracks through the use of ichnotaxa, and assigning track names depends on good preservation, among other factors.

Dinosaur tracks can tell us much about the behavior of both individual dinosaurs and groups of dinosaurs. Although some interesting behaviors for individual dinosaurs have been reflected by tracks, most locomotion simply consisted of walking. Group dynamics of dinosaurs are illustrated by multiple trackways, some of which may represent the simultaneous movement of several dinosaurs in unison, behaviors typical of herding or pack-hunting animals. Repeated movement by large groups of dinosaurs resulted in major changes in substrates and possibly imparted other effects on Mesozoic ecosystems during the Jurassic through to Cretaceous Periods. Tracks also are good cross-checks for skeletal data in biostratigraphy and paleobiogeography. A lack of bones in Mesozoic rocks does not necessarily mean that dinosaurs were not present in the sedimentary environments that produced the rocks.

Nests are important dinosaur trace fossils that, together with body fossils such as eggshells, embryos, juveniles, or adults, can give added insights into dinosaur reproductive behavior. Although still rarely interpreted, dinosaur nests are documented for one theropod, ornithopod, and sauropod. These nests were shallow, hollowed-out structures on a ground surface, are often rimmed, and in some cases are abundant in small areas. Dinosaur nests found so far are analogous to modern mound nests made by ground-dwelling birds and alligators, rather than hole nests made by other crocodilians and sea turtles.

Most other dinosaur trace fossils, toothmarks, gastroliths, and coprolites, are related to feeding. Toothmarks can show food preferences of some genera (sometimes species) and relationships of some dinosaurs to their food chains. Gastroliths can lend clues about how dinosaurs digested their food, including aspects of their soft-part anatomy. Coprolites are among the most valuable of all dinosaur trace fossils if they can be correctly attributed to specific dinosaurs. They provide information about paleodiets, paleoenvironmental conditions (habitats and climate), and diagenesis.

14

DISCUSSION QUESTIONS

1. Go to a sandy area where the surface can be smoothed, then perform the following behaviors in the area:
   a. Walking forward with your back parallel to the ground (stooped).
   b. Limping with an "injured" right leg.
   c. Walking backward with your feet pointing in the same direction as when you walked forward.
   d. Running full speed across the sand.
   e. Walking on hands and knees.
   What qualitative and quantitative differences did you see as a result of your experiments? Which results do you think would have low preservation potential, and why?

2. Calculate your own foot stress, using your weight in kilograms and measurements of your foot width and length in a shoe (right or left). After calculating that value, measure your foot without a shoe. How much does the foot-stress value change when you use the shoeless measurements?

3. Calculate the foot-stress values for a bipedal dinosaur track, assuming that the dinosaur weighed 83 kg and had a foot area of 374 cm^2. What are potential sources of error for your calculation, especially when considering the taphonomic factors that might have affected the track?

4. Calculate your leg length/foot length ratio by measuring each and calculating the value. For leg length, measure from where the femur inserts into the hip. Compare results with your classmates and calculate a mean ratio for the entire class. How different is your personal value from the class mean? How different is the class mean from that derived for dinosaurs?

5. You find a trackway of a probable ornithopod where it was moving bipedally. Only two tracks are preserved, the left and right footprints. What was its estimated speed, using a footprint length of 52 cm and pace of 204 cm? Calculate the variations in your answer by using the different ratios of 4.0, 5.0, and 5.5. Did the speeds become faster or slower? Explain why.

6. Think of examples of animals that move together as a coordinated group, whether on land, in the water, or in the air. How closely spaced are they to one another with regard to their body widths? How would you test the hypothesis that in a fossil trackway one individual (a "leader") initiated any changes in direction of the group?

7. What is an explanation for site fidelity of dinosaur nests, when dinosaurs were clearly capable of moving to a variety of environments for their egglaying? (Hint: Think about how known dinosaur nests were constructed.)

8. Dinosaur toothmarks that only penetrated flesh have not been described from the fossil record because of their low preservation potential. What parts of a prey animal (such as a herbivorous dinosaur) might have had the thickest amount of flesh covering the bones and thus

DISCUSSION QUESTIONS Continued

would not have preserved deep bites? Conversely, what parts of prey animals would have had the thinnest amount of flesh covering their bones and would more likely preserve toothmarks?

9. How would you explain why a suspected dinosaur coprolite does not show any coprophagy traces? Develop at least two hypotheses. What subsequent data could disprove each one?

Bibliography

Alexander, R. M. 1976. Estimates of speeds of dinosaurs. *Nature* **261**: 129–130.

Avanzini, M. 1998. Anatomy of a footprint: Bioturbation as a key to understanding dinosaur walk dynamics. *Ichnos* **6**: 129–139.

Bird, R. T. 1944. Did *Brontosaurus* ever walk on land? *Natural History* **53**: 60–67.

Bromley, R. G. 1996. *Trace Fossils: Biology, Taphonomy and Applications*. (2nd Edition). London: Chapman and Hall.

Brown, T., Jr. 1999. *Tom Brown's Science and Art of Tracking: Nature's Path to Spiritual Discovery*. New York: Berkley Books.

Campbell, H. W. 1972. Ecological or phylogenetic interpretation of crocodilian nesting habits. *Nature* **238**: 404–405.

Carpenter, K. 1999. *Eggs, Nests, and Baby Dinosaurs: A Look at Dinosaur Reproduction (Life of the Past)*. Bloomington, Indiana: Indiana University Press.

Carpenter, K., Hirsch K. F. and Horner, J. R. (Eds). 1994. *Dinosaur Eggs and Babies*. Cambridge, UK: Cambridge University Press.

Chiappe, L. M., Coria, R. A., Dingus, L., Jackson, F., Chinsamy, A. and Fox, M. 1998. Sauropod dinosaur embryos from the Late Cretaceous of Patagonia. *Nature* **396**: 258–261.

Chin, K. 1997. "What did dinosaurs eat? Coprolites and other direct evidence of dinosaur diets". *In* Farlow, J. O. and Brett-Surman, M. K. (Eds), *The Complete Dinosaur*. Bloomington, Indiana: Indiana University Press. pp. 371–382.

Chin, K., and Gill, B. D. 1996. Dinosaurs, dung beetles, and conifers: participants in a Cretaceous food web. *Palaios* **11**: 280–285.

Chin, K., Eberth, D. A. and Sloboda, W. J. 1999. Exceptional soft-tissue preservation in a theropod coprolite from the Upper Cretaceous Dinosaur Park Formation of Alberta. *Journal of Vertebrate Paleontolog*, **19**, **3**: Suppl. 37A.

Chin, K., Tokaryk, T. T., Erickson, G. M. and Calk, L. C. 1998. A king-sized theropod coprolite. *Nature* **393**: 680–682.

Cohen, A. S., Halfpenny, J., Lockley, M. G. and Michel, E. 1993. Modern vertebrate tracks from Lake Manyara, Tanzania and their paleobiological implications. *Paleobiology* **19**: 433–458.

Currie, P. J. 1997. "Gastroliths". *In* Currie, P. J. and Padian, K. (Eds), *Encyclopedia of Dinosaurs*. New York: Academic Press. p. 270.

Currie, P. J. and Jacobsen, A. R. 1995. An azhdarchid pterosaur eaten by a velociraptorine theropod. *Canadian Journal of Earth Sciences* **32**: 922–925.

Elbroch, M. and Marks, E. 2001. *Bird Tracks and Sign*. Mechanicsburg, Pennsylvania: Stackpole Books.

Erickson, G. M., Van Kirk, S. D., Jinntung S., Levenston, M. E., Caler, W. E. and Carter, D. R. 1996. Bite-force estimation for *Tyrannosaurus rex* from tooth-marked bones. *Nature* **382**: 706–708.

14

Farlow, J. O. and Chapman, R. E. 1997. "The scientific study of dinosaur footprints". *In* Farlow, J. O. and Brett-Surman, M. K. (Eds), *The Complete Dinosaur*. Bloomington, Indiana: Indiana University Press. pp. 519–553.

Gillette, D. D. 1995. True grit. *Natural History*, **104(6)**: 41–43.

Horner, J. R. 2000. Dinosaur reproduction and parenting. *Annual Review of Earth and Planetary Sciences*, **28**: 19–45.

Hunt, A. P., Chin, K. and Lockley, M. G. 1994. "The palaeobiology of vertebrate coprolites". *In* Donovan S. K. (Ed.), *The Palaeobiology of Trace Fossils*. Chichester: John Wiley & Sons. pp. 221–240.

Irby, G. V. 1996. "Paleoichnological evidence for running dinosaurs worldwide". *In* Morales, M. (Ed.), *The Continental Jurassic*. Bulletin 60, Museum of Northern Arizona, Flagstaff, Arizona, pp. 109–112.

Jacobsen, A. R. 1997. "Tooth marks". *In* Currie, P. J. and Padian, K. (Eds), *Encyclopedia of Dinosaurs*. New York: Academic Press: pp. 738–739.

Leonardi, G. (Ed.) 1987. *Glossary and Manual of Tetrapod Footprint Palaeoichnology*. Departmento Nacional da Producão Mineral, Brazil.

Lockley, M. G. 1990. *Tracking Dinosaurs: A New Look at an Ancient World*. Cambridge, UK: Cambridge University Press.

Lockley, M. G. 1997. "The paleoecological and paleoenvironmental utility of dinosaur tracks". *In* Farlow, J. O. and Brett-Surman, M. K. (Eds), *The Complete Dinosaur*. Bloomington, Indiana: Indiana University Press. pp. 554–578.

Lockley, M. G. and Hunt, A. P. 1995. *Dinosaur Tracks and Other Fossil Footprints of the Western United States*. New York: Columbia University Press.

Lull, R. S. 1953. *Triassic life of the Connecticut Valley*. State Geological and Natural History Survey of Connecticut Bulletin 81.

Manley, K. 1992. Surface polish measurements from bona fide and suspected sauropod dinosaur gastroliths, wave and stream transported clasts. *Ichnos* **2**: 167–169.

Mikhailov, K. E. 1997. "Eggs, eggshells, and nests". *In* Currie, P. J. and Padian, K. (Eds), *The Encyclopedia of Dinosaurs*. San Diego, California: Academic Press. pp. 205–209.

Norell, M. A., Clark, J. M., Chiappe, L. M. and Dashzeveg, D. 1995. A nesting dinosaur. *Nature* **378**: 774–776.

Over, D. J. 1995. An exercise on dinosaur trackways for introductory science courses. *Journal of Geological Education* **43**: 204–206.

Pemberton, S. G. and Frey, R. W. 1991. William Buckland and his "coprolitic" vision. *Ichnos* **1**: 317–325.

Rezendes, P. 1999. *Tracking and the Art of Seeing: How to Read Animal Tracks and Signs* (2nd Edition). Charlotte, Vermont: Camden House Publishing, Inc.

Sarjeant, W. A. S. 1990. "A name for the trace of an act: approaches to the nomenclature and classification of fossil vertebrate footprints". *In* Carpenter, K. and Currie, P. J. (Eds), *Dinosaur Systematics: Approaches and Perspectives*. Cambridge, U.K.: Cambridge University Press. pp. 299–307.

Stokes, W. L. 1987. Dinosaur gastroliths revisited. *Journal of Paleontology* **61**: 1242–1246.

Thulborn, R. A. 1990. *Dinosaur Tracks*. London: Chapman Hall.

Thulborn, R. A. 1991. Morphology, preservation and palaeobiological significance of dinosaur coprolites. *Palaeogeography, Palaeoclimatology, Palaeoecology* **83**: 341–366.

Varricchio, D. J. 1999. Gut contents for a Cretaceous tyrannosaur: implications for theropod digestive tracts. *Journal of Vertebrate Paleontology* **19**, 3: Suppl. 82A.

Varricchio, D. J., Jackson, F., Borlowski, J. and Horner, J. R. 1997. Nest and egg clutches for the theropod dinosaurs *Troodon formosus* and evolution of avian reproductive traits. *Nature* **385**: 247–250.

Chapter 15

During your study of dinosaurs you have frequently encountered the phrase "birds are dinosaurs." You may start to think of possible lines of evidence to disprove this statement, including body fossils, trace fossils, paleobiogeography, modern genetics, and behavioral ecology.

What information would convincingly falsify the currently reigning hypothesis about bird origins, which suggests that dinosaurs are still here today? Conversely, what types and amounts of evidence would convince you to conditionally accept the hypothesis? How do modern birds provide clues regarding their possible dinosaurian ancestry, and what behaviors do they show that are unknown in dinosaurs?

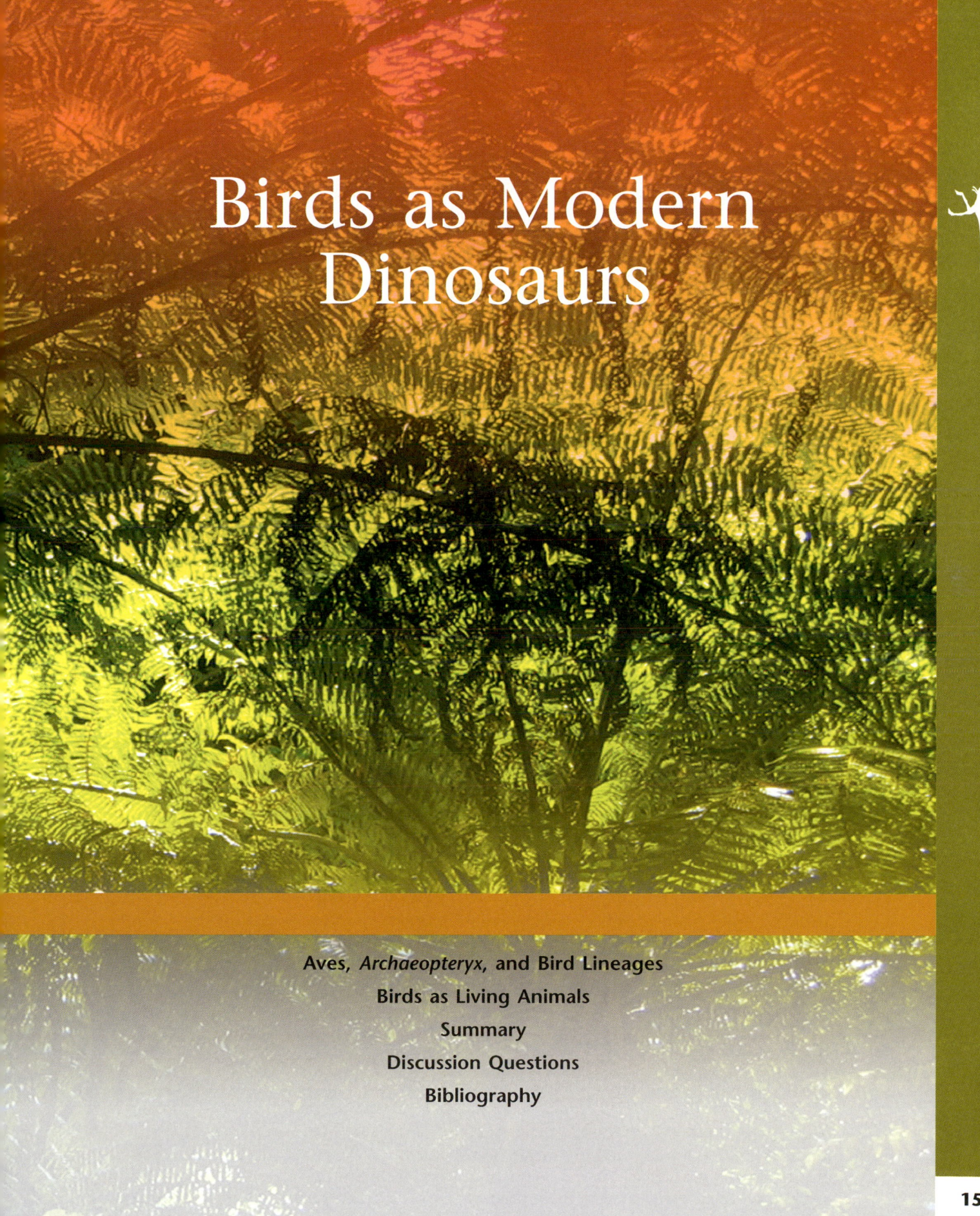

Birds as Modern Dinosaurs

15

Why Are Birds Dinosaurs?

The term "dinosaur" has been defined in several different ways in this book (Chapters 1 and 5). One of the recently proposed ones, which aids in interrelating most of the preceding chapters with this one, is:

> *An animal that is a member of the group descended from the most recent common ancestor of* Triceratops *and birds.*

The reason why *Triceratops*, a ceratopsian (Chapter 13), is singled out in this definition is because it represents the most advanced ornithischian, and birds represent the most advanced saurischians. Saurischians and ornithischians are different clades of dinosaurs, but they diverged from a common ancestor. Hence, whenever their most recent common ancestor lived (probably in the Middle to Late Triassic) is also when dinosaurs as a clade began. This application of phylogenetic methods results in a geologic range of dinosaurs from Late Triassic to the present, not from the Late Triassic to the end of the Cretaceous. Consequently, dinosaurs did not become extinct – they are still here today as birds.

The purpose of this chapter is to show that dinosaurs and birds are now intertwined topics that can provide perspectives on the past, present, and future of dinosaur studies. The long history of dinosaurs is intrinsically connected to modern birds because birds are dinosaurs just as much as humans are mammals. Moreover, observations of modern birds lend insights on how their extinct non-avian cousins may have lived and especially how they behaved. Yet another consideration is how the 145-million-year evolutionary history of birds is related to subjects other than dinosaurs, such as environmental changes during the latter part of the Mesozoic Era and mass extinctions at the end of the Cretaceous (Chapter 16). Consequently, birds comprise a crucial topic within dinosaur studies, one that helps to complete a picture of the evolutionary history of dinosaurs.

Aves, *Archaeopteryx*, and Bird Lineages

Birds share a large number of synapomorphies with non-avian dinosaurs, which means that their node-based clade, Aves, is within the Dinosauria, Saurischia, Theropoda, Tetanurae, Coelurosauria, Maniraptiformes, Maniraptora, and the stem-based clade Aviale (Fig. 15.1). Aviale is defined as including all living birds and maniraptorans more related to them than the dromaeosaur *Deinonychus*, so it includes a few non-avian theropods. This means that Dromaeosauridae and Aviale are sister

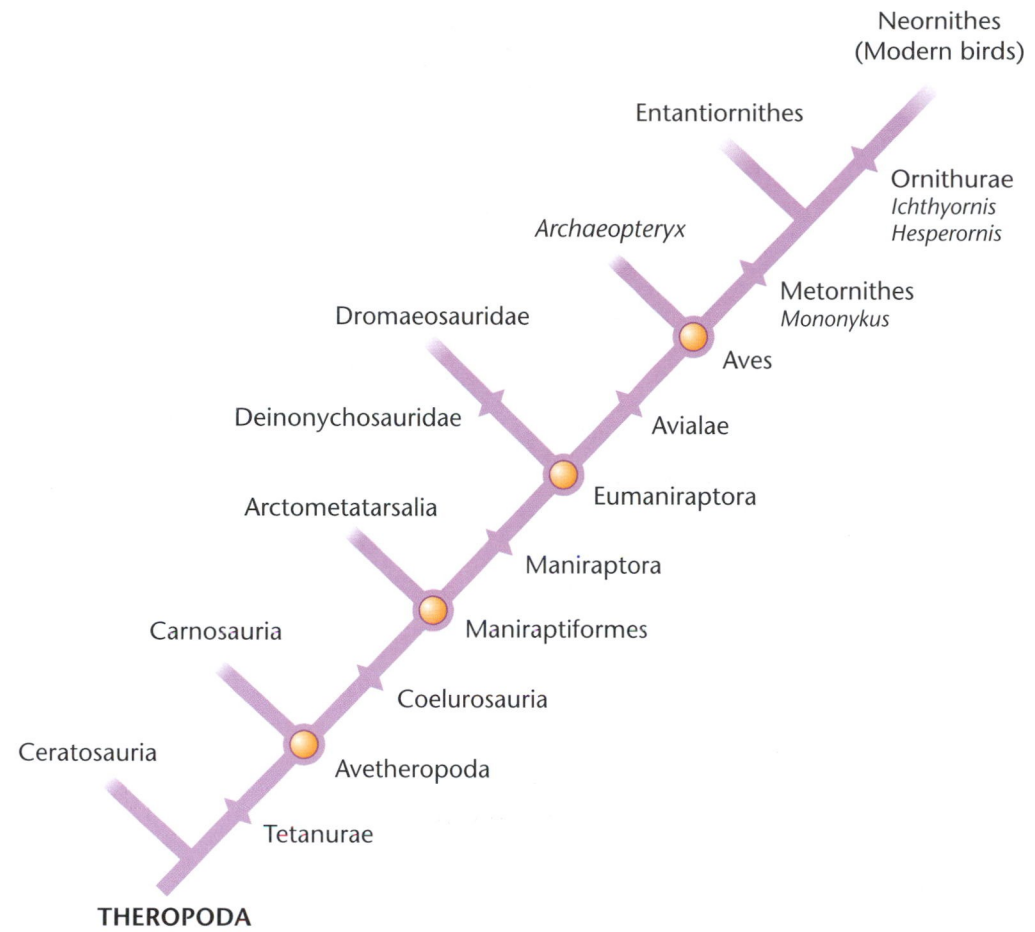

FIGURE 15.1 Cladogram showing the lineage within Theropoda leading to Aves (birds) and subsequent clades nested within Aves.

clades. More specifically, birds are defined cladistically as *Archaeopteryx lithographica* of the Late Jurassic (Chapters 3, 6, and 9) and all descendants of their most recent common ancestor. The important point about this definition is that *Archaeopteryx*, an exemplary "transitional" fossil (Chapter 6), is not considered to be the common ancestor of all birds, but rather is the most basal bird known. No other contenders for the superlative appellation of "oldest bird," also thought to have lived in the Late Jurassic, have been verified yet. However, tracks similar to those made by undoubted avians have been described from Late Triassic strata of Argentina. The large time gap between these tracks and *Archaeopteryx* suggests small non-avian theropods as the tracemakers, but research on this hypothesis was still being conducted at the time of this writing.

Characters of Aves (Fig. 15.2) include:

- Reduction of the number of caudal vertebrae to 23 or less. Their tails became shorter, which fused into a small structure called a **pygostyle** (although also present in a few non-avian dinosaurs).
- A forearm that is more than 90% of the length of the humerus and a fore limb length considerably longer (more than 120%) than that of the hind limb, which shows the tendencies of these limbs' use for flight.

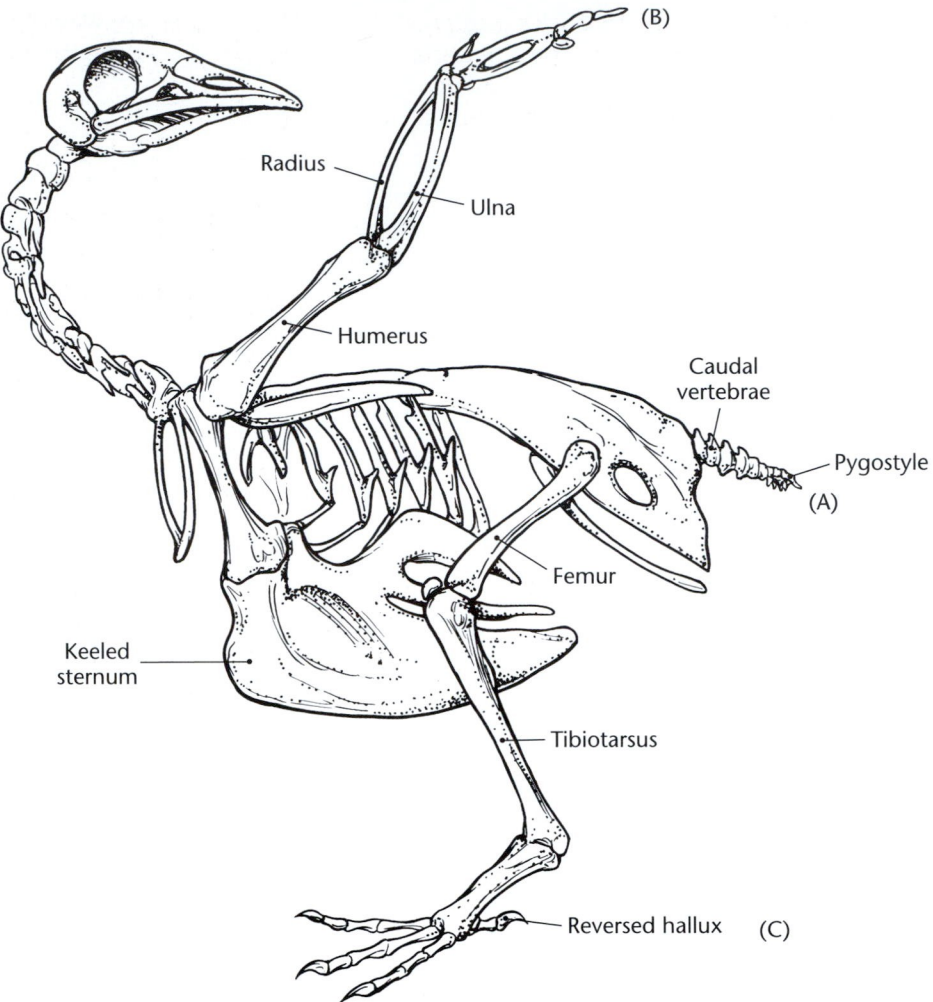

Radius

Ulna

(B)

Humerus

Caudal
vertebrae

Pygostyle

(A)

Femur

Keeled
sternum

Tibiotarsus

Reversed hallux   (C)

FIGURE 15.2 A few characters defining Aves (birds): (A) reduction of caudal vertebrae into a pygostyle; (B) forearm more than 90% of the length of the humerus and forelimb considerably longer (more than 120%) than the hindlimb; (C) anisodactyl foot, with a reversed hallux adapted for perching. Notice also the keeled (carinate) sternum and elongated coracoids.

■ An anisodactyl foot, with three forward-pointing digits (II through to IV) and a reversed hallux (digit I) adapted for perching.

Keep in mind that these characters are added to the previously mentioned characters of the theropod ancestors of birds, which means that this condensed list does not come close to describing what is defined as a bird. For example, the possession of feathers used to be a primary criterion for identification of an animal as a bird, especially under the Linnaean classification scheme, but numerous discoveries of non-avian feathered theropods have revoked this single-character identifier. Instead, feathers can be viewed as a possible plesiomorphy in birds and the few theropods that shared a common coelurosaurian ancestry. Feathered non-avian theropods (Chapter 9) include *Beipiaosaurus*, *Caudipteryx*, *Microraptor*, *Protarchaeopteryx*, *Sinornithosaurus*, *Sinosauropteryx*, and others, all from the Early Cretaceous of China. A more inclusive trait is the possession of low-density pneumatic bones, evident

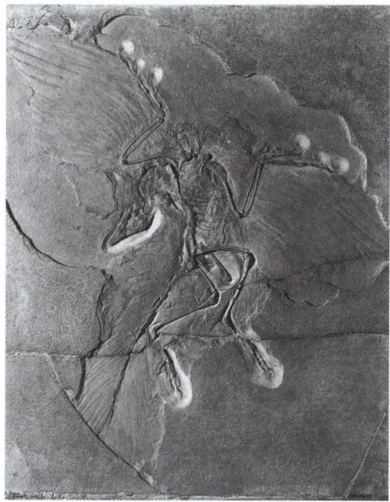

FIGURE 15.3 The Late Jurassic bird *Archaeopteryx lithographica* of the Solnhofen Limestone, Bavaria, Germany. Compare with Figure 2.4. Humboldt Museum für Naturkunde, Berlin/Peabody Museum of Natural History, Yale University.

in both theropods and birds (Chapter 9). Modern birds also have a number of air sacs throughout their bodies that help to lower their density and thus aid in flight.

As far as *Archaeopteryx* is concerned, it was small by dinosaurian standards, especially in the Late Jurassic when it shared the landscape with massive sauropods (Chapter 10). It is about 45 cm long, probably weighed less than half of a kilogram, and is comparable to the size of a large crow (Fig. 15.3). *Archaeopteryx* shows a mix of features normally associated with non-avian theropods and birds exclusively:

- A tail composed of an intermediate number of caudal vertebrae, which makes it short for a theropod but long for a bird.
- Claws on its fore limbs identical to those of some theropods, but not seen on most birds.
- Flight feathers connected to the fore limbs identical to those seen in most birds, but not in most theropods.
- Teeth in the jaws that are atypical of most birds, but non-serrated, which is atypical for theropods.
- A furcula ("wishbone") that represents a fusing of the clavicles, present only in a few theropods (such as some maniraptors), but in all birds.

Archaeopteryx also recently had its head examined and was found to be, for all practical purposes, "bird-brained." This investigation used CT scans (Chapter 4) to develop a finely resolved three-dimensional picture of the braincase for one specimen of *Archaeopteryx*, and the results showed that its brain was much more akin to that of a modern bird than a reptile.

Archaeopteryx was found in a fine-grained limestone, the Solnhofen Limestone of Bavaria, Germany. This deposit probably formed in a lagoon and dates to about 152 Ma in the Late Jurassic. Seven specimens and one feather represent *Archaeopteryx*; the feather, which was found in 1860, is presumed to be from this specimen because no other feather-bearing fossils have been found in the Solnhofen. The first complete skeleton with feather impressions was found in 1861, and its discovery contributed to then-raging debates about evolutionary theory prompted by Darwin's publication of *On the Origin of Species* in 1859 (Chapter 3). The other specimens were found at various times from 1877 to 1992. Interestingly, one of the specimens had been mislabeled in a museum collection as a pterosaur; John Ostrom realized its actual identity when he first saw it in 1970 (Chapter 3). In a similar manner, yet another specimen was discovered three years later when

15

FIGURE 15.4 Anatomy of a typical flight feather.

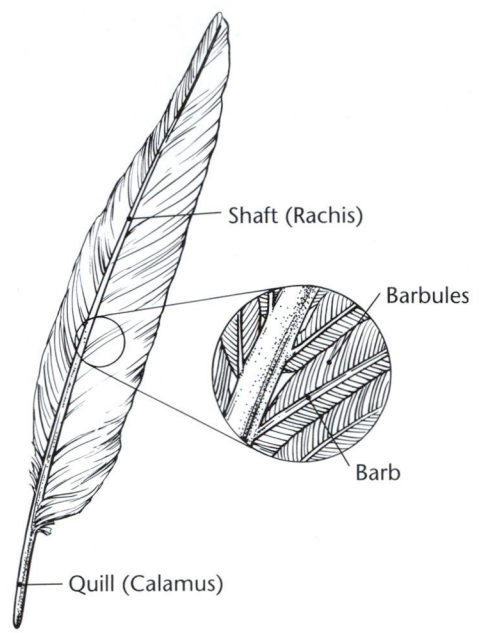

Shaft (Rachis)

Barbules

Barb

Quill (Calamus)

a closer examination of a skeleton initially identified as the small, Late Jurassic coelurosaur *Compsognathus* revealed that it was actually *Archaeopteryx*. Thus far, no other species of undoubted Late Jurassic birds are documented, making *Archaeopteryx* the sole known product of bird evolution in the Jurassic.

Of course, the most intriguing of the many interesting traits of *Archaeopteryx* are its feathers, which were originally recognized for their scientific significance and are still the subject of debate today. The reason for this interest is that the feathers of *Archaeopteryx* have the appearance of those seen in most modern birds. In modern birds, a feather is a keratinized integument that originates in a layer of skin below the surface of the animal; these feathers can be broadly categorized as **downy feathers** and **flight feathers** (Fig. 15.4). The primary function of downy feathers is insulation, a property exploited by humans who use these feathers as fill in winter coats and sleeping bags. In contrast, flight feathers are used mainly for aiding lift and descent by creating surfaces that move air in ways conducive to controlled flight. Some flightless birds still retain flight feathers as a primitive (but vestigial) trait, which indicates their descent from flighted ancestors.

A typical feather consists of a central, hollow **shaft** that terminates proximally into a **quill**. The shaft has **barbs** that branch from it at about 180° from one another (on opposite sides of the shaft). The barbs interlock through smaller **barbules** in a fashion similar to the teeth in a zipper. In a flight feather, the barbs collectively form a planar structure called a **vane**, creating air foils that help considerably in the aerodynamic ability of flighted birds. Downy feathers are relatively less organized, and the barbs will radiate in seemingly random directions, forming ill-defined vanes or no vanes at all.

Like any other structure, both types of feathers have more than one function. For example, bright colors and varied patterns contribute to intraspecies displays. The multiple uses of feathers can at least partially explain why flightless theropods such as *Caudipteryx* would have color banding evident in feathers that composed their tail fan. This banding is the only known direct evidence of coloration in dinosaurs. The hair-like dorsal fringe seen in *Sinosauropteryx* was probably composed of downy feathers, which were not used for flight either. This evidence suggests that the flight feathers seen in *Archaeopteryx* may have evolved for the purposes of display or insulation first, then flight later.

(A)

(B)

FIGURE 15.5 *Confuciusornis sanctus*, an Early Cretaceous bird from China. (A) Fossil specimen, with carbonized margin indicating presence of feathers. (B) Reconstruction of living animal. Note the prominent digits on the wings, indicating a primitive condition. Naturhistoriches Museum Basel, Basel, Switzerland.

Compared to *Archaeopteryx*, the most primitive bird interpreted so far is *Rahonavis ostromi*, found in Upper Cretaceous rocks of Madagascar. Its unusual mixture of maniraptoran and avian traits led to some controversy over whether it actually represents a bird or not, and its geologic age (well after the Late Jurassic) contributed to this skepticism. A clade of primitive birds that comprises a sister clade to all birds other than *Archaeopteryx* and *Rahonavis* is **Confuciusornithidae**. This clade is represented by abundant specimens of its namesake, *Confuciusornis sanctus* (Fig. 15.5) as well as *Changechengornis hengdaoziensis*. As might be surmised from their names, both species are from China, and they come from the same Lower Cretaceous formation.

Judging from the numerous and varied species found so far, Aves diversified considerably throughout the Cretaceous, suggesting that their evolution was relatively rapid and their ecological niches became more specifically definable. Some forms had definitely achieved full flight and probably had arboreal lifestyles, whereas others adapted to new habitats such as semi-arid inland areas and shallow marine regions. For example, avians are likely the tracemakers of bird-like tracks in some of the oldest Cretaceous strata of Spain, and these tracks are interpreted as having been made by shorebirds. This implies that birds had already radiated to such habitats

15

within 10 million years of *Archaeopteryx*. A few presumed dinosaur eggs and nests have also been suspected of actually belonging to avians; some Late Cretaceous eggshells closely match those known from birds. The oldest embryonic avian remains, found in Mongolia, are also Late Cretaceous and probably belong to the bird *Gobipteryx minuta*. Overall, the fossil record for birds improves dramatically in Cretaceous deposits, in comparison to their extreme rarity in Jurassic strata. Additionally, discoveries of the last 25 years in particular have added exponentially to unraveling the evolution of Cretaceous birds. At this writing, more than 50 genera of birds had been identified from Cretaceous strata, hailing from every continent, except Antarctica, and contributing to the ever more complicated cladograms which change with each new discovery (Table 15.1).

TABLE 15.1 Cretaceous birds, their approximate geologic ages, and general localities.

Genus	Age	Geographic Location
Alexornis	Late Cretaceous	Mexico
Apatornis	Late Cretaceous	Western USA
Apsaravis	Late Cretaceous	Mongolia
Avisaurus	Late Cretaceous	Western USA
Baptornis	Late Cretaceous	Western USA
Cathayornis	Early Cretaceous	China
Changchengornis	Early Cretaceous	China
Chaoyangia	Early Cretaceous	China
Confuciusornis	Early Cretaceous	China
Coniornis	Late Cretaceous	Western USA
Enaliornis	Early Cretaceous	UK
Eoalulavis	Early Cretaceous	Spain
Eocathayornis	Early Cretaceous	China
Eoenantiornis	Early Cretaceous	China
Gargantuavis	Late Cretaceous	France
Gobipterx	Late Cretaceous	Mongolia
Halomornis	Late Cretaceous	Eastern USA
Hesperornis	Late Cretaceous	Western USA
Iberomesornis	Early Cretaceous	Spain
Ichthyornis	Late Cretaceous	Western and Eastern USA
Jibenia	Early Cretaceous	China
Kizylkumavis	Late Cretaceous	Uzbekistan
Kuszholia	Late Cretaceous	Uzbekistan
Lectavis	Late Cretaceous	Argentina
Liaoningornis	Early Cretaceous	China
Nanantius	Early Cretaceous	Australia
Neuquenornis	Late Cretaceous	Argentina
Noguerornis	Early Cretaceous	Spain
Otogornis	Early Cretaceous	China
Parahesperornis	Late Cretaceous	Canada
Patagopteryx	Late Cretaceous	Argentina
Protopteyrx	Early Cretaceous	China
Rahonavis	Late Cretaceous	Madagascar
Sinornis	Early Cretaceous	China
Soroavisaurus	Early Cretaceous	Argentina

Although discussion still continues about whether Archaeopteryx was capable of self-powered flight (that is, flapping its wings instead of merely gliding), the Early Cretaceous Sinornis of China has characteristics closely associated with full flight.

One of the key features used to determine whether a bird flew is the degree of development seen in its sternum, to which the large flight muscles are attached. These sterna can be bony, cartilaginous, or completely absent; the latter two conditions are coincident with flightlessness. A bladed appearance to the middle of a sternum is a **keel** (also known as **carina**), analogous to the central ridge on the bottom of a boat. Once the keel is well-developed, the sternum is called **carinate**. *Archaeopteryx* has a mildly carinate sternum, whereas the Early Cretaceous *Sinornis* of China has a proportionally more carinate sternum. When coupled with elongated coracoids in the shoulder region (for the further attachment of flight muscles), this evidence advocates *Sinornis* as a flyer. Yet *Sinornis* and many other Cretaceous birds, such as the Early Cretaceous *Iberomesornis* of Spain, still retained teeth and other primitive traits that reflected their dinosaurian heritage.

Other Cretaceous birds varied from fully flighted to flightless varieties. For example, the very odd, wingless, cursorial *Mononykus* from the Upper Cretaceous of Mongolia was first classified as a non-avian theropod, but is now placed within Aviale and is considered more closely related to *Archaeopteryx* than to most other maniraptorans. Other flightless Late Cretaceous birds included the tern-like shorebird *Ichthyornis* and the toothed, marine diving bird *Hesperonis* (Fig. 15.6). Thus, birds certainly had taken to the air and shared flight time with pterosaurs during the Cretaceous, but they also ran in the same deserts as dromaeosaurs and swam in the same waters as plesiosaurs and other marine reptiles (Chapter 6). This expansion of habitats for birds is all the more remarkable because it happened during the last 70 million years of the Mesozoic, when most dinosaurs of the previous 90+ million years apparently spent all of their time firmly on the ground.

Of all the clades of Cretaceous birds, none of those with toothed species, such as **enantiornithines**, survived into the Tertiary Period. Although the fossil record for birds has improved considerably in recent years, inadequate information is available

FIGURE 15.6 Skeleton of *Hesperornis regalis*, a Late Cretaceous diving bird recovered from marine deposits in Kansas, and artistic reconstruction behind it. Note the vestigial wings, indicating secondary flightlessness in a Cretaceous bird. Sam Noble Oklahoma Museum of Natural History, Norman, Oklahoma.

to figure out when declines occurred in bird populations. Nevertheless, all lineages with the exception of the **neornithines**, known generally as "modern birds," went extinct either before or at the end of the Cretaceous Period. Why only these birds and no others made it into the Tertiary is unknown. One commonality of Late Cretaceous neornithines is that most were apparently shorebirds. This habitat preference may have had a survival advantage for whatever events happened toward the end of the Cretaceous and beginning of the Tertiary Periods (Chapter 16). Alternatively, the presence of these species in the fossil record may be a result of a preservation bias in the form of the more frequent burial of nearshore species. Taphonomy is the filter through which all interpretations of the fossil record necessarily must be made (Chapter 7), and birds in particular are difficult to preserve as body fossils because of their often small, hollow bones. As a result, the fossil record for birds is not expected to be very rich, making the discoveries of recent years from the Late Cretaceous all the more remarkable.

Bird Ancestors: Theropod Hypothesis and the Origin of Avian Flight

The most widely accepted hypothesis for bird ancestry in the Mesozoic is that certain lineages of small theropods, in combination with environmental factors that affected natural selection of these theropods, resulted in the evolution of birds by the Late Jurassic. As mentioned before, the shining example used as evidence in this evolutionary scenario is *Archaeopteryx*, known as the long-presumed link between reptilian ancestors and avian descendents. A theropod ancestry of *Archaeopteryx* is interpreted on the basis of its numerous anatomical features allied with theropods (Chapter 9):

- Both upper and lower jaws bearing pointed teeth.
- Tridactyl manus with digits I through to III, digit II the longest of the three, ending in claws.
- Semilunate (half-moon shaped) carpal in the wrist.
- V-shaped furcula.
- Ankle with differentiated (unfused) metatarsals II through to IV.
- Ascending process on the astragalus.
- Tridactyl pes symmetrical around digit III, with digit I retroverted (aniso-dactyl) and well-developed claws on all digits.
- Saurischian pelvis with long pubis.
- Gastralia.
- Six unfused sacral vertebrae.
- Moderately long tail (about 25 caudal vertebrae), with elongate processes (zygopophyses) that interlock to stiffen it.

Modern birds form a contrast:

- No teeth.
- Forearms where the carpals, metacarpals, and phalanges fuse into a **carpo-metatarsus** (the distal end of a chicken wing shows this structure quite well).
- A **synsacrum**, where the pelvic bones fuse with the sacral vertebrae.
- A pygostyle (although a few non-avian maniraptorans also have this feature).
- Fusion of the metatarsals (anklebones) into a **tarsometatarsus**.

Obviously, *Archaeopteryx* is not just an ordinary bird. Its teeth, unfused bones in the forearm, manus with phalanges and claws, unfused sacrum, long tail, and unfused bones in the ankle, along with many other traits, all point to its classification as a theropod. Yet it is also a bird because of modifications to this theropod body plan that represent novel traits.

On the basis of the previously mentioned traits for *Archaeopteryx*, Aves and Deinonychosauria are hypothesized as having a common ancestor from the node-based clade Coelurosauria. Depending on how a cladogram for birds and their theropod relatives is arranged, some predatory theropods, such as the Early Cretaceous *Deinonychus*, *Dromaeosaurus*, *Utahraptor*, and Late Cretaceous *Velociraptor*, are probably part of a sister clade to birds. A possible point of confusion is that, because of this common ancestry, both deinonychosaurs and primitive birds may have been feathered. This hypothesis is supported by the discovery of one feathered deinonychosaur, the Early Cretaceous *Sinornithosaurus* of China. This circumstance does not mean that birds were the ancestors of deinonochysaurs, but rather that they descended from the same ancestor and later became contemporaries. The largest problems with the theropod–bird hypothesis do not lie in working out whether birds evolved from theropods; a detailed comparative analysis reveals that more than 100 characters are shared by coelurosaurian dinosaurs and avians. The questions that are still unanswered are when and how birds evolved from theropod ancestors.

The "when" part of the theropod–bird question is probably easier to answer in a preliminary way. The stratigraphic position of *Archaeopteryx* indicates a minimum age of Late Jurassic for the evolution of birds. This suggests that the most immediate ancestors of birds may have originated during the Middle Jurassic or the earliest part of the Late Jurassic, with divergence from a hypothetical coelurosaur (probably maniraptoran) ancestor. A few fragmentary maniraptoran remains have been found in Late Jurassic deposits of North America, indicating a maniraptoran presence on two continents at that time. However, the lack of more complete, identifiable maniraptoran specimens is particularly vexing in this respect, because a gap results in the fossil lineage of coelurosaurs to avians. Nevertheless, the characters of *Archaeopteryx* with relation to other theropods and its stratigraphic position serve as a predictor not only for where in geologic time these avian ancestors lie, but also for what they should look like.

The question of how flighted birds evolved from flightless theropods is a rather contentious debate that may not be resolved in the near future. This pessimistic assessment acknowledges that the majority of data supporting the competing hypotheses are based on inferences gained from functional morphology and biomechanical analyses. Because the morphological features that define birds are intrinsically linked to adaptations for flight, the causes for the evolution of flight in certain theropods must be considered. As a result, two main hypotheses have been proposed for the origin of bird flight:

1 the **arboreal hypothesis**; and
2 **cursorial hypothesis**.

The arboreal hypothesis, known colloquially as the "trees-down" hypothesis, states that small, feathered theropods climbed into trees, evolved into gliding forms, and eventually gave rise to forms capable of full flight. The cursorial hypothesis, also known as the "ground-up" hypothesis, postulates that small, fast-moving, bipedal theropods flapped their feathered arms while pursuing and swatting at prey such as insects, which eventually led to jumping, short flights, then full flight in later descendants. Both hypotheses presume that feathers evolved from the skins of theropods in some rudimentary state as "proto-feathers" that were not associated with flight, thus meaning that the feathers were exaptations. Exaptations are inheritable traits that were already favorably adapted for a selective pressure before it happened (Chapter 6), meaning that these proto-feathers did not evolve specifically

15

for flight, but later helped with flying. After all, although feathers are useful for flighted birds today, they are not a prerequisite for flying ability in animals. For example, among tetrapods, pterosaurs and bats both independently developed full flying capabilities without the benefit of feathers. Consequently, theropods could have had feathers well before natural selection was applied in either an arboreal or cursorial scenario, and the feathers could have served an entirely different function.

Of these two hypotheses, the cursorial hypothesis was more popular with paleontologists until recently. It may have been more acceptable simply because it was less weak than the arboreal hypothesis. A turning point in this debate came about as a result of the recent discovery of three Early Cretaceous non-avian theropods that were apparently adapted for arboreal lifestyles. Two species, *Epidendrosaurus ninchengensis* and *Scansoriopteryx heilmanni*, are relatively small. They have an unusually long digit III on their hands, and long forearms, collectively. These traits would have served well for climbing trees. Another species, *Microraptor gui*, is also small (smaller than *Archaeopteryx*, in fact) and has feathers on all four limbs, which suggests that it was a glider. Adaptations for gliding are useless if the same animal also could not climb trees or similarly high objects, thus *Microraptor* is also assumed to have been a tree climber.

As just demonstrated, the debate over trees-down or ground-up origins for bird flight centers on functional morphology, and from such studies other apparent contradictions become apparent. For example, *Archaeopteryx* has claws on its manus that could have been used for climbing and feet that were seemingly adapted for perching. Nonetheless, its hallux is slightly too short for good perching, and it also has legs that were adapted for bipedal running. Likewise, its flight feathers are well developed, yet the bony sternum is poorly developed (in fact, only one specimen has a sternum preserved). Despite seven skeletons, some of them exquisitely preserved, and an inordinate amount of study by many careful and brilliant paleontologists over the course of 145 years, the exact details about how *Archaeopteryx* lived or got to where it was in its evolutionary history are still being debated.

Another problem with relying on functional morphology for either hypothesis is that few actual experiments or observations of modern analogues are included to test the assumptions. For example, no modern birds swat at insects or any other prey with their feathered arms for food gathering while they run, which disfavors the cursorial hypothesis. Similarly, the only modern analogue that favors the arboreal hypothesis is represented by one species of bird that has claws on its wings adapted for climbing trees, the hoatzin of South America (*Opisthocomus hoazin*). The hoatzin only has this ability in its juvenile state, as the adults have fused digits in their manus, just like any other bird.

Other independent data, such as trace fossils or facies associations, have not been integrated to any great extent into arguments for either hypothesis. For example, if theropod ancestors of birds were indeed cursorial before short flights, then preserved trackways of such behavior would help to confirm that this happened. Unfortunately, arboreal theropods would have left far fewer tracks, and scratch marks left on trees would have had low preservation potential. Another problem would be distinguishing non-avian theropod tracks from bird tracks in Jurassic rocks. So far no Jurassic bird tracks have been recognized, although the aforementioned avian-like tracks from the Late Triassic of Argentina are intriguing clues to animals that had feet similar to those of modern birds. With regard to *Archaeopteryx*, if it was at least partially ground dwelling, its short hallux means that it probably would not have left an impression of this digit. As a result, only a typical tridactyl and presumably non-avian theropod footprint would be evident.

One other piece of evidence relating to the trees-down or ground-up hypotheses is the paleoenvironmental context for *Archaeopteryx*. Its exclusive occurrence in lagoonal deposits of the Solnhofen Limestone means that it may have been flying far away from any forested areas, which would favor the cursorial hypothesis. Alternatively, all of the specimens in the Solnhofen also could have floated into the lagoon from forested areas (the "bloat-and-float" hypothesis explained in Chapter 7), which would not have negated the arboreal hypothesis. Other evidence supporting the latter scenario is the occurrence of the probable arboreal non-avian theropods, *Epidendrosaurus*, *Scansoriopteryx*, and *Microraptor*, in lake deposits, which means that they may have floated out into a water body before sinking to the bottom and becoming part of the fossil record.

The preceding discussion was presented in the context of an "either-or" argument, but in science alternative hypotheses do not have to be limited to just two. A third hypothesis, which is actually a variation of the cursorial one, is that ground running was helped along by vigorous flapping that increased theropod running speeds. Such an adaptation certainly would have aided in predator avoidance, which would have been particularly important for small theropods in the Late Jurassic. Under this hypothesis, natural selection of these flapping "pre-avian" theropods would have progressively led to full, self-powered flight. The contrast between this model and the previous cursorial model is that one would have been used for predation, whereas the other would have been used for avoiding it.

Yet another modification of the cursorial hypothesis is actually a neat synthesis of it with the arboreal hypothesis, which calls for the evolutionary development of **wing-assisted incline running**. This hypothesis differs from the others in that it has incorporated much experimental data from modern birds (partridges), rather than theorizing based on functional morphology. These experiments showed that juvenile partridges were capable of running up steep inclines, including vertical tree trunks, journeys that were made easier by an energetic flapping of their wings. This method also would have been an excellent method for predator avoidance, particularly if the predators were non-avian theropods with relatively short arms. The researcher who documented this behavior also tested the effects of feather area on incline running by trimming the feathers to half their length or cutting them off completely. Birds without feathers could not run up slopes greater than 60°, and the half-feathered individuals also were 10–20° behind the fully-feathered in climbing ability. This study thus helps to explain how a "half-wing" in a theropod would still have an evolutionary advantage over "no wing." The results also changed the perspectives of paleontologists who had not been studying extant avian dinosaurs for clues of their evolutionary history.

Although birds are theropods, the exact mechanisms responsible for the evolution of flightless theropod lineages into flighted birds are still poorly understood, although they are becoming clearer with each fossil discovery. Indeed, anatomical data derived from non-avian maniraptorans, *Archaeopteryx*, and other primitive birds have clearly demonstrated the clear progression of the fore limbs, chest, and shoulder girdles, adaptations favoring self-powered flight. For example, ratios of armspans to body lengths of the feathered non-avian theropods *Sinosauropteryx*, *Protarchaeopteryx*, and *Sinornithosaurus* show a potential progression from leapers to gliders (where wider armspans correlate with "wingspans"). Furthermore, *Archaeopteryx* may have been either a glider or used self-powered flight, but it may have been surpassed in the latter respect by the non-avian theropod *Cryptovolans*, which had a better developed keel (Chapter 9). Doubtless the steps of this evolutionary process and its contributing factors will gain even more clarity with further study and new fossil discoveries.

Bird Ancestors: "Thecodont" Hypothesis

Debates about bird ancestry included a hypothesis that birds originated from archosaur lineages separate from dinosaurs. A group of archosaurs, previously called

thecodonts, was considered as a common ancestral group to crocodilians, pterosaurs, and dinosaurs (Chapter 6). Based on cladistics, this grouping is now understood to be paraphyletic, and the recently held understanding is that Archosauria had at least two clades split from it, Crurotarsi and Ornithodira. Crurotarsans gave rise to some extinct crocodilian-like animals such as phytosaurs, as well as lineages that led to modern alligators and crocodiles. Ornithodirans gave rise to both pterosaurs and dinosaurs, both of which arrived on the Mesozoic scene by the Late Triassic.

Although these relationships seem clear now, this was not always the case. For example, under previous Linnaean classifications, thecodonts were more or less a conglomerate of crurotarsans and ornithodirans, and their evolutionary relationships were poorly understood. The first well-reasoned proposal of a thecodont origin for birds, stated in 1926 by German paleontologist Gerhard Heilmann, was based on the then-factual lack of clavicles (furcula) in theropods. In contrast, some primitive Triassic archosaurs did have clavicles. By Heilmann's logic, using a principle first articulated by Louis Dollo (Chapter 3), a structure that is lost is not re-acquired in an evolutionary lineage. This would have meant by default that non-dinosaur archosaurs were the ancestors of birds. However, when some maniraptors were later found to have furculas, the premise of Heilmann's hypothesis was negated and the theropod-origin hypothesis was correspondingly strengthened. Similarly, subsequent hypotheses of either thecodont or other archosaur ancestors have been based on only a few morphological traits shared by birds and the proposed bird ancestors. These attempts have not withstood critical scrutiny. Although the theropod–bird lineage admittedly has some gaps, a thecodont–bird lineage has chasms.

Evidence that added fuel to this still-simmering controversy was the announcement that feather-like structures were found on the dorsal surface of *Longisquama insignis*, a small Late Triassic archosaur from central Asia. The structures superficially do look like feathers in that they have central shafts with symmetrical branching forming elongated vanes, and they seem to originate from the body of the animal. *Longisquama*, which has been known to paleontologists for about 35 years, was interpreted as a gliding animal because of these unusual structures and its lightly built skeleton. However, these structures were only recently interpreted as feathers. As of this writing, the topic of whether the structures are indeed feathers, some odd type of scales, or some previously unknown structure is still unresolved. Regardless, preliminary claims of the presence of feathers on one specimen of a Late Triassic archosaur unrelated to dinosaurs do not automatically erase the enormous amount of character data that support dinosaur–bird ancestry. Feathers do not make a one-character "magic bullet" that proves a relationship of any given fossil to birds. By analogy, hair-like structures described in the wings of pterosaurs are not construed to imply that they were the ancestors of modern bats.

One consensus view is that birds represent convergent evolution from separate lineages. In this scenario, a thecodont ancestry provided bird descendants and theropod ancestry provided other bird descendants. This compromise nevertheless conjures a more complicated scenario than if birds had originated from just one lineage. It requires a sort of faith that such a unique body plan could have evolved independently from different lineages within approximately the same span of geologic time. Considering that the theropod hypothesis is backed by a robust data set and the competing ones are not, and that the coincidence of the more than 100 characters shared by theropods and birds is impossible to ignore, the most logical course is to conditionally accept the theropod hypothesis.

An interesting variation on the non-avian–avian theropod hypothesis is that some Cretaceous flightless and feathered "non-avian" theropods were actually descended from avian ancestors. As a result, the evolutionary sequence would have been:

1 non-avian flightless theropod;
2 avian flighted theropod; and
3 avian flightless theropod.

However, cladistic analyses of *Archaeopteryx* and other avians indicate that these are more derived forms than the flightless theropods conjectured as their descendants, therefore casting doubt on such an ancestor–descendant relationship. Regardless, the affirmation of this hypothesis would only modify the currently supported scenario for avian descent from non-avian theropods.

Nonetheless, a word of caution is warranted for any paleontologists who too quickly embrace a hypothesis as completely confirmed. Human prejudice can influence how evidence is viewed, fitting it to the hypothesis rather than cultivating awareness of how it may not fit. In other words, people see what they want to see. For example, a few years ago paleontologists named *"Archaeoraptor,"* a new Early Cretaceous genus of dinosaur from China, on the basis of a single specimen that had a blend of half-deinonychosaur and half-avian features. This specimen was hailed as further confirming dinosaur–bird links and was given much publicity in the popular press before undergoing peer review. However, a more careful examination later revealed that it was a chimera: the posterior half of a dromaeosaur, later identified as *Microraptor*, had been pasted to the anterior half of a fossil bird, *Yanornis martini* (Chapter 9). This two-for-one specimen was not a hoax perpetrated by the paleontologists who described it, but nevertheless it was embarrassing to them and the magazine that first announced it. The lesson from this mistake is that the euphoria surrounding a potentially important fossil find is understandable, but healthy skepticism helps to prevent hasty interpretations that just happen to reaffirm a currently reigning hypothesis.

Bird Evolution in the Cenozoic

As mentioned earlier, birds had already filled numerous niches, including those in aquatic environments, by the end of the Cretaceous. Thus, the extinction of both ground-dwelling dinosaurs and aerial pterosaurs opened many more niches for birds and mammals by the early part of the Cenozoic. In spite of the extinction of all bird clades, except the neornithines, birds diversified quickly in these niches during the first 10 million years or so of the Tertiary Period. Shorebirds, similar to (and probable ancestors of) modern flamingos, herons, and ducks, were particularly common early in the Tertiary. The largest group of modern birds, the **passerines**, otherwise known as songbirds (wrens, larks, sparrows, warblers, chickadees, crows, jays, magpies, and so on), occupied most other niches available in terrestrial environments.

Passerines comprise about 60% of all known species of modern birds, numbering nearly 6000 species. Their evolution was most likely dependent on their vocalizations, which are key novelties linked to their reproductive cycles. The sounds of lambeosaurines from the Late Cretaceous can be reasonably inferred on the basis of their huge nasal chambers (Chapter 11), but fossil passerines of the past 20 million years, with their tiny and delicate bones, are not amenable to a similar analysis. Because their vocalizations did not fossilize, passerines are poor subjects for cladistic analyses, demonstrating in this instance that cladistics is only one tool available to a paleontologist, not a panacea. The diversification of passerines not only changed the sound of terrestrial ecosystems, but also affected the biogeographic dispersal of flowering plants, as many of these birds ate fruit and had other interactions with flowering trees and shrubs.

Although fore-limb adaptations to flight comprise a hallmark of birds, some lineages show the evolution of secondary flightlessness during the early part of the Tertiary Period. This situation means that the inheritance of a lack of flying

FIGURE 15.7 Track of greater rhea (*Rhea americana*), a large ratite native to Patagonia, Argentina. Notice its close anatomical resemblance to Mesozoic theropod tracks depicted and described in previous chapters, with prominent digits II–IV, phalangeal pads, and well-developed claws.

ability recurred like so: flightless non-avian theropod → flighted bird → flightless bird. Modern birds can be broadly divided into **carinates** (flighted birds) and **ratites** (flightless birds), although exceptions to this dualistic classification are posed by a few species of flightless carinates. Among these birds are penguins, which probably evolved from flighted diving birds and now essentially "fly" through a liquid medium (using the same flapping motion) instead of the air.

Ratites, which include ostriches, emus, rheas, kiwis, and cassowaries, are of the most interest to dinosaur paleontologists because they are biomechanical and possible behavioral analogues to some flightless theropods of the Mesozoic. In fact, numerous studies of rhea and emu tracks have been used as modern analogues to Mesozoic theropod tracks (Fig. 15.7). A Tertiary ratite that might have been as terrifying as some Cretaceous dromaeosaurs to the mammals of its time was *Diatryma*, described by Edward Cope in 1876 (Fig. 15.8). *Diatryma* was a 2-meter tall bird that towered over most mammals about 50 million years ago. It probably weighed more than 150 kg and had a large head and beak adapted for meat eating. However, *Diatryma* was surpassed in mass by a bird that only recently went extinct, the herbivorous elephantbird (*Aepyornis maximus*). At nearly 3-meters tall and weighing 400 kg, the elephantbird was larger than any deinonychosaur, with the exception of *Utahraptor* (Chapter 9). It died out only about 1000 years ago, and its decline coincided with the arrival of humans in its habitats on Madagascar about 2000 years ago. Other ratites that went extinct within recent memory were the moas (e.g., *Dinornis*) of New Zealand (Fig. 15.9). *Dinornis maximus* was the tallest bird known (one specimen was 3.7 meters tall), and weighed more than 200 kg. However, other species of moas varied in size and most were considerably smaller. These ratites were probable victims of overhunting and habitat alteration by humans, and the last

FIGURE 15.8 Skeleton of the frightening Tertiary ratite *Diatryma* of North America. Be aware of its anatomical similarity to theropod skeletons from Chapter 9. Sam Noble Oklahoma Museum of Natural History, Norman, Oklahoma.

possible moa sighting was in 1947. Other modern ratites and many carinates similarly have been decreasing at rapid rates as a result of overhunting and human alterations of habitats, signaling the beginnings of a possible mass extinction for birds well into the Cenozoic (Chapter 16).

Carinates are by far the more diverse of the two groups and include the aforementioned passerines, but also (in general, and not cladistic categories):

1 waterbirds – albatrosses, boobies, cormorants, frigatebirds, gannets, grebes, loons, pelicans, petrels, and shearwaters;
2 wading birds – bitterns, cranes, egrets, herons, ibises, spoonbills, storks;
3 shorebirds – avocets, gulls, mudhens, oystercatchers, plovers, rails, sandpipers, stilts, and terns;
4 gamebirds – grouse, quails, and turkeys;
5 raptors – falcons, hawks, eagles, and owls, to name a few.

The preceding list is not meant to be memorized, but to impress that the variability of birds is almost taken for granted unless one starts to name all of them. This modern assortment put together with the Cenozoic fossil record of birds collectively point toward a post-Cretaceous success of birds that calls into question the popular appellation of the Cenozoic as the "Age of Mammals." In terms of shear numbers of species and individuals, it more arguably is the "Age of Birds."

15

(A)

(B)

FIGURE 15.9 *Dinornis*, a recently extinct genus of moas, which were a group of ratites native to New Zealand. (A) Skeleton of adult *Diornis maximus*; bust of Sir Richard Owen (Chapter 3) for scale. (B) Egg of *D. giganteus*, with a calculated volume of about four liters (!). Auckland Museum, Auckland, New Zealand.

Birds as Living Animals

Not surprisingly, bird reproduction is very similar to what has been interpreted for theropod dinosaurs, but other dinosaur clades may have shared behavioral traits with birds too.

Reproduction

Displays and courting behavior are common aspects of mating for birds, as well as auditory wooing through the use of songs. Displays can be made through colorful or prominent plumage (e.g., peacocks) as well as sometimes-complicated dances or songs performed for the benefit of receptive females, showing some parallels to some modern primates. Related to such pre-mating behaviors are territorial displays, where male birds will make aggressive movements or sounds that clearly communicate "stay away" to rivals. Interestingly, these behaviors rarely lead to actual fights between rival males, which have been conjectured for ceratopsians and pachycephalosaurs (Chapter 13).

Birds show a myriad of mating behaviors. Although they are often admired as paragons of "family values" because of the large number of species that have monogamous pair bonding, a significant number of species also slip in the occasional bird on the side. In such instances, bird pairs may be classified as **socially monogamous**, meaning that they help one another to raise their young, or **genetically monogamous**, where they are the only genetic parents of the young. Observations of bird behavior in recent years now suggest that the latter is actually rare, although most birds remain socially monogamous. Accordingly, many bird pairs will raise young that are not a result of their mating. In cases of nest parasitism (discussed in Chapter 8), some young that are raised may not even belong to the same species.

Nest building, now attributed to an ornithopod, *Maiasaura*, a few non-avian theropods such as *Troodon* and *Oviraptor* (Chapter 9), and at least one species of titanosaurid sauropod (Chapter 10), is common in modern birds, although not ubiquitous. Some birds do not nest at all, but lay their eggs on bare ground or rock. Similarly, a few bird nests consist of the barest scrape of a ground surface. However, others are among the most elaborate of any tetrapod-made structures, consisting of finely woven grasses or sticks, or borings made into hard soils augmented by vegetative material (Fig. 15.10A–C). Nests can be solitary or closely spaced in nesting colonies, the latter of which has been proposed for *Maiasaura*. Nesting colonies sometimes show regular spacing between individual nests, indicative of space requirements needed by parents for raising their respective broods (Fig. 15.10D).

Growth

Avian growth rates are often rapid, which is consistent with their endothermic physiology (Chapter 8). However, different groups of birds differ considerably in whether their young are born atricial or precocial, a consideration discussed for juvenile dinosaurs (Chapters 9 and 10). Most passerines, raptors, and herons have atricial juveniles, which means that they require much parental maintenance, including brooding that conserves body heat. In some instances, juveniles may stay in close proximity to their nests, even as they approach adult size (Fig. 15.11). On the other hand, most shorebirds and "game birds" (turkeys, grouse, quail) have precocial young that are active and somewhat self-sufficient soon after hatching. The latter situation enables parents to divide duties in raising the young, whereas the former almost necessitates that both parents are constantly around while their young develop.

Regardless of whether a species of bird is atricial or precocial in its juvenile stage, they all reach breeding age within a relatively short period of time compared to average lifespan, some as early as one year after hatching. Again, this is indicative of rapid growth rates relative to many mammals, and similar growth rates have been calculated for what are presumed as precocial juvenile theropods (Chapters 8 and 9). After they reach breeding age, most birds cease or otherwise slow their growth. The majority of bird species live less than 30 years (and some considerably less than that), but a few species of parrots can live more than 50 years in captivity.

Locomotion

The various ways that birds move are incredibly varied, going far beyond descriptions of merely "flying." Although most species of modern birds are indeed capable of

(A)

(B)

FIGURE 15.10 Variety of nests constructed and used by modern birds. (A) Ground scrape with a clutch of eggs on a sandy beach made by American oystercatcher (*Haematopus palliatus*), Georgia, USA. (B) Large and elaborate stick nest of osprey (*Pandion haliaetus*), Florida, USA. (C) Hole nest (burrow) in semi-consolidated sand with vegetation stuffed inside, made by kotare (kingfisher: *Halcyon santus*), North Island, New Zealand. (D) Nesting colony of takapu (Australasian gannet: *Morus serrator*) showing regularly spaced nest mounds formed by guano, North Island, New Zealand.

(C)

(D)

FIGURE 15.10 Continued

extended self-powered flight, they range from completely flightless (cursorial) to the fastest animals on Earth, once airborne (e.g., peregrine falcons, *Falco peregrinus*). As mentioned earlier, most flightless birds fit into a category of ratites, although a few flightless passerines evolved via the geographic isolation of New Zealand and other remote islands. Moreover, a few flighted birds also are maneuverable on the ground and can easily outrun their prey or predators. Other non-cursorial or non-aerial variations on locomotion include:

1 swimming on the surfaces of water bodies;
2 swimming under water surfaces;
3 diving;
4 burrowing; and
5 climbing.

FIGURE 15.11 Altricial juveniles of magnificent frigatebirds (*Fregata magnificens*), which have 1.7–2.4 m wingspans as adults but are completely dependent on their parents for the first year of life, despite approaching their sizes. Notice their eerie resemblance to non-avian theropods, downy feathers and all. San Salvador, Bahamas.

Because of this range of movements, birds can soar high above the Earth's surface (as much as 6000 m), dive as deep as 500 m, or live in nearly every other terrestrial and aquatic environment. This diversity of lifestyles far exceeds those known for non-avian dinosaurs, although avians had the advantage of more time to evolve them.

Perhaps the most noteworthy aspect of avian locomotion is how it is used for **migrations**, and in this respect birds are the most impressive of all tetrapods. A migration is the movement of birds between where they spend their winters and where they breed, hence these movements are seasonal and annual. Because reproduction often requires much caloric energy for mating, the development of eggs, and raising of young, birds will migrate away from their winter habitat to a place that has more calories and other nutrients available (Chapter 8), as well as adequate nesting habitats. Some of these migrations cover tens of thousands of kilometers, and even flightless birds, such as penguins, are known to migrate (via swimming, not waddling) hundreds of kilometers. Similar seasonal and annual migrations have been postulated for some dinosaurs that show large latitudinal variations, such as some hadrosaurids (Chapter 11), and some dinosaurs were clearly adapted for high-latitude (polar) environments as well (Chapter 8).

Feeding

Birds show a wide range of feeding strategies, from herbivorous (seeds, leaves, or fruits) to insectivorous to carnivorous, the latter manifested as either predation or scavenging. Darwin's original observations of finches in the Galapagos Islands noted

(A)

(B)

FIGURE 15.12 Wood-boring activities of birds related to nesting and feeding. (A) Hole nest in tree trunk made by pileated woodpecker (*Dryocopus pileatus*); Idaho, USA. (B) Oak acorns (*Quercus* sp.) wedged in holes made by acorn woodpecker (*Melanerpes formicivorus*) in trunk of ponderosa pine (*Pinus ponderosa*); California, USA.

how beak shapes for otherwise very similar species varied considerably in accordance with adaptations for food acquisition, for example, seed-crushing versus fruit eating versus insect nabbing. Not all birds are restricted to just one source of food, and some species switch from herbivory to insectivory to carnivory according to their needs. Even hummingbirds, which were always thought of as nectar eaters, prey upon and eat insects to supplement their diets. Other insectivorous birds, such as woodpeckers, have special adaptations to their bills and skulls for rapid hammering into wood in search of wood-boring insects, but they also can construct hole nests in tree trunks (Fig. 15.12A). A few shorebirds have long bills (partially to compensate for proportionally long legs) that are well suited for probing deeply into beach sands for crustaceans and molluscans. Of course, sharp beaks and talons associated with strong, grasping feet and rapid or near-silent flight, present in raptors and owls, are easily associated with predatory behaviors. However, one of the fiercest of predatory birds is the flightless cassowary (*Casuarius casuarius*) of northern Australia, which has been known to kill people with its vicious kicks. A similar kicking behavior has been conjectured for a few deinonychosaurs, which had large, sickle-like claws on digit II (Chapter 9).

Besides such obvious morphological adaptations, a few birds are the only tetrapods other than primates known to use tools for feeding. For example, some species of herons will hold feathers or similar lures in their beaks above a water surface to attract fish; nuthatches will use pieces of bark to force open bark on a tree trunk to acquire insects; and crows probe for insects using sticks or leaves. Some species of birds also show a feeding behavior markedly different from the vast majority of reptiles: **caching**, which is the storing of food for later consumption. One of the best examples of this type of behavior is in the acorn woodpecker (*Melanerpes formicivorus*), which drills holes into bark of trees and then tightly wedges acorns into these holes. Entire trees then become "grocery stores" for woodpeckers to visit later (Fig. 15.12B). Both tool use and caching are unknown in Mesozoic dinosaurs, and the evidence for such behavior is expected to be scanty, based on current analogues.

A noteworthy aspect of the interrelationships of bird feeding and flowering plants is their well-documented interdependence. Numerous flowering plants are dependent on birds for cross-pollination and seed dispersal, and likewise many birds are dependent on flowering plants for food and nesting materials. Indeed, some paleontologists have hypothesized that the near coincidence of the oldest flowering plants (Early Cretaceous) and oldest birds (Late Jurassic) in the geologic record possibly indicates a cause-and-effect relationship. Whether birds or pollinating insects played a role in the development of flowering plants is unknown, but the clear interconnections between birds, flowers, and fruits today argue for similar relationships in the geologic past.

Social Life

Modern birds are represented by nearly 10,000 species, hence their social lives are difficult to classify. The broadest categories that can be made for them are:

1 male–female pairs (discussed earlier); and
2 flocks.

Some male–female pairs rarely gather with others of their species; such spatial separation is probably related to male territoriality, food resource allocations, or other habitat requirements. Of course, any given flock of birds may be composed of a large number of male–female pairs, which increases the likelihood of gene mixing between pairs. The advantages of large flocks are numerous:

1 a collective protection of young ("strength in numbers");
2 finding food is easier with more eyes looking for it;
3 predators are more easily avoided for the same reasons as in (2); and
4 navigation during migrations.

Regardless of whether social behavior is limited to a few individuals or thousands in a breeding colony, much of it is facilitated by verbal and non-verbal communications. Non-verbal forms of communication include feather displays (plumage) and body movements; some of the latter consist of elaborate dances that either entice or intimidate. Verbal communications in birds are among the most complex of all tetrapods, but fall into two general categories, **calls** and **songs**. Calls are typically innate (not learned) and consist of brief vocalizations that express alarm, scold a predator or other intruder, signal other birds in a flock to stay together, or simply identify an individual so that another of its species knows its position. For example, blue jays will make sharp, loud calls that increase in number and tempo when a predator is in the vicinity of a nest. Geese will honk while in flight so that they can maintain their group formations. Crows will call to one another as a sort of linked chain as a flock moves over its territory. On the other hand, songs, which are normally learned, are often very complicated and can consist of numerous variations on a main theme. Songs are used for wooing or territorialism, and some birds even accomplish both tasks with the same song. The distinctiveness of most bird songs, particularly for passerines, enables carefully listening humans to distinguish species on the basis of sound alone. However, a few species of birds have song catalogues of thousands (e.g., brown thrashers), and others are excellent mimics of a large number of songs of other species (e.g., mockingbirds), which indicate a greater functionality to songs than mere flirting or fighting.

Male–female pairs are postulated for non-avian dinosaurs in some instances and are especially appealing hypotheses in cases where slightly different-sized and dual, parallel trackways might occur (Chapter 14). Flocking behavior, or at least the formation of large socially interacting groups of non-avian dinosaurs, is suggested by some ornithopod and sauropod nesting grounds, as well as monospecific bone beds of theropods, ornithopods, and ceratopsians (Chapters 9, 11, and 13). Vocalizations were likely in at least a few species of ornithopods, especially hadrosaurids with elaborate sinuses capable of directing air to make sounds (Chapter 11). Whether Mesozoic landscapes and seascapes were filled with the intermingled calls and songs of non-avian and avian dinosaurs is unknown.

Health

Although most bird species are very healthy, a few individuals suffer from ectoparasites and diseases (Chapter 7). Diseases in particular, whether fungal, bacterial, or viral, can spread quickly in some bird species because of close proximities of large numbers of individuals in flocks, exacerbated by the rapid movement of flighted birds. Avian diseases are receiving more attention in recent years because a few birds are recognized carriers of some diseases that also affect humans, such as salmonella and West Nile virus. Salmonella is contracted by contact or consumption of uncooked chicken eggs or chicken flesh. West Nile virus is transmitted through both mosquitoes and birds, causing a multiplicative effect that creates higher risk than if only one of these animals carried it.

Although most birds that reach adulthood seem outwardly healthy, injuries are common in those birds that spend a great deal of time on land or in the water. Cursorial birds might develop noticeable limps from skeletal or muscular maladies, which could have been caused by overuse or any number of other stresses. Missing feet or legs are a problem for waterbirds that rest on ocean surfaces, where their dangling feet tempt sharks and other predators (Fig. 15.13).

Injuries and deaths of birds from predation in terrestrial environments, although commonplace, are increased dramatically by habitat alterations that take away normal roosting spots or vegetative cover needed by birds to avoid detection. Such problems are compounded by the introduction of non-native predators that break the rules of the "Red Queen" (Chapter 6). In other words, these predators are from an evolutionary track in which their prey animals did not develop defenses against them. House cats in urbanized areas exemplify how habitat fragmentation and non-native predators combine to decimate songbird populations in North America; cats in the USA are estimated to kill hundreds of millions of birds each year. Migratory birds in particular are vulnerable to such risks, especially for ground-nesting species that subsequently experience increased juvenile mortality. Non-native organisms introduced to an environment, which have a disproportionately deleterious impact on native organisms, are termed **invasive species**. Invasive species, when co-occurring

FIGURE 15.13 Evidence of unsuccessful predation of a modern avian dinosaur: a footless laughing gull (*Larus altricilla*) on a beach in Georgia, USA. This observation was confirmed by examination of its trackways, which showed well-defined right-foot tracks alternating with impressions made by the metatarsal nub of the left leg.

with habitat alterations, have resulted in some species of birds becoming endangered or extinct.

The paleopathology of non-avian dinosaurs, covered previously (Chapter 7), indicates these animals certainly had some health problems, suffered injuries, or were subjected to predation. However, no evidence has been presented to suggest that habitat alterations combined with the introduction of invasive species accelerated dinosaur extinctions at any time. The latter factor is largely the product of human activities, although birds themselves are capable of transporting organisms long distances in short periods of time. Nevertheless, a large amount of information, multi-faceted and integrated, now adds up to a powerful argument that the end-Cretaceous extinction of non-avian dinosaurs was related to a sudden, catastrophic change in habitats. This mass extinction ensured the proliferation of birds in the Cenozoic, but it still inspires much curiosity: why did some dinosaurs make it past the Cretaceous, but others (including some avians), did not? This point of inquiry is the subject of the next and last chapter (Chapter 16).

SUMMARY

One of the most discussed topics in dinosaur studies is the evolution of dinosaurs into birds. Avians apparently began in the Middle to Late Jurassic, and the oldest known bird is *Archaeopteryx* from the Late Jurassic of Germany. *Archaeopteryx* shows a blend of features associated with non-avian theropods and birds, including flight feathers, thus it is often identified as a "transitional" fossil. Based on cladistic analyses of character traits, the most probable ancestor to *Archaeopteryx* was a maniraptoran, and deinonychosaurs comprise a sister clade to *Archaeopteryx* and other birds. Birds diversified considerably during the Cretaceous Period; more than 50 species are known from that period. Cretaceous bird evolution resulted in an impressive expansion of habitats, but most of their clades were extinct by the end of the Cretaceous. Subsequent diversification of birds has led to their inhabiting nearly all near-surface terrestrial and aquatic environments. Cenozoic birds can be broadly categorized as ratites ("flightless") and carinates ("flighted"), with ratites representing modern analogues to non-avian theropod dinosaurs. Modern birds are represented by more than 10,000 species, although a significant number of these have become extinct in the past several hundred years as a result of human overhunting and habitat alterations.

Two current hypotheses for the development of full flight in birds are the cursorial ("ground up") and arboreal ("trees down") hypotheses, with some variations on those themes. Although each hypothesis has its merits, they are largely based on functional morphology of non-avian and avian theropods. Recent insights into this realm, as well as experiments with living birds, have now generated variations of these hypotheses that could combine elements of each.

Birds are extremely diverse in their reproduction, growth, feeding, locomotion, social lives, and health, and in some instances their behaviors overlap with hypothesized dinosaur behaviors. Birds thus provide models of comparison for paleontologists interested in these facets of dinosaur behavior. Recent bird extinctions are largely the result of human-caused factors, such as habitat alterations that prevent adequate cover, food, and nesting material for birds, as well as invasive species of predators that decimate bird populations. Of these factors, habitat alterations have the most applicability to understanding non-avian dinosaur extinctions.

DISCUSSION QUESTIONS

1. What were the three definitions given for "dinosaur" in this book in Chapters 1, 5, and 15? How are they different or similar? Explain how this word can have different definitions, yet they still can convey the same concepts.

2. Review the characteristics of theropods, especially the lineage leading to maniraptorans, covered in Chapter 9, and compare them to what is described for *Archaeopteryx* in this chapter. How many matches did you find? What contrary evidence, if any, would sway you from accepting the hypothesis that birds originated from theropods?

3. Feathers in birds today have different (but often overlapping) purposes, such as insulation, display, and flight. In the case of flightless coelurosaurs, such as *Caudipteryx*, which of these functions was most likely and why?

4. Why is it that a bird has a saurischian hip (and is a saurischian), yet ornithischians are called the "bird-hipped" dinosaurs?

5. Look at Table 15.1 and eliminate all genera from China. What percentage of change happens to the number of genera? Also, considering that the oldest known bird is *Archaeopteryx* from Germany, what hypotheses could explain avian dispersal and diversifications without the Chinese finds?

6. Some carinates developed flightlessness in ecosystems that lacked appreciable numbers of mammals. What evolutionary factors may have resulted in the selection of reduced wings, which brought these birds back to a state similar to those of their theropod ancestors?

7. You are doing fieldwork in the southwestern USA on an excavation of Upper Triassic rocks when you discover what seems to be a bird skeleton. Which hypothesis of bird origins, theropod or "thecodont", would your discovery support and why? Would it necessarily only support one of the hypotheses? What supplementary evidence would support or disprove your initial identification of the skeleton as avian?

8. Which of the hypotheses for bird flight seem most plausible to you, based on the evidence presented here, and why? What body fossil or trace fossil evidence would be needed to support or disprove the hypothesis you currently favor?

9. Given the bird nests shown in Figure 15.10, arrange them in order of "most likely" to "least likely" to be preserved in the fossil record. What factors are involved in their preservation? How could some unusual conditions result in your changing this ranking?

10. If all birds went extinct tomorrow, how would the world be different? What ecosystems might be affected the most and in what ways? (Hint: Think of bird interactions with both insects and plants.)

Bibliography

Chatterjee, S. 1997. *The Rise of Birds: 225 Million Years of Evolution*. Baltimore, Maryland: Johns Hopkins University Press.

Chiappe, L. M. 1997. "Aves". *In* Currie, P. J. and Padian, K. (Eds), *The Encyclopedia of Dinosaurs*, San Diego, California: Academic Press. pp. 32–38.

Chiappe, L. M. and Witmer, L. M. (Eds). 2002. *Mesozoic Birds: Above the Heads of Dinosaurs*. Berkeley, California: University of California Press.

Chiappe, L. M. and Dyke, G. J. 2002. The Mesozoic radiation of birds. *Annual Review of Ecology and Systematics* **33**: 91–124.

Currie, P. J., Koppelhus, E. B., Shugar, M. A. and Wright, J. L. 2004. *Feathered Dragons: Studies on the Transition from Dinosaurs to Birds*. Bloomington, Indiana: Indiana University Press.

de Ricqles, A. J., Padian, K., Horner, J. R., Lamm, E. M. and Myhrvold. N. 2003. Osteohistology of *Confuciusornis sanctus* (Theropoda, Aves). *Journal of Vertebrate Paleontology* **23**: 373–386.

Dial, K. D. 2003. Wing-assisted incline running and the evolution of flight. *Science* **299**: 402–404.

Dingus, L. and Rowe, T. 1997. *The Mistaken Extinction: Dinosaur Evolution and the Origin of Birds*. New York: W. H. Freeman.

Elbroch, M. and Marks, E. 2001. *Bird Tracks and Sign: A Guide to North American Species*. Mechanicsburg, Pennsylvania: Stackpole Books.

Elphick, C., Dunning, J. B., Jr. and Sibley, D. A. 2001. *The Sibley Guide to Bird Life and Behavior* New York: Alfred A. Knopf.

Erickson, G. M., Rogers, K. C. and Yerby, S. A. 2001. Dinosaurian growth patterns and rapid avian growth rates. *Nature* **412**: 429–432.

Feduccia, A. 1999. *The Origin and Evolution of Birds* (2nd Edition). New Haven, Connecticut: Yale University Press.

Feduccia, A. 2002. Birds are dinosaurs: Simple answer to a complex problem. *Auk* **119(4)**: 1187–1201.

Gatesy, S. M. and Middleton, K. M. 1997. Bipedalism, flight, and the evolution of theropod locomotor diversity. *Journal of Vertebrate Paleontology* **17**: 308–329.

Melchor, R. N., de Valais, S. and Genise, J. F. 2002. Bird-like fossil footprints from the late Triassic. *Nature* **417**: 936–938.

Novas, F. E. and Puerta, P. F. 1997. New evidence concerning avian origins from the Late Cretaceous of Patagonia. *Nature* **387**: 390–392.

Padian, K. 2004. "Basal Aviale". *In* Weishampel, D. B., Dodson, P. and Osmólska, H. (Eds), *The Dinosauria* (2nd Edition). Berkeley, California: University of California Press. 210–231.

Padian, K., de Ricqles, A. J. and Horner, J. R. 2001. Dinosaurian growth rates and bird origins. *Nature* **412**: 405–408.

Paul, G. S. 1988. *Predatory Dinosaurs of the World*. New York: Simon & Schuster.

Paul, G. S. 2002. *Dinosaurs of the Air: The Evolution Loss of Flight in Dinosaurs*. Baltimore, Maryland: Johns Hopkins University Press.

Shipman, P. 1998. *Taking Wing:* Archaeopteryx *and the Evolution of Bird Flight*. New York: Simon & Schuster.

15

Chapter 16

Your nephew is having a problem accepting that dinosaurs are extinct, and your reassurances that they are still present today as birds does not satisfy him. "Why did they go extinct?" is a more difficult question to answer than "How did they go extinct?", so you decide to focus on the latter. You have a problem, partially because paleontologists seem to disagree on the causes of dinosaur extinctions, but mostly because their extinction is often compared to recent and probable near-future extinctions.

What are the postulated causes of dinosaur extinctions, and which seem most plausible in light of current geological and paleontological evidence? What are some of the factors associated with modern extinctions? What, if anything, do these modern extinctions share with extinctions interpreted from the fossil record? Do extinctions of the geologic past, including those of dinosaurs, provide any lessons applicable to a better understanding of modern extinctions?

Dinosaur Extinctions

Why Learn About Dinosaur Extinctions?

One of the enduring mysteries is why all dinosaurs that were not birds, which included all ornithischians and most saurischians, disappeared by the end of the Cretaceous Period. A few reports have been published about non-avian dinosaur skeletal material above the **Cretaceous–Tertiary (K–T) boundary** (where *K* is for Cretaceous and *T* is for Tertiary), but these may have been reworked inclusions from older, underlying deposits (Chapters 4 and 7). Consequently, as far as worldwide paleontological assessments are concerned, no dinosaurs lived into the Tertiary Period. One argument for this conclusion is that no tracks or other trace fossils attributable to non-avian dinosaurs have been found above the K–T boundary, either. This does not mean that no definitive non-avian dinosaur material will ever be found in Tertiary deposits, but such a find would be exceptional. Furthermore, based on current information, paleontologists predict that if non-avian dinosaur fossils are found above the boundary they will be restricted to the very earliest part of the Tertiary Period. How did a group of terrestrial animals that seemed to have dominated their landscapes for more than 160 million years vanish in only a few million years?

Coupled with this riddle of dinosaurian disappearances is the question of why only a few dinosaurs (birds) made it past the Cretaceous–Tertiary transition and are still here today (Chapter 15). The end of the Cretaceous is often referred to as a "Great Dying" by paleontologists and geologists, who recognize that it was a time of one of six worldwide **mass extinctions** in the history of the Earth. A mass extinction is defined as the disappearance of a large number of species within a relatively short period of time. Mass extinctions are interpreted in the geologic record by abrupt changes in vertical sequence of fossil assemblages that:

1 had a worldwide occurrence;
2 are noticeable in a variety of facies (different environments); and
3 seemed to cross many phylogenetic groups that lived in those environments, such as terrestrial plants, marine invertebrates, and terrestrial freshwater vertebrates.

One of these mass extinctions is probably happening now, which is almost entirely linked to human causes, as explained later.

Many clades composed of many species that lived in a number of environments during the Mesozoic were extinct by the end of the Cretaceous. Moreover, significant losses were incurred in other groups that did persist into the Tertiary Period. Among the casualties were:

1 some forms of planktonic algae;
2 marine invertebrates, such as the squid-like ammonoids and giant clams;

3 some flowering plants;

4 all marine reptiles, such as mosasaurs, ichthyosaurs, plesiosaurs, and pterosaurs (Chapter 6); and,

5 non-avian dinosaurs.

In other words, the world was a very different place at the end of the Cretaceous; nearly 50% of all taxa had disappeared over a few million years.

One argument posits that the K–T mass extinction opened to birds and mammals the many ecological niches that had been occupied by pterosaurs and dinosaurs. In this sense, extinctions are contributors to evolution, and ecosystems are changed when their biotic components change, so new opportunities are created for species to interact with their environments and one another (Chapter 6). Indeed, modern (avian) dinosaurs illustrate this generalization well. Birds constitute the most diverse group of terrestrial chordates, with about 10,000 species identified so far, and they live in virtually every near-surface ecosystem and latitude (Chapter 15). Because of their anatomical similarity to some theropods and their temporal overlap with non-avian dinosaurs, they represent a link to the Mesozoic that was largely severed by the mass extinction at the end of the Late Cretaceous. Because the termination of this mass extinction was only 65 million years ago (a mere 1.4% of geologic time; Fig. 1.2), paleontologists are also interested in finding out what factors caused it, whether these factors could arise again, and why some organisms that were alive during the last of the Mesozoic, such as birds, survived.

It is with some irony to note that the currently interpreted mass extinction is especially evident among birds. Judging by estimates of naturalists who have tabulated known bird extinctions since 1600, more than 100 species of previously well-known birds have become extinct as a result of human influences. Examples of these birds include the dodo, passenger pigeon, Carolina parakeet, and the moas of New Zealand (Chapter 15). (A notable exception was the re-discovery in early 2005 of the ivory-billed woodpecker in Arkansas of the USA, a bird that was thought extinct for more than 50 years.) This situation may mean that humans, despite their seeming never-ending fascination with dinosaurs, are causing a second extinction of them. Because birds have both aesthetic and utilitarian value to humans, their increasing losses are currently of some concern, but how their disappearances might affect human populations is uncertain. As a result, the lessons of extinctions of the past and the present may intersect through birds and other organisms that share a heritage with what happened at the end of the Mesozoic Era.

Definitions and Causes of Mass Extinctions

*An **extinction** of a species happens when the last individual of that species is prevented from mating and producing a viable offspring.*

Extinctions can be categorized as either **local extinctions**, where a population of a species dies out in one part of the world but is still present elsewhere, or **global extinctions**, where all populations of a species are gone from all habitats. Species do not evolve again or otherwise resurrect after they become extinct. This permanence is underscored by conservative estimates that more than 99% of all species that have ever lived on the Earth are now extinct. A small percentage of these species had their body parts preserved in the fossil record, along with the trace fossils of many not represented by body parts. Consequently, estimations of which species went extinct and when these occurred are admittedly biased and subject to some errors. Nevertheless, extinction is part of the history of life, and the sample of past life

FIGURE 16.1 The coelacanth *Latimeria chalumnae*, regarded as a "living fossil" because of how its lineage survived from the Cretaceous Period to now.

provided by the fossil record shows that the average fossil species lasted about 1 to 2 million years (with a large standard deviation).

How small errors may enter the databases of extinctions is exemplified by the occasional discovery of a so-called **living fossil**, an extant organism that represents an ancient lineage originally thought to have died out. In these cases, a supposed extinction of a species, which is based on a gap in the geologic record between its last known occurrence and the present, may not have actually happened. One example is represented by coelacanths, lobe-finned fish of the clade Actinistia, which shared time with the dinosaurs. Coelacanths were thought to have become extinct during the Cretaceous, well before the end of the period (about 75 Ma), but in 1938 some specimens were discovered living off the coast of Madagascar as an extant species, *Latimeria chalumnae* (Fig. 16.1). This species was also found recently thousands of kilometers to the east in the Pacific Ocean, indicating that it is geographically widespread. The explanation for a nearly 75-million-year gap in the geologic record for members of this clade is that coelacanths, which were shallow-marine species, had adapted to deep-oceanic environments by the end of the Cretaceous. Most geologists and paleontologists, in contrast, concentrate their efforts on shallow-marine or continental deposits. As a result, no fossil coelacanths were found in deposits less than 75 million years old, and researchers accordingly, but mistakenly, interpreted their extinction in the Cretaceous.

Coelacanths also illustrate how a local extinction can become regarded as a global extinction in the geologic record. Similarly, some species may disappear from the geologic record in a stratigraphic interval, only to be found later by geologists in overlying, younger deposits, and such fossils have been nicknamed **Lazarus species**. Lazarus species are explainable by ecological principles, such as the temporary displacement of a species from an ecosystem by some changes in that ecosystem (e.g., temperature, nutrient availability, habitats). Although such taxa appeal to the imagination and give hope of "lost worlds" where supposedly extinct organisms dwell out of the sight of humans, they actually are rare. Thorough checking

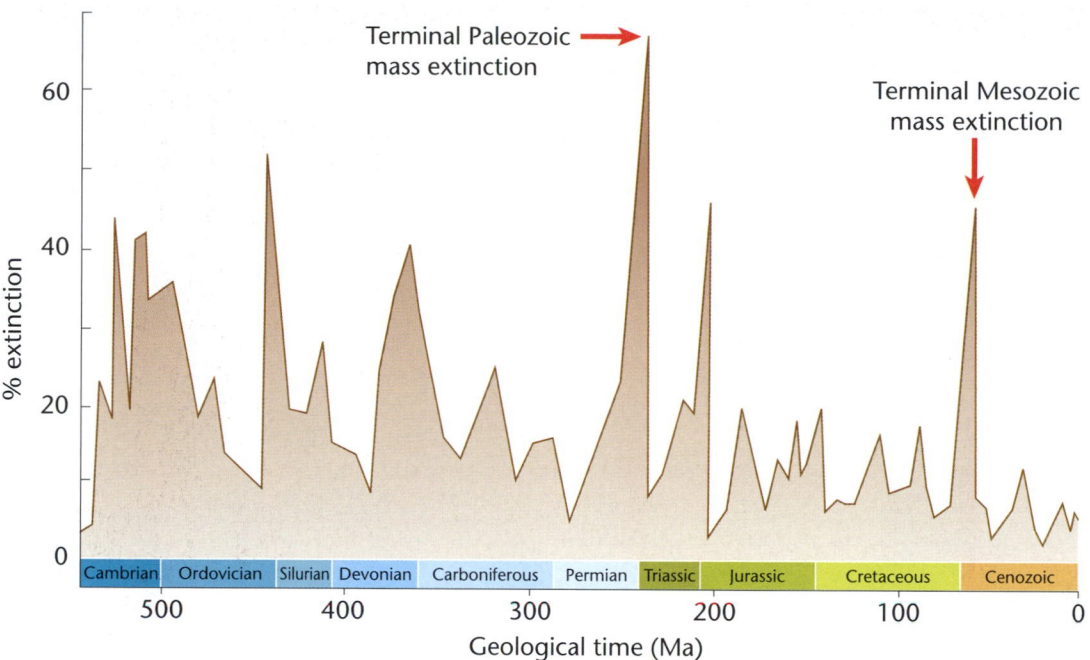

FIGURE 16.2 Mass extinctions of marine genera during the Phanerozoic Eon, plotted as percentage of genus extinctions versus geologic time. Adapted from data by Sepkoski (1994).

and re-checking of the geologic record by geologists and paleontologists has resulted in an assessment of geologic ranges for fossil species that becomes more precise over time, not less.

Five mass extinctions have been recognized in the fossil record: the end-Ordovician, end-Devonian, end-Permian, end-Triassic, and end-Cretaceous (Fig. 16.2). Of these, by far the greatest was the end-Permian, in which an estimated 95% of all species died out. Other extinctions of lesser magnitude are in between these mass extinctions. A striking difference in fossil content above and below a given horizon with worldwide consistency, however, is what constitutes the scientific basis for originally drawing such boundaries. Indeed, the principle of biological succession was recognized and articulated early in the nineteenth century because of the repeated observations by field geologists of fossil lineages that had limited geologic ranges (Chapter 4). In fact, the establishment of time divisions between the eras of the Paleozoic, Mesozoic, and Cenozoic Eras reflects mass extinctions at the end of the Paleozoic (end-Permian) and Mesozoic (end-Cretaceous).

Some proposed causes of mass extinctions include:

- Global climate changes (cooling or warming), which were initiated by extended periods of volcanism or variations in the Earth's rotation and orbit.
- Changes in positions of the continents, which would have affected oceanic circulation patterns or the amount of shallow-marine habitats available for marine life.
- Impact of **bolides**, which are extraterrestrial bodies such as comets or meteorites.

These bolides certainly could have caused local extinctions through a direct hit. Moreover, they may have contributed to global extinctions through ejected debris blocking sunlight, which would have disrupted photosynthesis in plants and thus deprived animals of food.

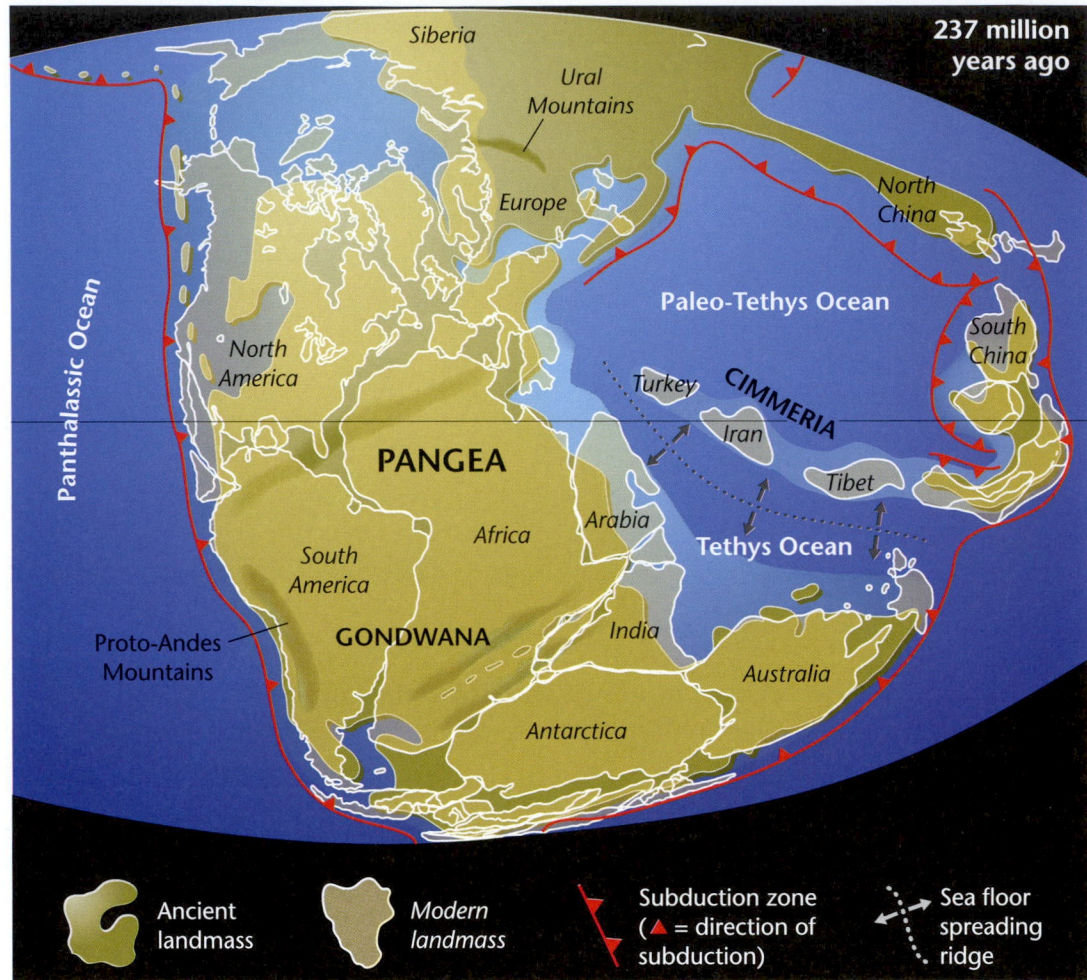

FIGURE 16.3 Pangea, the supercontinent caused by the uniting of Laurasia and Gondwana at the end of the Permian Period, which was a time of the greatest of all mass extinctions. Reprinted by permission from Scotese, C. R., 2001. Atlas of Earth History, Volume 1, Paleogeography, PALEOMAP Project, Arlington, Texas, 52 pp.

With these possibilities in mind, a look at the end of the Permian Period (about 245 Ma) is warranted because it is considered the worst-case scenario among all of the known mass extinctions. A review of this mass extinction should make the later discussion of the more famous K–T mass extinction easier to understand. Based on present geological and paleontological information derived worldwide from above and below the Permian–Triassic boundary, the Earth had a set of conditions that adversely affected biodiversity over the course of a few million years. Several factors have been proposed as related to the end-Permian extinction:

- The continents were all massed together into one "supercontinent" called Pangea at that time. Large continents typically have challenging environmental conditions develop in their interiors, such as the Gobi Desert of Asia and the tundra of Siberia, which have lower biodiversity than other terrestrial ecosystems (Fig. 16.3).
- Shallow marine environments became limited as sea level dropped, which reduced the available habitats for marine organisms that breed and otherwise dwell in such environments.
- Both the north and south ends of Pangea were situated in polar regions, which restricted terrestrial organisms to its center, which had tundras,

deserts, mountains, and other inhospitable habitats that typically have low biodiversity.

■ Pangea blocked oceanic circulation patterns, particularly around the equator, so marine organisms were hindered in migrating or dispersing their larvae to alternative habitats.

■ Extensive periods of volcanism happened in present-day Siberia, causing the release of expansive and thick deposits of basalt and so altering vast stretches of habitats. This volcanism also may have contributed a great deal of CO_2 to the atmosphere. Because CO_2 is a greenhouse gas, it consequently may have triggered global warming.

The end-Permian extinction is so far attributed to a combination of these factors, not any single cause. Attempts to find evidence of a large bolide impact at the end of the Permian are inconclusive to date.

Two of the five mass extinctions happened during the Mesozoic Era, specifically the end-Triassic and the end-Cretaceous. The first extinction, at the end of the Late Triassic, probably aided in the ascendancy of dinosaurs in the latter part of the Mesozoic, whereas the second at the end of the Late Cretaceous resulted in the demise of all dinosaurs except birds. The Late Triassic mass extinction is interesting in that the large terrestrial vertebrates of the Triassic became extinct, such as some amphibians (metoposaurs), therapsids, rauisuchians, and phytosaurs. These animals shared some of the same environments as theropods, prosauropods, and a few ornithischians (Chapter 6). The demise of other large archosaurs in apparent inverse proportion to the ascendancy of the dinosaurs was at one time attributed to the adaptive superiority of upright posture, bipedalism, and possible endothermy in the latter group. A significant number of both invertebrate and vertebrate marine species, however, also went extinct toward the end of the Late Triassic. Furthermore, land-dwelling species seemed to have undergone two stages of extinctions separated by about 10 million years. These data suggest climatic fluctuations were at least partially responsible for the extinctions, rather than interspecific competition. Most recently, evidence of a bolide impact at the end of the Triassic has been added as yet another factor that negatively affected floral and faunal assemblages. As a result, causes of the end-Triassic mass extinction are now starting to converge with those proposed for the end-Cretaceous mass extinction.

Although mass extinctions have received a tremendous amount of attention, they have been happening for as long as multicellular life has existed. To show how extinctions are also a part of the history of dinosaurs, extinctions within each of the major clades of dinosaurs will be reviewed here. Such a review should aid in understanding some of the commonalities of extinctions in general, and how some dinosaurs and their ecosystems were affected.

Extinctions of Major Dinosaur Clades

Non-Avian Theropods

Non-avian theropods first show up in the geologic record in the Late Triassic, represented by both skeletal material and tracks. They then seem to have been affected by the end-Triassic extinction, exemplified by the disappearance of herrerasaurids and most ceratosaurs. Species that were common during the Late Triassic (e.g. *Coelophysis bauri*) apparently died out, and theropod tracks in Lower Jurassic strata change dramatically in both types and sizes. Larger ceratosaurs subsequently replaced relatively smaller ones in the Early Jurassic (e.g., *Dilophosaurus* and *Syntarsus*), and tetanurans made their first appearance by the Middle Jurassic. However, ceratosaurs declined considerably throughout the rest of the Jurassic, with a few exceptions in the Early Cretaceous (e.g., *Giganotosaurus* and *Carcharodontosaurus*). Tetanurans, such as therizinosaurs, troodontids, oviraptorids, and dromaeosaurs, then dominated theropod assemblages in the latter part of the Cretaceous.

Whether a few theropods made it past the Cretaceous Period is subject to controversy because a few exceptional occurrences of theropod hard parts, normally represented as teeth, were found in Tertiary deposits immediately above what geologists regard as the boundary. Taking taphonomic considerations into account, one interpretation could be that these remains may have been eroded from older (underlying) deposits, transported, and redeposited in the younger deposits. This exemplifies the role and importance of included particles and superposition for relative age determination (Chapter 4). Experiments with theropod teeth show that they can be tumbled artificially for long distances and incur few signs of wear from transport. These results suggest that a pristine condition for theropod teeth is not necessarily a sign that they are either locally derived or contemporaneous with surrounding sediment. In other words, teeth found in unusual strata may have been transported over both time and space. Theropod eggshell fragments also have been found just above the K–T boundary, but they were subject to the same possibility of transport, and without accompanying embryonic material their theropod affinity is indefinite. The overall conclusion reached by most paleontologists is that non-avian theropods did not make the journey past the K–T boundary.

Nevertheless, the preceding statement could be disproved in at least one way. Probably the most difficult to refute datum for the hypothesis of a non-avian theropod surviving into the Tertiary would be a footprint, particularly from a distinctive tetanuran species such as *Tyrannosaurus*. Because tracks are not easily transported and thus represent an *in-situ* presence of dinosaurs (Chapter 14), they are considered as more reliable indicators of dinosaurs co-existing with the environments than body fossils. Along these lines, the only known example of a probable tyrannosaurid footprint (see Fig. 9.12A) was found only several meters below the K–T boundary. This shows that at least one large non-avian theropod was alive and walking near the end of the Cretaceous, but it still did not quite make it to the Cenozoic. Otherwise, no undoubted non-avian theropod tracks have been found closer to the K–T boundary or in the Tertiary.

Sauropodomorphs

Prosauropods were among the first dinosaurs in the Late Triassic, but they died out by the end of the Early Jurassic. This extinction happened either despite their dominance as the largest terrestrial herbivores of their time or because of it. One hypothesis for extinctions is that a lack of resources will favor smaller individuals with lower nutritional needs, as explained later. Sauropods, in contrast to prosauropods, lasted from the Early Jurassic until the Late Cretaceous. Sauropods, however, seemed to have undergone local extinctions by the Early Cretaceous. This circumstance highlights one of the most intriguing problems in the study of sauropods: the timing of their distribution in the Late Jurassic to Early Cretaceous transition. For example, the large, abundant, and successful sauropods in western North America, such as *Apatosaurus*, *Barosaurus*, *Brachiosaurus*, *Camarasaurus*, and *Diplodocus*, were extinct by the end of the Jurassic. Only a few sauropod species with rare occurrences (such as *Pleurocoelus*) occur in any Early Cretaceous rocks of that same region, but large sauropods then show up again near the end of the Cretaceous.

South America and a few other places in the world were playing host to large titanosaurids such as *Argentinosaurus*, *Saltasaurus*, and *Titanosaurus* during the Late Cretaceous. Titanosaurids were in fact the most common sauropods throughout the Cretaceous. Interestingly, titanosaurids were nearly unknown in North America until the end of the Late Cretaceous. One hypothesis to explain this gap in sauropod fossils in North America during much of the Cretaceous is that a local extinction of sauropods occurred on that continent. Titanosaurids then later migrated north from South America into niches devoid of sauropods by the Late Cretaceous. Maps of the global distribution of sauropods, on the basis of both body and trace fossils in Cretaceous rocks, show that they were still apparently thriving in some environments, however patchy their distribution, in South America, Africa, and Asia, until the end of the Mesozoic.

Ornithopods

Ornithopods, which originated from ornithischian ancestors in the Early Jurassic, comprised a successful clade up until the end of the Cretaceous. A few clades within Ornithopoda nevertheless underwent their own respective extinctions before the end of the Cretaceous. Perhaps the most striking extinction was of the heterodontosaurids, which only lived during the Early Jurassic and then only in southern Africa. Other non-iguanodontians, such as hypsilophodontids, were more successful, and some species, such as *Thescelosaurus*, retained primitive features of their clade well into the Late Cretaceous.

It was iguanodontians, however, that included some of the most diverse and abundant assemblages of all dinosaurs. Iguanodontia as a clade was long-lived and apparently thrived throughout the entirety of the Cretaceous. Hadrosaurids, in particular, proliferated during the Late Cretaceous and are regarded as iconic images of herbivorous dinosaur success. *Edmontosaurus*, *Hadrosaurus*, *Maiasaura*, *Parasaurolophus*, and many others are examples of Cretaceous hadrosaurids, which are well represented in the fossil record. For such profusion to end at the termination of the Cretaceous is all the more striking and provokes more questions than answers. This apparently rapid decline in numbers and diversity of ornithopods, and other herbivorous dinosaurs at the end of the Cretaceous, suggests that their extinctions were a result of some abrupt changes in their ecosystems that affected their food supply: terrestrial plants.

Thyreophorans

Stegosaurs comprise the first major clade of thyreophorans to become extinct. They were at their most prolific in the Jurassic, particularly the Late Jurassic, exemplified by *Stegosaurus* of North America and *Kentrosaurus* of Tanzania, as well as some stegosaur genera in China. Nevertheless, they were substantially reduced by the end of the Early Cretaceous and effectively disappeared long before the end of the Late Cretaceous. This puzzling and seemingly premature extinction of what was a successful group has led to a hypothesis that these large herbivores were displaced from their niches by other herbivores that diversified considerably by the Early Cretaceous, such as ornithopods (Chapter 11) and ceratopsians (Chapter 13). Furthermore, the demise of stegosaurs coincided with the well-documented propagation of flowering plants through the Early Cretaceous, which irrevocably changed floral patterns in terrestrial ecosystems. The difference no doubt had a ripple effect on biota throughout the Cretaceous; stegosaurs may not have been able to adapt to the new foodstuffs or habitats caused by the radical shift from spore-bearing plants and conifers to those that bore fruit.

On the other hand, ankylosaurs, represented by both ankylosaurids and nodosaurids, flourished during the same time interval and were relatively common well into the Late Cretaceous in North America and Asia. Their herbivorous habits were apparently not adversely affected by the advent of angiosperms; indeed, their success suggests that they may have been caused by it. In the end, however, their highly developed armor and other defenses did not aid their survival, and like all other non-avian dinosaurs they were extinct by the end of the Cretaceous.

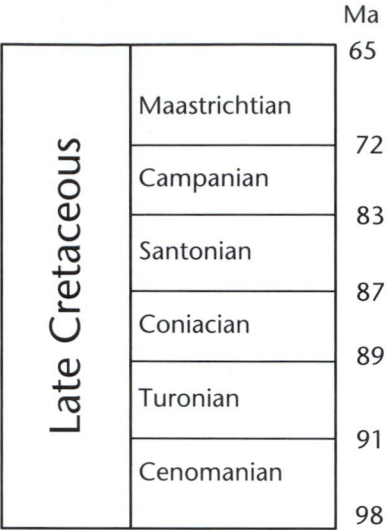

FIGURE 16.4 Age subdivisions of the Late Cretaceous, terminated by the Maastrichtian Age.

Marginocephalians

Marginocephalians were relative latecomers on the Mesozoic scene in comparison to other dinosaur clades, with primitive forms evident in Lower Cretaceous strata but none known from the Jurassic. Consequently, extinctions that occurred in primitive (basal) lineages of marginocephalians are poorly-defined, although a few conservative forms in both pachycephalosaur and ceratopsian lineages persisted until the **Maastrichtian Age**, the latest part of the Late Cretaceous (Fig. 16.4). The Maastrichtian is dated as having lasted about 6 million years (71–65 Ma). Strata representing this time interval are among the most thoroughly scrutinized in the geologic record for dinosaur body fossils. As a result, if marginocephalians are present in these rocks, persistent paleontologists should have found them by now.

Based on analyses of strata formed during the Maastrichtian, some of the advanced marginocephalians, and ceratopsids in particular, were among the last dinosaurs to walk the Earth. This conclusion is based on the excellent fossil record for ceratopsids, which show a considerable decline in species closer to the K–T boundary. In contrast, the extinction of pachycephalosaur lineages is poorly understood. Some genera, such as *Pachycephalosaurus*, *Stegoceras*, and *Stygimoloch*, occur in strata constituting the very last of the Cretaceous, but their remains are relatively rare in all strata, therefore little is known as to whether their numbers declined toward the K–T boundary. Nevertheless, they did not make it into the Tertiary Period. Likewise, no ceratopsians made it past the K–T boundary, although bones of *Triceratops* are common in strata just below it; less common are those of other ceratopsids *Leptoceratops* and *Torosaurus*. Analyses of dinosaur faunas in the latest part of the Cretaceous indicate that chasmosaurines, a formerly successful clade, were already extinct before the end of the Cretaceous, accompanied by hadrosaurids and nodosaurids (Chapters 11 and 12).

What happened to marginocephalians, as well as all of the other dinosaurs? The currently accepted answer is that the last marginocephalians died out from causes common to all dinosaurs toward the end of the Maastrichtian, with the exception of a few avian theropods. The last part of this chapter deals with how those avian theropods made it past one of the largest extinctions witnessed in Earth history, as well as what made all other dinosaurs become part of the fossil record.

The Dinosaur Fossil Record and the K–T Boundary

The phrase **paradigm shift** generally denotes a revision or complete overturning of a long-held hypothetical or theoretical framework. In the long-term perspective on the history of scientific endeavors, the following may qualify as examples of paradigm shifts in science:

- The change from the geocentric (Earth-centered) to the heliocentric (sun-centered) model for the solar system, influenced by the theories of the sixteenth century by Nicolas Copernicus (1473–1543).
- Publication of *On the Origin of Species* by Charles Darwin in 1859, which revolutionized the biological sciences.
- Discovery of radioactivity in 1896 by Antoine Henri Becquerel (1852–1908), which helped lead to the development of atomic theory in the twentieth century.
- Development of the theories of special and general relativity, articulated by Albert Einstein (1879–1955) early in the twentieth century, which among other things showed the interrelationships between time, space, matter, and energy.
- The outgrowth of plate tectonics as a theory, from a hypothesis of continental drift proposed in 1912 by Alfred Wegener (1880–1930).

Unfortunately for the process of science, scientists nowadays have greater access to the modern mass media and can announce paradigm shifts to a large number of people before their hypotheses have gone through initial peer review. In these instances, any scientific discovery that seems to disagree with an already established hypothesis or theory can be proclaimed as a paradigm shift. Not coincidentally, such announcements garner much publicity in the popular press while the many rebuttals and clarifications proceed among scientists. The mainstream media, of course, depends upon such excitement and discord for increased circulation and revenues. Nonetheless, the use of mass media for promoting scientific ideas and egos is not unique to modern tabloid newspapers, television, and Web sites. Indeed, in the late nineteenth century, Cope and Marsh publicly sniped at one another via newspaper editorials, and telegrams telling of dinosaur discoveries provided instant gratification and glory to these and other paleontologists more than 100 years before e-mail (Chapter 3).

If one paleontology paradigm shift could be said to have occurred in the latter half of the twentieth century, it was the realization that catastrophes happen. This specific shift in scientific thinking demonstrates that "what goes around, comes around," that is, science does not necessarily travel in a straight line of progress but rather will revisit old ideas in new ways. **Catastrophism**, a belief system that the Earth underwent cataclysmic events or upheavals in the past, that wiped out most or all of life, has undergone a revival of sorts. However, the current flavor of catastrophism, informally termed **neo-catastrophism**, is not tied in with religious beliefs, as was the case with the various cultures that have flood myths or other stories of past global woe. In paleontology it is based on extensive data sets and hypotheses that have been rigorously examined, and in some cases falsified, by the worldwide scientific community (Chapter 2).

In this instance, neo-catastrophism is only about 25 years old. In 1980, four researchers, Luis Alvarez (1911–88), a former Nobel Prize winner in physics; his son Walter Alvarez, a geologist; and Frank Asaro and Helen Michel, both chemists, published what turned out to be a seminal paper in the peer-reviewed journal *Science*. In this paper, they proposed that the Earth was hit by a 10-km-wide bolide 65 million years ago, and that the dire consequences of this impact were directly

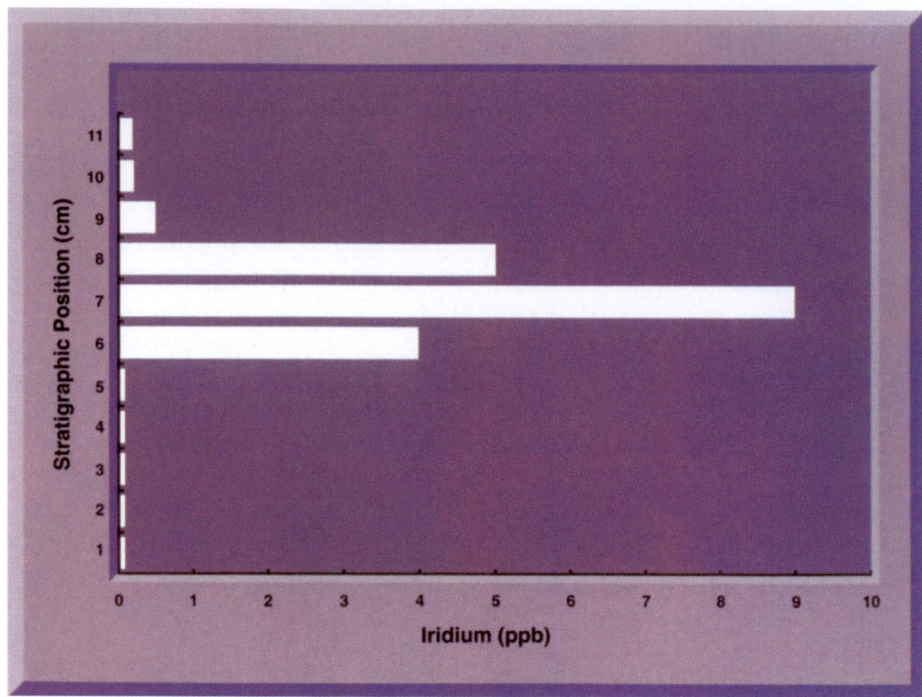

FIGURE 16.5 Iridium-bearing clay layer and its geochemical signature from the K–T boundary at Gubbio, Italy. Iridium concentration in parts per billion (ppb) throughout the clay layer. After Cowen (2000), *History of Life*, 3e, Blackwell Science, Inc., Malden, MA, p. 284, fig. 18.1.

responsible for the mass extinction evident in the geologic record at that time. This hypothesis generated a huge amount of debate and publicity. The resultant fanfare was at least in part because this extinction event coincided with the last known stratigraphic occurrence of non-avian dinosaurs.

Evidence supporting their hypothesis was based primarily on a geochemical anomaly, an unusually high concentration of iridium (^{192}Ir) in a 2.5-cm-thick clay layer in Gubbio, Italy. This clay layer separates underlying Cretaceous strata from overlying Tertiary Period strata in that area (the K–T boundary) and was interpreted as a depositional record of the very end of the Cretaceous Period. Iridium, a rare element in the Earth's crust that is chemically similar to platinum (^{195}Pt), is normally present in an abundance of less than 1 part per billion (ppb), about 0.3 ppb. In some meteorites, however, it has been measured at close to 3000 ppb, which is 10,000 times greater than background levels. In the K–T boundary clay at Gubbio, the iridium concentration was recorded as 9 ppb, which is about 30 times higher than background levels (Fig. 16.5). Later investigations by the research team revealed that other localities in the world, with a preserved K–T boundary, also had unusually high amounts of iridium.

After eliminating other potential sources of iridium, such as from volcanism, the researchers concluded that only a bolide impact could have been the source of such a large amount of iridium and distributed it worldwide within the short period of geologic time represented by the thin clay layer. Using an iridium concentration typical for a meteorite, they projected the mass of the meteorite needed for dispersal of the iridium found at the K–T boundary in different places in the world. These calculations yielded a hypothetical meteorite that was 10 km in diameter with a mass of about 5.0×10^{12} metric tons. Once caught by the Earth's gravity, the meteorite would have been moving at about 20 to 25 km/s, at a trajectory angle of 20 to 30° degrees in relation to the Earth's surface. On impact, it would have released kinetic energy of about 1.0×10^{23} J. This amount of energy is nearly 5 billion times the energy released by the nuclear bomb used by the USA on Hiroshima, Japan, near the end of World War II. The crater that should have resulted from the impact was predicted to have been more than 100 km wide. An impact of this size would have had devastating effects on life during the Late Cretaceous:

- The living and non-living matter in the direct impact area would have vaporized, instantly causing a local extinction of all organisms.
- The ground displacement in a radius of several hundred kilometers around the impact site would have been severe, causing major earthquakes.
- High-amplitude and fast-moving sea waves, **tsunamis**, would have engulfed nearby coastal areas if the meteorite hit the ocean, and earthquakes also caused by the impact would have caused their own tsunamis. For example, an earthquake off the west coast of Indonesia generated a devastating tsunami that killed more than 100,000 people on December 26, 2004.
- The heat wave associated with the release of energy would have ignited vast forest fires in any nearby forested ecosystems; the air displacement would have knocked down trees hundreds to thousands of kilometers away from the impact site.
- Any carbonate or sulfur-rich rocks (such as limestone or gypsum, respectively) near such a meteor strike would have combusted. In combination with water, the resulting chemicals would have produced long periods of acid rain, which in turn would have acidified soils for years afterwards.
- Dust ejected into the atmosphere, plus soot from forest fires, would have darkened the sky and blocked out sunlight for more than a year, hampering photosynthesis and sending global ecosystems into chaos because of the lack of food for animals. This proposed situation is nicknamed "nuclear winter," because a similar effect has been projected for a nuclear war.

The bolide impact hypothesis has been tested rigorously since it was proposed in 1980 and so far has been strengthened and refined by additional data. Moreover, alternative hypotheses, what few there were, now look weaker in the face of more data. For example, since the early 1980s, additional testing has revealed more than 100 sites in the world where the K–T boundary shows a higher-than-background concentration of iridium (Fig. 16.6). The relatively higher concentrations of iridium were proximal to North America, implying that the impact site was close to these sites of highest concentrations.

Other pieces of evidence for an impact have come from the presence at the K–T boundary of **microtektites**, which are tiny glass spheres caused by high temperatures associated with meteorite impacts. Moreover, an actual meteorite fragment was found at the K–T boundary on the seafloor of the Atlantic Ocean. **Shocked quartz**, which is a form of quartz created by extreme high-pressure conditions, is also abundant at the K–T boundary in some sites in North America. **Tsunamites**, which are sedimentary deposits formed by tsunamis, have been interpreted from both the seafloor and land deposits associated with the K–T boundary rimming the Caribbean. These independently derived forms of evidence therefore collectively argue in favor of the impact hypothesis.

The so-called "smoking gun" was announced in 1990 – researchers believed they might have located the impact structure, in its predicted location and fitting its predicted size. The **Chicxulub crater** (named after a Mexican town near the site) is about 170 km wide and located on the seafloor northeast of the Yucatan Peninsula of Mexico (Fig. 16.7). Radiometric dates taken from rocks at the bottom of the impact structure derived an age of 65.07 ± 0.1 Ma by the use of $^{40}Ar/^{39}Ar$

FIGURE 16.6 K–T boundary in Recife, Brazil, one of more than 100 such boundaries in the world that show elevated amounts of iridium, yet far removed from Gubbio, Italy.

ratios (Chapter 3). A comparison of this age to the dates that have been calculated for the K–T boundary for North America is favorable: for example, two K–T boundary dates are 65.0 ± 0.04 Ma and 65.4 ± 0.1 Ma. Although the resolution of radiometric age dates should never be expected to yield a perfect fit, the calculated age range for the crater and the K–T boundary, when combined with all of the other evidence, constitutes a very persuasive hypothesis that the Earth was hit by a large bolide at the end of the Cretaceous Period.

Nevertheless, the second part of the hypothesis, that the impact caused the mass extinction of the end-Cretaceous, is still under debate. This uncertainty is because the exact effects of the impact on global biota are still not well defined and its effects cannot be separated easily from whatever extinctions may have been happening already before the impact. In terms of previously existing adaptations, certainly very few could have prepared a species for surviving the abnormal environmental stresses that would have been imposed by a 10-km-wide meteorite impact. Nevertheless, extinctions have many causes and it is the coincident timing of many of them that defines a mass extinction, not the causes themselves.

Thus, causes of the extinctions that happened near and at the end of the Cretaceous have been the subject of intensive study, mainly focused on the North American K–T boundary. In this case, experts have been working on the same stratigraphic sections of the Upper Cretaceous **Hell Creek Formation** in Montana and the Dakotas. Continuing a long tradition in scientific research, different research groups working on the same problem disagree with one another about what

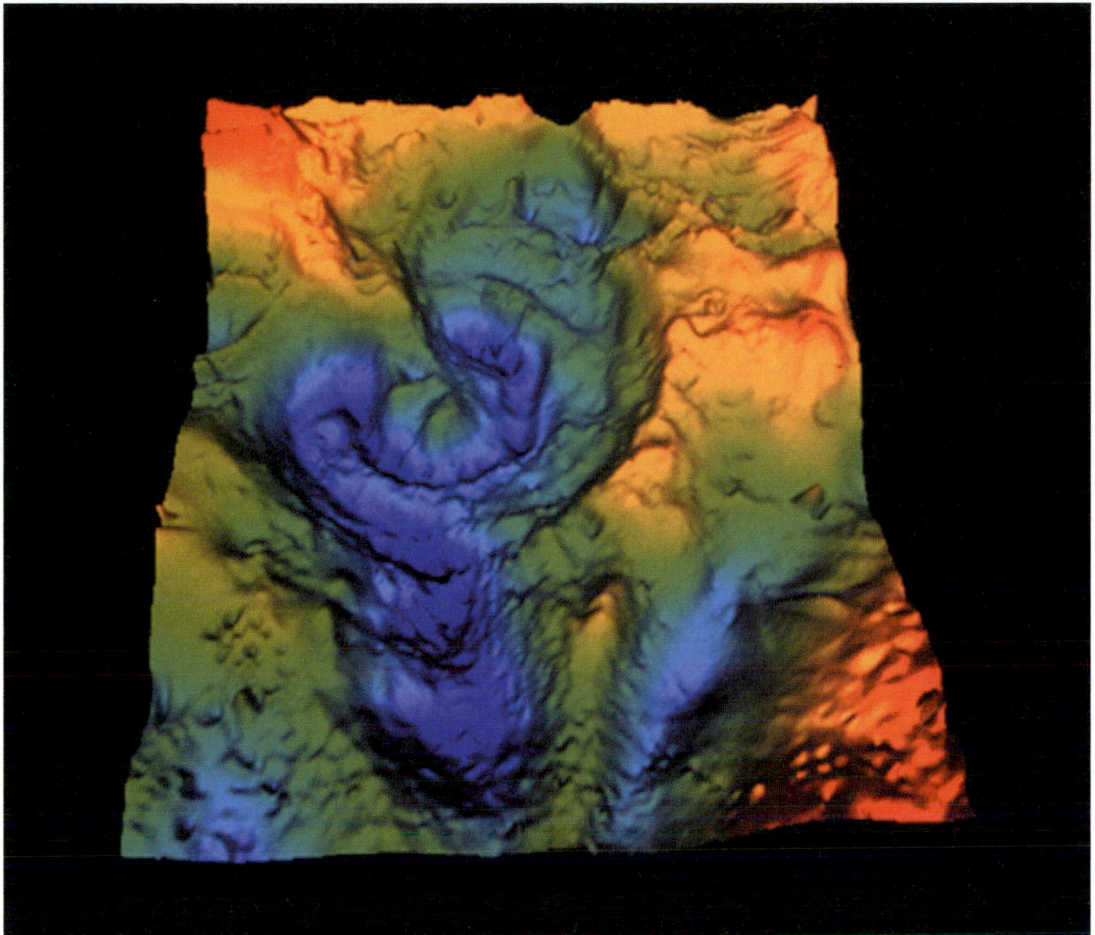

FIGURE 16.7 Map, based on a gravity field, of probable impact structure associated with the end of the Cretaceous Period, Chicxulub, Mexico. Reprinted courtesy of Virgil L. Sharpton, University of Alaska, Fairbanks.

the rocks tell them. In this instance, they disagree about whether the terminal Cretaceous extinction of dinosaurs was gradual or sudden.

Various sources of fossil and geological information indicate that some gradual changes in global systems, which included the biosphere, were already occurring before the end of the Cretaceous. Climate is largely driven by oceanic changes, and sea level was becoming lower during the last few million years of the Cretaceous. This lowering would have decreased the available habitats for marine life but also affected oceanic productivity. If organic materials from the deep ocean were not circulated sufficiently to the ocean surface (a process called **upwelling**), much planktonic algae would have died. This circumstance would have caused a deleterious ripple effect in oceanic ecosystems that depended on the algae for food. The oceanic record of the Late Cretaceous indeed shows that planktonic organisms in general, such as algae and the one-celled, shelled amoebae called **foraminifera**, decreased in numbers just below the K–T boundary. Some species vanish at the boundary and new species are found above the boundary. Geochemical evidence from Late Cretaceous marine rocks also suggests slower oceanic circulation and lower productivity before the K–T boundary. Additionally, the continents were in new positions relative to earlier in the Cretaceous, which also would have affected oceanic currents (Fig. 16.8). The exact effects of these positions on those circulation patterns, however, are uncertain.

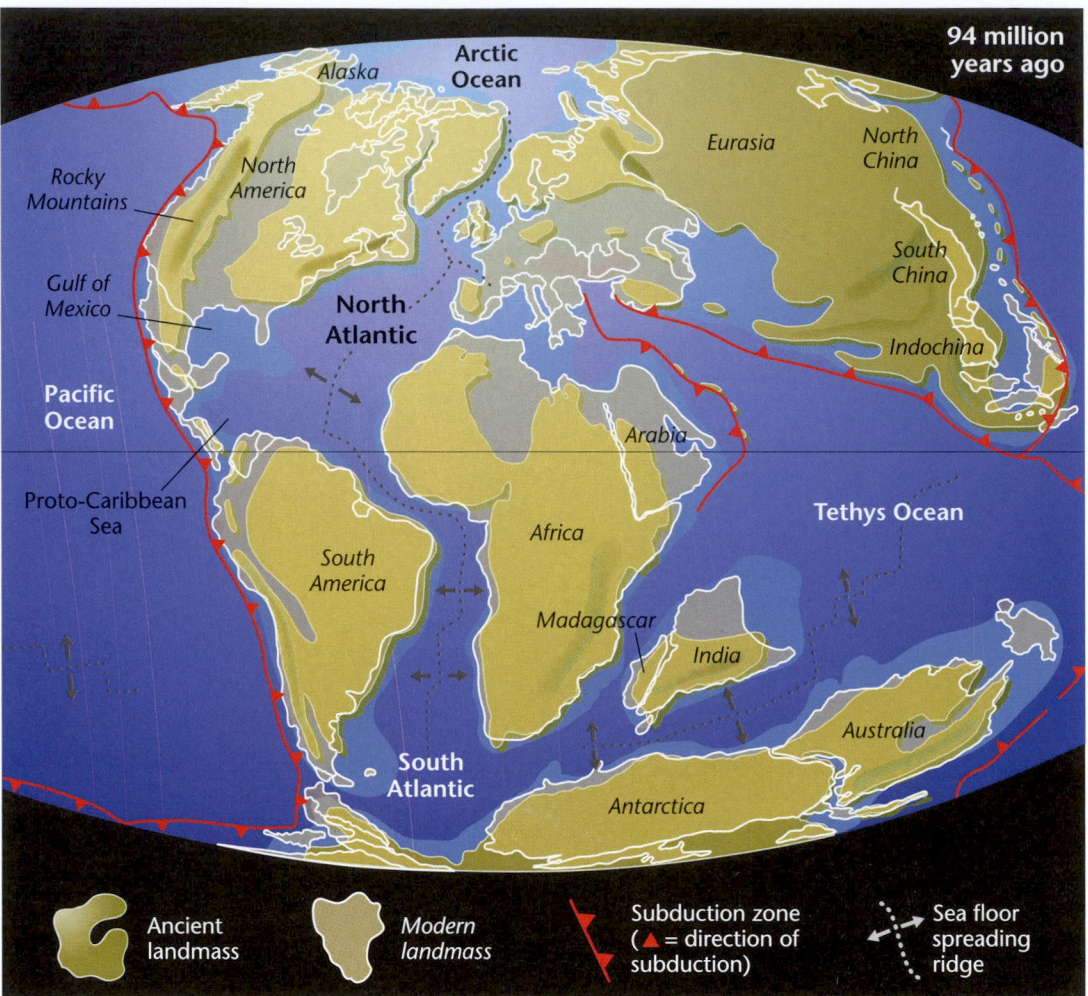

FIGURE 16.8 Paleogeographic map of the end-Cretaceous, showing continental landmasses relative to one another at the time the dinosaurs went extinct. Reprinted by permission from Scotese, C. R., 2001. Atlas of Earth History, Volume 1, Paleogeography, PALEOMAP Project, Arlington, Texas, 52 pp.

On land, extensive volcanism began in India about a million years before the K–T boundary, forming a widespread (500,000 km²) and as much as 2-km-thick deposit of basalts called the **Deccan Traps**. This volcanism was proposed as an alternative villain to a meteorite for the observed effects of a global catastrophe, including the iridium anomaly. Volcanic rocks are the only other potential sources of iridium, although in quantities much less than found in meteorites. For this and other reasons, the volcanic eruption as a single cause for the end-Cretaceous extinction has since been abandoned as a hypothesis. Nevertheless, the volcanism was massive enough to have had some effect on climate. Two possible effects are actually opposite in their results:

1 increased volcanism produced greater amounts of greenhouse gases, such as CO_2, which would have contributed to global warming; or
2 increased volcanism produced greater amounts of particulate matter (ash) in the atmosphere, which would have partially blocked sunlight and thus aided in global cooling.

Whether one, the other, or both process happened as a result of the Deccan basalts is still debated.

Also in terrestrial environments, angiosperms showed some gradual declines during the Maastrichtian in some locations, though in other places they were not affected. Their pollen record underwent a sharp decrease at the K–T boundary, corresponding with an increased abundance of fern spores just above the boundary. Independent studies of fossil plant remains also show a dramatic extinction of flowering plants at the boundary. Interestingly, angiosperms bounced back rapidly after the K–T boundary, and their pollen became proportionally more abundant in comparison to fern spores in the earliest Tertiary deposits. Ferns typically are pioneer species in terrestrial ecosystems following a natural disaster, so their abundance following a global catastrophe, followed by a gradual return of flowering plants, is expected. This pattern is consistent with the impact hypothesis, but of course may be related to another, unknown cause.

The marine and terrestrial vertebrate groups that died out by the end of the Cretaceous were varied. Among the marine reptiles, ichthyosaurs actually became extinct millions of years before the K–T boundary, followed by plesiosaurs and mosasaurs (Chapter 6). Pterosaurs, which showed up at about the same time as dinosaurs in the Late Triassic, had also been declining in both number and diversity before the end of the Cretaceous. Among terrestrial groups of vertebrates, the greatest losses in species were in birds and marsupials. Fish, amphibians, crocodiles, snakes, lizards, turtles, and mammals other than marsupials incurred casualties, but their species losses were no more than 40% by some estimates.

Interestingly, the extinctions did not seem to discriminate against vertebrates based on thermoregulation (Chapter 8). Some sea-dwelling vertebrates that were presumably ectotherms, such as mosasaurs, plesiosaurs, and ichthyosaurs, became extinct, but sea turtles did not. On land, endotherms such as pterosaurs, birds, and marsupials were badly affected, but others were not. Ectotherms, such as fish, amphibians, and so-called "modern" reptiles, made it into the Tertiary. One of the few patterns noticeable in vertebrate extinctions is that the larger animals became extinct, whereas the smaller ones survived. Why this happened is unknown, but it may have been related to food availability (and organic productivity in general) after the Cretaceous, a time when ecosystems may not have been able to sustain large animals. Another ecological trait shared by terrestrial vertebrates that survived the K–T boundary is their living in freshwater habitats; in one study, vertebrate clades interpreted to have lived in such ecosystems had a nearly 90% survival rate.

Most importantly, from the perspective of dinosaur studies, all non-avian dinosaurs died at the end of the Cretaceous. Maastrichtian dinosaurs are represented by some familiar forms: ankylosaurids, nodosaurids, hypsilophodontids, hadrosaurids, pachycephalosaurids, ceratopsians, titanosaurids, dromaeosaurids, ornithomimids, and tyrannosaurids (Fig. 16.9). This long list partially illustrates how their erasure by the end of that 6-million-year timespan was significant. In accordance with the concept of a global extinction, non-avian dinosaurs also vanished everywhere, leaving no traces. No *in-situ* dinosaur bones, tracks, eggs, or nests are found above the K–T boundary. With regard to the latter, dinosaur eggs are very abundant as whole specimens and eggshell fragments in Upper Cretaceous strata of southeastern China, but they are completely absent from overlying Tertiary strata. The totality and finality of dinosaur extinction, when these animals had enjoyed a 165-million-year history, is the main evidence for the K–T extinction.

As mentioned earlier, two hypotheses have been proposed for the timing of dinosaur extinctions toward the end of the Cretaceous: gradual and sudden extinction. Both proponents of the gradual and sudden extinction hypotheses sampled the last 10 million years of the Cretaceous, as represented in the Hell Creek Formation. Supporters of the gradual-extinction hypothesis presented data showing that dinosaur

FIGURE 16.9 Maastrichtian dinosaurs that lived during that age, all of which became extinct by its end.

genera gradually diminished in diversity over the last 8 million years of the Cretaceous, from 28 genera (at about 74 Ma) to none by the end of the Cretaceous. The same studies showed that mammal diversity, based on the number of fossil mammal genera, increased dramatically in the same area after the end of the Cretaceous. In contrast, more recent studies in the same area but by different researchers concluded that dinosaur fauna were abundant until the K–T boundary, then suddenly absent from the geologic record after that boundary. According to these studies, some of the dinosaur bones are as abundant in the last 3-meter interval of Hell Creek as they are lower in the section, thus favoring a sudden extinction.

Without denigrating the data or the tremendous effort behind its collection, problems with accepting either hypothesis on the basis of work done so far should be noted. The majority of information for these hypotheses was derived from relatively few skeletal remains deposited in terrestrial environments. These remains are fragmentary in some cases, consisting of only one specimen per species, or they could have been reworked from underlying deposits. This situation complicates strict age assignments given to the geologic ranges of dinosaur genera found at specific horizons. Also, how the data are plotted or what statistical techniques are used to show either gradual or sudden extinction can create much dissension. For example, in some studies the data were plotted as percentage extinctions of families, using the Linnaean classification system, which may not be as instructive as genera or species lists. In at least two instances, only a single specimen of a single species, representing a Linnaean family of dinosaurs, is recorded as having gone extinct. Taxonomic hindrances to extinction analyses have been alleviated somewhat by applying cladistics in recent years, but the small number of dinosaur specimens renders Linnaean versus cladistic arguments moot.

Researchers freely acknowledge the problems involved with their data. The paleontologists and their volunteers working on the Hell Creek Formation have been incredibly thorough in their searches for fossil remains. Thus, the current disagreement between them is certainly not over the fieldwork or data collected, but rather on the interpretation of such data, a common source of contention in paleontology (Chapter 2). Statistical analyses of the relatively meager data gained so far suggest that as much as 40% of the dinosaur faunas in the studied Hell Creek interval could have become extinct before the paleontologists would actually see any broad-scale change. As a result, only further work and agreement on the emergence of common patterns of dinosaur extinctions will help to resolve whether dinosaurs died out gradually or suddenly.

As discussed previously, bones have their advantages and disadvantages as sources of paleontological data. For example, the occurrence of dinosaur bones in Tertiary deposits is attributed by most paleontologists to reworking of the bones from older Cretaceous deposits. In other words, bones do not tell us when the last dinosaur walked on the Earth. This question can be answered by dinosaur tracks, which are invariably *in-situ* records of when dinosaurs were alive. So far, the youngest dinosaur track found, made by an adult hadrosaurid, occurs only 37 cm below the K–T boundary in southern Colorado. Older tracks of ceratopsians (Chapter 13), and a probable *Tyrannosaurus* track (Chapter 9), occur only a few meters below the boundary. Although depositional rates vary, a reasonable estimate is that the 37-cm interval only represents a few thousand years, meaning that at least one dinosaur was alive soon, geologically speaking, before the end of the Cretaceous.

Modern Extinctions

In a literal interpretation of phylogenetic relationships, dinosaurs did not go extinct because they are still around today as birds and are possibly more diverse now than they ever were during the Mesozoic. However, because some modern bird species have become so reduced in numbers that they are less likely to breed and produce viable offspring, they are also undergoing a possible mass extinction. What this extinction means in terms of how much ecosystems will change is unknown, but the world would certainly be a very different one without them (Chapter 15).

The bulk of recent scientific evidence has led to the hypothesis that humans are currently observing a sixth mass extinction in the Phanerozoic Eon. Moreover, this mass extinction has been correlated with human alteration of habitats through:

1 introduced species;
2 resource acquisition;
3 pollution; and
4 global climate change.

Documenting to what degree human impact is causing extinctions is actually the only controversial point within the scientific community regarding the modern mass extinction. In this case, separating so-called normal extinctions that would have happened without human influence, such as those at the end of the Permian or Cretaceous Periods, from those caused by humans, is problematic. However, so far the only species in the history of the Earth shown to have been ultimately responsible for the extinction of another species is *Homo sapiens*. The recognized human causes of extinctions include the following factors, introduced earlier:

■ Pollution of ecosystems with chemicals that disrupt reproductive cycles, documented particularly in predatory birds such as eagles, hawks, and other raptors (Fig. 16.10).
■ Alteration of habitats through deforestation or filling wetland areas for development.
■ Harvesting, hunting, or purposeful elimination of organisms targeted as pests.
■ Changing the biogeographic distribution of previously isolated populations, resulting in competition for resources, parasitism, and predation where it did not exist before the introduction of the imported species.

FIGURE 16.10 Bald eagle (*Haliaeetus leucocephalus*), a modern predatory theropod of North America that was considered an endangered species and is now protected in the USA. PhotoDisk.

Even in cases when humans were not the species that caused the ecological disturbance that resulted in local extinctions, humans were still ultimately responsible for those extinctions. Humanly introduced species are called **exotic**, meaning that humans purposefully or accidentally brought them to a new ecosystem. Exotic species can then qualify as invasive species if they displace other native species in an ecosystem or otherwise prompt major imbalances. Examples of invasive species that proliferated in their new terrestrial ecosystems and encouraged adverse interspecific competition include:

1 lagomorphs (rabbits and hares) in Australia, which have overpopulated and subsequently denuded vegetation;
2 the American gray squirrel (*Sciurus carolinensis*), which is competitively displacing the native European red squirrel (*Sciurus vulgaris*) in England; and
3 the prolific creeping vine nicknamed "kudzu" (*Pueraria lobata*), the bane of all geologists in the southeastern USA who prefer to see rocks not covered by vegetation.

Surprisingly, probably the most devastating invasive species in terrestrial ecosystems is the beloved domestic cat, *Felis domestica* (Fig. 16.11). As mentioned in Chapter 15, both feral and human-owned cats are responsible for killing more than an estimated 100 million songbirds each year; feral cats in particular have an adverse effect on ground-nesting birds in grasslands and forests that previously lacked such predators. The shrinking of proper habitats for migrating and nesting birds exacerbates this problem, as the limited choices tend to overlap with urbanized areas that contain large cat populations.

FIGURE 16.11 Invasive species in terrestrial habitats thought to be responsible for decimating modern bird populations, the mammal *Felis domestica*, which is proliferated in those habitats by another species of mammal, *Homo sapiens*.

Local extinctions, which may then have later translated into global extinctions, are not a recent phenomenon. Many local extinctions are correlated with the arrival of human populations. Notable historical examples of local extinctions include the large marsupials and birds in Australia about 40,000 years ago, the large placental mammals in North America and Europe about 10,000 years ago, the fauna on Madagascar about 2000 years ago, and the moas of New Zealand starting about only 800 years ago. The last two of these local extinctions resulted in the loss of the elephantbird, the largest avian theropod, and a species of moa that was the tallest avian theropod, respectively (Chapter 15).

Did Dinosaurs Become Extinct?

After all of the global change (whether gradual and catastrophic) associated with the end of the Cretaceous had stabilized, the plant and animal composition of terrestrial ecosystems was forever changed and much of what we consider as the modern world had arrived. Some ferns, conifers, angiosperms, placental mammals, crocodiles, lizards, snakes, amphibians, and fish made it into the Tertiary Period, but marine reptiles and pterosaurs were gone. The only dinosaurs that survived were the avian theropods, specifically the neornithines.

Still, in a philosophical sense, dinosaurs are very much alive in our imaginations, as is evident by literature, art, and technological attempts to re-create their visages and behavior through computer animations and other special effects. Paleontologists who work with dinosaur body or trace fossils also try to re-create dinosaurs in their minds as living animals, rather than as static museum displays. The ubiquitous images of dinosaurs in popular culture and children's enduring fascination with them should also ensure that dinosaurs should live in this sense as long as humans live.

The long history of dinosaurs and their impact on human life continues, but will be changed with future discoveries and insights. Our scientific understanding of how dinosaurs originated, lived, and functioned as transitory elements of Mesozoic landscapes helps us to better understand the evolution of life. Thinking about how they became extinct or how some of these dinosaurs may be with us today are also important to our understanding them. Most importantly, the process we use to understand dinosaurs, composed of scientific inquiry and imagination, provides a framework for understanding other aspects of our natural world and our role in its continuing evolution.

SUMMARY

Mass extinctions have happened five times in the Phanerozoic Eon before humans emerged. Although the end-Permian extinction event was the greatest of these in the number of species lost, the end-Cretaceous mass extinction is the most famous, mostly because it involved non-avian dinosaurs. However, part of the fame associated with the end-Cretaceous extinction is also related to its possible causes; one hypothesis is a catastrophic meteorite impact. This impact, once hypothesized on only a few lines of evidence, was continually verified through subsequent research that eventually resulted in the discovery of the probable impact structure. Other contributing factors in the extinctions could have been global climate change, which included a drop in worldwide sea level, as well as extensive volcanism in India. Dinosaurs were not the only creatures to become extinct; numerous planktonic algal and foraminiferan species, as well as some invertebrates, vanished from the oceans. Among vertebrates, all large marine and flying reptiles disappeared, as well as many birds and marsupial species. The survivors included various plant groups (ferns, conifers, and flowering plants), other mammals, reptiles (crocodiles, snakes, lizards, and turtles), and neornithines (modern birds).

Current evidence indicates that another mass extinction is occurring now and it is affecting the numbers and diversity of bird species. Ironically, considering the public fascination with dinosaurs of the Mesozoic, this extinction represents a second one for dinosaurs and it is largely attributable to human alterations of terrestrial and marine ecosystems. Human-caused factors that contribute to local and global extinctions include the introduction of invasive species, changing of habitats, overharvesting, and pollution. As a result, lessons can be learned through this overlapping of the geologic past and present associated with the extinctions of avian and non-avian dinosaurs.

DISCUSSION QUESTIONS

1. This chapter mentioned that the states of thermoregulation, endothermic or ectothermic, did not seem to play a role in which groups of vertebrates became extinct at the end of the Cretaceous. Some of the evidence presented for thermoregulation of dinosaurs suggested that some of them had an intermediate form of thermoregulation that was between endothermy and ectothermy during their growth. How might such a thermoregulation method have been a disadvantage for some dinosaurs at the end of the Cretaceous Period, thus selecting against their survival?

2. Review the possible causes of extinctions at the end Permian and compare them to those proposed for the end of the Cretaceous. What similarities or differences did you notice? Would the extinction still have been noticeable if a meteorite had not hit the Earth at the end of the Cretaceous?

3. The author states that a mass extinction is happening now and implies that it is comparable to the mass extinctions of the geologic past. How could you disprove this statement? What evidence would have to be compiled to contradict what is defined as a mass extinction, whether it is in the geologic past or present?

4. Review the major dinosaur clades of Chapters 9–13 and chart when they went extinct. Which ones were extinct by the end of the:
   a. Triassic,
   b. Jurassic,
   c. Early Cretaceous, and
   d. Late Cretaceous?

5. How could a local extinction in the Hell Creek fauna of North America be related to the K–T bolide impact that happened in the Gulf of Mexico? Explain how a local extinction can then be unrepresentative of a global extinction. What new information would test the hypothesis that the Hell Creek fauna represents a global extinction?

6. Why would freshwater vertebrates have a survival advantage over land-dwelling vertebrates? Think of and list all of the ecological factors that would affect either group and look for contrasts.

7. Why would smaller terrestrial vertebrates have a survival advantage over larger terrestrial vertebrates? Think of and list all the ecological factors that would affect either group and look for contrasts, especially in the light of thermoregulation (as mentioned in Question 1).

8. Explain how both marine regressions and extensive volcanism may have contributed to dinosaur extinctions before a bolide impact. Could the impact have been just another factor in extinctions, rather than a single cause? Explain why or why not.

9. Think of some invasive species in your local area and list them. What are their effects on local extinctions? What could be done to decrease their abundance or adverse effects on native plants and animals?

10. Find out which rare (endangered) species of birds live in your area. What are the factors interpreted to have caused their rarity?

Bibliography

Alvarez, L. W., Alvarez, W., Asaro, F. and Michel, H. V. 1980. Extraterrestrial cause for the Cretaceous-Tertiary boundary extinction. *Science* **208**: 1095–1108.

Alvarez, W. 1997. T. Rex *and the Crater of Doom*. Princeton, N.J.: Princeton University Press.

Archibald, J. D. and Fastovsky, D. E. 2004. "Dinosaur extinction". *In* Weishampel, D. B., Dodson, P. and Osmólska, H. (Eds), *The Dinosauria* (Second Edition). Berkeley, California: University of California Press. pp. 672–684.

Bourgeois, J. T., Hansen, T. A., Wilberg, P. L. and Kauffman, E. G. 1988. A tsunami deposit at the Cretaceous-Tertiary boundary in Texas. *Science* **241**: 567–570.

Brett, R. 1992. The Cretaceous-Tertiary extinction: A lethal mechanism involving anhydrite target rocks. *Geochemica et Cosmochimica Acta* **56**: 3603–3606.

Courtillot, V. and McClinton, J. 1999. *Evolutionary Catastrophes: The Science of Mass Extinction*. Cambridge, UK: Cambridge University Press.

d'Hondt, S., Pilson, M. E. Q., Sigurdsson, H., Hanson, A. K. and Carey, S. 1994. Surface water acidification and extinction at the Cretaceous-Tertiary boundary. *Geology* **22**: 983–986.

Dingus, L. W. 1984. Effects of stratigraphic completeness on interpretations of extinction rates across the Cretaceous-Tertiary boundary. *Paleobiology* **10**: 420–438.

Duncan, R. A. and Pyle, D. G. 1988. Rapid eruption of the Deccan flood basalts at the Cretaceous-Tertiary boundary. *Nature* **333**: 841–843.

Eaton, J. G., Kirkland, J. I. and Doi, K. 1989. Evidence of reworked Cretaceous fossils and their bearing on the existence of Tertiary dinosaurs. *Palaios* **4**: 281–286.

Eldredge, N. 1991. *The Miner's Canary: Unraveling the Mysteries of Extinction*. Princeton, N.J.: Princeton University Press.

Eldredge, N. 1998. *Life in the Balance: The Biodiversity Crisis*. Princeton, N.J.: Princeton University Press.

Erwin, D. H. 1994. The Permo-Triassic extinction. *Nature* **367**: 231–236.

Fastovsky, D. E. and Dott, R. H., Jr. 1986. Sedimentology, stratigraphy, and extinctions during the Cretaceous-Paleogene transition at Bug Creek, Montana. *Geology* **14**: 279–282.

Florentin, J.-M., Maurrasse, R. and Sen, G. 1991. Impacts, tsunamis, and the Haitian Cretaceous-Tertiary boundary layer. *Science* **252**: 1690–1693.

Frankel, C. 1999. *The End of the Dinosaurs: Chicxulub Crater and Mass Extinctions*. Cambridge, UK: Cambridge University Press.

Hallam, A. 1990. The end-Triassic mass extinction event. *Geological Society of America Special Paper* **247**: 577–583.

Hildebrand, A. R., Penfield, G. T., Kring, D. A., et al. 1991. Chicxulub Crater: A possible Cretaceous-Tertiary boundary impact crater on the Yucatan peninsula. *Geology* **19**: 867–871.

Hurlbert, S. H. and Archibald, J. D. 1995. No statistical support for sudden (or gradual) extinction of dinosaurs. *Geology* **23**: 881–884.

Kring, D. A. 1995. The dimensions of the Chicxulub impact crater and impact melt sheet. *Journal of Geophysical Research* **100**: 16979–16986.

Kyte, F. T. 1998. A meteorite from the Cretaceous/Tertiary boundary. *Nature* **396**: 237–239.

Leakey, R. and Lewin, R. 1995. *The Sixth Extinction: Patterns of Life and the Future of Humankind*. New York: Doubleday Books.

Lofgren, D. L., Hotton, C. L. and Runkel, A. C. 1990. Reworking of Cretaceous dinosaurs into Paleocene channel deposits, upper Hell Creek Formation, Montana. *Geology* **18**: 874–877.

Powell, J. L. 1998. *Night Comes to the Cretaceous: Comets, Craters, Controversy, and the Last Days of the Dinosaurs*. New York: W. H. Freeman.

Raup, D. 1991. *Extinction: Bad Genes or Bad Luck?* New York: W. Norton & Company, Inc.

Retallack, G. J. Leahy, G. D. and Spoon, M. D. 1987. Evidence from paleosols for ecosystem changes across the Cretaceous-Tertiary boundary in Montana. *Geology* **15**: 1090–1093.

Sharpton, V. L. and Ward, P. E. (Editors, and all authors therein). 1990. Global catastrophes in Earth history. *Geological Society of America* Special Paper 247.

Sharpton, V. L., Dalrymple, G. B., Marin, L. E., Ryder, G., Schuraytz, B. and Urrutia, F. J. 1992. New links between Chicxulub impact structure and the Cretaceous-Tertiary boundary. *Nature* **359**: 819–821.

Sheehan, P. M. and Fastovsky, D. E. 1992. Major extinctions of land-dwelling vertebrates at the Cretaceous-Tertiary boundary, eastern Montana. *Geology* **20**: 556–560.

Sheehan, P. M., Fastovsky, D. E., Barreto, C. and Hoffmann, R. G. 2000. Dinosaur abundance was not declining in a "3 m gap" at the top of the Hell Creek Formation, Montana and North Dakota. *Geology* **28**: 523–526.

Signor, P. W. and Lipps, J. H. 1982. "Sampling bias, gradual extinction patterns and catastrophes in the fossil record". *In* Silver, L. T. and P. H. Schultz (Editors), Geological Implications of Impacts of Large Asteroids and Comets on the Earth. *Geological Society of America Special Paper* **190**: 291–296.

Sigurdsson, H., S. d'Hondt and S. Carey. 1992. The impact of the Cretaceous/Tertiary bolide on evaporite terrane and generation of major sulfuric acid aerosol. *Earth and Planetary Science Letters* **109**: 543–559.

Smit, J. and van der Kaars, S. 1984. Terminal Cretaceous extinctions in the Hell Creek area, Montana: Compatible with catastrophic extinction. *Science* **223**: 1177–1179.

Stimmesbeck, W., Barbarin, J. M., Keller, G., et al. 1993. Deposition of channel deposits near the Cretaceous-Tertiary boundary in northeastern Mexico: Catastrophic or normal sedimentary deposits? *Geology* **21**: 797–800.

Swisher, C. C., Grajales, N. J. M., Montanari, A., et al. 1992. Coeval Ar-Ar ages of 65 million years ago from Chicxulub crater melt-rock and Cretaceous-Tertiary boundary tektites. *Science* **257**: 954–958.

Ward, P. D. 1993. *On Methuselah's Trail: Living Fossils and the Great Extinctions*. New York: W. H. Freeman.

White, P. D., Fastovsky, D. E. and Sheehan, P.M. 1998. Taphonomy and suggested structure of the dinosaurian assemblage of the Hell Creek Formation (Maastrichtian), eastern Montana and western North Dakota. *Palaios* **13**: 41–51.

Wignall, P. B. and Hallam, A. 1997. *Mass Extinctions and Their Aftermath*. Oxford, UK: Oxford University Press.

Williams, M. E. 1994. Catastrophic versus noncatastrophic extinction of the dinosaurs: Testing, falsifiability, and the burden of proof. *Journal of Paleontology* **68**: 183–190.

Wolbach, W. L., Gilmour, I., Anders, E., Orth, C. J. and Brooks, R. R. 1988. A global fire at the Cretaceous/Tertiary boundary. *Nature* **334**: 665–669.

Wolfe, J. A. and Upchurch, G. R. 1987. Leaf assemblages across the Cretaceous-Tertiary boundary in the Raton Basin, New Mexico and Colorado. *National Academy of Sciences Proceedings* **84**: 5096–5100.

16

Glossary

A

Acetabulum   Opening (hole) on each side of a pelvis that allows for the insertion of a ball-like proximal end of each femur. A distinguishing character of a dinosaur.

Adaptation   Physical attribute of an organism that can help it to survive at least long enough to successfully reproduce.

Aerodynamics   Physics of air flow. Important for interpreting sedimentary environments and taphonomy.

Allantois   Sac that develops between the eggshell and amnion for respiration of the embryo in a cleidoic egg.

Allochthonous   In taphonomy, refers to a body that has been moved (perhaps very far) from where it originally died.

Allometry   Study of size and how it changes with growth of an organism in various dimensions.

Allosauridae   Clade of saurischian dinosaurs, placed with clades Theropoda, Tetanurae, Avetheropoda, and Carnosauria.

Altricial   Behavioral reference to state of juveniles not capable of moving and fending for themselves soon after birth, requiring much parental care. (Contrast with precocial.)

Amino acid   Organic compound that forms the basis for much soft tissue in an animal and helps to facilitate biochemical reactions. Must have a carboxyl group ($COOH$) and amino group (NH_2), in which the carboxyl performs as an acid and the amino as a base.

Amnion   Fluid-filled sac surrounding a developing embryo, characteristic of amniotes.

Amniota   Clade of tetrapods that reproduce through by enclosed eggs with an amnion. Members are amniotes.

Amphibia   Paraphyletic group of chordates and tetrapods (formerly Class Amphibia under gradistic classification) that normally are dependent on water bodies for reproduction. Contrast with amniotes.

Anachronism   Juxtaposition of items or situations that belong to different and separate time periods, such as *Stegosaurus* (of the Jurassic Period) with *Tyrannosaurus* (of the Cretaceous Period).

Ankylosauria   Clade of ornithischian dinosaurs, placed with clade Thyreophora, that includes clades Ankylosauridae and Nodosauridae.

Ankylosauridae   Clade of ornithischian dinosaurs, placed with the clades Thyreophora and Ankylosauria, that shares common ancestor with clade Nodosauridae.

Anterior   Reference to the front part of an animal or front surface of a part.

Appendicular   Anatomical reference to the appendages (limbs) of an animal.

Arboreal hypothesis   Explanation of origin of flight in theropods from tree-dwelling species. Also known as the "trees down" hypothesis.

Archaeology Study of human artifacts and other traces of human behavior. Someone who studies artifacts is an archaeologist.

Archosauria Clade of diapsids, characterized by a minimum of the following traits: openings anterior to the orbits (antorbital fenestrae), teeth with serrations compressed laterally and none on the palate, dentary fenestrae, differently shaped calcaneum, and elongated ilium and pubis. Part of lineage of dinosaurs.

Arctometatarsalia Clade of saurischian dinosaurs, placed with clades Theropoda, Tetanurae, Coelurosauria, and Maniraptoriformes. Named for how the middle metatarsal is "pinched" between metatarsals on either side of it. Includes clades Troodontidae, Ornithomimosauria, and Tyrannosauridae.

Asthenosphere Hot, plastically flowing, and partially molten portion of the upper part of the mantle that theoretically interacts with the lithosphere to cause plate tectonic movement.

Autochthonous In taphonomy, refers to a body that is in the same place where it originally died.

Aves Clade of saurischian dinosaur, placed with the clades Theropoda, Tetanurae, and Avetheropoda, Coelurosauria, Maniraptoriformes, Maniraptora, and Avialae. Colloquially known as birds.

Avetheropoda Clade of saurischian dinosaurs, placed with clades Theropoda and Tetanurae. Includes clades Carnosauria, Sinraptoridae, Allosauridae, and Coelurosauria.

Axial Anatomical reference to the axis of an animal, such as its vertebrae.

B

Binomial Nomenclature In Linnaean classification, method that names a species through combined use of a genus and trivial name, e.g., *Stegosaurus stenops*.

Biocoenosis Assemblage of organisms in the fossil record that is autochthonous and thus represents the living community.

Biogeochemistry Study of chemical processes caused by organisms in geologic media and how elements are cycled.

Biologic Succession, (Principle of) Observation that fossil assemblages may change in a vertical sequence of rocks (succeed one another). Explained through extinctions and evolution that happened through over time.

Biological Evolution, (Theory of) Generally accepted explanation for observations of organisms that are (or were) modified through descent.

Biomechanics Study of how living systems, such as animal bodies, perform work.

Biometry Study of life through measurements and statistical methods.

Biomineralization Process for formation of hard parts (shells, bones, teeth) in organisms.

Biomolecule Combinations of elements that only could have been formed only by organisms. Examples include nucleic acids, lipids (fatty molecules), carbohydrates (also known as sugars), and proteins.

Biostratinomy Processes that affect an organism between its death and final burial; an important part of taphonomy.

Bioturbation Process in which an organism causes mixing of a sediment. Product of bioturbation is a bioturbate texture.

Bipedal Using of two legs for locomotion.

Bird General term applied to members of clade Aves, which includes *Archaeopteryx* and all of its descendants.

Body Fossil   Any direct evidence of ancient life as represented by bodily remains. Includes actual or altered body parts (such as bone), impressions of any body part (skin, muscles, feathers), and eggs (which are body parts of embryos).

Bone   Biomineralized skeletal tissue in vertebrates composed of dahllite but often in combination with softer organic tissue in varying proportions. Can be either cancellous ("spongy" or low-density) or compact (high-density).

Brachiosauridae   Clade of saurischian dinosaurs, placed with clades Sauropodomorpha and Sauropoda, which had more rounded skulls more rounded than diplodocids and nares were positioned more anteriorly (and below) the orbitals.

Brooding   Behavior where eggs or juveniles are provided with care by the parents, such as through protection or insulation.

C

Camarasauridae   Clade of saurischian dinosaurs, placed with clades Sauropodomorpha and Sauropoda. They have more rounded skulls than diplodocids, spoon-like teeth, and smaller numbers of their cervical, dorsal, and caudal vertebrae.

Carbonization   Fossilization process where volatile elements, such as carbon, nitrogen, and oxygen, depart from an organism's body and leave a carbon impression. Common mode of preservation for soft tissues in both plants and animals.

Carina, (of Tooth)   Narrow, blade-like or ridge-like part of a tooth that may have serrations or other denticles.

Carina, (of Sternum)   Bladed middle part of the sternum to which large flight muscles are attached in avians. Also known as the keel.

Carnivory; (Carnivorous)   Meat eating, but more generally the eating of any animal, whether through predation or scavenging.

Carnosauria   Clade of saurischian dinosaurs, placed with clades Theropoda, Tetanuare, and Avetheropoda. Previously was a general category for all large meat-eating dinosaurs.

Cartilage   Mass of protein collagen arranged as parallel, linear fibers, forming a part of a flexible, non-ossified connective tissue in a chordate.

Cast   Positive (convex) feature made from what was originally a negative (concave) feature, such as for preservation of dinosaur skin and tracks.

Caudal   Anatomical position referring to the tail of an animal, such as caudal vertebrae.

Centrosaurine   Clade of ornithischian dinosaurs, placed with clades Marginocephalia, Ceratopsia, and Neoceratopsia, characterized by short squamosals.

Cerapoda   Node-based clade of ornithischian dinosaurs. Includes sister clades of Marginocephalia and Ornithopoda.

Ceratopsia   Clade of ornithischian dinosaurs, placed with clade Marginocephalia, characterized by a minimum of the following traits: rostral bone anterior to the maxilla that paired with a predentary to form a sharp beak, frill formed by the parietals that hung past the rest of the skull, cheeks that extend laterally and posteriorly, and a palate positioned high in the skull. Includes sister clades of Psittacosauridae and Neoceratopsia.

Ceratopsidae   Clade of ornithischian dinosaurs, placed with clades Marginocephalia, Ceratopsia, and Neoceratopsia.

Ceratosauria   Clade of saurischian dinosaurs, placed with clade Theropoda, characterized by a minimum of the following traits: fusion of bones in ankle and feet, sacrum fused to ilium and ribs, two fenestrae on the pubis, four digits, but with digit IV reduced so that digits I through III were the most functional, and two pairs of cavities (pleurocoels) in the cervical vertebrae.

Cervical   Any body part with reference to the neck, such as cervical vertebrae.

Character   Trait or characteristic of an organism's anatomy distinctive enough to use for classification, such as in cladistics.

Chasmosaurine   Clade of ornithischian dinosaurs, placed with clades Marginocephalia, Ceratopsia, and Neoceratopsia, characterized by long squamosals.

Chordate   Clade of animal distinguished by a notochord, dorsal nerve cord, and pharyngeal gill slits. Placed with this clade is the clade Vertebrata.

Chromosome   Packet of genetic material in a cell, contains DNA.

Clade   Group of organisms defined on the basis of common ancestry as indicated by shared, derived characters (synapomorphies).

Cladistics   Classification system that places organisms into clades based on shared derived traits (synapomorphies). (See also phylogenetic classification.)

Cladogram   Diagram showing relatedness of clades to one another; serves as a visual display of a hypothesis on evolutionary relationships between clades.

Cleidoic Egg   Enclosed structure in combination with an embryo that provides a food supply (through a yolk sac) and a membrane for respiration, temperature maintenance, and waste disposal.

Cloaca   Organ in some birds that serves a dual purpose as the end orifice for the alimentary canal and oviduct.

Clutch   One or more eggs laid in a single egg-laying episode by a mother.

Coelophysoidea   Clade of saurischian dinosaurs, dinosaurs placed with clades Theropoda and Ceratosauria. One of the first theropod clades in the Late Triassic.

Coelurosauria   Clade of saurischian dinosaurs, placed with clades Theropoda, Tetanurae, and Avetheropoda. Includes other clades far too numerous to mention here, but with all of its members are united by possession of a semilunate carpal.

Cold-blooded   Popularized description of ectothermy.

Compression Shape   Overall geometric outline of a track. Useful for identifying a track made in substrates other than sand or mud, and a common mode of preservation for undertracks.

Continental Drift, (Hypothesis of)   First substantial hypothesis, stated originally by Alfred Wegener, explaining the geographic distribution of similar fossils, rocks, and rock structures in widely separated continents that also have closely fitting coastlines. Seminal hypothesis for later development of the theory of plate tectonics.

Coprolite   Trace fossil representing the end product of digestion by an animal. Fossilized equivalent of feces.

Cranial   Anatomical position referring to the head of an animal, such as cranial bones.

Cretaceous Period   Time interval in the geologic time scale spanning from about 140 to 65 million years ago. Last period of the Mesozoic Era.

Cretaceous-Tertiary (K-T) Boundary   Division between Cretaceous and Tertiary Periods (from about 65 million years ago) marking a mass extinction that resulted in the end of the non-avian dinosaurs and many of their contemporary species.

Crop   Muscular organ anterior to the stomach and gizzard in the alimentary canal of some certain herbivorous animals, such as some birds, and presumed to have existed in some herbivorous dinosaurs.

Crust   Solid and less-dense upper part of the lithosphere.

Cursorial Hypothesis   Explanation of the origin of flight in theropods from ground-dwelling species. Also known as the "ground up" hypothesis.

D

Dahllite   Mineral composed primarily of calcium phosphate and described by approximately the same formula as the mineral apatite, $Ca_{10}(PO_4)_6(OH, CO_3, F)_2$. Forms the main framework of hard parts (bones and teeth) in chordates.

Darwinism   Term used to describe older version of hypothesis of natural selection, as co-authored by Charles Darwin and Alfred Wallace.

Diapsida   Clade of amniotes with paired temporal fenestrae on each side of the skull. Part of lineage for dinosaurs.

Deinonychosauria   Clade of saurischian dinosaurs, placed with clades Theropoda, Tetanurae, Avetheropoda, Coelurosauria, Maniraptiformes, and Maniraptora.

Dental Battery   Assemblage of interlocking teeth that form broad grinding or shearing surfaces for chewing of plant material. Typical in hadrosaurids and ceratopsians.

Dentition   Sum of an animal's teeth in its jawbones.

Diagenesis   Biological, chemical, and physical processes in a sediment or sedimentary rock that occur after those sediments have been deposited. Important factor in taphonomy.

Digitigrade   Locomotion that involves contact only through the digits. (Contrast with plantigrade.)

Dinosaur   (1) A reptile- or bird-like animal with an upright posture that spent most (perhaps all) of its life on land and lived only during the Mesozoic Period. (2) An animal that had a minimum of the following synapomorphies: three or more sacral vertebrae, shoulder girdle with backward-facing (caudally pointing) glenoid, asymmetrical manus with less than or equal to three phalanges on digit IV, acetabulum with open medial wall, tibia with cnemial crest, astragalus with a long ascending process that fits into the anterior part of the tibia, sigmoidally shaped third metatarsal, postfrontal absent, humerus with long deltopectoral crest, and femur with ball-like head on proximal end. (3) The closest common ancestor to *Triceratops* and modern birds.

Dinosauria   Clade of archosaurs characterized by certain synapomorphies. (See dinosaur.) Term originally coined by Sir Richard Owen.

Diplodocidae   Clade of saurischian dinosaurs, placed with clades Sauropodomorpha and Sauropoda, distinguished by more cervical and caudal vertebrae than other sauropods, teeth restricted to the anterior portion of the skull, and nares dorsal to the orbitals.

Diploid   Number of chromosomes that results from the uniting of two haploid gametes.

DNA   Deoxyribonucleic acid.

Dorsal   Reference referring to the top surface of a horizontally oriented animal.

Dromaeosauridae   Clade of saurischian dinosaurs, placed with clades Theropoda, Tetanurae, Avetheropoda, Coelurosauria, Maniraptoriformes, Maniraptora, and Deinonychosauria.

E

Ecological Community   Collective group of organisms in an ecosystem.

Ecology   Study of interrelationships between organisms and their environments.

Ecosystem   Environment where organisms are interacting with both one another and abiotic (non-biological) factors.

Ectotherm   Animal that is dependent on the ambient temperature outside of its body for maintaining internal body temperature. (Contrast with endotherm.)

Egg (Cleidoic)   Permeable structure enclosing and interacting with an embryo. Is a body part of the embryo, hence its fossil equivalent is a body fossil.

Egg (Gamete)   Sex cell with only a haploid set of chromosomes supplied by the female of a species.

Eggshell Types   Categories of morphologically distinctive mineralized remains of cleidoic eggs. Examples include geckoid, testuoid, crocodiloid, dinosauroid spherulitic, dinosauroid prismatic, and ornithoid.

Eisospherite Layer   Inner, organic shell membrane of an egg.

Ellipsoid   In geometry, a body where all plane sections are either circles or ellipses. Used to describe geometry of most dinosaur eggs.

Embryo   Stage of development in ontogeny of an organism between a zygote (fertilized egg) and juvenile. In amniotes, associated with an amnion and (in cases of oviparous amniotes) an egg.

Enantiornithines   Clade of toothed avians that went extinct by the end of the Cretaceous Period.

Encephalization Quotient (EQ)   Ratio of cerebral cortex mass:totalbrain mass. Used as measure of relative "braininess" of an animal.

Endotherm   Animal that is dependent on its body for maintaining internal temperature rather than the ambient temperature outside of its body. (Contrast with ectotherm).

Entomology   Study of insects.

Ethics   Set of principles of conduct or behavior in human society and how that behavior affects people's relationships with one another.

Euornithopoda   Clade of ornithischian dinosaurs, placed with clade Ornithopoda, includes clades Hypsilophodontidae and Iguanodontia.

Evolution   Change in a population between generations, or descent with modification. See also biological evolution, (theory of).

Exospherite Layer   Crystalline (mineralized) exterior of an egg.

Extinction   Permanent cessation of propagation of a species, which can be either local or global. Occurs when the last individual of a species is prevented from reproducing.

Extinction, (Mass)   Near-simultaneous extinction of many different (unrelated) species.

F

Fabrosaurids   Paraphyletic group of primitive ornithiscian dinosaurs, most closely aligned with clade Ornithopoda.

Facies   All of the characteristics imparted by an environment to its sediment at approximately the same time; can be represented by biofacies (organismal remains), ichnofacies (traces left by organismal behavior), and lithofacies (composition and proportions of sediments).

Fact   Phenomenon that has an actual, objective existence, regardless of whether it was observed or not.

Feather   Integument associated with avian and known in a few non-avian theropods, which can function as flight or downy feathers but also is used for display and insulation. Consists of central shaft with barbs emanating from it, forming a vane in flight feathers.

Formation   Mappable unit of rock, given a formal name that typically refers to a place with a type section of the formation.

Fractionation   Fractional change in the amount of stable isotopes, where one becomes depleted while the other is enriched.

G

Gamete Sex cell (egg or sperm) containing a haploid complement of chromosomes, that for better or worse will combine with another (complementary) gamete to form a zygote.

Gastrolith Literally "stomach stones," refers to stones ingested by a vertebrate to use for either digestion or ballast. The fossil equivalent of this is a trace fossil.

Genasauria Node-based clade that includes clades Thyreophora and Ornithopoda.

Gene Nucleotide sequence in a DNA molecule that provides a code for a protein or part of a protein. Can be either dominant or recessive.

Genome Sum total of genes, conveyed in a DNA molecule and encompassing coding for all of an organism's proteins. Represents the genetic potential of an organism.

Genotype Genetic expression of an organism. A pair of genes at a locus on a chromosome. Contrast with phenotype.

Genus Name applied as first part of binomial nomenclature used in species name. Can be used by itself but also represents a broader category that may include several species.

Geochemistry Study of chemistry in regarding how it pertains to earth processes and geologic media.

Geographic Information Systems (GIS) Computer programs, facilitated through computer hardware, that integrate spatial data with other forms of information.

Geography Study of the earth's surface, typically facilitated through the use of maps.

Geologic Map Graphical, two-dimensional representation of the outcrop patterns of rock units (formations) on the land surface. Typically has topographic information, (such as elevation changes) superimposed on it.

Geologic Time Scale Standard description of time intervals in the history of the earth, based on a combination of relative dating and absolute dating criteria.

Geophysics Study of how basic physical principles are used to better understand the earth, particularly its interior.

Gizzard Muscular organ anterior of the stomach in the alimentary canal. In herbivorous dinosaurs, former presence is indicated by gastroliths.

Grades Levels applied to classifying organisms, which are, (in order of most to least inclusive,) kingdom, phylum, class, order, family, genus, and species. In botany, the equivalent grade to a phylum is a division.

Gradistics Classification system that places organisms into grades (levels) that become more inclusive based on anatomical similarity.

Gravitation, (Theory of) Generally accepted explanation for observations of the attraction of matter for matter.

H

Hadrosauridae Clade of ornithischian dinosaurs, placed with clades Ornithopoda, Euornithopoda, and Iguanodontia, characterized by a minimum of the following traits: long and wide anterior portion of skull ("duckbill"), well-developed dental batteries, loss of antorbital and surangular fenestrae, increase of vertebrae to at least 8 sacral and 12 cervical vertebrae, loss of digit I on the manus, development of prominent unguals on the pes, and long forelimbs relative to hindlimbs.

Haploid Half of the normal complement of chromosomes in a somatic (body) cell. Characteristic of a gamete.

Herbivory; (Herbivorous) Plant-eating.

Herpetology Study of amphibians and reptiles.

Herrerasauridae Clade of saurischian dinosaurs, often placed within clade Theropoda but sometimes placed outside of the Dinosauria. Consists of a minimum of the following traits: long pubis with relation to its femur, three sacral vertebrae, semiperforate to open acetabulum with a well-developed medial wall, femur length nearly twice that of the humerus, elongate skull nearly equal in length with its femur, serrated and recurved conical teeth, long and equally sized metatarsals I and V on the pes, and manus with five digits but digits IV and V reduced and without unguals. Considered one of the more primitive clades of dinosaurs.

Heterodont Dentition that shows a variety of teeth adapted for different functions.

Heterodontosauridae Clade of ornithischian dinosaurs, placed with clades Ornithopoda, and limited to the Early Jurassic of South Africa. Identified by heterodont dentition.

Homocephalidae Clade of ornithischian dinosaurs, placed with clade Marginocephalia and Pachycephalosauria. Often known as "flat-headed" dinosaurs.

Homeotherm Animal that maintains a near-constant internal body temperature, regardless of whether it is endothermic or ectothermic.

Homologue Body parts that are the same in different animals, although they may have a different morphology because they have been modified with descent as adaptations.

Hydrodynamics Physics of water flow. Important for interpreting sedimentary environments and taphonomy.

Hypothesis Conditional explanation of an observation or series of observations that typically proposes a cause for the observations. Must be testable and falsifiable.

Hypsilophodontidae Paraphyletic group of ornithischian dinosaurs within clade Ornithopoda and Euornithopoda. Among first dinosaurs to achieve pleurokinesis, a jointing between the premaxilla and the rest of the skull (including the premaxilla) that caused the maxilla to shift outward when the mouth closed.

I

Ichnofabric Resultant patterns or texture imparted to a substrate (either unconsolidated or solid) as a result of organismal behavior, such as trampling of sediment by dinosaurs.

Ichnology Study of traces left by organisms as a result of their behavior.

Ichnotaxon Name given to a trace fossil on the basis of its form (not its tracemaker). A type of parataxonomy, where binomial nomenclature is used for both ichnogenera and ichnospecies.

Igneous Rock Rock formed from originally molten material (magma), can be either plutonic (cooling far below the earth's surface) or volcanic (cooling on or near the earth's surface). Minerals from these rocks often are used for calculating radiometric ages.

Iguanodontia Clade of ornithischian dinosaurs, placed with clades Ornithopoda and Euornithopoda. Includes Hadrosauridae.

Ilium One of the hip bones, paired and lateral to the sacral vertebrae of the axial skeleton.

Inclusions, (Principle of) Basic geologic principle used in determining relative ages of rocks; particles of a pre-existing rocks incorporated into a sediment must be older than the rock including them.

Insectivory; (Insectivorous) Insect-eating habit. Considered as a hypothetical feeding strategy for some theropods (therizinosaurs).

Integument Skin and its derivatives, such as scales or feathers.

Interspecific Between different species, such as in most parasitic or predator-prey relations.

Intraspecific Within the same species, such as in intraspecific competition for mates.

Ischium One of the hip bones, posterior to both the pubis and ilium.

Isotope Variation on atomic weight of an element, can be either radioactive (undergoing decay) or stable (not decaying).

J

Jurassic Period Time interval in the geologic time scale spanning about 206 to 140 million years ago. Middle period of the Mesozoic Era.

L

Lateral Anatomical reference to the side of an organism, or of any body part, farther away from a midline.

Lineage "Line of descent," defined by populations that went through generations, from ancestors leading to descendants.

Lines of Arrested Growth (LAG's) Lines recorded in dinosaur bones that represent periods of interrupted growth, which can be attributed to yearly cycles in growth and thus suggestive of ectothermy.

Linnaean Classification Hierarchical classification system of organisms based on grades (levels), also known as gradistics. Defined by Carl von Linnž, (also known as Carolus Linneaus).

Lithosphere Cool, rigid exterior of the earth that incorporates the crust and upper part of the mantle. Composes plates (which can have continents) that interact with one another as a result of processes emanating from the asthenosphere.

M

Mammalia Clade of amniotes characterized by hair and mammary glands, among other traits. Descended from synapsid ancestors during the Late Triassic.

Maniraptoriformes Clade of saurischian dinosaurs, placed with clades Theropoda, Tetanuare, Avetheropoda, and Coelurosauria.

Mantle Relatively more dense and thickest layer of the earth, immediately below the crust, and forming the lower part of the lithosphere and all of the asthenosphere.

Manus Hand, associated with the forelimbs. Contrast with pes.

Marginocephalia Clade of ornithischian dinosaurs, considered a sister clade to Ornithopoda. Characterized by an abbreviated posterior portion of the premaxillary as it contributes to the palate and a shortened pubis accompanied by widely spaced hip sockets. Includes clades Pachycephalosauria and Ceratopsia.

Medial Middle part of an organism, especially definable in those organisms with bilateral symmetry, but also can refer to the middle portion of a body part.

Meiosis Splitting of diploid cells into haploid cells, which produces gametes.

Megalosauridae Clade of saurischian dinosaurs, placed with clades Theropoda and Tetanurae. Considered as an outgroup of the latter.

Mesozoic Era Division of geologic time scale that is defined as lasting from about 245 to 65 million years ago, consisting of the Triassic, Jurassic, and Cretaceous Periods.

Metamorphic Rock   Rock formed from heat and pressure that changes the original minerals or texture of a previous rock, such as another metamorphic rock, igneous rock, or sedimentary rock.

Mineral   Solid, naturally occurring, inorganic substance with a definite chemical composition and an ordered internal arrangement of atoms expressed as a crystal habit.

Mitosis   Splitting of diploid cells to form more diploid cells.

Mold   Impression made as a result of a body or body part contacting with a substrate. Can be either external or internal. Contrast with cast.

Monophyletic   Group of organisms with a common ancestor.

Morphology   Study of form in an organism, but also refers to the overall shapes of an organism and products of its behavior, i.e., track morphology and egg morphology.

Morrison Formation   Well-known Upper Jurassic rock unit made famous for its abundant dinosaur body fossils and some trace fossils. Named after its type area in Morrison, Colorado.

Mummification   Mode of preservation that involves desiccation (dehydration) of a body.

Mutualism   Form of commensalism, where two of or more species benefit from associating with one another.

N

Natural Selection   Hypothesis that is part of the theory of biological evolution. States that species have genetic variation within their populations. These variations may constitute adaptations with relation to intraspecific competition or environmental factors, and those better adapted individuals survive long enough to reproduce, thus selectively passing on their inheritable traits. Older version called Darwinism, new version called Neo-Darwinism.

Necrolysis   Decomposition of a body after it dies. Important process in taphonomy.

Neoceratopsia   Clade of ornithischian dinosaurs, placed with clades Marginocephalia and Ceratopsia.

Neo-Darwinism   Term used to describe newer version of Darwinism, which includes concepts of modern genetics.

Neornithines   Clade of avians represented by modern birds, originated in the Cretaceous Period. Can be divided into carinates (flighted birds) and ratites (flightless birds).

Nest   Biogenic structure that typically (but not always) contains a clutch. Best represented by an arrangement of eggs or eggshells in definite patterns, but can also be denoted as a semicircular depression with a raised rim that was originally used to hold eggs or young (mound nest), or a hollow, subsurface chamber for holding eggs (hole nest).

Node-based Clade   Clade that has all of the descendants of the most recent common ancestor for two groups, where the common ancestor forms the node.

Nodosauridae   Clade of ornithischian dinosaurs, placed with clades Thyreophora and Ankylosauria. Had laterally placed nares, and lacked horns and tail club.

Nucleic Acids   Biomolecules such as RNA and DNA that are chains of nucleotides.

O

Occlusion   Closing of an animal's mouth so that the teeth from the upper and lower jaws come together on a surface (occlusal surface).

Omnivory; (Omnivorous) Eating of both plants and animals.

Ontogeny Growth history (development) of an organism during its lifetime.

Opinion Idea that may or may not be based on factual information, but more on how a person feels.

Original Horizontality Basic geologic principle that sediments are originally deposited in more-or-less horizontal layers.

Ornithischia Clade placed with the Dinosauria that is one of the two defining all dinosaurs. Characterized by a hip arrangement that has the pubis together with the ischium and pointing posteriorly.

Ornithomimosauridae Clade of saurischian dinosaurs, placed with clades Theropoda, Tetanurae, Avetheropoda, Coelurosauria, Maniraptiformes, and Arctometatarsalia. Also known as the "ostrich dinosaurs."

Ornithopoda Clade of ornithischian dinosaurs, characterized by a minimum of the following traits: offset tooth row, where the maxillary teeth are more dorsal than those in the premaxillary, (but teeth in the latter might be missing altogether), occlusal surface more dorsal than jaw joint, crescent-shaped paraoccipital process, and premaxilla with an elongate process that touches either, (or both,) the prefrontal or lachrymal.

Osteoderm Bony outgrowth of skin. Typical trait in titanosaurids and thyreophorans.

Oviduct In female birds and reptiles, the passageway between uterus and cloaca, colloquially called a "birth canal." Paired in some crocodilians.

Oviparous Amniote that reproduces by laying an enclosed egg on land.

Oviraptorisauria Clade of saurischian dinosaurs, placed with clades Theropoda, Tetanurae, Avetheropoda, Coelurosauria, and Maniraptoriformes.

P

Pace Angulation In a trackway, angle described by the pace of one side in comparison to the overall stride.

Pace Length In a trackway, the distance between two successive tracks made by appendages from opposite sides, such as right-left or left-right.

Pachycephalosauria Clade of ornithischian dinosaurs, placed with clade Marginocephalia. Characterized by fusing together the frontals and parietals into a thick deposit of bone, thus the popular nickname of "bonehead" dinosaurs.

Pachycephalosauridae Clade of ornithischian dinosaurs, placed with clades Marginocephalia and Pachycephalosauria. Also known as the "dome-headed" dinosaurs.

Paleobiogeography Study of how maps can be used to describe the geographic distribution of organisms during the geologic past.

Paleoecology Study of interrelationships between ancient organisms and their original environments.

Paleontology Study of ancient life. Can include invertebrate paleontology (animals without backbones), vertebrate paleontology (animals with backbones), micropaleontology (one-celled organisms), and paleobotany (plants).

Paleopathology Study of sickness, injuries, and other abnormalities in the health of ancient organisms.

Paraphyletic A group of related organisms but excluding some of their descendants that might be much different. (See monophyletic.)

Parasitism   Form of symbiosis where one organism is living at the expense of another host organism. Parasitic organisms can be either endoparasites (within the host) or ectoparasites (outside the host).

Parataxonomy   Naming of either trace fossils or some body fossils, (such as eggs,) similar to that for biological species, using binomial nomenclature to identify a specific morphology.

Permineralization   Filling of pores in a fossil by minerals precipitated from solution. Common mode of preservation for petrified wood and bones.

Pes   Foot, associated with the hindlimbs. Contrast with manus.

pH   Negative logarithm of the hydrogen ion concentration in a solution, measured on a scale from 1 (most acidic) to 14 (most basic).

Phalangeal Formula   Count of phalanges associated with each digit in a manus or pes.

Phalanges (plural of phalanx)   Distal bones associated with the digits of either a manus or pes.

Phylogenetic Classification   System used for classifying organisms on the basis of their inferred phylogeny. Also known as cladistics.

Phylogeny   Evolutionary history of an organism.

Piscivory; (Piscivorous)   Fish-eating habit. Proposed as a hypothetical feeding strategy for some theropods, such as spinosaurs.

Plantigrade   Locomotion that involves contact of the digits and more proximal bones of the limb, such as tarsals and metatarsals. Also known as "flat-footed." (Contrast with digitigrade.)

Plate Tectonics, (Theory of)   Generally accepted explanation for observed earthquakes, volcanoes, and some other geologic phenomena that occur in definite places on the earth.

Plesiomorphy   Character in an ancestor that is also observable in a descendant. Considered as a primitive trait, such as ziphodont teeth.

Pleurocoels   Cavities within vertebrae of some dinosaurs, such as ceratosaurs.

Pneumatic (Bones)   Air-filled or otherwise less dense bones. Typical trait of both avian and non-avian theropods.

Polyphyletic   Group of organisms that had separate ancestors.

Population   Group of organisms interbreeding with one another, presumably representing a species.

Posterior   Anatomical reference to the rear of an animal or any part.

Precocial   Behavioral reference to juveniles capable of moving and fending for themselves soon after birth, with little or no parental care.

Pressure-release Structure   Deformity associated with a track caused by the application or release of pressure by a foot.

Prosauropoda   Clade of saurischian dinosaurs, placed with clade Sauropodomorpha, characterized by a minimum of the following traits: atlas and a total of 10 cervical vertebrae, 15 dorsal, three sacral (attached to the pelvic bones), and nearly 50 caudal vertebrae, phalangeal formula on pes of 2-3-4-5-1, with small unguals, and phalanx on digit V seemingly vestigial, phalangeal formula on manus of 2-3-4-3-2, with unguals present on digits I through III, enlarged ungual on digit I, which is deviated from the rest of the manus.

Protein   Combination of any 20 amino acids into a compound that facilitates biochemical reactions or provides structural support for an organism. Examples are albumin, collagen, hemoglobin, and osteocalcin.

Protoceratopsidae Uncertainly assigned as a clade of ornithischian dinosaurs, placed with clades Marginocephalia, Ceratopsia, and Neoceratopsia, among the most primitive of neoceratopsians.

Proximal Anatomical reference to a body part that is close to an organism's midline. Contrast with distal.

Psittacosauridae Clade of ornithischian dinosaurs, placed with clades Marginocephalia and Ceratopsia, considered as the oldest clade of ceratopsians.

Pubis One of the hip bones, anterior and medial with relative to the ischium and ilium.

Q

Quadrupedal Using four legs for locomotion.

Quantum Mechanics, (Theory of) Generally accepted explanation for observed behavior of atomic and subatomic particles.

R

Replacement Process where a body part of an organism is replaced by a material with a different composition from the original material.

Reproductive Isolation Separateness of a species from another, represented by an inability to reproduce with one another to produce viable offspring.

Reptilia; (Reptiles) General term for a polyphyletic group of amniotes that includes modern snakes, lizards, crocodillans, and turtles. Sometimes synonymized with clade Eureptilia, but the latter excludes clade Anapsida, which includes turtles.

Resonating Chambers Tube-like structures that connected with nasal cavities in some hadrosaurids. Hypothetically used for changing sounds produced by moving air through the skull.

Respiratory Turbinates Spaces within nasal cavities that accommodate folded bony or cartilaginous structures, often lined with mucous membranes. Considered as an indicator of endothermy.

RNA Ribonucleic acid.

Ruminant Herbivore that uses a multi-chambered stomach for the digestion of its food, which is physically mashed into a compacted mass called a bolus, (or cud).

S

Sacral Body parts referring to the hip region, such as sacral vertebrae.

Saurischia Clade placed with the clade Dinosauria that is one of the two defining all dinosaurs. Characterized by a hip arrangement that has the pubis pointing anterior and separate from the ischium.

Sauropoda Clade of saurischian dinosaurs, placed with clade Sauropodomorpha, characterized by a minimum of the following traits: 12 or more cervical vertebrae, neural arches and other processes smaller but forming better support for musculature as struts, sacrum with four or more vertebrae, large and separate pelvic bones, including an ilium with an expanded anterior end, minimum of 45 to more than 80 caudal vertebrae.

Sauropodomorpha Clade of saurischian dinosaurs, characterized by a minimum of the following traits: distal part of the tibia covered by an ascending process of the astralagus, short hindlimbs in comparison to the torso length, thin and flat (spatula-like) teeth with bladed and serrated crowns, minimum of 10 cervical vertebrae that are typically elongated and a total of 25 presacral vertebrae, and large digit I on the manus, pes with five digits and unguals on digits I–III, manus with five digits with smaller phalanges on digits II–V, ungual on digit I of manus. Includes clades Prosauropoda and Sauropoda.

Sediment   Unconsolidated material, typically formed as a result of weathering from a previously existing rock.

Sedimentary Environment   Place where sediments are deposited.

Sedimentary Rock   Rock formed through consolidation of sediment, forming either chemical or clastic sedimentary rocks.

Sedimentology   Study of sediments, specifically how they are formed, transported, buried, and altered with burial.

Semi-erect Posture   Body alignment used by an animal that is in between and an upright and sprawling posture, with its legs directly underneath its torso.

Serrations   Square-, triangular-, oval-, or rectangular-shaped denticles on carinae separated by narrow indentations (cellae) along a tooth surface.

Sexual Selection   Choosing of mates on the basis of attractive traits possessed by that mate. An important part of natural selection.

Sister Group   Taxon that had the same "parent," or ancestral group, as another taxon and split from that ancestral group, e.g., Sauropodomorpha is a sister group to Theropoda because they both have a saurischian ancestor.

Site Fidelity   Habitual (annual) visitation and inhabitation of an area by a species of animal, typically for breeding and brooding purposes.

Speciation   Evolution of one species into another species. Has been observed in some organisms, thus is part of factual basis of the theory of biological evolution.

Species (Biological Concept)   Population of organisms that can interbreed and produce offspring that can also reproduce with one another. A closed gene pool.

Species (Name)   Taxon for biological species based on binomial nomenclature, italicized and using genus and trivial name; e.g., *Triceratops horridus*, *Tyrannosaurus rex*.

Sperm   Male gamete containing a haploid complement of chromosomes. Although outnumbered by millions of its compatriots, one will combine with female gamete (egg) to form a zygote. (Everyone should know this, but if not, this textbook was worth writing in order to pass on that knowledge alone.)

Sprawling Posture   Body alignment used by an animal that has its legs lateral to (outside of) the plane of its torso.

Stable Isotope   Variation of an element based on differing numbers of neutrons, which affect its mass (isotope) that does not undergo radioactive decay. Ratios of stable isotopes are used for interpreting temperatures in ancient ecosystems.

Stegosauria   Clade of ornithischian dinosaurs, placed with the clade Thyreophora, characterized by osteoderms evident as spikes on the shoulder regions (parascapular spines) and dermal armor restricted to two rows of vertically oriented plates, parallel to but also lateral to the axial skeleton (parasagittal plates) or spikes.

Stem-based Clade   Clade that has a shared common ancestor more recent than that of another group.

Straddle   Width of a trackway, measured across the diameter of the trackway.

Stratigraphy   Study of rock layers, (particularly sedimentary rocks:), how they were formed, and mapping of their geometry (areal distribution and thickness).

Stride Length   In a trackway, the distance between two successive steps made by the same foot.

Superposition, (Principle of) Basic geologic principle that originally horizontal layers have the oldest layers on the bottom and youngest layers on the top of a sequence, i.e., whatever is on the top is younger and vice versa.

Symbiosis Two or more organisms of different species living together or interacting in a way that one or all are dependent on the relationship for survival. Parasitism is a form of symbiosis, as is mutualism.

Synapomorphy Character, shared between two or more groups of organisms, that was derived from earlier characters. Evolutionarily derived ("new" or "novel") anatomical trait. Used in cladistic (phylogenetic) classification system.

Synapsida Clade of amniotes characterized by single temporal foramen on each side of the skull. Clade Mammalia is placed with this clade.

Synonymy Different species name assigned to an organism that has already been given a species name (a common problem in naming of dinosaur species).

T

Taphonomy Study of all processes that affect an organism after it dies, (such as scavenging, tissue degradation, transport of a body, and burial) and its fossilization potential. Also can be applied to the preservation of trace fossils.

Taxonomy Classification system in which names are applied to organisms or groups of organisms with regard to defining features; name given is a taxon. System applied to naming traces of behavior by organisms or their eggs is parataxonomy.

Tetanurae Clade of saurischian dinosaurs, placed with clade Theropoda, characterized by a minimum of the following traits: dentition in the maxilla only anterior to the orbital, antorbital and maxillary fenestrae, accompanied by increased pneumaticity of the skull, manus with digits I through III but digit III is absent in some cases, tibia that overlapped a reduced fibula, development of a large notch on the ischium, well-developed stiffening of the caudal vertebrae by processes (zygopophyses) that extended anterior and posterior from the neural arches.

Theory Hypothesis, or set of related hypotheses, that withstood repeated testing to the point of widespread acceptance by the scientific community.

Thanatocoenosis Assemblage of organisms in the fossil record that may be composed of only allochthonous species or a mixture of allochthonous and autochthonous species. Also known as a "death assemblage."

Therizinosauridae Clade of saurischian dinosaurs, placed with clades Theropoda, Tetanurae, Avetheropoda, Coelurosauria, and Maniraptiformes. Considered as the most unusual of theropods, especially in their dentition. They possibly were insectivorous, omnivorous, or herbivorous.

Thermodynamics Study of heat and its relationship to work.

Theropoda Clade of saurischian dinosaurs, has minimum of following traits: lachrymal bone prominently exposed on dorsal surface of skull, minimum of five sacral vertebrae, manus with claws (unguals) and reduction or loss of digits IV and V, slightly curved femur, which is also more than twice as long as the humerus, pes with digits II through IV, digit I separate from pes, pes length greater than width, bilaterally symmetrical from digit III, and a well-defined processes on cervical and caudal vertebrae. Generally known as "carnivorous dinosaurs" but probably with some exceptions.

Thoracic Any body part referring to the thorax (torso), such as thoracic vertebrae.

Thyreophora Clade of ornithischian dinosaurs, characterized by dermal armor, typically present as osteoderms in rows parallel to the midline of the body and a well-developed postorbital process associated with the jugal and a palpebral (supraorbital bone). Includes clades Stegosauria and Ankylosauridae.

Titanosauridae Clade of saurischian dinosaurs, placed with clades Sauropodomorpha and Sauropoda, distinguished by dermal armor (osteoderms) and procoelous caudal vertebrae.

Tooth Bone-like projection from the jaw of a chordate used primarily for grasping, biting, or chewing food. If found in the fossil record is considered as a body fossil.

Toothmark Trace left on a substrate, either as a puncture or scrape mark, that represents behavior by an animal using its teeth. If found in the fossil record is considered as a trace fossil.

Topographic Map Graphical, two-dimensional representation of differences in elevation in a specific area as well as its surface features, such as forested areas, roads, and cities.

Trace Element Element that is normally in very small amounts in the earth's crust or in a body.

Trace Fossil Any indirect evidence of ancient life as represented by the effects of behavior. Includes tracks, trails, burrows, toothmarks, gastroliths, coprolites, and nests.

Track Trace left by appendage of an animal, typically as a result of its locomotion. If found in the fossil record is considered as a type of trace fossil. A series of two or more successive tracks made by the same foot is a trackway.

Trail Path worn by repeated movement of animals along a route, typically the result of numerous trackways.

Triassic Period Time interval in the geologic time scale spanning from about 245 to 206 million years ago. Dinosaurs had evolved from archosaur ancestors by the latter part of this period, including the clades Theropoda and Sauropodomorpha.

Tridactyl Three-toed track or foot, often affiliated with either theropods or most ornithopods, (especially hadrosaurids).

Trivial Name Second part of a species name (in binomial nomenclature) that must always be used in combination with and preceded by a genus name. Is always lowercase and italicized.

Troodontidae Clade of saurischian dinosaurs, placed with clades placed with clades Theropoda, Tetanurae, Avetheropoda, Coelurosauria, Maniraptoriformes, and Arctometatarsalia.

Type Specimen Individual organism or parts of an organism used to define a species.

Tyrannosauridae Clade of saurischian dinosaurs, placed with clades Theropoda, Tetanurae, Avetheropoda, Coelurosauria, Maniraptoriformes, and Arctometatarsalia.

U

Undertrack Track preserved below the actual surface on which an animal walked as a result of compression of underlying layers.

Ungual Nail or hoof, located at the distal ends of phalanges.

Upright posture Body alignment used by an animal that stands and walks with its legs directly underneath its torso.

V

Ventral Bottom side of a horizontally oriented organism or a body part; contrast with dorsal.

Vertebrata, (Vertebrates) Clade with synapomorphy of a series of bones (vertebrae) forming the main axial elements in the dorsal part of the skeleton. Placed with clade Chordata.

Vertebrae (plural of Vertebra) Repeated and interconnected bones that formed the main axial elements in the dorsal part of a skeleton. Can be classified as cervical, dorsal, sacral, and caudal, in order from cranially to caudally, and with each composed of a centrum, neural arch, and nerve canal.

Viviparous Amniote that reproduces by giving "live birth," where its juveniles are retained in its body just before birth, rather than being born from cleidoic eggs.

W

Warm-blooded Popularized description of endothermy.

Y

Yolk Sac Food supply in a cleidoic egg.

Z

Ziphodont Curved and serrated teeth; considered as a plesiomorphic trait in dinosaurs and present in most theropods.

Zygopophyses Processes emanating from the caudal vertebrae that, when well-developed, helped to stiffen the tail.

Zygote Female gamete (egg) that has combined its haploid complement of chromosomes with that of a male gamete (sperm); a fertilized egg.

Index

Page numbers in *italics* refer to figures; those in **bold** to tables.